Microorganisms for Sustainability

Volume 34

Microorganisms have been in existence since the origin of life on earth and can survive the most extreme habitats or conditions on earth. Microorganisms are involved in regulating biogeochemical cycles, maintaining plant and animal health, and sustaining the global food chain. Moreover, they play crucial roles in addressing the challenges of climate change and achieving the targets of Sustainable Development Goals (SDGs).

This multidisciplinary book series captures the role of microbes towards building a sustainable world, while encompassing cutting-edge technologies and current needs across various fields such as agriculture sustainability, bioremediation, restoration of degraded habitats and wastelands, and food security. Additionally, this series explores microbial applications in industries, and building their utilization in clean and green energy solutions. Furthermore, themes like microbial secondary metabolites, extremophilic microbes and modern omics, including next generation sequencing and metagenomics, are also covered in this series.

With contributions from researchers across the globe, this series addresses the important call of 'One Planet-One Health-One Future'. It comprises a collection of diverse volumes that provides insights for scientists, young researchers, educators and decision-makers in the government, private sector, and non-governmental organizations, empowering their efforts to achieve the global goals.

The series invites, evaluates, and accepts book proposals to ensure a diverse, inclusive, and evolving program. The final decision regarding acceptance rests with the series editor.

Peer-review

This book series follows a stratified review process. Proposals for individual volumes are reviewed by the series editor and then the editorial board members. On a case-to-case basis, external reviewers are also invited for further evaluation of the book proposal. Review of the chapters is the responsibility of the volume editor(s). A manuscript submission platform has been recently made available to the authors, volume editors and the series editor.

Naveen Kumar Arora • Brahim Bouizgarne
Editors

Microbial Biotechnology for Sustainable Agriculture Volume 2

Springer

Editors
Naveen Kumar Arora
Department of Environmental Science
Babasaheb Bhimrao Ambedkar University
Lucknow, Uttar Pradesh, India

Brahim Bouizgarne
Department of Biology
Ibn Zohr University
Agadir, Morocco

ISSN 2512-1901 ISSN 2512-1898 (electronic)
Microorganisms for Sustainability
ISBN 978-981-97-2354-6 ISBN 978-981-97-2355-3 (eBook)
https://doi.org/10.1007/978-981-97-2355-3

This Springer imprint is published by the registered company Springer Nature Singapore Pte Ltd.
The registered company address is: 152 Beach Road, #21-01/04 Gateway East, Singapore 189721, Singapore

Preface

Microbial technology has become pivotal in sustainable agriculture, providing innovative solutions to pressing farming challenges. The sustainable and quickly developing microbial technologies for agricultural ecosystems are not only helping in improvement of crop productivity but also in reducing harmful impacts on the environment. By harnessing the power of microbes through their bioinoculants, farmers can optimize nutrient uptake, suppress diseases, improve productivity in stressed agro-ecosystems, and promote eco-friendly practices. These advancements signify a transformative shift toward a more sustainable and efficient agricultural future.

In this comprehensive volume, *Microbial Technology for Sustainable Agriculture* Volume 2, multifaceted roles of microbial bioinoculants and their applications in fostering agricultural sustainability are explored and explained. With a keen focus on the dynamic interrelationship between microorganisms and various facets of plant health and growth, this book delves into the advances and future perspectives in the field of microbial technology involving development of novel bioinoculants.

The chapters in this book serve as evidence of the diverse and profound impact of microbial technology on sustainable agricultural practices. The volume explores the applications of varied types of microbe-based inoculants in solving the issues related to agro-ecosystems. From the exploration of the role of arbuscular mycorrhizal fungi (AMF) in enhancing plant tolerance to abiotic stresses to the elucidation of the intricate plant-microbe interactions in the management of plant diseases, each chapter covers different aspects of microbial technology in agriculture.

Furthermore, the book delves into futuristic applications of plant growth-promoting microbes, such as the potential impact of engineered nanoparticles on rhizosphere microbes, and the application of nano-biofungicides as effective tools for the control of fungal pathogens. With a keen emphasis on the pivotal role of transcriptomics in explaining the mechanisms of abiotic stress tolerance in plants, this volume offers a comprehensive overview of the intersection between innovative technologies and sustainable agricultural practices. The chapters in the volume also explain the role of beneficial soil microbes, including endophytes, in the bioremediation of contaminated and stressed agro-ecosystems, suggesting the development of novel bioinoculants for the future.

By bringing together a diverse array of perspectives from leading experts in the field, *Microbial Technology for Sustainable Agriculture* Volume 2 aims to serve as a comprehensive guide for researchers, agronomists, and policymakers seeking to harness the transformative potential of microbial technology to build a more sustainable and resilient agricultural future.

The editors extend their sincere gratitude to all the authors for generously sharing their expertise and research contributions in this volume. Additionally, our heartfelt thanks go to the dedicated team at Springer Nature, especially Dr Sanchi Bhimrajka Associate Editor, Books, Life Sciences, Springer Nature, India, and Dr Naren Aggarwal, Editorial Director, Medicine, Biomedical and Life Sciences, Books Asia, Springer Nature, India, for their unwavering support.

NKA would also like to express profound gratitude to Vice-Chancellor, BBA University, Lucknow, UP, India, for his invaluable support. Moreover, special acknowledgment goes to the team of dedicated research scholars, namely Dr Tahmish Fatima, Ms Priya Mishra, Ms Priyanka Verma, Ms Ankita Bhattacharya, and Ms Chanda Bharti, for their diligent efforts in manuscript editing. Our aspiration is that this book serves as a valuable resource, inspiring new ideas and innovations in the field of sustainable agriculture. Furthermore, we hope that this volume contributes significantly to the attainment of sustainable development goals (SDGs) by advancing the knowledge and practices crucial for a more sustainable agricultural future.

Lucknow, Uttar Pradesh, India — Naveen Kumar Arora
Agadir, Morocco — Brahim Bouizgarne

About the Book

This volume discusses the significant impact of microbial technology in promoting agricultural sustainability and a circular economy. The book covers the latest developments in microbial bioinoculants including both fungi and bacteria, their applications in remediating barren fields, enhancing crop productivity, and protecting plants against abiotic and biotic stresses.

Plants are associated with diverse groups of microbes residing in either the rhizosphere, endosphere, or phyllosphere and are being commonly addressed as "Plant Growth-Promoting Microorganisms" (PGPM). Microbial inoculants facilitate plant growth and soil remediation through various mechanisms including participation in biogeochemical cycling, protection against stresses, synthesis of plant growth regulators (like phytohormones), and modulation of secondary metabolites, antibiotics, plant hormones, and various signal compounds. In recent times, the increasing demand of microbial products has led to designing of novel bioformulations for specific conditions of diverse agro-ecosystems. Modern omic-based techniques have further eased the processes through better understanding of action mechanisms and thereby developing tailor-made bioinoculants. A better understanding of the mechanisms has also advanced application strategies, and new engineered techniques such as the use of nanoparticles augment the effects of PGPM and ensure the successful, slow delivery of inoculants to soil or plants. Commercialization of microbial inoculants still requires attention and targeting the stressed, infertile lands can sustain local/national economy and replace chemicals that are worsening the situation both on land and in water.

This book provides latest information about microbial inoculants, which are of interest to faculty, agronomists, microbiologists, plant pathologists, researchers, and students. The book is an excellent study material for the undergraduate or postgraduate students of agriculture, biotechnology, agronomy, microbiology, biotechnology, soil and environmental sciences.

About the Book

Contents

Contributors

Murtaza Abid Department of Botany, University of Lucknow, Lucknow, Uttar Pradesh, India

M. M. Abid Ali Khan Department of Botany, Shia Post Graduate College, Lucknow, Uttar Pradesh, India

Department of Botany, University of Lucknow, Lucknow, Uttar Pradesh, India

Rachid Aboutayeb Environment and Sustainable Management of Natural Resources Research Unit, Regional Center of Agricultural Research of Settat, National Institute of Agricultural Research, Rabat, Morocco

Mohamed Ait-El-Mokhtar Laboratory of Biochemistry, Environment & Agri-food URAC 36, Department of Biology, Faculty of Science and Techniques – Mohammedia, Hassan II University of Casablanca, Mohammedia, Morocco

Laboratory of Agro-Food, Biotechnologies and Valorization of Plant Bioresources (AGROBIOVAL), Department of Biology, Faculty of Science – Semlalia, Cadi Ayyad University, Marrakesh, Morocco

Centre d'Agrobiotechnologie et Bioingénierie, Unité de Recherche Labellisée CNRST (Centre AgroBiotech-URL-CNRST-05), Cadi Ayyad University, Marrakesh, Morocco

Youssef Ait-Rahou Laboratory of Agro-Food, Biotechnologies and Valorization of Plant Bioresources (AGROBIOVAL), Department of Biology, Faculty of Science – Semlalia, Cadi Ayyad University, Marrakesh, Morocco

Centre d'Agrobiotechnologie et Bioingénierie, Unité de Recherche Labellisée CNRST (Centre AgroBiotech-URL-CNRST-05), Cadi Ayyad University, Marrakesh, Morocco

Higher Institute of Nursing Professions and Health Techniques, Ministry of Health, Ouarzazate, Morocco

Fatima Zahra Aliyat Environment and Valorization of Microbial and Plant Resources Unit, Faculty of Sciences, Moulay Ismail University, Meknes, Morocco

African Genome Center, Mohammed VI Polytechnic University (UM6P), Ben Guerir, Morocco

Mohamed Anli Laboratory of Agro-Food, Biotechnologies and Valorization of Plant Bioresources (AGROBIOVAL), Department of Biology, Faculty of Science – Semlalia, Cadi Ayyad University, Marrakesh, Morocco

Centre d'Agrobiotechnologie et Bioingénierie, Unité de Recherche Labellisée CNRST (Centre AgroBiotech-URL-CNRST-05), Cadi Ayyad University, Marrakesh, Morocco

Department of Life, Earth and Environmental Sciences, University of Comoros, Patsy University Center, Moroni, Comoros

Naveen Kumar Arora Department of Environmental Science, School of Earth and Environmental Sciences, Babasaheb Bhimrao Ambedkar University, Lucknow, Uttar Pradesh, India

Pranay Arora Department of Earth and Environmental Science, University of Manchester, Manchester, UK

Khalid Azim Integrated Crop Production Research Unit, Regional Center of Agricultural Research of Agadir, National Institute of Agricultural Research, Rabat, Morocco

Mohamed Bakki Faculty of Sciences, Department of Biology, University Ibn Zohr (UIZ), Agadir, Morocco

Marouane Baslam Laboratory of Agro-Food, Biotechnologies and Valorization of Plant Bioresources (AGROBIOVAL), Department of Biology, Faculty of Science – Semlalia, Cadi Ayyad University, Marrakesh, Morocco

Centre d'Agrobiotechnologie et Bioingénierie, Unité de Recherche Labellisée CNRST (Centre AgroBiotech-URL-CNRST-05), Cadi Ayyad University, Marrakesh, Morocco

Department of Life and Food Sciences, Graduate School of Science and Technology, Niigata University, Niigata, Japan

Raja Ben-Laouane Laboratory of Agro-Food, Biotechnologies and Valorization of Plant Bioresources (AGROBIOVAL), Department of Biology, Faculty of Science – Semlalia, Cadi Ayyad University, Marrakesh, Morocco

Centre d'Agrobiotechnologie et Bioingénierie, Unité de Recherche Labellisée CNRST (Centre AgroBiotech-URL-CNRST-05), Cadi Ayyad University, Marrakesh, Morocco

Laboratory of Environment and Health, Department of Biology, Faculty of Science and Techniques, Boutalamine, Errachidia, Morocco

Chanda Bharti Department of Environmental Science, School of Earth and Environmental Sciences, Babasaheb Bhimrao Ambedkar University, Lucknow, Uttar Pradesh, India

Ankita Bhattacharya Department of Environmental Science, School of Earth and Environmental Sciences, Babasaheb Bhimrao Ambedkar University, Lucknow, Uttar Pradesh, India

Brahim Bouizgarne Laboratory of Plant Biotechnology, Department of Biology, Faculty of Sciences, Ibn Zohr University (UIZ), Agadir, Morocco

Abderrahim Boutasknit Laboratory of Agro-Food, Biotechnologies and Valorization of Plant Bioresources (AGROBIOVAL), Department of Biology, Faculty of Science – Semlalia, Cadi Ayyad University, Marrakesh, Morocco

Centre d'Agrobiotechnologie et Bioingénierie, Unité de Recherche Labellisée CNRST (Centre AgroBiotech-URL-CNRST-05), Cadi Ayyad University, Marrakesh, Morocco

Department of Biology, Multidisciplinary Faculty of Nador, Mohamed First University, Nador, Morocco

Bhavin Changela Department of Biomedical Engineering, A.V. Parekh Technical Institute, Rajkot, Gujarat, India

Disha Changela Department of Biochemistry, Saurashtra University, Rajkot, Gujarat, India

Kruti Dangar Department of Biochemistry, Saurashtra University, Rajkot, Gujarat, India

Nandita Das Department of Microbiology, Assam University, Silchar, India

Abdelhi Dihazi Laboratory of Agrobiotechnology and Bioengineering, Cadi Ayyad University, Faculty of Sciences and Technology, Marrakech, Morocco

Mohammed El Guilli National Institute of Agricultural Research, Regional Center for Agricultural Research, Kenitra, Morocco

Fouad Elame Integrated Crop Production Research Unit, Regional Center of Agricultural Research of Agadir, National Institute of Agricultural Research, Rabat, Morocco

Yamna Elbakhouch Integrated Crop Production Research Unit, Regional Center of Agricultural Research of Agadir, National Institute of Agricultural Research, Rabat, Morocco

Laboratory of Agrobiotechnology and Bioengineering, Cadi Ayyad University, Faculty of Sciences and Technology, Marrakech, Morocco

Abdessamad Fakhech Laboratory of Agro-Food, Biotechnologies and Valorization of Plant Bioresources (AGROBIOVAL), Department of Biology, Faculty of Science – Semlalia, Cadi Ayyad University, Marrakesh, Morocco

Juhi Goyal Department of Biotechnology, Mohanlal Sukhadia University, Udaipur, Rajasthan, India

Jamal Ibijbijen Environment and Valorization of Microbial and Plant Resources Unit, Faculty of Sciences, Moulay Ismail University, Meknes, Morocco

Nileshwariba Jadeja Department of Biochemistry, Saurashtra University, Rajkot, Gujarat, India

Preet Jain Department of Biotechnology, Mohanlal Sukhadia University, Udaipur, Rajasthan, India

Megha Kaviraj ICAR-National Rice Research Institute, Cuttack, Odisha, India

Upendra Kumar ICAR-National Rice Research Institute, Cuttack, Odisha, India

Vivek Kumar Himalayan School of Biosciences, Swami Rama Himalayan University Jolly Grant, Dehradun, India

Islamul Hoque Laskar Department of Biotechnology, Assam University, Silchar, India

Shivani Mahra Department of Biotechnology, Motilal Nehru National Institute of Technology Allahabad, Prayagraj, India

El Hassan Mayad Laboratory of Biotechnology and Valorisation of Natural Resources, Faculty of Sciences of Agadir, Ibn Zohr University, Agadir, Morocco

Pranab Bihari Mazumder Department of Biotechnology, Assam University, Silchar, India

Abdelilah Meddich Laboratory of Agro-Food, Biotechnologies and Valorization of Plant Bioresources (AGROBIOVAL), Department of Biology, Faculty of Science – Semlalia, Cadi Ayyad University, Marrakesh, Morocco

Centre d'Agrobiotechnologie et Bioingénierie, Unité de Recherche Labellisée CNRST (Centre AgroBiotech-URL-CNRST-05), Cadi Ayyad University, Marrakesh, Morocco

Isha Mishra IITians4Nation Foundation, Delhi, India

Priya Mishra Department of Environmental Science, School of Earth and Environmental Sciences, Babasaheb Bhimrao Ambedkar University, Lucknow, India

Toshiaki Mitsui Laboratory of Biochemistry, Faculty of Agriculture, Niigata University, Niigata, Japan

Department of Life and Food Sciences, Graduate School of Science and Technology, Niigata University, Niigata, Japan

Laila Nassiri Environment and Valorization of Microbial and Plant Resources Unit, Faculty of Sciences, Moulay Ismail University, Meknes, Morocco

Rajeev Padbhushan Department of Soil Science and Agricultural Chemistry, Bihar Agricultural University, Sabour, Bihar, India

Piyush Pandey Department of Microbiology, Assam University, Silchar, India

S. N. Pandey Department of Botany, University of Lucknow, Lucknow, Uttar Pradesh, India

Ved Prakash Department of Biotechnology, Motilal Nehru National Institute of Technology Allahabad, Prayagraj, India

Nitish Rai Department of Biotechnology, Mohanlal Sukhadia University, Udaipur, Rajasthan, India

Department of Zoology, University of Lucknow, Lucknow, Uttar Pradesh, India

Padmaja Rai Department of Biotechnology, Motilal Nehru National Institute of Technology Allahabad, Prayagraj, India

Ketaki Ramani Department of Microbiology, School of Science, RK University, Rajkot, Gujarat, India

Shweta Rana Department of Physical and Natural Sciences, FLAME University, Pune, India

Mihir Raval Department of Pharmaceutical Sciences, Sardar Patel University, Vallabh Vidhyanagar, Gujarat, India

Mohd. Zahid Rizvi Department of Botany, Shia Post Graduate College, Lucknow, Uttar Pradesh, India

Snehasini Rout ICAR-National Rice Research Institute, Cuttack, Odisha, India

Samarth Sharma Department of Biotechnology, Motilal Nehru National Institute of Technology Allahabad, Prayagraj, India

Shivesh Sharma Department of Biotechnology, Motilal Nehru National Institute of Technology Allahabad, Prayagraj, India

Archana Singha ICAR-National Rice Research Institute, Cuttack, Odisha, India

Ilyass Tabrika Research Unit "Integrated Plant Production" Regional Centre of Agricultural Research, Agadir, Morocco

Laboratory of Mechanic, Process Energy and Environment, National School of Applied Sciences, Ibn Zohr University, Agadir, Morocco

Kavita Tiwari Department of Biotechnology, Motilal Nehru National Institute of Technology Allahabad, Prayagraj, India

Sneha Tripathi Department of Biotechnology, Motilal Nehru National Institute of Technology Allahabad, Prayagraj, India

Priyank Upadhyay Department of Biotechnology, Mohanlal Sukhadia University, Udaipur, Rajasthan, India

Udaya Kumar Vandana Department of Biotechnology, Assam University, Silchar, India

Priyanka Verma Department of Environmental Science, School of Earth and Environmental Sciences, Babasaheb Bhimrao Ambedkar University, Lucknow, Uttar Pradesh, India

Said Wahbi Laboratory of Agro-Food, Biotechnologies and Valorization of Plant Bioresources (AGROBIOVAL), Department of Biology, Faculty of Science – Semlalia, Cadi Ayyad University, Marrakesh, Morocco

Centre d'Agrobiotechnologie et Bioingénierie, Unité de Recherche Labellisée CNRST (Centre AgroBiotech-URL-CNRST-05), Cadi Ayyad University, Marrakesh, Morocco

Priti Yadav Department of Biotechnology, Mohanlal Sukhadia University, Udaipur, Rajasthan, India

Mina Zaafrani Laboratory of Mechanic, Process Energy and Environment, National School of Applied Sciences, Ibn Zohr University, Agadir, Morocco

Chapter 1
Role of Microbial Bio-inoculants in Sustainable Agriculture

Islamul Hoque Laskar, Udaya Kumar Vandana, Nandita Das, Piyush Pandey, and Pranab Bihari Mazumder

Abstract The increasing global demand for food has led agricultural crop producers to rely heavily on the use of chemical fertilizers to enhance production yields. However, the unregulated use of these chemicals has resulted in negative consequences for both human health and soil fertility. Biofertilizers have emerged as an alternative to these chemicals. Microbial inoculants used as biofertilizers and biopesticides have proven to be effective and environment-friendly approaches towards sustainable agriculture, without causing any known adverse effects or pollution. Recently, concerns have been raised regarding the impact of climate change on the beneficial traits of these bio-inoculants in nature. In order to preserve and more efficiently utilize plant-associated beneficial microbes, genetic tools have been employed to modify them and prevent them from altering their essential phenotypes. In this chapter, we have evaluated various genera of microbes that are useful in crop production, as well as genetic approaches to enhance their efficiency. The use of genetic engineering and nanotechnologies has been shown to yield precise benefits to agriculture and has therefore been discussed in the context of the benefits associated with microbes in agriculture. Furthermore, strategies have been elaborated upon to expand the production of bio-inoculants from their current bottleneck status in laboratory research to large-scale industrial production, with challenges for routine agricultural application.

Keywords Chemical fertilizers · Cyanobacteria · PGPR · Mycorrhiza · Climate change · Genetic engineering

---

I. H. Laskar · U. K. Vandana · P. B. Mazumder (✉)
Department of Biotechnology, Assam University, Silchar, India

N. Das · P. Pandey
Department of Microbiology, Assam University, Silchar, India

N. K. Arora, B. Bouizgarne (eds.), *Microbial Biotechnology for Sustainable Agriculture Volume 2*, Microorganisms for Sustainability 34,
https://doi.org/10.1007/978-981-97-2355-3_1

1.1 Introduction

Food is a crucial component for the survival of all heterotrophic organisms on Earth, including human beings. The dependence of humans on agriculture has increased significantly due to the growing demand for food (Koyande et al. 2019). However, conventional farming practices predominantly involve the use of chemical fertilizers that are rich in inorganic materials such as nitrogen and phosphorus. Such practices have been found to cause pollution to the environment and lead to the degradation of soil fertility over time. This, in turn, can cause environmental pollution such as eutrophication of water bodies and nitrate pollution of groundwater, ultimately hampering soil homeostasis (Pathak and Kumar 2016). Excessive use of urea as a source of nitrogen for crops has been identified as a significant cause of nitrate pollution, which can lead to diseases like methemoglobinemia and hypertension in infants, ultimately causing severe disabilities. Furthermore, the widespread use of inorganic potassium in agriculture has raised concerns about the depletion of reserves of rock phosphates. A report by the Food and Agricultural Organization (FAO) in 2017 revealed that out of the total potassium fertilizers used in agriculture annually, only 0.2% get utilized by plants, and the rest is precipitated by metal cations in the soil (Soumare et al. 2020). In addition to chemical fertilizers, the extensive use of pesticides in agriculture has been linked to various health problems, including cancer, asthma, neurological defects, neonatal disorders, as well as reproductive health in human beings. These observations highlight the dire consequences of conventional agriculture practices on human health and the environment, and the need for more sustainable and environmentally friendly agricultural practices (Alori and Babalola 2018). Alternative methods such as organic farming, which avoids the use of synthetic fertilizers and pesticides, and focuses on building healthy soil, can help overcome the demand for food while preserving the environment and human health.

Biofertilizers have emerged as a promising alternative to chemical fertilizers due to their eco-friendly nature and lack of observed adverse effects on human health and the environment. Biofertilizers, also known as bio-inoculants, consist of active or latent microbial strains, typically bacteria, sometimes in combination with other microorganisms like fungi and algae, which are formulated to stimulate microbial activity and enhance the mobilization of soil nutrients (Suyal et al. 2016). These bio-inoculants are designed to promote plant growth and enhance productivity by colonizing the root system of plants, both internally and externally, competing with native microorganisms in the soil to establish a positive interaction with the plant through rhizocompetence traits (Romano et al. 2020). Beneficial microbial groups used as bio-inoculants in agriculture include cyanobacteria, mycorrhizal fungi, rhizobia, and plant growth-promoting rhizobacteria (PGPR) (Pandey and Maheshwari 2007). Although there are many bacterial and fungal strains that are useful to plants, PGPR have shown the most promising traits and are commonly used in agriculture. PGPR can provide benefits to plants through various mechanisms, such as phytohormone synthesis, soil nutrient mobilization, mitigation of abiotic stresses, and protection or induction of the plant's defence system against biotic stresses (Garcia-fraile et al. 2015). These traits are primarily influenced by environmental factors and are susceptible to alteration based on changes in the environment.

Global climate change has the capacity to impact every aspect of living world. Soil microbes in concern have also shown signs of alterations with changing environment. The respiration rate of soil microbes is getting changed with increase in global temperature as the process of soil respiration is temperature sensitive. Apart from that 5 °C increase in temperature has shown to decline the relative abundance of soil bacteria as well as alter the bacterial to fungal community ratio in soil (Classen et al. 2015). The global temperature rise is majorly due to the increase in CO_2 emissions in atmosphere; and increase in CO_2 prefers the fungal development over bacteria in soil. Arbuscular mycorrhizal fungi and decomposers were found to attain enhanced rate of decomposition of cellulose in elevated CO_2 conditions. But molecular-based mechanisms related to these observed changes in relation to climate change are not yet known wholly (de Vries and Griffiths 2018). However, it can be suggested that as environmental factors have influence on phenotype, the genotype over the course of time may as well get altered to accommodate the changing environmental variables. Thus, to enhance or preserve the microbial beneficial traits, advanced genetic engineering approaches are being tested (Shamim et al. 2022) though it is still in experimentation due to ethical questions but in future it may gain accessibility.

Genetic engineering is a promising approach to develop microbial inoculants for agriculture, as it allows for the targeted modification of microbial traits related to plant growth promotion (Mojica and Rodriguez-Valera 2016). However, the effective delivery of these essential microbes to the soil rhizosphere remains a critical challenge. To address this issue, nanotechnology has emerged as an important tool for enhancing the delivery and efficacy of biofertilizers in agriculture. Specifically, the encapsulation of biofertilizers in nanoparticles can improve their precision and effectiveness in promoting plant growth (Kalia and Kaur 2019). This approach, known as nanobiofertilization, uses nanomaterials such as polymers and gold to effectively bind microbial inoculants and enhance their growth-promoting effects on plants (Shang et al. 2019).

1.2 Role of Microbes as Bio-inoculants

The rhizomicrobiome associated with a particular plant is predominantly controlled by that plant through the exudation of specialized root compounds. Additionally, plants also produce signaling molecules to recruit specific species of microbes, influencing their genetic and biochemical expression (Backer et al. 2018). The microbiota residing in the root system can also self-regulate by producing quorum sensing molecules, which facilitate communication and influence physiological shifts. Conversely, plants respond to quorum sensing signals from microbes by producing analogues that regulate their rhizomicrobiome composition (Smith et al.

2015). Recent studies have revealed that not all microbes within a phytomicrobiome have equal effects on the host plant, and that certain hub or core microbiota are more intimately associated with the plant and exert greater influence over its development (Lyu et al. 2021). The plant core microbiome comprises members of the microbiota that are consistently and systematically associated with the plant throughout various phases and under different environmental conditions. Hub microbiota, on the other hand, are members of the core microbiome that exhibit the most mutualistic and facilitative interactions with the host plant and are central to the assembly of its rhizomicrobiome (Noman et al. 2023).

The rhizomicrobiome encompasses a diverse array of microorganisms that play crucial roles in various plant functions, including nutrient acquisition, mineral transportation, phytohormone synthesis, antibiotic production, secondary metabolite secretion, soil structure improvement, and modulation of various signal molecules, ultimately resulting in enhanced plant growth and development (Backer et al. 2018). Specifically, plant-beneficial microbes such as cyanobacteria, plant growth-promoting rhizobacteria (PGPR), and mycorrhizal fungi have been utilized as biofertilizers and biopesticides in agriculture to augment crop growth and mitigate the effects of biotic and abiotic stress (Pathak and Kumar 2016). These microbial groups possess distinct characteristics and mechanisms for plant growth promotion, such as cyanobacteria's ability to fix atmospheric nitrogen and PGPR's ability to solubilize phosphorus and produce plant growth regulators. Furthermore, mycorrhizal fungi establish a symbiotic relationship with the plant, enhancing nutrient uptake and promoting stress tolerance (Ahemad and Kibret 2014; Debnath et al. 2022). The utilization of these microbial groups in agriculture has shown promising results, with potential implications for sustainable and environmentally friendly farming practices, as listed in Table 1.1 and Fig. 1.1.

1.2.1 Cyanobacteria

Cyanobacteria are one of the primitive photosynthetic microorganisms capable of carbon dioxide and nitrogen fixation. Cyanobacteria are useful in sustainable agriculture due to their ability to tolerate extreme environments and do not compete for arable land (Kuraganti et al. 2020; Pandey and Dheeman 2022). Their cultivation in non-arable land makes the land fertile due to their production and release of exopolysaccharides (EPS). The EPS are able to exert this benefit by gluing the soil particles together and enhancing the organic nutrient accumulation as well as water holding capacity. Thus, increase in organic content and soil moisture enables the growth and development of other microbes like PGPR and eventually alters the physico-chemical properties of the inhabited land (Pathak et al. 2018).

Cyanobacteria can fix atmospheric nitrogen in both symbiotic and free living forms. The fixed nitrogen is released into the environment in the form of ammonia, amino acids, polypeptides, vitamins, and auxin-like compounds (Kollmen and Strieth 2022). *Anabaena variabilis*, *Aulosira fertilisima*, *Nostoc linkia*, *Calothrix*

Table 1.1 Important microbial genera utilized in agricultural practices

Microbial groups	Microbes genera	Beneficial role; Microbial group specific	Reference
Cyanobacteria	*Anabaena, Anabaenopsis, Aluosira, Boryanum, Camptylonema, Calothrix, Chroococcidiopsis, Chthonoplastes, Chlorogloea, Chlorogloeopsis, Dermocarpa, Fischerella, Gloeocapsa, Gloeotrichia, Haplosiphon, Lyngbya, Mastigocladus, Microcoelus, Myxosarcina, Nodularia, Nostoc, Nostochopsis, Oscillatoria, Pleurocapsa, Rivularia, Scytonema, Scytonematopsis, Schizothrix, Synecococcus, Tolypothrix, Trichodesmium, Westiella, Westieollopsis, Wollea, Xenococcus*	• Nitrogen fixation • Increases water holding capacity • CO_2 sequestration • Improvement in soil fertility • Biocontrol • CH_4 oxidation • Bioremediation • Bio-fuel production	Pathak et al. (2018) Chittora et al. (2020) Yadav et al. (2017)
Plant growth-promoting rhizobacteria (PGPR)	*Azospirillum, Azorhizobium, Sinorhizobium, Bradyrhizobium, Mesorhizobium, Allorhizobium, Enterobacter, Frankia, Herbaspirilium, Azospirilium, Burkholderia, Azotobacter, Azospirillum, Azorhizobium, Bacillus, Herbaspirillum, Klebsiella, Pseudomonas, Paenibacillus, Rhizobium*	Direct benefits • Nutrient acquisition • Nitrogen fixation • Phosphate solubilization • Potassium solubilization • Phytohormone production • Siderophore production • Exopolysaccharide production Indirect benefits • Stress management (abiotic and biotic) • Antibiotic production • Induced systemic resistance	Gouda et al. (2018) Kaymak (2011) Jha and Saraf (2015) Kotoky and Pandey (2018) Vandana et al. (2021)

(continued)

Table 1.1 (continued)

Microbial groups	Microbes genera	Beneficial role; Microbial group specific	Reference
Mycorrhizal fungi	*Acaulospora*, *Alternaria*, *Entrophospora*, *Gigaspora*, *Glomus*, *Sclerocystis*, *Funneliformis*, *Glomus aggregatum*, *Glomus claroideum*, *Glomus clarum*, *Glomus etunicatum*, *Glomus fasciculatum*, *Geosporum*, *Gigaspora*, *Glomus intraradices*, *Glomus mosseae*, *Paraglomus occultum*, *Scutellospora*	• Selective absorption of immobile and mobile elements • Enhances water uptake by plants • Tolerance to soil stress (salinity, heavy metals, drought, and nutrient imbalance) • Increases root development during vegetative propagation • Increases plant resistance to pathogens and pests	Giovannini et al. (2020) Siddiqui et al. (2008)

sp., *Scytonema* sp., and *Tolypothrix* sp. are some of the important nitrogen-fixing cyanobacteria (Chittora et al. 2020). Apart from that cyanobacteria can reduce soil salinity and produce many useful by-products. Cyanobacteria can reduce greenhouse gas emission and play an important role in bio-fuel production (Singh et al. 2016). The successful applications of these cyanobacteria were reported in barley, chili, lettuce, maize, oats, radish, sugarcane, and tomatoes (Thajuddin and Subramanian 2005).

1.2.2 Plant Growth-Promoting Rhizobacteria (PGPR)

PGPR are a diverse group of microorganisms that have been shown to confer numerous beneficial effects on plants. These microorganisms utilize various mechanisms to promote plant growth and development (Bhattacharyya and Jha 2012). Direct mechanisms include the production of phytohormones such as auxins, gibberellins, and cytokinins, as well as mineral solubilization, nitrogen fixation, and phosphate solubilization. Indirect mechanisms include the production of siderophores, antimicrobial enzymes, and the induction of systemic resistance against various stresses. Furthermore, PGPR can also produce antioxidant enzymes that scavenge reactive oxygen species, thereby protecting plants from oxidative damage (Devi et al. 2017; Zandi and Schnug 2022; Maheshwari et al. 2023). The use of PGPR as biofertilizers has been found to be a reliable source of promoting plant growth and increasing crop yield. PGPR-induced phytohormone production helps to overcome seed dormancy and promote germination in a shorter time span. Inorganic phosphate solubilization is a crucial activity of PGPR, where the bacteria release specific organic acids into the soil to solubilize phosphate complexes and convert them to

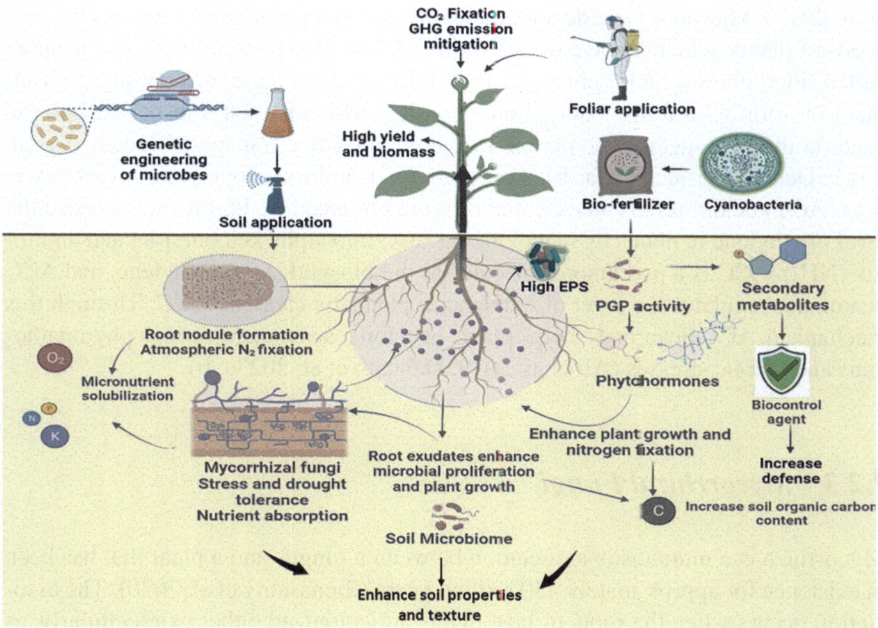

Fig. 1.1 Role of different biofertilizers in the enhancement of plant growth and development. Increase in nitrogen fixation, and micronutrient solubilization occur due to the application of biofertilizers

ortho-phosphates, which can be taken up by plants (Maheshwari et al. 2013a, b; Rajkumari et al. 2022). Moreover, the use of PGPR as nitrogen fixers has gained acceptability due to their ability to fix free nitrogen. Several rhizobial genera, including *Rhizobium*, *Azorhizobium*, *Sinorhizobium*, *Bradyrhizobium*, *Mesorhizobium*, *Allorhizobium*, and other nodule forming bacteria such as *Frankia*, have been reported to act as biological nitrogen fixers for plants. Rhizobacterial strains of *Herbaspirilium*, *Azospirilium*, *Bacillus*, *Herbaspirillum*, *Klebsiella*, and *Burkholderia* are associated with phosphate solubilization and nitrogen fixation in rice crops; they are proposed as an alternative to urea in wheat production (Maheshwari et al. 2013a, b). The use of PGPR as biofertilizers is a promising approach to sustainable agriculture that has the potential to increase crop yield and reduce environmental impacts (Shah et al. 2021).

PGPR can also antagonize the effects of metal stress or toxicity and hence provide enhanced tolerance (Nadeem et al. 2014). They protect the plant from toxicity through immobilization of heavy metals and restricting their translocation by precipitation in plants. *Pseudomonas aeruginosa* and *Burkholderia gladioli* were reportedly able to mitigate cadmium stress in 10 days old seedling of *Lycopersicon esculentum* (Khanna et al. 2019; Kumar et al. 2009). Iron chlorosis is a common disorder in plants that results from iron deficiency, leading to a decrease in micronutrient availability and increased susceptibility to microbial infections (Maheshwari et al. 2013). Microbes have developed highly evolved pathways to address the iron needs of plants, which involve the utilization of low molecular weight iron chelators called siderophores. Siderophores can solubilize iron from the surrounding environment by forming a ferric-siderophore complex, which is then transported into the bacteria through specific membrane receptors via active transport (Beneduzi et al. 2012; Debnath et al. 2023a, b). Furthermore, 1-aminocyclopropane-1-carboxylic acid (ACC) deaminase is an essential enzyme produced by PGPR that regulate the level of ethylene in plants by metabolizing ACC into alpha ketobutyrate and ammonia (NH_3). ACC is a precursor molecule for the biosynthesis of ethylene, and ACC deaminase regulates the level of ethylene in plants by cleaving ACC. Through this mechanism, ACC deaminase helps plants develop resistance to various phytopathogens and abiotic stresses (Ali et al. 2019; Debnath et al. 2023a, b).

1.2.3 Mycorrhizal Fungi

Mycorrhiza is a mutualistic association between a fungus and a plant that has been in existence for approximately 450 million years (Ramasamy et al. 2020). The association occurs when the roots of host plants are colonized either extracellularly, as in ectomycorrhiza, or intracellularly, as in endomycorrhiza. Mycorrhizal fungi have been found to colonize the roots of most plants, including bryophytes, pteridophytes, gymnosperms, and angiosperms. The fungi involved in this mutualistic relationship are mostly from Ascomycetes, Basidiomycetes, and Zygomycetes (Madouh and Quoreshi 2023). It is estimated that approximately 95% of vascular plants have a symbiotic association with mycorrhizal fungi.

Ectomycorrhiza (ECM) involves a wide variety of species, around 5000, and most of them belong to Basidiomycetes, with few being in Ascomycetes and Zygomycetes class of fungi. The families of plants inhabited by ECM involve *Betulaceae*, *Fagaceae*, *Pinaceae*, and *Dipterocarpaceae* (Argiroff et al. 2022). These economically important plants cover wide areas of temperate and boreal forests. The ECM make association with plant root tips by making a sheath (mantle) around it. The ECM traverse into the cortex of root cells to form a specialized structure called "hartig net" (Futai et al. 2008). This hartig net emerges outside of soil to absorb nutrients and water. Thus plant absorbs water and nutrients from the outside through this structure and in return provides fungi with photosynthetic products and other essential materials (Bucking et al. 2012).

Endomycorrhiza or the arbuscular mycorrhiza (AM) are characterized by the presence of specialized structures called arbuscules. These structures grow within the root cells and ramify like a tree within the cell. AM develops and extends its extraradical mycelium up to 6–10 centimeters outwards into the surface of soil and supplies water and nutrients to the plant, mostly phosphate ions, and in exchange plant provides it with carbon molecules like sugars and other carbohydrates (Etesami and Alikhani 2022). AM fungi association mostly occurs through the genera *Acaulospora*, *Entrophospora*, *Gigaspora*, *Glomus*, and *Sclerocystis*. The fungi secrete enzymes that enable their hyphae to penetrate the fleshy cortical and epidermal layers of the root cells. After 2–3 days of inhabitation, they form arbuscules to facilitate nutrient transfer (Ramasamy et al. 2020). Root colonization of plants by these fungi may alter the root exudation pattern of the plant and eventually may alter the root microbiome structure of the rhizosphere. The hyphae of these endomycorrhizae secrete a glycoprotein called glomalin that increases the water stable aggregates (WSA) by accumulating the soil particles, thus improving the inhabited soil structure (Kennedy and de Luna 2005). Apart from that, most AM fungi have vesicular structures for storage of oil-rich products. These vesicular structures in some species function as propagules though in most species AM propagates through a multilayered cell wall containing body called spores. These spores contain lipids that act as storage materials and can survive for several years and therefore, are a viable agent for establishing new colonies. There are also other variations in terms of forming new colonies, for instance, the suborder *Glomineae* can propagate through spores, vesicles, or fragments of hyphae, confirming the means of propagation are not restricted to spores or vesicles only (Müller et al. 2017).

1.3 Why Biofertilizers Are Necessary?

Biofertilizers are defined as products containing live or dormant microorganisms that enhance plant growth by providing essential nutrients like nitrogen, phosphorus, and other minerals to the crops, seeds, or soil (Reddy and Saravanan 2013). Biofertilizers are eco-friendly and are capable of mitigating the impact of abiotic and biotic stressors on crops without adversely affecting soil chemistry or integrity. As a result, they are a valuable tool for promoting sustainable agriculture (Bhardwaj et al. 2014). In contrast, chemical fertilizers work by directly supplying inorganic nutrients to plants, which are effective in the short term. However, in the long run, unused nutrients accumulate in the soil and degrade its quality. In this regard, biofertilizers act as a precise nutrient uptake mechanism from the soil, and they can reduce the gap between nutrient supply and crop acquisition (Debnath et al. 2019). Despite the benefits of biofertilizers, the lack of standardization regarding their optimal environmental conditions and storage methods may make them less effective than chemical fertilizers, which are standardized for application and can be stored for a longer period. Additionally, chemical fertilizers are favored by farmers because they are easy to apply and maintain their quality during application. However, several studies have shown that the combination of biofertilizers and measured amount of chemical fertilizers can yield significantly better results (Soumare et al. 2020). Therefore, although chemical fertilizers have detrimental effects on soil, human health, and the environment (Chandini et al. 2019), until biofertilizer production is increased to meet the demand, and their application is standardized with respect to particular environments, chemical fertilizers will remain a viable option for crop producers.

1.3.1 Commercially Available Biofertilizers

Marketed biofertilizers are mostly utilized or limited to certain consumers due to a lack of understanding among common masses about their efficacy and also variation in efficacy of the same product in different climates or environmental conditions (Owen et al. 2015). Even the production and commercialization of biofertilizers are limited to Asia, Europe, and North America. In Asia, particularly India has a certain level of commercial presence of biofertilizers; rest of the continent does not have any significant presence in terms of commerce (Mishra et al. 2016, 2020). Some of the commercially available biofertilizers have been listed in Table 1.2. Poor quality and low adaptability of biofertilizers are discouraging users to use them extensively. In addition, low funding in research of these marketed products is also putting obstacles in standardization and optimization of this product for its field application in different environmental conditions. Apart from this, varied inferences over research studies of the same product are also demotivating the biofertilizer companies to opt for further research of their marketed products (Stamenković et al. 2018). So, to circumvent the shortcomings of commercial biofertilizers, industries and researchers need to collaborate efficiently to carry out further extensive research on implementation of biofertilizers on various environmental conditions for their effective application in agriculture (Soumare et al. 2020).

1.4 Genetic Engineering to Enhance the Bio-inoculant Effectiveness

The sensitivity of microbial inoculants to environmental conditions can be improved to some extent by introducing stress-tolerant genetic variants of microbial species or strains (Phillips 2008). The same can be applied to enhance the growth promotion activity of any microorganism in a particular rhizosphere.

Genetic diversity over the course of evolution in microbes is attained through genetic exchanges, mutation, and selection to achieve adaptation in a particular ecological niche (Francino 2012). The various selective traits particular to soil microbes, which are beneficial to plants, are needed to be magnified to attain better outcome in a continuously changing environment. The advent of recombinant technology has opened the door to sharing these beneficial traits among microbes, which might lack these traits but are habitants of a particular rhizosphere (Ashajyothi et al. 2022). In this regard, many genetically modified microorganisms (GMM) have been

Table 1.2 Some of the commercially available biofertilizers

Product name	Manufacturer	State	Benefit	Microbes used	Reference
Phosphobacteria and Phosphatika	TNAU Agritech Portal (India)	Powder/ liquid	Increase yield 5–30%	2 bacterial & 2 fungal species	http://agritech.tnau.ac.in
Phosphobacteria	Monarch Bio-Fertilisers and Research Centre (India)	Powder	Dissolve 30–50 kg of phosphorous per hectare	Bacteria	http://www.monarchbio.co.in/bio_fertilizers.html
Phosphobacterium	SAFS Organic Enterprises (India)	Powder/ liquid	Solubilizes insoluble phosphates and makes it available to the plant	Bacteria	https://www.indiamart.com/safsorganicenterprises/bio-fertilizer.html#bio-fertilizer-phosphobacterium
Phosphobacteria	Agro Bio-Tech Research Centre LTD (India)	Powder/ liquid	Phosphate solubilization for the plant	*Bacillus megaterium var. phosphaticum*	http://www.abtecbiofert.com/products.htm
Phosphofix	International Panaacea Limited (India)	Liquid	Reduces 25–30% phosphatic fertilizer requirement	Bacteria	https://www.iplbiologicals.com/
Phosphomax	Varsha Bioscience and Technology India Private Limited (India)	Powder	Increases crop yield by 15–25%	*Bacillus megatherium*	http://www.varshabioscience.com/products/phosphomax.html
Symbion-P	T. Stanes and Company Limited (India)	Liquid	Solubilizes fixed phosphorus in the soil and makes it available to plants	*Bacillus megaterium var. phosphaticum*	http://www.tstanes.com/products-symbion-p.html
Jumpstart LCO	Novozymes Biologicals Limited (Canada)	Powder/ Granular	Phosphate solubilizing bio-inoculants	*Penicillium bilaii*	https://www.novozymes.com/en/advance-yourbusiness/agriculture/crop-production/jumpstart

(continued)

Table 1.2 (continued)

Product name	Manufacturer	State	Benefit	Microbes used	Reference
P Sol B® - BM	AgriLife (India)	Powder (spore)	The bacterium propagates on its own upon application using carbon source of plant and secretes gluconic acid, citric acid, succinic acid, etc. which eventually solubilize insoluble phosphates for plant uptake	*Bacillus megaterium*	http://www.agrilife.in/bioferti_psolb_bm.htm
BiotaMax	CustomBiologicals Inc. (USA)	Powder (spore)	Increases root mass Processes nutrients efficiently Degrades organic material. Produces plant growth hormones Reduces root oxidation	*Bacillus subtilis*, *B. megaterium*, *B. licheniformis*, *B. pumilus*, *B. laterosporus*, *Paenibacillus polymyxa*, *Trichoderma harzianum*, *T. viride*, *T. polysporum*, *T. koningii*	http://www.biotamax.com
Jumpstart®	Novozymes Biologicals Limited (Canada)		Increased root development Improved nitrogen fixation in legume crops Improved stress tolerance Improved seed quality Lower environmental impact and higher yield	*Penicillium bilaiae*	http://www.novozymes.com/en/solutions/agriculture/bioag-in-australia
Custom B5™	Custom Biologicals Inc. (USA)		Enhances soil productivity	*Bacillus subtilis*, *B. laterosporus*, *B. licheniformis*, *B. megaterium*, *B. pumilus*	http://www.biotamax.com
Ovalis Rhizofertil	Agriculture XPRT	Powder	Improves soil quality by mineral amendment and stimulates plant growth	*Pseudomonas putida* I-4163	https://www.agriculture-xprt.com/products/ovalis-rhizofertil biofertilizers-518517

PONCHO/ VOTiVO™ mix	Bayer CropScience LP (USA)		Provide plant protection against insects and nematodes	*Bacillus firmus* mixed with clothianidin	http://fs1.agrian.com/pdfs/Poncho_VOTiVO_Labelnewa.pdf
Custom N_2	Custom Biologicals Inc. (USA)	Liquid	Improves plant's nitrogen utilization	*Paenibacillus polymyxa*	http://www.biotamax.com

developed under laboratory conditions and were tested for their efficacy in the field in a controlled environment (Morrissey et al. 2002). However, it is important to monitor field application of GMMs, not only to observe its beneficial effects on growth and development of plants but also to ensure the stability of engineered traits as well as its fitness in natural conditions. Apart from that, effect on other microbes inhabiting along with introduced modified microbes should also be noted (Ryan et al. 2009). There is an EU directive regarding the stepwise release of any genetically modified organisms (GMO) into environmental conditions. A notable example in this case is the *Pseudomonas putida* WCS358r strain, which was released in the Netherlands for a field trial of 5 years from 1997 to 2002 following the EU directive of safety regarding GMOs (Sharma and Shukla 2022). The bacterium was genetically modified for biocontrol activity against pathogenic fungi *Rhizoctonia solani* and *Gaeumannomyces graminis* var. *triticion* on wheat. Cluster of genes which encode anti-fungal metabolites phenazine carboxylic acid (PCA) and 2,4-diacetyl phloroglucinol (DAPG) was acquired from *Pseudomonas fluorescens* and was introduced into the genome of WCS358r strain through mini-Tn5 transposon system. Field trial GMMs are mostly modified through mini-transposons that lack transposase activity, providing stable integration into the recipient's chromosome and reducing the risk of gene transfer in nature (Glandorf 2019).

Genetic engineering has enabled the modification of microbes to acquire new traits, which can be useful for various purposes, such as enhancing plant growth, biocontrol against phytopathogens, and bioremediation (Smith et al. 1992; Aeron et al. 2011). However, the potential risk of horizontal gene transfer from genetically modified microbes to natural microflora raises concerns about the safety of environmental release. Therefore, field trials of genetically modified bacteria should be conducted in a secluded area with a low risk of widespread transmission of the modified traits (Glick and Bashan 1997). In this context, Table 1.3 outlines some of the microbes that have been genetically modified to transfer essential traits to those lacking them in their wild form (Rosenblueth et al. 2018). Nevertheless, before implementing such modifications in nature, it is crucial to establish a consensus within the scientific community on the safety of environmental release (Amarger 2002). While some previous studies have reported no evidence of gene transfer from genetically modified microbes, more research is needed to address the safety concerns associated with the release of genetically engineered microorganisms into the environment (Kong et al. 2016). Therefore, it is important to approach this issue with caution and take into account all aspects of safety linked to the environmental release of genetically modified microbes (Stemke 2004).

Table 1.3 List of some of the genetically modified microbes for agricultural use

Genetically modified microbe (GMM)	Introduced gene	Obtained from	Beneficial role	Field trail	Reference
Pseudomonas fluorescens and Agrobacterium radiobacter	*tox*	*Bacillus thuringiensis* HD-1	Successful expression of tox gene and found to have insecticidal activity against *Manduca sexta*	Not done	Obukowicz et al. (1986)
Escherichia coli HB101	*mps*	*Erwinia herbicola*	Ability to solubilize hydroxyapatite (HAP) under the condition of phosphate starvation	Not done	Goldstein and Liu (1987)
Pseudomonas cepacia 526	*cryIA(C)*	*Bacillus thuringiensis* HD-1	Expression of insecticidal crystal protein was observed and upon treatment of tobacco plants with the bacterium showed insecticidal activity against *Manduca sexta*	Not done	Stock et al. (1990)
Rhizobium leguminosarum	Δ-endotoxin gene	*Bacillus thuringiensis subsp. tenebrionis*	Expression of δ-endotoxin gene which was found to be toxic against larvae of *Sitonalepidus*	Not done	Skøt et al. (1990)
Escherichia coli JM109	*gabY*	*Pseudomonas cepacia*	Participates in phosphate solubilization and production of gluconic acid	Not done	Babu-khan et al. (1995)
Escherichia coli JM109	*pKKY*	*Enterobacter agglomerans*	Increase in phosphate solubilization but no significant change in pH	Not done	Kim et al. (1997)
Escherichia coli DH5α and HB101	*pqq/D,E*	*Rahnella aquatilis*	Increased activity of phosphate solubilization and gluconic acid production	Not done	Kim et al. (1998)
Pseudomonas fluorescens and Herbaspirillum seropedicae	*cry1Ac7*	*Bacillus thuringiensis* 234	Expression of the Cry protein and with proper promoter, both the modified strain produced enough Bt toxins to kill the *Eldana saccharina* larvae	Not done	Downing et al. (2000)

(continued)

Table 1.3 (continued)

Genetically modified microbe (GMM)	Introduced gene	Obtained from	Beneficial role	Field trail	Reference
Burkholderia cepacia IS-16 and *Pseudomonas* sp. PSS	*pqq*	*Erwinia herbicola*	Solubilization of tricalcium phosphate	Not done	Rodríguez et al. (2001)
Azospirillum brasilense Cd and Sp245	ACC deaminase	*Enterobacter cloacae* UW4	Enhanced root growth in seedlings inoculated with the modified bacterium	Not done	Holguin and Glick (2001)
Escherichia coli DH5α	PQQ biosynthesis	*Serratia marcescens* ER2	Capable of gluconic acid production and tricalcium phosphate solubilization	Not done	Krishnaraj and Goldstein (2001)
Burkholderia cepacia IS-16	*NapA* acid phosphatase gene	*Morganella morganii*	Extracellular enzymatic activity was enhanced in the recombinant strain	Not done	Fraga et al. (2001)
Escherichia coli DH5α	*pqq/A,B,C,D,E,F*	*Enterobacter intermedium* 60-2G	Expression of endogenous glucose dehydrogenase, which secreted gluconic acids that solubilized the insoluble phosphates	Not done	Kim et al. (2003)
Azospirillum sp.	*pqq*	*Erwinia herbicola*	Ability to solubilize dicalcium phosphate	Not done	Vikram et al. (2007)
Rhizobium spp. ST1 and IC3123	*fhuA*	*Escherichia coli*	Successful expression of *FhuA* and was able to utilize ferri-ferrichrome as iron source	Not done	Rajendran et al. (2007)
Rhizobium sp. TAL1145	*acdS*	*Sinorhizobium* sp. BL3	Enhanced expression of ACC deaminase and compared to wild type BL3 showed greater activity	Not done	Tittabutr et al. (2008)
Azospirillum brasilense Sp245	*ipdC*	*Escherichia coli* DHα	Seed inoculated with genetically modified bacterium showed enhanced shoot biomass	Not done	Baudoin et al. (2010)

(continued)

Table 1.3 (continued)

Genetically modified microbe (GMM)	Introduced gene	Obtained from	Beneficial role	Field trail	Reference
Enterobacter asburiae PSI3	*gad*	*Pseudomonas putida* KT2440	The transformant showed better MPS phenotype, i.e., better phosphate solubilization as well as 2 Ketogluconic acid productions	Not done	Kumar et al. (2013)
Pseudomonas fluorescens pf36	*pase4*	Metagenomic fosmid library	Expression of protease PASE4 was observed and found to have the ability to degrade *Radopholus similis* tissues	Not done	Chen et al. (2018)

1.5 Nano-Biofertilizers: An Approach to Precision Agriculture

The challenges associated with short shelf life, desiccation, high dosage, field stability, and improper storage of biofertilizers are the major issues which limit the efficacy of biofertilizers (Mishra et al. 2017). Interestingly, nanoencapsulation of chemical fertilizers with nanoparticles like silver, zinc oxide, and titanium dioxide has shown promising results in terms of its slow release and reduction in precipitation of chemical fertilizers by soil and thus opened a new arena of precision agriculture (Mittal et al. 2020). The precision agriculture has the benefit of limited use of fertilizers and pesticides with greater impacts as well as reducing the cost involved in chemical management and restrains unnecessary loss (Zhou et al. 2023; Singha et al. 2021). This inspiring new measure has also been applied to biofertilizers to overcome the possible drawbacks of biofertilizers in field application. In this context, the nanobiofertilizer is a hybrid of nanotechnology and biofertilizers to attain the benefit of both approaches in a cocktail manner for greater efficacy in field application of biofertilizers in a sustainable eco-friendly way (Kumari and Singh 2020). Nanoencapsulation of biofertilizers with silver, gold, zeolite, chitosan, and polymers has provided the fertilizers with better activity and stability due to the increased surface area and surface tension of the nanoparticles, which hold nutrients or other useful materials firmly compared to any conventional delivery materials (Iqbal et al. 2019; Duhan et al. 2017). Nanoencapsulation allows biofertilizers to enhance the shelf life and controlled release of microbes (Vejan et al. 2016). It helps the microbes to hold firmly to the plant and perform efficient nutrient uptake through slow and steady release of the nutrients (Gouda et al. 2018).

To protect biofertilizers from desiccation in field application, polymer emulsion has proven to be a useful measure. Polymer emulsion like water in oil preparation can have the capacity to trap water into an oil layer (Rojas-Sánchez et al. 2022). The water in this case contains the necessary microbes to be utilized in the field, which remains protected from water evaporation. The microbes in this preparation have longer shelf life and remain more viable. This preparation has longer storage period and acts effectively during field distribution (VanderGheynst et al. 2006). An experiment with gold nanoparticles encapsulated with certain PGPR applied in horticultural plants of Fabaceae has also provided impressive inferences. The PGPR - *P. fluorescens, Bacillus subtilis,* and *Paenibacillus elgii* had shown significant plant growth promotion when applied to their respective host plants (Shukla et al. 2015; Duhan et al. 2017). Therefore, nanofertilizers can be an astounding approach to augment the activities of available biofertilizers though it is still in experimentation phase (Danish and Hussain 2019). Hence a suitable combination of nanoparticles and microbes can be a useful tool towards sustainable agriculture.

1.6 Strategies to Expand the Bottleneck Status of Biofertilizers

Modern agricultural practices towards sustainable and eco-friendly agriculture necessitate the integration of various approaches, such as biofertilizers, nanotechnology, biotechnology, and microbiological methods. Any individual branch of study alone will not be sufficient to efficiently address the drawbacks faced by crop producers during farming (El-Ramady et al. 2018). It is imperative to prioritize the use of indigenous efficient microbes to enhance the growth and development of specific plants. Furthermore, the core and hub microbiome knowledge for a particular plant, in addition to the knowledge of indigenous microbial communities, is essential to ensure that any biofertilizer applied is adaptable to the plant's rhizosphere (Jones et al. 2019). Without ensuring the stability and adaptability of particular microbial inoculants, the outcomes will not be satisfactory. While genetic manipulation of better-performing microbes may be risky due to limited knowledge about their activity in field conditions and the risk of conjugation by genetically modified microbes among related and unrelated microbes (Nambisan 2017), manipulation of the rhizomicrobiome presents a possible option for the current situation. Rhizosphere engineering, coupled with rhizomicrobiome manipulation, may offer a better outcome. In this approach, volatile organic carbon compounds of plant origin can be applied to shape the rhizomicrobiome to desired characteristics (Zhan et al. 2022). Thus, by obtaining specific root metabolites necessary for influencing particular rhizobacteria, the rhizomicrobiome can be manipulated towards better growth-promoting microbes for better results. The genetic traits responsible for maintaining the root secretions can be improved, leading to the development of better crop varieties that can potentially influence beneficial root microbiota through engineered root metabolite secretions (Ryan et al. 2009; Schulz-Bohm et al. 2018).

Moreover, optimizing the delivery systems of biofertilizers is critical. In this regard, nanomaterials offer a better option. The use of nanotechnology may significantly enhance the activity of biofertilizers, but extensive evaluation of this approach is also necessary (Iqbal et al. 2019). Thus, a consortium of efforts will be required to overcome the limitations of biofertilizers, and once these limitations are addressed, the industrial sector will likely adopt biofertilizers as a potential product to meet the growing agricultural demands of fertilizers and pesticides.

1.7 Future Prospects

The use of microbial bio-inoculants in sustainable agriculture has been increasingly recognized as an environmentally friendly and cost-effective approach to promote plant growth, yield, and soil health (Nosheen et al. 2021). Bio-inoculants contain beneficial microorganisms such as bacteria, fungi, and actinomycetes that colonize plant roots and promote nutrient uptake, reduce stress, and enhance plant growth (Saeed et al. 2021). Moreover, bio-inoculants have been shown to contribute to soil fertility, soil structure, and crop resilience, reducing the need for synthetic fertilizers and pesticides (Maitra et al. 2022).

Recent advancements in biotechnology and genetic engineering have opened up new opportunities for developing highly effective and specialized bio-inoculants. For example, genetic modification of rhizobacteria to enhance plant growth promotion has shown promising results (Aslani borj et al. 2022). Similarly, the use of nanotechnology in the delivery systems of bio-inoculants has enabled improved efficacy and better targeted delivery of microorganisms to the plant roots (Shang et al. 2019). Bio-inoculants are being used for a wide range of crops, including cereals, legumes, fruits, and vegetables, and have been shown to increase crop productivity and quality, while reducing environmental impacts (O'Callaghan et al. 2022). Therefore, the future prospects of microbial bio-inoculants in sustainable agriculture are highly promising, and their role in addressing global challenges of food security and environmental sustainability is becoming increasingly important. However, further research and development is required to fully realize the potential of microbial bio-inoculants in sustainable agriculture (Suman et al. 2022).

Microbial inoculants offer several advantages over agrochemicals, such as sustainability, environmental safety, and improved soil health. However, the widespread adoption of microbial inoculants is still impeded by certain challenges (Canfora et al. 2021). While research on the application of microbial inoculants in the form of liquids, sprays, dips, or drenches has yielded promising results, their implementation on a large scale is impractical due to the high quantities of inoculants required (O'Callaghan et al. 2022). Additionally, microbial inoculants, such as PGPR, are highly selective and targeted, unlike broad-spectrum chemical inputs, and their efficacy under field conditions may be inconsistent due to the presence of various competing microorganisms (Naik et al. 2019). The variable shelf life of microbial inoculants is another factor that hinders their widespread adoption in agriculture. Additionally, some microbial inoculants can be extremely hazardous when used as biocontrol agents because some of them are toxic and have adverse effects on species that are not their intended targets (Kumari et al. 2023). Therefore, extensive research and development are needed to address these challenges and make microbial inoculant technology a viable option for sustainable agriculture practices.

1.8 Conclusion

Soil microbes possessing plant growth-promoting traits can offer significant benefits to plants, making them a valuable tool for sustainable and eco-friendly agriculture. However, the translation of predicted benefits from lab to field conditions is limited, hindering their widespread use in agriculture. To address this issue, the standardization of microbial inoculum and optimization of delivery methods can enhance their efficacy in natural settings. Moreover, successful manipulation of the rhizomicrobiome offers the possibility of enhanced activity of plant-beneficial microbes for promoting plant growth. While this endeavor may provide fruitful results in field conditions, it is important to address the associated risks. In addition, nanotechnology applications in the delivery system of biofertilizers hold great potential for enhancing the activities of bio-inoculants. Furthermore, biochemical engineering of suitable root metabolites may attract particular microbes for rhizosphere colonization, providing another promising avenue for the development of bio-inoculants. While individual approaches in this regard may appear advantageous, a conglomeration of these approaches holds the most promising outcome. Successful implementation may significantly reduce the use of chemical fertilizers and contribute to environmental restoration.

Acknowledgments IH, UB, ND, PBM, and PP are grateful to the Department of Biotechnology, Ministry of Science and Technology, Govt of India, for financial support.

Conflict of Interest Author(s) declares no conflict of interest.

References

Aeron A, Dubey RC, Maheshwari DK, Pandey P, Bajpai VK, Kang SC (2011) Multifarious activity of bioformulated *Pseudomonas fluorescens* PS1 and biocontrol of *Sclerotinia sclerotiorum* in Indian rapeseed (*Brassica campestris* L.). Eur J Plant Pathol 131(1):81–93. https://doi.org/10.1007/s10658-011-9789-z

Ahemad M, Kibret M (2014) Mechanisms and applications of plant growth promoting rhizobacteria: current perspective. J King Saud Univ Sci 26(1):1–20. https://doi.org/10.1016/j.jksus.2013.05.001

Ali J, Sharma DC, Bano A et al (2019) Chapter 29—Exploiting microbial enzymes for augmenting crop production. In: Kuddus M (ed) Enzymes in food biotechnology. Academic press, pp 503–519. https://doi.org/10.1016/B978-0-12-813280-7.00029-3

Alori ET, Babalola OO (2018) Microbial inoculants for improving crop quality and human health in Africa. Front Microbiol 9. https://doi.org/10.3389/fmicb.2018.02213

Amarger N (2002) Genetically modified bacteria in agriculture. Biochimie 84(11):1061–1072. https://doi.org/10.1016/S0300-9084(02)00035-4

Argiroff WA, Zak DR, Pellitier PT, Upchurch RA, Belke JP (2022) Decay by ectomycorrhizal fungi couples soil organic matter to nitrogen availability. Ecol Lett 25(2):391–404. https://doi.org/10.1111/ele.13923

Ashajyothi M, Charishma K, Patel A et al (2022) Rhizosphere microbiome: significance in sustainable crop protection. In: Singh UB, Sahu PK, Singh HV, Sharma PK, Sharma SK (eds) Rhizosphere microbes: biotic stress management. Springer Nature, Singapore, pp 283–309. https://doi.org/10.1007/978-981-19-5872-4_14

Aslani borj M, Etesami H, Alikhani HA (2022) Silicon improves the effect of phosphate-solubilizing bacterium and arbuscular mycorrhizal fungus on phosphorus concentration of salinity-stressed alfalfa (*Medicago sativa* L.). Rhizosphere 24:100619. https://doi.org/10.1016/j.rhisph.2022.100619

Babu-Khan S, Yeo TC, Martin WL, Duron MR, Rogers RD, Goldstein AH (1995) Cloning of a mineral phosphate-solubilizing gene from *Pseudomonas cepacia*. Appl Environ Microbiol 61(3):972–978. https://doi.org/10.1128/aem.61.3.972-978.1995

Backer R, Rokem JS, Ilangumaran G et al (2018) Plant growth-promoting Rhizobacteria: context, mechanisms of action, and roadmap to commercialization of biostimulants for sustainable agriculture. Front Plant Sci 9. https://doi.org/10.3389/fpls.2018.01473

Baudoin E, Lerner A, Mirza MS et al (2010) Effects of *Azospirillum brasilense* with genetically modified auxin biosynthesis gene ipdC upon the diversity of the indigenous microbiota of the wheat rhizosphere. Res Microbiol 161(3):219–226. https://doi.org/10.1016/j.resmic.2010.01.005

Beneduzi A, Ambrosini A, Passaglia LMP (2012) Plant growth-promoting rhizobacteria (PGPR): their potential as antagonists and biocontrol agents. Genet Mol Biol 35(4 SUPPL):1044–1051. https://doi.org/10.1590/S1415-47572012000600020

Bhardwaj D, Ansari MW, Sahoo RK, Tuteja N (2014) Biofertilizers function as key player in sustainable agriculture by improving soil fertility, plant tolerance and crop productivity. Microb Cell Factories 46(1). https://doi.org/10.1186/1475-2859-13-66

Bhattacharyya PN, Jha DK (2012) Plant growth-promoting rhizobacteria (PGPR): emergence in agriculture. World J Microbiol Biotechnol 28(4):1327–1350. https://doi.org/10.1007/s11274-011-0979-9

Bucking H, Elliot L, Prashant A (2012) The role of the mycorrhizal symbiosis in nutrient uptake of plants and the regulatory mechanisms underlying these transport processes. In: Nabin Kumar D, Sudam Charan S (eds) Plant Science. IntechOpen, Rijeka, p Ch. 4. https://doi.org/10.5772/52570

Canfora L, Costa C, Pallottino F, Mocali S (2021) Trends in soil microbial inoculants research: a science mapping approach to unravel strengths and weaknesses of their application. Agriculture 11(2):158. https://www.mdpi.com/2077-0472/11/2/158

Chandini KR, Kumar R, Prakash O (2019) The impact of chemical fertilizers on our environment and ecosystem. Res Trends Environ Sci 2:71–86

Chen D, Wang D, Xu C et al (2018) Nematicidal protease genes screened from a soil metagenomic library to control *Radopholus similis* mediated by *Pseudomonas fluorescens* pf36. Appl Microbiol Biotechnol 102(7):3301–3314. https://doi.org/10.1007/s00253-018-8869-9

Chittora D, Meena M, Barupal T, Swapnil P, Sharma K (2020) Cyanobacteria as a source of biofertilizers for sustainable agriculture. Biochem Biophys Rep 22:100737. https://doi.org/10.1016/j.bbrep.2020.100737

Classen AT, Sundqvist MK, Henning JA, Newman GS, Moore JA, Cregger MA et al (2015) Direct and indirect effects of climate change on soil microbial and soil microbial-plant interactions: what lies ahead? Ecosphere 6(8):1–21. https://doi.org/10.1890/ES15-00217.1

Danish M, Hussain T (2019) Nanobiofertilizers in crop production. In: Panpatte DG, Jhala YK (eds) Nanotechnology for agriculture: crop production & protection. Springer Singapore, Singapore, pp 107–118. https://doi.org/10.1007/978-981-32-9374-8_6

de Vries FT, Griffiths RI (2018) Chapter five—impacts of climate change on soil microbial communities and their functioning. In: Horwath WR, Kuzyakov Y (eds) Developments in soil science, vol 35. Elsevier, pp 111–129. https://doi.org/10.1016/B978-0-444-63865-6.00005-3

Debnath S, Rawat D, Mukherjee AK, Adhikary S, Kundu R (2019) Applications and constraints of plant beneficial microorganisms in agriculture. In: Mirmajlessi SM, Radhakrishnan R (eds) Biostimulants in plant science. IntechOpen, Rijeka., p Ch. 3. https://doi.org/10.5772/intechopen.89190

Debnath S, Das N, Maheshwari DK, Pandey P (2022) Interactions of rhizobia with nonleguminous plants: a molecular ecology perspective for enhanced plant growth. In: Maheshwari DK, Dobhal R, Dheeman S (eds) Nitrogen fixing bacteria: sustainable growth of non-legumes. Springer Nature Singapore, Singapore, pp 23–64. https://doi.org/10.1007/978-981-19-4906-7_3

Debnath S, Chakraborty S, Langthasa M et al (2023a) Non-rhizobial nodule endophytes improve nodulation, change root exudation pattern and promote the growth of lentil, for prospective application in fallow soil. Front Plant Sci 14. https://doi.org/10.3389/fpls.2023.1152875

Debnath S, Das A, Maheshwari DK, Pandey P (2023b) Treatment with atypical rhizobia, *Pararhizobium giardinii* and *Ochrobactrum* sp. modulate the rhizospheric bacterial community, and enhances Lens culinaris growth in fallow-soil. Microbiol Res 267:127255. https://doi.org/10.1016/j.micres.2022.127255

Devi KA, Pandey G, Rawat AKS, Sharma GD, Pandey P (2017) The endophytic symbiont—*Pseudomonas aeruginosa* stimulates the antioxidant activity and growth of *Achyranthes aspera* L. Front Microbiol 8. https://doi.org/10.3389/fmicb.2017.01897

Downing KJ, Leslie G, Thomson JA (2000) Biocontrol of the sugarcane borer *Eldana saccharina* by expression of the *Bacillus thuringiensis* cry1Ac7 and *Serratia marcescens* chiA genes in sugarcane-associated bacteria. Appl Environ Microbiol 66(7):2804–2810. https://doi.org/10.1128/AEM.66.7.2804-2810.2000

Duhan JS, Kumar R, Kumar N, Kaur P, Nehra K, Duhan S (2017) Nanotechnology: the new perspective in precision agriculture. Biotechnol Rep 15:11–23. https://doi.org/10.1016/j.btre.2017.03.002

El-Ramady H, El-Ghamry A, Mosa A, Alshaal T (2018) Nanofertilizers vs. biofertilizers: new insights. Env Biodivers Soil Secur 2:51–72. https://doi.org/10.21608/jenvbs.2018.3880.1029

Etesami H, Alikhani HA (2022) Silicon improves the effect of phosphate-solubilizing bacterium and arbuscular mycorrhizal fungus on phosphorus concentration of salinity-stressed alfalfa (*Medicago sativa* L.). Rhizosphere 24:100619

Fraga R, Rodríguez H, González T (2001) Transfer of the gene encoding the NapA acid phosphatase of *Morganella morganii* to a *Burkholderia cepacia* strain. Acta Biotechnol 21(4):359–369. https://doi.org/10.1002/1521-3846(200111)21:4<359::AID-ABIO359>3.0.CO;2-B

Francino MP (2012) The ecology of bacterial genes and the survival of the new. Int J Evol Biol 394026. https://doi.org/10.1155/2012/394026

Futai K, Taniguchi T, Kataoka R (2008) Ectomycorrhizae and their importance in Forest ecosystems. In: Siddiqui ZA, Akhtar MS, Futai K (eds) Mycorrhizae: sustainable agriculture and forestry. Springer, Netherlands, Dordrecht, pp 241–285. https://doi.org/10.1007/978-1-4020-8770-7_11

Garcia-fraile P, Menendez E, Rivas R (2015) Role of bacterial biofertilizers in agriculture and forestry. AIMS Bioeng 2(3):183–205. https://doi.org/10.3934/bioeng.2015.3.183

Giovannini L, Palla M, Agnolucci M et al (2020) Arbuscular mycorrhizal fungi and associated microbiota as plant biostimulants: research strategies for the selection of the best performing Inocula. Agronomy 10(1):106. https://www.mdpi.com/2073-4395/10/1/106

Glandorf DCM (2019) Re-evaluation of biosafety questions on genetically modified biocontrol bacteria. Eur J Plant Pathol 154(1):43–51. https://doi.org/10.1007/s10658-018-1598-1

Glick BR, Bashan Y (1997) Genetic manipulation of plant growth-promoting bacteria to enhance biocontrol of phytopathogens. Biotechnol Adv 15(2):353–378. https://doi.org/10.1016/S0734-9750(97)00004-9

Goldstein AH, Liu ST (1987) Molecular cloning and regulation of a mineral phosphate solubilizing gene from *Erwinia Herbicola*. Bio/Technology 5(1):72–74. https://doi.org/10.1038/nbt0187-72

Gouda S, Kerry RG, Das G, Paramithiotis S, Shin H-S, Patra JK (2018) Revitalization of plant growth promoting rhizobacteria for sustainable development in agriculture. Microbiol Res 206:131–140. https://doi.org/10.1016/j.micres.2017 08.016

Holguin G, Glick BR (2001) Expression of the ACC deaminase gene from *Enterobacter cloacae* UW4 in *Azospirillum brasilense*. Microb Ecol 41(3):281–288. http://www.jstor.org/stable/4251821

Iqbal M, Umar S, Mahmooduzzafar (2019) Nano-fertilization to enhance nutrient use efficiency and productivity of crop plants. In: Husen A, Iqbal M (eds) Nanomaterials and plant potential. Springer International Publishing, Cham, pp 473–505. https://doi.org/10.1007/978-3-030-05569-1_19

Jha CK, Saraf M (2015) Plant growth promoting rhizobacteria (PGPR): a review. J Agric Res Dev 5(2):108–119. https://doi.org/10.13140/RG.2.1.5171.2164

Jones P, Garcia BJ, Furches A, Tuskan GA, Jacobson D (2019) Plant host-associated mechanisms for microbial selection. Front Plant Sci 10. https://doi.org/10.3389/fpls.2019.00862

Kalia A, Kaur H (2019) Nano-biofertilizers: harnessing dual benefits of nano-nutrient and biofertilizers for enhanced nutrient use efficiency and sustainable productivity. In: Pudake RN, Chauhan N, Kole C (eds) Nanoscience for Sustainable Agriculture. Springer International Publishing, Cham, pp 51–73. https://doi.org/10.1007/978-3-319-97852-9_3

Kaymak HC (2011) Potential of PGPR in agricultural innovations. In: Maheshwari DK (ed) Plant growth and health promoting bacteria. Springer, Berlin, Heidelberg, pp 45–79. https://doi.org/10.1007/978-3-642-13612-2_3

Kennedy AC, de Luna LZ (2005) Rhizosphere. In: Hillel D (ed) Encyclopedia of soils in the environment. Elsevier, pp 399–406. https://doi.org/10.1016/B0-12-348530-4/00163-6

Khanna K, Jamwal VL, Gandhi SG, Ohri P, Bhardwaj R (2019) Metal resistant PGPR lowered cd uptake and expression of metal transporter genes with improved growth and photosynthetic pigments in *Lycopersicon esculentum* under metal toxicity. Sci Rep 9(1):5855. https://doi.org/10.1038/s41598-019-41899-3

Kim KY, McDonald GA, Jordan D (1997) Solubilization of hydroxyapatite by *Enterobacter agglomerans* and cloned *Escherichia coli* in culture medium. Biol Fertil Soils 24(4):347–352. https://doi.org/10.1007/s003740050256

Kim KY, Jordan D, Krishnan HB (1998) Expression of genes from *Rahnella aquatilis* that are necessary for mineral phosphate solubilization in *Escherichia coli*. FEMS Microbiol Lett 159(1):121–127. https://doi.org/10.1111/j.1574-6968.1998.tb12850.x

Kim CH, Han SH, Kim KY et al (2003) Cloning and expression of pyrroloquinoline quinone (PQQ) genes from a phosphate-solubilizing bacterium *Enterobacter intermedium*. Curr Microbiol 47(6):457–461. https://doi.org/10.1007/s00284-003-4068-7

Kollmen J, Strieth D (2022) The beneficial effects of cyanobacterial co-culture on plant growth. Life 12(2):223. https://www.mdpi.com/2075-1729/12/2/223

Kong HG, Kim NH, Lee SY, Lee SW (2016) Impact of a recombinant biocontrol bacterium, *Pseudomonas fluorescens* pc78, on microbial community in tomato rhizosphere. Plant Pathol J 32(2):136–144. https://doi.org/10.5423/ppj.Oa.08.2015.0172

Kotoky R, Pandey P (2018) Plant-microbe symbiosis as an instrument for the mobilization and removal of heavy metals from contaminated soils-a realistic approach. Curr Biotechnol 7(2):71–79. https://doi.org/10.2174/2211550106666170321104354

Koyande AK, Chew KW, Rambabu K, Tao Y, Chu D-T, Show P-L (2019) Microalgae: a potential alternative to health supplementation for humans. Food Sci Human Wellness 8(1):16–24. https://doi.org/10.1016/j.fshw.2019.03.001

Krishnaraj PU, Goldstein AH (2001) Cloning of a *Serratia marcescens* DNA fragment that induces quinoprotein glucose dehydrogenase-mediated gluconic acid production in *Escherichia coli* in the presence of stationary phase *Serratia marcescens*. FEMS Microbiol Lett 205:215–220. https://doi.org/10.1111/j.1574-6968.2001.tb10950.x

Kumar T, Wahla V, Pandey P, Dubey RC, Maheshwari DK (2009) Rhizosphere competent *Pseudomonas aeruginosa* in the management of *Heterodera cajani* on sesame. World J Microbiol Biotechnol 25(2):277–285. https://doi.org/10.1007/s11274-008-9890-4

Kumar C, Yadav K, Archana G, Naresh Kumar G (2013) 2-Ketogluconic acid secretion by incorporation of *Pseudomonas putida* KT 2440 gluconate dehydrogenase (gad) operon in *Enterobacter asburiae* PSI3 improves mineral phosphate solubilization. Curr Microbiol 67(3):388–394. https://doi.org/10.1007/s00284-013-0372-z

Kumari R, Singh DP (2020) Nano-biofertilizer: an emerging eco-friendly approach for sustainable agriculture. Proc Natl Acad Sci, India, Sect B Biol Sci 90(4):733–741. https://doi.org/10.1007/s40011-019-01133-6

Kumari M, Swarupa P, Kesari KK, Kumar A (2023) Microbial inoculants as plant biostimulants: a review on risk status. Life 13(1):12. https://www.mdpi.com/2075-1729/13/1/12

Kuraganti G, Edla S, Pallaval VB (2020) Cyanobacteria as biofertilizers: current research, commercial aspects, and future challenges. In: Yadav AN, Rastegari AA, Yadav N, Kour D (eds) Advances in plant microbiome and sustainable agriculture: functional annotation and future challenges. Springer Singapore, Singapore, pp 259–278. https://doi.org/10.1007/978-981-15-3204-7_11

Lyu D, Zajonc J, Pagé A et al (2021) Plant Holobiont theory: the Phytomicrobiome plays a central role in evolution and success. Microorganisms 9(4):675. https://www.mdpi.com/2076-2607/9/4/675

Madouh TA, Quoreshi AM (2023) The function of arbuscular mycorrhizal fungi associated with drought stress resistance in native plants of Arid Desert ecosystems: a review. Diversity 15(3):391. https://www.mdpi.com/1424-2818/15/3/391

Maheshwari DK, Saraf M, Aeron A (2013) Bacteria in agrobiology. Disease management. Springer, Berlin. https://doi.org/10.1007/978-3-642-37241-4

Maheshwari DK, Saraf M, Aeron A (2013a) Bacteria in agrobiology. Disease management. Springer Nature Singapore. https://doi.org/10.1007/978-3-642-33639-3

Maheshwari DK, Saraf M, Aeron A (2013b) Bacteria in agrobiology: crop productivity. Springer, Berlin, Heidelberg. https://books.google.co.in/books?id=HTbiDfTCuSsC

Maheshwari DK, Das A, Dheeman S, Pandey P (2023) An overall insight into the attributes, interactions, and future applications of "microbial consortium" for plant growth promotion with contemporary approaches. In: Maheshwari DK, Dheeman S (eds) Sustainable agrobiology: design and development of microbial consortia. Springer Nature Singapore, Singapore, pp 3–22. https://doi.org/10.1007/978-981-19-9570-5_1

Maitra S, Brestic M, Bhadra P et al (2022) Bioinoculants—natural biological resources for sustainable plant production. Microorganisms 10(1):51. https://www.mdpi.com/2076-2607/10/1/51

Mishra J, Prakash J, Arora NK (2016) Role of beneficial soil microbes in sustainable agriculture and environmental management. Clim Chang Environ Sustain 4(2):137–149. https://doi.org/10.5958/2320-642x.2016.00015.6

Mishra S, Keswani C, Abhilash PC, Fraceto LF, Singh HB (2017) Integrated approach of agri-nanotechnology: challenges and future trends. Front Plant Sci 8. https://doi.org/10.3389/fpls.2017.00471

Mishra J, Dutta V, Arora NK (2020) Biopesticides in India: technology and sustainability linkages. 3 Biotech 10:210. https://doi.org/10.1007/s13205-020-02192-7

Mittal D, Kaur G, Singh P, Yadav K, Ali SA (2020) Nanoparticle-based sustainable agriculture and food science: recent advances and future outlook. Front Nanotechnol 2:579954. https://doi.org/10.3389/fnano.2020.579954

Mojica FJM, Rodriguez-Valera F (2016) The discovery of CRISPR in archaea and bacteria. FEBS J 283(17):3162–3169. https://doi.org/10.1111/febs.13766

Morrissey JP, Walsh UF, O'Donnell A, Moënne-Loccoz Y, O'Gara F (2002) Exploitation of genetically modified inoculants for industrial ecology applications. Antonie Van Leeuwenhoek 81(1):599–606. https://doi.org/10.1023/A:1020522025374

Müller A, Ngwene B, Peiter E, George E (2017) Quantity and distribution of arbuscular mycorrhizal fungal storage organs within dead roots. Mycorrhiza 27(3):201–210. https://doi.org/10.1007/s00572-016-0741-0

Nadeem SM, Ahmad M, Zahir ZA, Javaid A, Ashraf M (2014) The role of mycorrhizae and plant growth promoting rhizobacteria (PGPR) in improving crop productivity under stressful environments. Biotechnol Adv 32(2):429–448. https://doi.org/10.1016/j.biotechadv.2013.12.005

Naik K, Mishra S, Srichandan H, Singh PK, Sarangi PK (2019) Plant growth promoting microbes: potential link to sustainable agriculture and environment. Biocatal Agric Biotechnol 21:101326. https://doi.org/10.1016/j.bcab.2019.101326

Nambisan P (2017) An introduction to ethical, safety and intellectual property rights issues in biotechnology. Elsevier Science. https://books.google.co.in/books?id=rikUDgAAQBAJ

Noman M, Ahmed T, Shahid M, Mudassir Nazir M, Azizullah, Li D, Song F (2023) Salicylic aciddoped iron nano-biostimulants potentiate defense responses and suppress Fusarium wilt in watermelon. J Adv Res S2090-1232(23):00174–00171. Advance online publication. https://doi.org/10.1016/j.jare.2023.06.011

Nosheen S, Ajmal I, Song Y (2021) Microbes as biofertilizers, a potential approach for sustainable crop production. Sustain For 13(4):1868. https://www.mdpi.com/2071-1050/13/4/1868

O'Callaghan M, Ballard RA, Wright D (2022) Soil microbial inoculants for sustainable agriculture: limitations and opportunities. Soil Use Manag 38(3):1340–1369. https://doi.org/10.1111/sum.12811

Obukowicz MG, Perlak FJ, Kusano-Kretzmer K, Mayer EJ, Watrud LS (1986) Integration of the delta-endotoxin gene of *bacillus thuringiensis* into the chromosome of root-colonizing strains of pseudomonads using Tn5. Gene 45(3):327–331. https://doi.org/10.1016/0378-1119(86)90031-4

Owen D, Williams AP, Griffith GW, Withers PJA (2015) Use of commercial bio-inoculants to increase agricultural production through improved phosphrous acquisition. Appl Soil Ecol 86:41–54. https://doi.org/10.1016/j.apsoil.2014.09.012

Pandey P, Dheeman S (2022) Conclusions: the rhizobial eminence for biological nitrogen fixation biological nitrogen fixation—revisited and refined. In: Maheshwari DK, Dobhal R, Dheeman S (eds) Nitrogen fixing bacteria: sustainable growth of non-legumes. Springer Nature Singapore, Singapore, pp 379–382. https://doi.org/10.1007/978-981-19-4906-7_17

Pandey P, Maheshwari DK (2007) Bioformulation of *Burkholderia* sp. MSSP with a multispecies consortium for growth promotion of *Cajanus cajan*. Can J Microbiol 53(2):213–222. https://doi.org/10.1139/w06-118. %M 17496969

Pathak DV, Kumar M (2016) Microbial inoculants as biofertilizers and biopesticides. In: Singh DP, Singh HB, Prabha R (eds) Microbial inoculants in sustainable agricultural productivity: Vol 1: Research Perspectives. Springer India, New Delhi, pp 197–209. https://doi.org/10.1007/978-81-322-2647-5_11

Pathak J, Rajneesh, Maurya PK, Singh SP, Häder D-P, Sinha RP (2018) Cyanobacterial farming for environment friendly sustainable agriculture practices: innovations and perspectives. Front Environ Sci 6. https://doi.org/10.3389/fenvs.2018.00007

Phillips T (2008) Genetically modified organisms (GMOs): transgenic crops and recombinant DNA technology. Nat Educ 1(1):213

Rajendran G, Mistry S, Desai AJ, Archana G (2007) Functional expression of *Escherichia coli* fhuA gene in *Rhizobium* spp. of *Cajanus cajan* provides growth advantage in presence of $Fe3^{+}$: ferrichrome as iron source. Arch Microbiol 187(4):257–264. https://doi.org/10.1007/s00203-006-0191-8

Rajkumari J, Katiyar P, Dheeman S, Pandey P, Maheshwari DK (2022) The changing paradigm of rhizobial taxonomy and its systematic growth upto postgenomic technologies. World J Microbiol Biotechnol 38(11):206. https://doi.org/10.1007/s11274-022-03370-w

Ramasamy M, Geetha T, Yuvaraj M (2020) Role of biofertilizers in plant growth and soil health. In: Everlon Cid R, Ademar Pereira S (eds) Nitrogen fixation. IntechOpen, Rijeka., p Ch. 1. https://doi.org/10.5772/intechopen.87429

Reddy CA, Saravanan RS (2013) Polymicrobial multi-functional approach for enhancement of crop productivity. Adv Appl Microbiol 82:53–113. https://doi.org/10.1016/b978-0-12-407679-2.00003-x

Rodríguez H, Gonzalez T, Selman G (2001) Expression of a mineral phosphate solubilizing gene from *Erwinia herbicola* in two rhizobacterial strains. J Biotechnol 84(2):155–161. https://doi.org/10.1016/s0168-1656(00)00347-3

Rojas-Sánchez B, Guzmán-Guzmán P, Morales-Cedeño LR et al (2022) Bioencapsulation of microbial inoculants: mechanisms, formulation types and application techniques. Appl Biosci 1(2):198–220. https://www.mdpi.com/2813-0464/1/2/13

Romano I, Ventorino V, Pepe O (2020) Effectiveness of plant beneficial microbes: overview of the methodological approaches for the assessment of root colonization and persistence. Front Plant Sci 11. https://doi.org/10.3389/fpls.2020.00006

Rosenblueth M, Ormeño-Orrillo E, López-López A et al (2018) Nitrogen fixation in cereals. Front Microbiol 9. https://doi.org/10.3389/fmicb.2018.01794

Ryan P, Dessaux Y, Thomashow L, Weller D (2009) Rhizosphere engineering and management for sustainable agriculture. Plant Soil 321:363–383. https://doi.org/10.1007/s11104-009-0001-6

Saeed Q, Xiukang W, Haider FU et al (2021) Rhizosphere bacteria in plant growth promotion, biocontrol, and bioremediation of contaminated sites: a comprehensive review of effects and mechanisms. Int J Mol Sci 22:19. https://doi.org/10.3390/ijms221910529

Schulz-Bohm K, Gerards S, Hundscheid M, Melenhorst J, de Boer W, Garbeva P (2018) Calling from distance: attraction of soil bacteria by plant root volatiles. ISME J 12(5):1252–1262. https://doi.org/10.1038/s41396-017-0035-3

Shah A, Nazari M, Antar M et al (2021) PGPR in agriculture: a sustainable approach to increasing climate change resilience. Front Sustain Food Syst 5. https://doi.org/10.3389/fsufs.2021.667546

Shamim K, Khan SA, Ahmad S (2022) Consumers' understanding of nutrition labels for ultra-processed food products. J Public Aff 22(1):e2398

Shang Y, Hasan MK, Ahammed GJ, Li M, Yin H, Zhou J (2019) Applications of nanotechnology in plant growth and crop protection: a review. Molecules 24(14):2558. https://www.mdpi.com/1420-3049/24/14/2558

Sharma B, Shukla P (2022) Futuristic avenues of metabolic engineering techniques in bioremediation. Biotechnol Appl Biochem 69(1):51–60. https://doi.org/10.1002/bab.2080

Shukla SK, Kumar R, Mishra RK et al (2015) Prediction and validation of gold nanoparticles (GNPs) on plant growth promoting rhizobacteria (PGPR): a step toward development of nano-biofertilizers. Nanotechnol Rev 4(5):439–448. https://doi.org/10.1515/ntrev-2015-0036

Siddiqui ZA, Akhtar MS, Futai K (2008) Mycorrhizae: sustainable agriculture and forestry. Springer Nature, Singapore. https://doi.org/10.1007/978-1-4020-8770-7

Singh JS, Kumar A, Rai AN, Singh DP (2016) Cyanobacteria: a precious bio-resource in agriculture, ecosystem, and environmental sustainability. Front Microbiol 7. https://doi.org/10.3389/fmicb.2016.00529

Singha KM, Singh B, Pandey P (2021) Host specific endophytic microbiome diversity and associated functions in three varieties of scented black rice are dependent on growth stage. Sci Rep 11(1):12259. https://doi.org/10.1038/s41598-021-91452-4

Skøt L, Harrison SP, Nath A, Mytton LR, Clifford BC (1990) Expression of insecticidal activity in *Rhizobium* containing the δ-endotoxin gene cloned from *Bacillus thuringiensis* subsp. *tenebrionis*. Plant Soil 127:285–295

Smith E, van Elsas JD, van Veen JA (1992) Risks associated with the application of genetically modified microorganisms in terrestrial ecosystems. FEMS Microbiol Rev 8(3–4):263–278. https://doi.org/10.1111/j.1574-6968.1992.tb04992.x

Smith D, Subramanian S, Lamont J, Bywater-Ekegärd M (2015) Signaling in the phytomicrobiome: breadth and potential. Front Plant Sci 6. https://doi.org/10.3389/fpls.2015.00709

Soumare A, Boubekri K, Lyamlouli K, Hafidi M, Ouhdouch Y, Kouisni L (2020) From isolation of phosphate solubilizing microbes to their formulation and use as biofertilizers: status and needs. Front Bioeng Biotechnol 7. https://doi.org/10.3389/fbioe.2019.00425

Stamenković S, Beškoski V, Karabegović I, Lazić M, Nikolić N (2018) Microbial fertilizers: a comprehensive review of current findings and future perspectives. Span J Agric Res 16:1. https://doi.org/10.5424/sjar/2018161-12117

Stemke DJS (2004) Genetically modified microorganisms. In: Parekh SR (ed) The GMO handbook: genetically modified animals, microbes, and plants in biotechnology. Humana Press, Totowa, NJ, pp 85–130. https://doi.org/10.1007/978-1-59259-801-4_4

Stock CA, McLoughlin TJ, Klein JA, Adang MJ (1990) Expression of a *bacillus thuringiensis* crystal protein gene in *Pseudomonas cepacia* 526. Can J Microbiol 36(12):879–884. https://doi.org/10.1139/m90-152

Suman J, Rakshit A, Ogireddy SD, Singh S, Gupta C, Chandrakala J (2022) Microbiome as a key player in sustainable agriculture and human health. Front Soil Sci 2. https://doi.org/10.3389/fsoil.2022.821589

Suyal DC, Soni R, Sai S, Goel R (2016) Microbial inoculants as biofertilizer. In: Singh DP, Singh HB, Prabha R (eds) Microbial inoculants in sustainable agricultural productivity: Vol 1: Research Perspectives. Springer India, New Delhi, pp 311–318. https://doi.org/10.1007/978-81-322-2647-5_18

Thajuddin N, Subramanian G (2005) Cyanobacterial biodiversity and potential applications in biotechnology. Curr Sci 89(1):47–57. http://www.jstor.org/stable/24110431

Tittabutr P, Awaya JD, Li QX, Borthakur D (2008) The cloned 1-aminocyclopropane-1-carboxylate (ACC) deaminase gene from *Sinorhizobium* sp. strain BL3 in *Rhizobium* sp. strain TAL1145 promotes nodulation and growth of *Leucaena leucocephala*. Syst Appl Microbiol 31(2):141–150. https://doi.org/10.1016/j.syapm.2008.03.001

Vandana UK, Rajkumari J, Singha LP et al (2021) The endophytic microbiome as a hotspot of synergistic interactions, with prospects of plant growth promotion. Biology 10(2):101. https://www.mdpi.com/2079-7737/10/2/101

VanderGheynst JS, Scher H, Guo H-Y (2006) Design of formulations for improved biological control agent viability and sequestration during storage. Ind Biotechnol 2(3):213–219. https://doi.org/10.1089/ind.2006.2.213

Vejan P, Abdullah R, Khadiran T, Ismail S, Nasrulhaq Boyce A (2016) Role of plant growth promoting rhizobacteria in agricultural sustainability—a review. Molecules 21(5):573. https://www.mdpi.com/1420-3049/21/5/573

Vikram A, Alagawadi AR, Krishnaraj PU, Mahesh Kumar KS (2007) Transconjugation studies in *Azospirillum* sp. negative to mineral phosphate solubilization. World J Microbiol Biotechnol 23(9):1333–1337. https://doi.org/10.1007/s11274-007-9365-z

Yadav S, Rai S, Rai R, Shankar A, Singh S, Rai LC (2017) Cyanobacteria: role in agriculture, environmental sustainability, biotechnological potential and agroecological impact. In: Singh DP, Singh HB, Prabha R (eds) Plant-microbe interactions in agro-ecological perspectives: volume 2: microbial interactions and agro-ecological impacts. Springer Singapore, Singapore, pp 257–277. https://doi.org/10.1007/978-981-10-6593-4_10

Zandi P, Schnug E (2022) Reactive oxygen species, antioxidant responses and implications from a microbial modulation perspective. Biology 11(2):155. https://www.mdpi.com/2079-7737/11/2/155

Zhan C, Matsumoto H, Liu Y, Wang M (2022) Pathways to engineering the phyllosphere microbiome for sustainable crop production. Nat Food 3:997. https://doi.org/10.1038/s43016-022-00636-2

Zhou S, Qing C, He J, Xu D (2023) Impact of agricultural division of labor on fertilizer reduction application: evidence from Western China. Int J Environ Res Public Health 20(5):3787. https://www.mdpi.com/1660-4601/20/5/3787

Web Pages

http://agritech.tnau.ac.in
http://fs1.agrian.com/pdfs/Poncho_VOTiVO_Labelnewa.pdf
http://www.abtecbiofert.com/products.htm
http://www.agrilife.in/bioferti_psolb_bm.htm
http://www.biotamax.com
http://www.monarchbio.co.in/bio_fertilizers.html
http://www.novozymes.com/en/solutions/agriculture/bioag-in-australia
http://www.tstanes.com/products-symbion-p.html
http://www.varshabioscience.com/products/phosphomax.html
https://www.agriculture-xprt.com/products/ovalis-rhizofertilbiofertilizers-518517
https://www.indiamart.com/safsorganicenterprises/bio-fertilizer.html#bio-fertilizer-phosphobacterium
https://www.iplbiologicals.com/
https://www.novozymes.com/en/advance-yourbusiness/agriculture/crop-production/jumpstart

Chapter 2
Microbe-Based Biostimulants: Latest Developments and Future Perspectives

Ankita Bhattacharya, Priya Mishra, Isha Mishra, Pranay Arora, and Naveen Kumar Arora

Abstract The agricultural sector needs to achieve the demand of increasing population in an ecofriendly and sustainable manner. Microbe-based biostimulants are proving to be good replacement for chemical fertilizers and pesticides, which have many ill effects on the environment and human health. Microbes have the natural ability to elevate plant growth, inhibit plant pathogens, combat abiotic stresses, and improve soil quality, thus enhancing overall plant and soil health. Plant growth-promoting microorganisms (PGPMs) reside in rhizosphere, within plant tissues (endophytes), over plant surface (epiphytes), or found free living in the soil ecosystem. In the last few decades, commercialization of microbial biostimulants has increased manifold with the awareness in society about their benefits and negative impact of agrochemicals. Plant growth-promoting rhizobacteria (PGPR) belonging to *Bacillus*, *Pseudomonas*, *Rhizobium*, *Azotobacter*, and *Azospirrilum* species, plant growth-promoting fungi (PGPF) including genera of *Trichoderma*, *Aspergillus*, *Piriformospora*, *Penicillium*, and *Phoma*, and mycorrhizal species of genera *Glomus*, *Entrophospora*, *Acaulospora*, and *Sclerocystis* are most commonly available microbial inoculants dominating the market currently that are reported to boost the growth in several plant varieties. PGPMs are well known for fixing atmospheric nitrogen, solubilizing complex bound nutrients like phosphorus, potassium, zinc,

A. Bhattacharya · P. Mishra · N. K. Arora (✉)
Department of Environmental Science, School of Earth and Environmental Sciences, Babasaheb Bhimrao Ambedkar University, Lucknow, India

I. Mishra
IITians4Nation Foundation, Delhi, India

P. Arora
Department of Earth and Environmental Science, University of Manchester, Manchester, United Kingdom

N. K. Arora, B. Bouizgarne (eds.), *Microbial Biotechnology for Sustainable Agriculture Volume 2*, Microorganisms for Sustainability 34,
https://doi.org/10.1007/978-981-97-2355-3_2

and other elements to simple forms that can be taken up by plant roots, and producing metal chelating compounds like siderophores for uptake of iron. In the past few years, microbial biostimulants have emerged to be useful in soil remediation by removal of toxic heavy metals and pesticides, and removal of excess sodium ions from saline soils along with enhancing plant growth. They are being used in wastewater treatment plants and ecological restoration of forests, wetlands, and waterbodies. Hence, with all these benefits PGPMs are efficient tools for the development of cost-effective bioformulations, which help to improve the growth and yield of plants in a sustainable manner. Emerging technologies like metagenomics and metabolomics can be used to find new insights in molecular and metabolic cross talks between plants and microbes. Such insights may result in better products having biostimulatory effects on crops.

Keywords Biostimulants · Plant growth-promoting microbes · Nutrient solubilization · Sustainable agriculture

2.1 Introduction

Human population has increased from 1.5 billion at the end of nineteenth century to around 6 billion at the beginning of this century,[1] and has touched the mark of 7.8 billion in March 2020 (Worldometers 2020). With such increase in the trend of population comes the need to overcome hunger. The agricultural sector tries its best to fulfill the demand, and various unsustainable approaches are implemented to combat food insecurities. The Green Revolution resulted in increased fertilizer and pesticide production and usage in agro-systems. Although the use of agrochemicals has many instant advantages as they are relatively inexpensive and have high nutrient content that are easily absorbed by plants, all this comes at the expense of implications to humans and environmental health. Agrochemicals have proven to ultimately degrade the quality of land by nutrition loss, soil salinization, acidification, desertification, contamination of ground and surface water, and loss of ecologically important microbial communities and biodiversity (Han et al. 2016; Arora 2019). With the increase in research and awareness about the injudicious use of these chemical inputs, emphasis is now being given on the use of microbe-based biostimulants (Yakhin et al. 2017; Arora et al. 2020a, b). Microbe-based biostimulants provide a sustainable and sound replacement as they have natural ecological power to help plants in nutrient acquisition through biological nitrogen fixation and phosphate, potassium, and micronutrient uptake and mineralization (Suhag 2016; Devi et al. 2022).

Kauffman et al. (2007) were the first to define the term "biostimulants" as materials, other than fertilizers, that promote plant growth when applied in low quantities. This definition demarcates the difference between fertilizers and biostimulants

[1] https://www.un.org/en/global-issues/population

by the quantity of them applied in the agricultural system. Fertilizers are used in bulk as their main constituents are simple forms of nutrients that plants can take directly, whereas biostimulants are substances that enhance the ability of plants in some way so that they can enhance nutrient uptake, combat abiotic stresses like high salinity, drought conditions, etc., hence helping in the overall growth performance of plants. The European Biostimulants Industry Council (EBIC) defines plant biostimulants as "any substance(s) and/or micro-organisms whose function when applied to plants or the rhizosphere is to stimulate natural processes to enhance/benefit nutrient uptake, nutrient efficiency, tolerance to abiotic stress, and crop quality." EBIC also excludes biostimulants from the pesticide category as they do not directly control pests. Biostimulants can broadly be categorized into substances and microorganisms; substances include materials like humic and fulvic acids, protein hydrolysates, N-containing compounds, seaweed extracts, botanicals, chitosan, and other biopolymers and inorganic compounds. Microflora that can show plant growth promotion includes fungi and bacteria residing in rhizosphere, within plant system or freely in soil (Du Jardin 2015; Olowe et al. 2020).

Plants are naturally surrounded by an array of micro living systems, which help them in various ways and are known as plant growth-promoting microorganisms (PGPMs). Rhizosphere is considered to be the most active region to find these minute creatures as the microbes get carbohydrates and other plant synthesized materials in the form of root exudates for their growth and multiplication (Glick 2014; Gupta et al. 2015; Goswami and Deka 2022). The PGPMs help plants in acquiring nutrients, enhancing plant growth, combating various environmental stresses, and protecting against pathogens. These microorganisms can additionally improve soil structural and chemical properties like texture, aeration, water retention, cation exchange capacity, organic matter content, lowering the toxicity, balancing pH, salinity, temperature, etc., thus rendering a versatile and sustainable approach to have quality agro-ecosystems. A diversity of microbial genera are found to coexist in soil, rhizosphere, and within the plant system. Some of the genera that have shown to work as plant growth-promoting rhizobacteria (PGPR) include *Azospirillum*, *Azotobacter*, *Arthrobacter*, *Burkholderia*, *Bacillus*, *Enterobacter*, *Klebsiella*, *Pseudomonas*, *Serratia*, and *Rhizobium*. Some common bacterial endophytic genera are *Azorhizobium*, *Azospirillum*, *Bradyrhizobium*, *Gluconacetobacter*, *Methylobacterium*, *Rhizobium*, *Agrobacterium*, *Proteobacteria*, *Burkholderia*, *Enterobacter*, *Klebsiella*, *Pantoea*, *Pseudomonas*, *Bacillus*, and *Arthrobacter* (Arora et al. 2013; Verma et al. 2019). Similarly, there are a number of fungal genera reported either in rhizosphere or as endophytes, acting as PGPMs.

History of symbiosis of plants and microbes dates back to the time probably when terrestrial plants arrived on earth. Plant fungi colonize almost 90% of terrestrial plant roots and most of the time it is an obligatory relationship. Both plants and fungi benefit as plants receive nutrition and other necessary requirements from the hyphal network while fungi obtain plant made nutrition. Most successful fungal symbionts are known to be ectomychorrhiza and arbuscular mycorrhiza (Behie and Bidochka 2014; Hagh-Doust et al. 2022). These microbes also have the ability to help plants in avoiding various abiotic stresses. According to FAO (2017), 6–20%

(upto USD 120 billion) of total agricultural production worldwide is estimated to be lost by abiotic stresses like salinity, droughts, floods, frosts, nutrient deficiencies, and various soil and air toxicities. Microbes can help plants cope with salinity (Arora et al. 2020a, b; Mishra et al. 2022), drought (Vurukonda et al. 2016; Ashry et al. 2022), and enhance plant growth even in stressed environments by various mechanisms like by producing exopolysaccharides (EPS), accumulation of osmolytes, production of reactive oxygen species (ROS), scavenging enzymes, phytohormone production, gene regulation, and many more (Sandhya and Ali 2015; Mishra et al. 2021). This chapter will mainly focus on the types of microorganisms that are currently used as plant stimulants and how they help in the acquisition of various essential nutrients.

2.1.1 Microbial Biostimulants

Microbe-based biostimulants have emerged as a promising technology for increasing the productivity in the era of sustainable agriculture (Rouphael and Colla 2020). In the last few decades, their commercialization has also increased. According to MarketsandMarkets's™ (a market research platform), the global market of biostimulants (microbial amendments, humic substances, amino acids, and seaweed extracts) has reached USD 2.6 billion in 2019, which is expected to reach USD 4.9 billion by 2025 at a compound annual growth rate (CAGR) of 11.24% (MNM 2020). According to Fortune Business Insight (2019), the global market of biostimulants (microbial and non-microbial) was found to reach USD 2.24 billion in 2018 and projected to increase to USD 5.69 billion till 2026 at a CAGR of 12.4%. Among microbe-based bioinoculants, bacteria, fungi, and mycorrhiza have shown great demand and are being used in enhancing plant growth via the production of plant hormones, improving nutrient availability and elevating environmental stresses such as drought and salinity. Various microbial biostimulants reported to enhance productivity in crops through nutrient acquisition and other mechanisms are mentioned in Table 2.1.

2.1.2 Bacterial Biostimulants

There are diverse types of PGPR that are currently commercialized in the form of biostimulants (Ruzzi and Aroca 2015; Arora et al. 2020a, b). Major bacterial biostimulants belong to *Bacillus* spp., *Pseudomonas* spp., *Rhizobium* spp., *Bradyrhizobium* spp., *Serratia* spp., *Azospirrilum* spp., *Azotobacter* spp., *Arthobacter*, *Burkholderia* spp., and *Frankia* spp. (Kloepper et al. 1989; Etesami and Adl 2020).

Table 2.1 Microbial biostimulants for enhancing crop productivity through nutrient acquisition and other mechanisms

S. no.	Microbial biostimulants	Nutrient	Crop	Mechanism	Reference
1.	Coinoculation of *Glomus intraradices* and *Trichoderma atroviride*	Many micro- and macronutrients like P, Mg, Fe, Zn, and B	Vegetable crops (lettuce, melon, pepper, tomato, and zucchini)	Various, including increasing root surface area, producing diverse enzymes, organic molecules, siderophores	Colla et al. (2015)
2.	*Trichoderma asperellum*	P, K, Mg, and Zn	Tomato	By reducing the severity of disease Fusarium wilt	Li et al. (2018)
3.	*Trichoderma harzianum*	P and microelements (Fe, Mn, Cu, and Zn)	Tomato	Solubilizing sparingly soluble minerals by acidification by organic acids, chelation by siderophores, redox by ferric reductase, and hydrolysis by phytase	Li et al. (2015)
4.	*Trichoderma harzianum*	N, P, K	Tomato	Enhancement of the rhizospheric microflora	Cai et al. (2015)
5.	*Glomus versiforme* (arbuscular mycorrhiza)	Mg	*Citrus sinensis* and *Citrus reticulate*	Enhanced Mg distribution in all plant parts	Xiao et al. (2014)
6.	*Funneliformis mosseae* (arbuscular mycorrhiza)	P	*Asparagus officinalis* L.	Increased uptake of P	Ping et al. (2014)
7.	*Glomus clarum* (arbuscular mycorrhiza)	N, P, K	*Capsicum annuum* L.	Increased uptake of N, P, K under salinity stress	Kaya et al. (2009)
8.	*Glomus mosseae* (arbuscular mycorrhiza)	P, K, Ca, and Mg	*Capsicum annuum* L.	Reducing Cu accumulation hence increased uptake of essential nutrients	Abdel Latef (2011)
9.	Indigenous AM consortia (*Glomus*, *Gigaspora*, and *Scutellospora* spp.)	P and K	Ornamental plants, *Petunia hybrida*, *Callistephus chinensis*, and *Impatiens balsamina*	Increased uptake of P and K, especially in shoot	Gaur et al. (2000)

(continued)

Table 2.1 (continued)

S. no.	Microbial biostimulants	Nutrient	Crop	Mechanism	Reference
10.	*Glomus fasciculatus* and *G. monosporus* (vesicular-arbuscular mycorrhizal fungi)	P	*Allium cepa* L. (Onion)	Increased uptake of P in deficient saline soil	Ojala et al. (1983)
11.	*Rhizophagus intraradices* (arbuscular mycorrhizal fungi)	P, K, Fe, Mn, Cu, Zn, Mg	*Lactuca sativa* L.	Enhancement of solubilization and uptake of nutrients	Baslam et al. (2013)
12.	*Bradyrhizobium yuanmingense*	Nitrogen	*Cajanus cajan* L.	Symbiotic nitrogen fixation	Araujo et al. (2015)
13.	*Pseudomonas putida* and *Pantoea dispersa*	Iron	Mungbean	Biofortification by siderophore production	Patel et al. (2018)
14.	*Providencia* sp. (PGPR) and *Anabaena* sp., *Calothrix* sp., and *Anabaena* sp. (Cyanobacteria)	Nitrogen and micronutrients	Wheat	Enhancing biofortification of micronutrients and protein synthesis by cyanobacteria as nitrogen assimilation	Rana et al. (2012)
15.	*Azotobacter coroocoocum*, *Azospirillum brasilens*, *Pseudomonas putida*, *Bacillus lentus* (PGPR)	P	*Zea mays* L.	Mobilization of phosphate	Yazdani et al. (2009)
16.	*Arthrobacter* sp. and *Bacillus* sp.	P	Tomato	Stress-induced phosphate solubilization	Banerjee et al. (2010)
17.	Two *Bacillus* strains OSU-142 and M3	N, P, K, Ca, Mg, Fe, Mn, Zn	Raspberry	Uptake and solubilization	Orhan et al. (2006)
18.	*Mesorhizobium* sp. and PGPR *Pseudomonas aeruginosa*	*N*, *P*, and *Fe*	Chickpea (*Cicer arietinum* L.)	Nitrogen acquisition by *Mesorhizobium* sp., and P and Fe uptake by *Pseudomonas aeruginosa*	Verma et al. (2013)
19.	*Pseudomonas fluorescens*	Zn	Wheat	Zn accumulation and solubilization	Sirohi et al. (2015)
20.	*P. fluorescens*	Phosphate	*Pisum sativum* L.	Phosphate solubilization	Otieno et al. (2015)
21.	*Bacillus aryabhattai*	Zinc	*Zea maize*	Zinc solubilization	Mumtaz et al. (2017)

(continued)

Table 2.1 (continued)

S. no.	Microbial biostimulants	Nutrient	Crop	Mechanism	Reference
22.	*Bacillus* sp.	Phosphate	*Oryza sativa*	Phosphate solubilization and organic acid production	Panhwar et al. (2013)
23.	*P. fluorescence*	Fe	Wheat	Siderophore production	Gull and Hafeez (2012)
24.	*Aneurinibacillus aneurinilyticus*, *Aeromonas* sp., and *Pseudomonas* sp.	Fe	Wheat	Siderophore production	Kumar et al. (2018)
25.	*Pantoea ananatis*, *Rahnella aquatilis*, and *Enterobacter* sp.	K and P	*Oryza sativa*	Potassium and phosphate solubilization	Bakhshandeh et al. (2017)
26.	*Paenibacillus polymyxa*	N	*Zea mays*	Nitrogen fixation	Padda et al. (2017)

Among bacterial biostimulants, *Rhizobium* has well-established global market and accounted for maximum use in the form of nitrogen (N) biofertilizer. Not only for enhancing N uptake, rhizobial biostimulants are now also recognized for several other PGP activities useful in the cultivation of legume and non-legume crops (Volpiano et al. 2018). Different species of rhizobia, i.e., *Rhizobium laguerreae*, *Rhizobium leguminosarum*, *Bradyrhizobium diazoefficiens*, *Bradyrhizobium japonicum*, *Sinorhizobium meliloti*, *Rhizobium phaseoli*, *Mesorhizobium ciceri*, and *Azorhizobium caulinodans*, have been reported to promote growth in legume and non-legume crops such as *Spinacia oleracea*, *Phaseolus vulgaris*, *Glycine max*, *Zea mays*, *Medicago truncatula* (Jiménez-Gómez et al. 2018; Jones 2019; Purwaningsih et al. 2019; Speck et al. 2019; Sun et al. 2020). Currently, several rhizobial biostimulants are available in global market for improving nodulation, plant growth, and yield of leguminous crops. The bioformulations of these biostimulants are either in liquid or in solid form (Arora et al. 2017). For example, 1495MAR is a liquid bioformulation of *Bradyrhizobium* sp. produced by Marondera SPRL, Zimbabwe; Legumefix is a peat-based bioformulation of *Bradyrhizobium* sp. manufactured by Legume technology, UK. In India, the market of rhizobial biostimulants is also prevalent and robust.

The role of fluorescent pseudomonads as biostimulants has also been found to enhance productivity in several crops. Fluorescent pseudomonads are well known for their root colonization ability, PGP activity, and antagonistic traits useful for deterring phytopathogens. Keerthana et al. (2022) documented that fluorescent pseudomonads are versatile PGPR having multifarious PGP and biocontrol activities. Arora and Mishra (2016) reviewed the application of fluorescent pseudomonads in agro-ecosystems. In the last few years, application of fluorescent pseudomonads-based biostimulants has been found to enhance the quality of crop produce. For example, in a study Jiménez et al. (2020) found that the application of *Pseudomonas fluorescence* has not only increased biomass but also enhanced total oil content and lipid composition in oilseed crops. Apart from *P. fluorescence*, there are several other species of fluorescent pseudomonads that are identified as biostimulants. Currently, *Pseudomonas putida*, *Pseudomonas entomophila*, and *Pseudomonas nitroreducens* are being commercialized in the form of biostimulants (Shah et al. 2017; Ansari and Ahmad 2018; Trinh et al. 2018; Fatima and Arora 2021; Mishra et al. 2020; Khare and Arora 2021). The role of *Pseudomonas aeruginosa* as a biostimulant is also reported by some workers (Gupta et al. 2020; Ghadamgahi et al. 2022), but due to its potential pathogenicity, application of cell-based bioformulation is not recommended in this case (Mishra and Arora 2018). In such circumstances, secondary metabolites of fluorescent pseudomonads could be utilized in the preparation of novel biostimulants having broad spectrum agricultural applications (Mishra et al. 2018). In a study, Tewari and Arora (2018) revealed the application of secondary metabolites (salicylic acid and EPS) from *P. aeruginosa* for ameliorating salinity stress, productivity loss and biocontrol of phytopathogenic fungi *Macrophomina phaseolina* in sunflower.

Members of the genus *Bacillus* are also very well-known bacterial biostimulants having applications in sustainable agriculture. There are several species of *Bacillus* such as *B. pumilus*, *B. sonorensis*, *B. velezensis*, *B. amyloliquefaciens*, *B. flexus*, *B. oryzicola*, and *B. megaterium* that are commercialized in the form of biostimulants (Upadhyay et al. 2019; Borriss 2020; Bhatt and Maheshwari 2020; Luo et al. 2022). Recently, *Bacillus*-based biostimulants are also linked in enhancing alkaloid content in spices and medicinal plants. For instance, application of *Bacillus halotolerans* was found to increase phenolic compounds in *Coriandrum sativum* L. (Jiménez-Gómez et al. 2020). In another study, *Bacillus safensis* increased stevioside content in *Stevia rebaudiana* (Prakash and Arora 2020).

Apart from the above-mentioned bacterial biostimulants, there are some other bacteria that also play an important role in sustainable agriculture. For example, *Paenibacillus mucilaginosus*, *Burkholderia cenocepacia*, *Burkholderia seminalis*, *Kocuria turfanensis*, *Delftia tsuruhatensis*, *Arthrobacter nitroguajacolicus*, *Arthrobacter enclensis*, *Azotobacter chroococcum*, *Serratia marcescens*, and *Serratia glossina* are reported for their ability to promote the growth of *Vigna radiata*, *Zea mays*, *Lycopersicon esculatum*, *Capsicum annuum*, *Rauwolfia serpentina*, *Triticum* sp., button mushroom, cherry tomato, *Oryza sativa* and *Sesamum indicum*, respectively (Jung et al. 2017; Safdarian et al. 2019; Moon and Yoon 2019; Kotoky et al. 2019; You et al. 2020; El-Beltagi et al. 2022). Puri et al. (2016) found that the application of *Paenibacillus polymyxa* improved N fixation and enhanced seedling length and biomass in *Zea mays*. *B. seminalis* isolated as plant growth-promoting endophyte promoted growth in many plant species including *Arabidopsis*, ching chiang pak choi, pak choi, varieties of lettuce, and Chinese amaranth (Hwang et al. 2021). Similarly, application of a phosphate-solubilizing *Serratia* sp. in the form of a biostimulant improved growth in maize (Ludueña et al. 2017).

2.1.3 Fungal Biostimulants

Application of plant growth-promoting fungi (PGPF) in the form of biostimulants has also been reported and used widely. Among PGPF, *Trichoderma*, *Aspergillus*, *Piriformospora*, *Penicillium*, and *Phoma* are commonly used as biostimulants (Elsharkawy et al. 2017). *Trichoderma* spp. are free-living soil/rhizosphere fungi, known for beneficial activities towards plants Among all known fungal biostimulants *Trichoderma* spp. are extensively used for diverse crops. In the last few decades global application of *Trichoderma* spp.-based products has increased tremendously (Singh et al. 2014; Woo et al. 2023). Several species of *Trichoderma* such as *T. longibrachiatum*, *T. asperellum*, *T. harzianum*, *T. virens*, and *T. atroviride* are now considered as an important segment of fungal-based bioinoculants (Singh et al. 2014; Lucini et al. 2019; Anam et al. 2019; Visconti et al. 2020). There are several studies where the application of *Trichoderma*-based biostimulants was found to be effective in enhancing plant growth and suppressing plant disease as well. In a study, tomato seed priming with *Trichoderma aggressivum*, *Trichoderma saturnisporum*, and *T. longibrachiatum* was reported to increase seedling height, shoot fresh and dry weight, and root fresh weight and dry weight (Maurya et al. 2019). Application of *Trichoderma brevicrassum* was reported to not only promote the growth of cucumber but also found to be very useful in the management of cucumber disease caused by *Rhizoctonia solani* (Zhang and Zhuang 2020). Similarly, two strains of *Trichoderma* (*T. asperelloides* and *T. lentiforme*) not only promoted the growth of cotton plant but also showed biocontrol against the fungal pathogen *Sclerotinia sclerotiorum* (Silva et al. 2022). *Aspergillus* spp. are another type of fungal biostimulant that are well known for their PGP activity. Several species such as *A. niger*, *A. tubingensis*, *A. japonicus*, *A. terreus*, and *A. flavus* have applications as potential fungal biostimulants (Zhao et al. 2018; Jiang et al. 2018; Hamayun et al. 2019; Galeano et al. 2021). Baron et al. (2018) for the first time reported the bio-stimulating properties of *Aspergillus sydowii* and *Aspergillus brasiliensis*. Yin et al. (2017) found that phosphate-solubilizing fungus *Aspergillus aculeatus P93* produces organic acids and phytohormones to avail phosphorus and enhance growth in maize plant. Recently, *Aspergillus brunneoviolaceus* HZ23 found to have promoted the growth of leafy vegetable Pakchoi in newly reclaimed mountainous soil; additionally the PGPF improved soil properties, rhizosphere bacterial community structure, and metabolites of pakchoi rhizosphere soil samples (Li et al. 2023). The use of *Penicillium* spp. as PGPF is also reported. There is a vast array of secondary metabolites including plant growth regulators, which are produced by several species of *Penicillium*. For example, *P. verruculosum*, *P. bilaii*, *P. viridicatum*, *P. commune*,

P. simplicissimum, and *P. oxalicum* are reported to increase nutrient uptake, synthesize phytohormones, and produce defense-related secondary compounds in several crops (Sindhu et al. 2018; Banerjee and Dutta 2019; Chalfoun et al. 2019; Ai et al. 2023).

2.1.4 Mycorrhizal Biostimulants

Apart from free-living fungal biostimulants, there is another important group of fungi termed as mycorrhizae, which form mutualistic association with approximately 80% of plants and play important roles in soil biochemistry. This symbiotic association is formed with higher plants including crop plants, for example, cereals, pulses, tomatoes, onions, etc., as well as with wild plant communities. These associations are being exploited for commercial development of mycorrhiza-based biostimulants. The fungal genera such as *Glomus*, *Entrophospora*, *Acaulospora*, *Gigaspora*, *Sclerocystis*, and *Scutellospora* are known for their intricate symbiotic association with plants. However, among them, *Glomus* is the important genus that is recommended for biostimulant application. Currently, *G. mosseae*, *G. intraradices*, *G. fasciculatum*, and *G. microcarpum* are being used as biostimulants for promoting vegetative growth, yield, and quality of various crops (Sohrabi et al. 2018; Vani et al. 2018; Sahodaran et al. 2019; Sellappan et al. 2023). *Funneliformis mosseae* is another important mycorrhizal fungus that is known to be used as a biostimulant. The fungus has been used for improving nutritional status and confers protection against phytopathogens in crops such as banana and grapes (Shahabivand et al. 2018; Valat et al. 2018). Researchers have also reported plant beneficial properties of *Acaulospora spinosa*, *Gigaspora margarita*, and *Gigaspora rosea*, and recommended their use as biostimulants for several crops (Ke et al. 2016; Xie et al. 2016; Sakamoto and Kaji 2017; Juntahum et al. 2022).

2.2 Microbial Biostimulants for Providing Nutrients to Plants

According to the quantity required plant nutrients are classified as macronutrients and micronutrients. Elements that are categorized as macronutrients are carbon (C), hydrogen (H), oxygen (O), nitrogen (N), phosphorus (P), potassium (K), calcium (Ca), magnesium (Mg), and sulfur (S). On the other side micronutrients include zinc (Zn), copper (Cu), iron (Fe), manganese (Mn), boron (B), molybdenum (Mo), chlorine (Cl), and nickel (Ni) (Fageria et al. 2002; Fan et al. 2021). Some plants also essentially require silicon (Si), sodium (Na), vanadium (V), and cobalt (Co) (Fageria et al. 2008; Choudhary et al. 2020). Among these, plants acquire C, H, and O from hydrological and atmospheric cycles; rest of the elements have to be applied externally to have good productivity.

2.2.1 Microbial Biostimulants for Macronutrients

Nitrogen (N) is a building block of living molecules like proteins and nucleic acids, and is required mainly by plants for growth. Plants are able to utilize many chemical forms of N, be it simple inorganic N compounds such as NH_4^+, NO_3^- to organic forms (Trenkel 2021). But presently, focus is mainly on inorganic forms because of their success and ease of production at the cost of environmental degradation. Excessive use of these chemicals causes issues like eutrophication, reactive N gases in the troposphere, acid rain, groundwater contamination, and contribute to global warming and climate change. These chemical manufacturing processes are highly energy driven with the requirement of large amounts of fossil fuels. Moreover, chemical fertilizers affect the soil microbiota by dis-balancing the natural processes revolving around them (Robertson and Vitousek 2009; Fox et al. 2016). There are mainly two natural pathways for fixing nitrogen; first, atmospheric fixation, in which gaseous nitrogen is converted to plant usable ammonia with the help of lightening, combustion, and volcanic activity, and second is through microbial fixation mainly by symbiotic and free-living nitrogen fixing microbes (diazotrophs). Microbial ability to fix N_2 is due to nitrogenase enzyme with the requirement of 16 ATP molecules for each ammonia molecule formed. Symbiotic association can further be divided into legume and non-legume N_2 fixation. Non-symbiotic fixation is through various blue green algae and aerobic and anaerobic bacteria (De Bruijn 2015; Hill and Jones 2019).

Microbial inoculants provide a healthier and environmentally sound alternative to the use of chemical N_2 fertilizers. Rhizospheric and endophytic microbes can be isolated and can be checked for the ability of N_2 fixation commonly either by the investigation for the presence of *nif* gene or by checking the growth in nitrogen free media (Bhattacharjee et al. 2008; Nagrale et al. 2023). *Rhizobium* is the most common legume-specific symbiotic microbe; *R. japonicum*, *R. leguminosarum*, *R. lupine*, *R. meliloti*, *R. phaseolus*, and *R. trifolii* are some commonly used biofertilizers. Free-living bacteria such as *Azospirillum* species including *A. lipoferum* and *A. brasilense* have proven to be good biofertilizers. *Azotobacter* includes species like *A. chroococcum*, *A. vinelandii*, *A. beijerinckii*, *A. insignis*, and *A. macrocytogenes* (Zaim et al. 2017; Yang et al. 2022). Studies have shown that the inoculation of strains of *Azospirillum* increased the yield of wheat crop by up to 36% through its ability of nitrogen fixation (Piccinin et al. 2013; Chhetri et al. 2022). Cyanobacteria are most commonly found in rice fields, hence also known as "paddy organisms." Rice requires high nitrogen input for production; *Azolla*, a free-floating fern when applied with BGA (*Anabaena*), has proven effective in nitrogen fixation (García-Fraile et al. 2015; Adhikari et al. 2020). The boundaries of diazotrophs have become limitless, and many studies are being done showing the non-conventional strains and plant varieties performing biological N_2 fixation. The list of N_2-fixing bacteria associated with non-legumes includes species of *Achromobacter*, *Alcaligenes*, *Arthrobacter*, *Acetobacter*, *Azomonas*, *Beijerinckia*, *Bacillus*, *Clostridium*, *Enterobacter*, *Erwinia*, *Derxia*, *Desulfovibrio*, *Corynebacterium*, *Campylobacter*, *Herbaspirillum*, *Klebsiella*, *Lignobacter*, *Mycobacterium*, *Rhodospirillum*,

Rhodopseudomonas, *Xanthobacter*, *Mycobacterium*, and *Methylosinus* (Bhat et al. 2015; Dheeman and Maheshwari 2022; Jalal et al. 2022; Katiyar et al. 2022). Ambrosini et al. (2016) have shown that a strain of *Bacillus mycoides*, which was isolated from sunflower rhizosphere, was able to fix nitrogen in sunflower plant. There are studies which have shown that major cereal crops like wheat and maize can also be benefited with inoculation of less familiar nitrogen fixing bacterial strains like *Klebsiella pneumonia* and *Pseudomonas protegens* (Fox et al. 2016; Liu et al. 2019).

Phosphorus (P) is deficient in tropical and subtropical soils due to high reactivity of soluble P with other elements and compounds. Phosphorus is available to plants in inorganic and organic forms. Chemical fertilizers of P have taken up the market after the Green Revolution, but excessive use to the threshold limit causes no further increase in crop growth. This situation calls for an urgent need for some sustainable strategies. The use of phosphate solubilizing microbes (PSM) is helpful for this situation. Rhizosphere has high proportion of these phosphate solubilizing microbes. Arbuscular mycorrhiza and other fungi along with bacteria have created an understanding of how to have a long-term ecological and cost-effective system to overcome phosphate deficiency. The ability of rhizospheric bacteria and fungi for solubilizing phosphate is known since long. *Bacillus* and *Pseudomonas* are the most promising genera for phosphate solubilization (Amy et al. 2022; Sarmah and Sarma 2022) and important fungal genera include *Trichoderma*, *Aspergillus*, and *Penicillium* (Li et al. 2022; Kour and Yadav 2022). Nowadays, attention has also been given to phosphate-solubilizing plant growth-promoting rhizobacteria. In many studies, it has been proved that bacterial consortia can also enhance phosphate solubilization. For example, Li et al. (2017) have shown that species of *Paenibacillus* and *Pseudomonas* isolated from maize rhizosphere can efficiently solubilize organic and inorganic forms of phosphate. Similarly species of *Bacillus* and *Alcaligenes* have been found to promote phosphate solubilization by production of organic acids in sunflower plants (Shahid et al. 2015). Also, a combination of mycorrhizae and PGPR has shown excellent results. Inoculation of a arbuscular mycorrhiza along with strains of PGPR of genus *Bacillus* and *Streptococcus* have shown to enhance phosphate solubilization (Wahid et al. 2016). PGPMs are also able to solubilize phosphate under biotic and abiotic stresses such as salinity, pH, temperature, and heavy metals. For example, phosphate solubilization is found to be enhanced by inoculation of a *Trichoderma* species under abiotic stress of heavy metal cadmium and fluctuation of pH and temperature (Rawat and Tewari 2011). Phosphate-solubilizing bacterium and arbuscular mycorrhizal fungus in combination enhanced P uptake in salinity-stressed alfalfa (*Medicago sativa* L.) (Etesami and Alikhani 2022).

Potassium (K) is among the three most essential elements required by plants, inadequacy of which causes poor growth of roots, seeds, and low resistance to pathogens. Although soil is rich in K containing minerals, they are often unavailable to plants. Potassium solubilizing microbes help in solubilizing K from the rocks of mica, orthoclase, illite etc. (Meena et al. 2014; Velázquez et al. 2016). Species of genus *Bacillus*, *Burkholderia*, *Agrobacterium*, *Achromobacter Klebsiella*, *Enterobacter*, *Pantoea* and *Pseudomonas* have the ability to solubilize K (Sattar et al. 2019). Growth promotion in potato plants was observed, which were fertilized with K-feldspar and inoculated with PGPR *Bacillus cereus* (Ali et al. 2021). Also, indigenous species of potassium solubilizing bacteria have been found to enhance crop productivity and quality of tea (Bagyalakshmi et al. 2012). Foliar treatment of microbial biostimulant has proved to increase growth and uptake efficiency of K in almond plants (Saa et al. 2015). Potassium dissolution ability of PGPR belonging to the species of *Enterobacter* and *Pseudomonas* was utilized in wheat cultivation (Khani et al. 2019).

2.2.2 *Microbial Biostimulants for Micronutrients*

Enriching any staple crop with micronutrients by using the best traditional breeding practices and modern biotechnology is known as biofortification (Nestel et al. 2006). Although the concept is great, micronutrients like Fe and Zn are added to the daily intake of food or as food supplements, but success by using breeding and biotechnological techniques is not consistent (Murgia et al. 2012). Hence more attention must be given to naturally available options such as microbes. There are several works that have been done with PGPR and fungi relating to solubilization and uptake of micronutrients like Fe, Zn, Cu, manganese, boron, etc. Three rhizobacterial strains of *Bacillus*, *Providencia*, and *Brevundimonas* species were found to enhance the micronutrient uptake of wheat plants (Rana et al. 2012). Similarly, the content of micronutrients like Fe, Cu, Mn, and Zn in wheat has been found to increase after combined application of indigenous PGPR like *B. megaterium*, *Arthrobacter chlorophenolicus*, and *Enterobacter* (Kumar et al. 2014). Verma et al. (2020) reported Zn and phosphate-solubilizing *Rhizobium radiobacter* species that enhanced the productivity of loose leaf lettuce under salinity stress. Promising Zn fortifying PGPR strains showing biofortification improved up to sixfold as compared with uninoculated *Triticum aestivum* (Shaikh and Saraf 2017). In a study *Bacillus subtilis* GB03 was shown to increase the accumulation of Fe in plants (Zhang et al. 2009). *Bacillus* species isolated from cotton rhizosphere were found to be excellent Zn and P solubilizers and cotton growth promoters (Ahmad et al. 2021). Mumtaz et al. (2017) showed that Zn solubilizing *Bacillus* spp. are potential candidates for biofortification in maize. Foliar application of a biostimulant derived from microbial fermentation of cereal grains enhanced Zn uptake in sunflower (Tian et al. 2015). Ercoli et al. (2017) have reported in a study that arbuscular mycorrhiza can show enhanced uptake of Zn and Fe in wheat. Dhaked et al. (2022) reported K and Zn solubilizing bacteria along with other plant growth-promoting properties (PGPR) that were isolated from different rhizospheric soil. Recently, Mishra et al. (2023) reviewed in detail the use and role of beneficial soil microorganisms in biofortification of micronutrients and trace elements in diverse crops.

2.3 Emerging Areas of Microbial Biostimulants

In recent past, microbial biostimulants have also emerged as novel tools for environment management. Nowadays they are being utilized in wastewater treatment, heavy metal remediation, azo dyes decolorization, pesticides degradation, etc. (Fig. 2.1). Currently, the role of PGPR in environmental nutrient recycling and water pollution abatement is being studied. For example, the plant growth-promoting *Pseudomonas* sp. was reported to remediate P and ammonium from wastewater. About 98% of total P was accumulated and removed from synthetic wastewater, and this PGPR-based technology can be further utilized for wetland remediation (Daly et al. 2019). PGP bacteria are also reported to have application in phytodepuration systems like constructed wetlands. The use of specific bacterial biostimulants in microbe-assisted phytodepuration decreases stress by enhancing plant growth and degrading phytotoxic compounds present in constructed wet lands. The additional profit of utilization of PGP bacteria in phytodepuration is that they can be used for

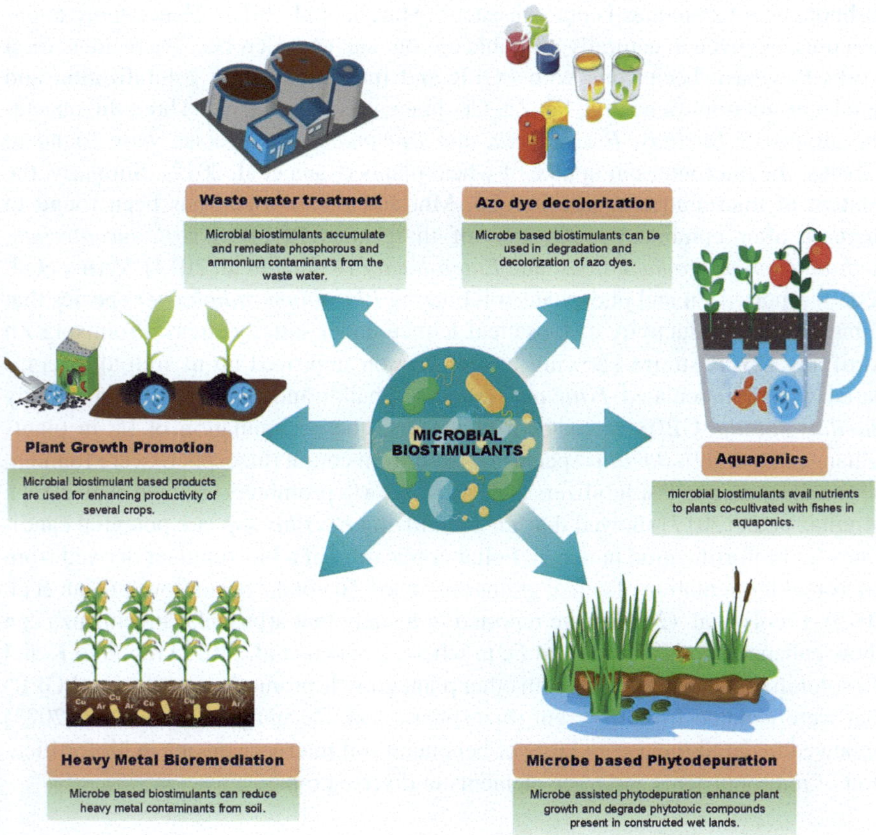

Fig. 2.1 Applications of microbial bioinoculants in various areas

the degradation of micropollutants present in constructed wetlands (Riva et al. 2020). PGPMs are documented to show a great opportunity as microbial biostimulants in aquaponics as they can prove to be efficient substitutes of chemicals used in aquaponics (Bartelme et al. 2018).

The microbial biostimulants can also be used for azo dye decolorization. For instance, plant growth-promoting *Bacillus* sp. was reported to decolorize azo dyes (i.e., reactive black-5, reactive red-120, direct blue-1, and congo red) through NADH-ubiquinone: oxidoreductase activity along with phosphate solubilization, and ACC deaminase production in high metal and salt levels. Thus microbial bioinoculants could be a powerful tool for bioremediating textile wastewater and wastewater contaminated soil (Mahmood et al. 2017). In a similar study, the plant growth-promoting bacteria *Rhodobacter erythropholi*, *Azotobacter vinelandii*, *Rhizobium meliloti*, and *Bacillus megaterium* are reported to adsorb and decolorize dye Direct Red 73 (DR73) on sugar cane bagasse (Kadam et al. 2014).

Another emerging area where microbe-based biostimulants have proven their excellence is related to heavy metal bioremediation. There is evidence that biofilm producing *Pseudomonas* sp. show high metal tolerance and can efficiently reduce heavy metal contamination from soil (Meliani and Bensoltane 2016). Similarly, *Ochromobacrtum* and *Bacillus* sp. are reported to ameliorate lead, cadmium, and arsenic toxicity along with plant growth promotion (Pandey et al. 2013). Nitrogen fixing bacterium *A. chroococcum* is reported to produce melanin, which has potential to chelate heavy metals (Banerjee et al. 2014). The role of symbiotic nitrogen fixing *Rhizobium* strains in remediation of heavy metal is also described by various researchers (Nocelli et al. 2016; Lebrazi and Fikri-Benbrahim 2018; Jach et al. 2022).

Microbial bioinoculants are now being developed to play diverse roles and have immense capabilities in solving the key environmental challenges of global concern in future, particularly those related to soil and agriculture.

2.4 Future Perspectives

The emerging application of microbe-based bioinoculants in the agricultural sector for better plant growth and disease resistance has opened a new gateway towards cost-effective and environmentally sustainable approach. Plant microbiome is a huge bionetwork and further elucidation of molecular mechanisms and characterization of bioactive compounds from potential strains under optimal and suboptimal conditions via multidisciplinary techniques can be helpful in designing novel biopreparations (Rouphael and Colla 2020). Unraveling the molecular mechanisms of plant microbiome can be a promising approach to tackle multiple environmental stresses across various crops and will help in developing climate-smart cropping (Arora 2019; Kerchev et al. 2020). In this regard, metagenomics, metabolomics, high-throughput sequencing and phenotyping are some important biotechnological tools in order to check the impact and develop future biostimulants having a positive role in agro-ecosystems in diverse ways (de Frias et al. 2018; El Boukhari et al.

2020; Arora et al. 2020a, b; Graziano et al. 2022). Recently, a new bio-encapsulation technique such as nano- or micro-encapsulation (use of polymeric nanoparticle coating) has been discovered to provide a much favorable micro-niche for the microbial cells as well as to provide them physical and chemical protection to maintain their viability (Vassilev et al. 2020). Moreover, a detailed understanding of carrier additives in biostimulants is necessary to determine the overall functional properties of the product and finalize strategies needed for their application and improve efficiency. Apart from this, careful and extensive field trials and market surveys should be carried out for the successful technology transfer from lab to field to strengthen farmers' acceptance to new techniques (Tabassum et al. 2017). Also, partnerships and collaborations between industries, academia, and research organizations should be promoted to increase knowledge and awareness regarding product development and its process (Backer et al. 2018). Lastly, government authorities, NGOs, researchers, and agriculturists should come forward in formulating policies and encourage farmers' participation through training programs to boost up the use of these green products, which are only a miniscule part of the global market (Arora and Mishra 2019). Together, all these efforts can serve as a prerequisite to minimize our reliance on the use of injudicious chemical inputs and maximize crop production to meet food demands of ever growing population.

2.5 Conclusion

Microbial biostimulants have arisen as an economically sound and sustainable technology for replacing agrochemicals. The chapter has documented important roles of PGPMs in overall plant and soil health focusing on their role in nutrient solubilization, phytohormone production, biotic and abiotic stress tolerance, and many more. The role of PGPMs in bioremediation of water and soil contaminants is emerging as a successful approach. In future, usage of biotechnological tools can help in reclamation and redesigning of degraded ecosystems with the help of these beneficial microbes. Commercialization of this green technology will help us to achieve the goals of global food security and environmental sustainability, which is the need of the hour in the era of unprecedented anthropogenic activities.

References

Abdel Latef AAH (2011) Influence of arbuscular mycorrhizal fungi and copper on growth, accumulation of osmolyte, mineral nutrition and antioxidant enzyme activity of pepper (*Capsicum annuum* L.). Mycorrhiza 21(6):495–503

Adhikari K, Bhandari S, Acharya S (2020) An overview of azolla in rice production: a review. Rev Food Agric 2(1):04–08

Ahmad I, Ahmad M, Hussain A, Jamil M (2021) Integrated use of phosphate-solubilizing *Bacillus subtilis* strain IA6 and zinc-solubilizing *Bacillus* sp. strain IA16: a promising approach for improving cotton growth. Folia Microbiol 66:115–125

Ai M, Han F, Yang X, Chu H, Luo C, Tan S et al (2023) Endophytic *Penicillium oxalicum* CX-1 prevented *Phytophthora cactorum* blight on *salvia miltiorrhiza* and promoted plant growth. J Appl Microbiol 134:lxad010

Ali AM, Awad MY, Hegab SA, Gawad AMAE, Eissa MA (2021) Effect of potassium solubilizing bacteria (*Bacillus cereus*) on growth and yield of potato. J Plant Nutr 44(3):411–420

Ambrosini A, de Souza R, Passaglia LM (2016) Ecological role of bacterial inoculants and their potential impact on soil microbial diversity. Plant Soil 400:193–207

Amy C, Avice JC, Laval K, Bressan M (2022) Are native phosphate solubilizing bacteria a relevant alternative to mineral fertilizations for crops? Part I. When rhizobacteria meet plant P requirements. Rhizosphere 21:100476

Anam GB, Reddy MS, Ahn YH (2019) Characterization of *Trichoderma asperellum* RM-28 for its sodic/saline-alkali tolerance and plant growth promoting activities to alleviate toxicity of red mud. Sci Total Environ 662:462–469

Ansari FA, Ahmad I (2018) Biofilm development, plant growth promoting traits and rhizosphere colonization by *Pseudomonas entomophila* FAP1: a promising PGPR. Adv Microbiol 8:235–251

Araujo J, Díaz-Alcántara CA, Velázquez E, Urbano B, González-Andrés F (2015) *Bradyrhizobium yuanmingense* related strains form nitrogen-fixing symbiosis with *Cajanus cajan* L. in Dominican Republic and are efficient biofertilizers to replace N fertilization. Sci Hortic 192:421–428

Arora NK (2019) Impact of climate change on agriculture production and its sustainable solutions. Environ Sustain 2:95–96

Arora NK, Mishra J (2016) Prospecting the roles of metabolites and additives in future bioformulations for sustainable agriculture. Appl Soil Ecol 107:405–407

Arora NK, Mishra I (2019) United Nations sustainable development goals 2030 and environmental sustainability: race against time. Environ Sustain 2:339–342

Arora NK, Tewari S, Singh R (2013) Multifaceted plant-associated microbes and their mechanisms diminish the concept of direct and indirect PGPRs. In: Arora N (ed) Plant microbe symbiosis: fundamentals and advances. Springer, New Delhi, pp 411–449

Arora NK, Verma M, Mishra J (2017) Rhizobial bioformulations: past, present and future. In: Mehnaz S (ed) Rhizotrophs: plant growth promotion to bioremediation. Microorganisms for sustainability, vol 2. Springer, Singapore, pp 69–99

Arora NK, Fatima T, Mishra I, Verma S (2020a) Microbe-based inoculants: role in next green revolution. In: Shukla V, Kumar N (eds) Environmental concerns and sustainable development. Biodiversity, soil and waste management, vol 2. Springer, Singapore, pp 191–246

Arora NK, Fatima T, Mishra J, Mishra I, Verma S, Verma R et al (2020b) Halo-tolerant plant growth promoting rhizobacteria for improving productivity and remediation of saline soils. J Adv Res 26:69–82

Ashry NM, Alaidaroos BA, Mohamed SA, Badr OA, El-Saadony MT, Esmael A (2022) Utilization of drought-tolerant bacterial strains isolated from harsh soils as a plant growth-promoting rhizobacteria (PGPR). Saudi J Biol Sci 29(3):1760–1769

Backer R, Rokem JS, Ilangumaran G, Lamont J, Praslickova D, Ricci E et al (2018) Plant growth-promoting rhizobacteria: context, mechanisms of action, and roadmap to commercialization of biostimulants for sustainable agriculture. Front Plant Sci 9:1473

Bagyalakshmi B, Ponmurugan P, Marimuthu S (2012) Influence of potassium solubilizing bacteria on crop productivity and quality of tea (*Camellia sinensis*). Afr J Agric Res 7(30):4250–4259

Bakhshandeh E, Pirdashti H, Lendeh KS (2017) Phosphate and potassium-solubilizing bacteria effect on the growth of rice. Ecol Eng 103:164–169

Banerjee S, Dutta S (2019) Plant growth promoting activities of a fungal strain *Penicillium commune* MCC 1720 and its effect on growth of black gram. Pharma Innov 8(12):121–127

Banerjee S, Palit R, Sengupta C, Standing D (2010) Stress induced phosphate solubilization by *'Arthrobacter'* Sp. and *'Bacillus'* sp. isolated from tomato rhizosphere. Aust. J Crop Sci 4(6):378

Banerjee A, Supakar S, Banerjee R (2014) Melanin from the nitrogen-fixing bacterium *Azotobacter chroococcum*: a spectroscopic characterization. PLoS One 9(1):e84574

Baron NC, Costa NTA, Mochi DA, Rigobelo EC (2018) First report of *Aspergillus sydowii* and *Aspergillus brasiliensis* as phosphorus solubilizers in maize. Ann Microbiol 68:863–870

Bartelme RP, Oyserman BO, Blom JE, Sepulveda-Villet OJ, Newton RJ (2018) Stripping away the soil: plant growth promoting microbiology opportunities in aquaponics. Front Microbiol 9:8

Baslam M, Garmendia I, Goicoechea N (2013) The arbuscular mycorrhizal symbiosis can overcome reductions in yield and nutritional quality in greenhouse-lettuces cultivated at inappropriate growing seasons. Sci Hortic 164:145–154

Behie SW, Bidochka MJ (2014) Nutrient transfer in plant–fungal symbioses. Trends Plant Sci 19(11):734–740

Bhat TA, Ahmad L, Ganai MA, Khan OA (2015) Nitrogen fixing biofertilizers; mechanism and growth promotion: a review. J Pure Appl Microbiol 9(2):1675–1690

Bhatt K, Maheshwari DK (2020) Zinc solubilizing bacteria (*Bacillus megaterium*) with multifarious plant growth promoting activities alleviates growth in *Capsicum annuum* L. 3 Biotech 10:36

Bhattacharjee RB, Singh A, Mukhopadhyay SN (2008) Use of nitrogen-fixing bacteria as biofertiliser for non-legumes: prospects and challenges. Appl Microbiol Biotechnol 80(2):199–209

Borriss R (2020) Phytostimulation and biocontrol by the plant-associated *Bacillus amyloliquefaciens* FZB42: an update. In: Kumar M, Kumar V, Prasad R (eds) Phyto-microbiome in stress regulation. Environmental and microbial biotechnology. Springer, Singapore, pp 1–20

Cai F, Chen W, Wei Z, Pang G, Li R, Ran W et al (2015) Colonization of *Trichoderma harzianum* strain SQR-T037 on tomato roots and its relationship to plant growth, nutrient availability and soil microflora. Plant Soil 388(1–2):337–350

Chalfoun SM, Angélico CL, Resende MLV, Moraes GE (2019) Selection of fungal isolates with potential for phosphate solubilization and formulation of inoculant for coffee crops. Coffee Sci 14(3):315–325

Chhetri G, Kim I, Kang M, So Y, Kim J, Seo T (2022) An isolated *Arthrobacter* sp. enhances rice (*Oryza sativa* L). plant growth. Microorganisms 10(6):1187

Choudhary S, Zehra A, Wani KI, Naeem M, Hakeem KR, Aftab T (2020) The role of micronutrients in growth and development: transport and signalling pathways from crosstalk perspective. In: Aftab T, Hakeem KR (eds) Plant Micronutrients. Springer, Cham, pp 73–81

Colla G, Rouphael Y, Di Mattia E, El-Nakhel C, Cardarelli M (2015) Co-inoculation of *Glomus intraradices* and *Trichoderma atroviride* acts as a biostimulant to promote growth, yield and nutrient uptake of vegetable crops. J Sci Food Agric 95(8):1706–1715

Daly I, Jellali S, Mehri I, Reis MA, Freitas EB, Oehmen A et al (2019) Phosphorus and ammonium removal characteristics from aqueous solutions by a newly isolated plant growth-promoting bacterium. Environ Technol 41(20):959–3330

De Bruijn FJ (2015) The quest for biological nitrogen fixation in cereals: a perspective and prospective. In: De Bruijn FJ (ed) Biological nitrogen fixation, vol 2. Wiley Online Library, pp 1087–1101

de Frias UA, Pereira GKB, Guazzaroni ME, Silva-Rocha R (2018) Boosting secondary metabolite production and discovery through the engineering of novel microbial biosensors. Biomed Res Int 2018:7021826

Devi R, Kaur T, Kour D, Yadav A, Yadav AN, Suman A et al (2022) Minerals solubilizing and mobilizing microbiomes: a sustainable approach for managing minerals' deficiency in agricultural soil. J Appl Microbiol 133(3):1245–1272

Dhaked BS, Koli GK, Triveni S, Reddy RS, Jaiswal A, Koli DK (2022) Screening of potassium and zinc solubilizing bacteria for plant growth promoting properties (PGPR) from different rhizospheric soil. Pharm Innov J 11(2):1315–1319

Dheeman S, Maheshwari DK (2022) Ecology of nitrogen-fixing bacteria for sustainable development of non-legume crops. In: Maheshwari DK, Dobhal R, Dheeman S (eds) Nitrogen fixing bacteria: sustainable growth of non-legumes. Microorganisms for sustainability, vol 36. Springer, Singapore, pp 301–315

Du Jardin P (2015) Plant biostimulants: definition, concept, main categories and regulation. Sci Hortic 196:3–14

El Boukhari MEM, Barakate M, Bouhia Y, Lyamlouli K (2020) Trends in seaweed extract based biostimulants: manufacturing process and beneficial effect on soil-plant systems. Plan Theory 9:359

El-Beltagi HS, Ahmad I, Basit A, Abd El-Lateef HM, Yasir M, Tanveer Shah S et al (2022) Effect of *azospirillum* and *azotobacter* species on the performance of cherry tomato under different salinity levels. Gesunde Pflanz 74(2):487–499

Elsharkawy MM, Abass JM, Kamel SM, Hyakumachi M (2017) The plant growth promoting fungus *Penicillium* sp. GP16-2 enhances the growth and confers protection against cucumber mosaic virus in tobacco. J Virol Sci 1:145–154

Ercoli L, Schüßler A, Arduini I, Pellegrino E (2017) Strong increase of durum wheat iron and zinc content by field-inoculation with arbuscular mycorrhizal fungi at different soil nitrogen availabilities. Plant Soil 419:153–167

Etesami H, Adl SM (2020) Plant growth-promoting Rhizobacteria (PGPR) and their action mechanisms in availability of nutrients to plants. In: Kumar M, Kumar V, Prasad R (eds) Phytomicrobiome in stress regulation. Environmental and Microbial Biotechnology. Springer, Singapore, pp 147–203

Etesami H, Alikhani HA (2022) Silicon improves the effect of phosphate-solubilizing bacterium and arbuscular mycorrhizal fungus on phosphorus concentration of salinity-stressed alfalfa (*Medicago sativa* L.). Rhizosphere 24:100619

Fageria NK, Baligar VC, Clark RB (2002) Micronutrients in crop production. Adv Agron 77:185–268

Fageria NK, Baligar VC, Li YC (2008) The role of nutrient efficient plants in improving crop yields in the twenty first century. J Plant Nutr 31(6):1121–1157

Fan X, Zhou X, Chen H, Tang M, Xie X (2021) Cross-talks between macro-and micronutrient uptake and signaling in plants. Front Plant Sci 12:663477

FAO (2017) The future of food and agriculture—trends and challenges, Rome

Fatima T, Arora NK (2021) *Pseudomonas entomophila* PE3 and its exopolysaccharides as biostimulants for enhancing growth, yield and tolerance responses of sunflower under saline conditions. Microbiol Res 244:126671

Fortune business insight (2019) Biostimulants market size share and industry analysis by source (Microbial and Non-microbial), by active ingredients (Seaweed extracts, humic substances, vitamins& amino acids, microbial amendments and others), by application (Foliar treatment, soil treatment and seed treatment), by crop (row crops, fruits and vegetables, turf and ornamentals) and regional forecast 2019–2026. Available online https://www.fortunebusinessinsights.com/industry-reports/biostimulants-market-100414

Fox AR, Soto G, Valverde C, Russo D, Lagares A Jr, Zorreguieta Á et al (2016) Major cereal crops benefit from biological nitrogen fixation when inoculated with the nitrogen-fixing bacterium *Pseudomonas protegens* Pf-5 X940. Environ Microbiol 18(10):3522–3534

Galeano RMS, Franco DG, Chaves PO, Giannesi GC, Masui DC, Ruller R et al (2021) Plant growth promoting potential of endophytic *Aspergillus niger* 9-p isolated from native forage grass in Pantanal of Nhecolândia region, Brazil. Rhizosphere 18:100332

García-Fraile P, Menéndez E, Rivas R (2015) Role of bacterial biofertilizers in agriculture and forestry. AIMS Bioeng 2(3):183–205

Gaur A, Gaur A, Adholeya A (2000) Growth and flowering in *Petunia hybrida, Callistephus chinensis* and *Impatiens balsamina* inoculated with mixed AM inocula or chemical fertilizers in a soil of low P fertility. Sci Hortic 84(1–2):151–162

Ghadamgahi F, Tarighi S, Taheri P, Saripella GV, Anzalone A, Kalyandurg PB et al (2022) Plant growth-promoting activity of *Pseudomonas aeruginosa* FG106 and its ability to act as a biocontrol agent against potato, tomato and taro pathogens. Biology 11(1):140

Glick BR (2014) Bacteria with ACC deaminase can promote plant growth and help to feed the world. Microbiol Res 169(1):30–39

Goswami M, Deka S (2022) Rhizodeposits: an essential component for microbial interactions in rhizosphere. In: Singh UB, Rai JP, Sharma AK (eds) Re-visiting the rhizosphere eco-system for agricultural sustainability. Rhizosphere Biology. Springer, Singapore, pp 129–151

Graziano S, Caldara M, Gullì M, Bevivino A, Maestri E, Marmiroli N (2022) A metagenomic and gene expression analysis in wheat (*T. durum*) and maize (*Z. mays*) biofertilized with PGPM and biochar. Int J Mol Sci 23(18):10376

Gull M, Hafeez FY (2012) Characterization of siderophore producing bacterial strain *Pseudomonas fluorescens* Mst 8.2 as plant growth promoting and biocontrol agent in wheat. Afr J Microbiol Res 6:6308–6631

Gupta G, Parihar SS, Ahirwar NK, Snehi SK, Singh V (2015) Plant growth promoting rhizobacteria (PGPR): current and future prospects for development of sustainable agriculture. J Microb Biochem Technol 7(2):096–102

Gupta V, Kumar GN, Buch A (2020) Colonization by multi-potential *Pseudomonas aeruginosa* P4 stimulates peanut (*Arachis hypogaea* L.) growth, defence physiology and root system functioning to benefit the root-rhizobacterial interface. J Plant Physiol 248:153144

Hagh-Doust N, Färkkilä SM, Moghaddam MSH, Tedersoo L (2022) Symbiotic fungi as biotechnological tools: methodological challenges and relative benefits in agriculture and forestry. Fungal Biol Rev 42:34–55

Hamayun M, Hussain A, Afzal Khan S, Iqbal A, Lee IJ (2019) *Aspergillus flavus* promoted the growth of soybean and sunflower seedlings at elevated temperature. Biomed Res Int 2019:1295457

Han SH, An JY, Hwang J, Kim SB, Park BB (2016) The effects of organic manure and chemical fertilizer on the growth and nutrient concentrations of yellow poplar (*Liriodendron tulipifera* Lin.) in a nursery system. For Sci Technol 12(3):137–143

Hill PW, Jones DL (2019) Plant–microbe competition. New Phytol 221(2):796–806

Hwang HH, Chien PR, Huang FC, Hung SH, Kuo CH, Deng WL et al (2021) A plant endophytic bacterium, *Burkholderia seminalis* strain 869T2, promotes plant growth in *Arabidopsis*, pak choi, Chinese amaranth, lettuces, and other vegetables. Microorganisms 9(8):1703

Jach ME, Sajnaga E, Ziaja M (2022) Utilization of legume-nodule bacterial symbiosis in phytoremediation of heavy metal-contaminated soils. Biology 11(5):676

Jalal A, Filho MCMT, da Silva EC, da Silva Oliveira CE, Freitas LA, do Nascimento V (2022) Plant growth-promoting bacteria and nitrogen fixing bacteria: sustainability of non-legume crops. In: Maheshwari DK, Dobhal R, Dheeman S (eds) Nitrogen fixing bacteria: sustainable growth of non-legumes. Microorganisms for sustainability, vol 36. Springer, Singapore, pp 233–275

Jiang H, Yin Z, Shi F, Liu C, Fan B (2018) Isolation, identification of a salt-tolerant, phosphate-solubilizing and crop-growth promoting *Aspergillus japonicus*. Acta Microbiol Sin 5:862–881

Jiménez JA, Novinscak A, Filion M (2020) *Pseudomonas fluorescens* LBUM677 differentially increases plant biomass, total oil content and lipid composition in three oilseed crops. J Appl Microbiol 128:1119–1127

Jiménez-Gómez A, Flores-Félix JD, García-Fraile P, Mateos PF, Menéndez E, Velázquez E et al (2018) Probiotic activities of *Rhizobium laguerreae* on growth and quality of spinach. Sci Rep 8:1–10

Jiménez-Gómez A, García-Estévez I, García-Fraile P, Escribano-Bailón MT, Rivas R (2020) Increase in phenolic compounds of *Coriandrum sativum* L. after the application of a *Bacillus halotolerans* biofertilizer. J Sci Food Agric 100(6):2742–2749

Jones KM (2019) Features of Sinorhizobium meliloti exopolysaccharide succinoglycan required for successful invasion of *Medicago truncatula* nodules. In: de Bruijn F (ed) The Model Legume *Medicago truncatula*. Wiley Online Library, pp 571–578

Jung BK, Khan AR, Hong S-J, Park G-S, Park Y-J, Kim H-J et al (2017) Quorum sensing activity of the plant growth-promoting rhizobacterium *Serratia glossinae* GS2 isolated from the sesame (*Sesamum indicum* L.) rhizosphere. Ann Microbiol 67:623–632

Juntahum S, Ekprasert J, Boonlue S (2022) Efficiency of arbuscular mycorrhizal fungi for the growth promotion of sugarcane under pot conditions. Sugar Tech 24(6):1738–1747

Kadam AA, Kulkarni AN, Lade HS, Govindwar SP (2014) Exploiting the potential of plant growth promoting bacteria in decolorization of dye disperse red 73 adsorbed on milled sugarcane bagasse under solid state fermentation. Int Biodeterior Biodegrad 86:364–371

Katiyar P, Kumar S, Arora NK (2022) Interactions of nitrogen-fixing bacteria and cereal crops: an important dimension. In: Maheshwari DK, Dobhal R, Dheeman S (eds) Nitrogen fixing bacteria: sustainable growth of non-legumes. Microorganisms for sustainability, vol 36. Springer, Singapore, pp 169–194

Kauffman GL, Kneivel DP, Watschke TL (2007) Effects of a biostimulant on the heat tolerance associated with photosynthetic capacity, membrane thermostability, and polyphenol production of perennial ryegrass. Crop Sci 47:261–267

Kaya C, Ashraf M, Sonmez O, Aydemir S, Tuna AL, Cullu MA (2009) The influence of arbuscular mycorrhizal colonisation on key growth parameters and fruit yield of pepper plants grown at high salinity. Sci Hortic 121(1):1–6

Ke LI, Jian-jun XU, Lu-yao W, Chu-nan W, Chen YU, Jian-hua GUO (2016) Effects of *Acaulospora spinosa* on plant growth and lycopene related genes (psy1 and psy2) ex-pression of tomato. Ying Yong Sheng Tai Xue Bao 27(2):499–503

Keerthana U, Prabhukarthikeyan SR, Baite MS, Yadav MK, Kumar RN, Kumar AM et al (2022) Fluorescent pseudomonads: a multifaceted biocontrol agent for sustainable agriculture. In: Singh H, Vaishnav A (eds) New and future developments in microbial biotechnology and bioengineering. Elsevier, pp 83–92

Kerchev P, van der Meer T, Sujeeth N, Verlee A, Stevens CV, Breusegem FV et al (2020) Molecular priming as an approach to induce tolerance against abiotic and oxidative stresses in crop plants. Biotechnol Adv 40:107503

Khani AG, Enayatizamir N, Norouzi Masir M (2019) Impact of plant growth promoting rhizobacteria on different forms of soil potassium under wheat cultivation. Lett Appl Microbiol 68(6):514–521

Khare E, Arora NK (2021) Biosurfactant based formulation of *Pseudomonas guariconensis* LE3 with multifarious plant growth promoting traits controls charcoal rot disease in *Helianthus annus*. World J Microbiol Biotechnol 37:1–14

Kloepper JW, Lifshitz R, Zablotowicz RM (1989) Free-living bacterial inocula for enhancing crop productivity. Trends Biotechnol 7:39–44

Kotoky R, Nath S, Maheshwari DK, Pandey P (2019) Cadmium resistant plant growth promoting rhizobacteria *Serratia marcescens* S2I7 associated with the growth promotion of rice plant. Environ Sustain 2:135–144

Kour D, Yadav AN (2022) Stress adaptive phosphorus solubilizing microbiomes for agricultural sustainability. J Appl Biol Biotechnol 10(6):1–3

Kumar A, Maurya BR, Raghuwanshi R (2014) Isolation and characterization of PGPR and their effect on growth, yield and nutrient content in wheat (*Triticum aestivum* L.). Biocatal Agric Biotechnol 3(4):121–128

Kumar P, Thakur S, Dhingra G, Singh A, Pal MK, Harshvardhan K et al (2018) Inoculation of siderophore producing rhizobacteria and their consortium for growth enhancement of wheat plant. Biocatal Agric Biotechnol 15:264–269

Lebrazi S, Fikri-Benbrahim K (2018) *Rhizobium*-legume symbioses: heavy metal effects and principal approaches for bioremediation of contaminated soil. In: Meena R, Das A, Yadav G, Lal R (eds) Legumes for soil health and sustainable management. Springer, Singapore, pp 205–233

Li RX, Cai F, Pang G, Shen QR, Li R, Chen W (2015) Solubilisation of phosphate and micronutrients by *Trichoderma harzianum* and its relationship with the promotion of tomato plant growth. PLoS One 10:6

Li Y, Liu X, Hao T, Chen S (2017) Colonization and maize growth promotion induced by phosphate solubilizing bacterial isolates. Int J Mol Sci 18(7):1253

Li YT, Hwang SG, Huang YM, Huang CH (2018) Effects of *Trichoderma asperellum* on nutrient uptake and Fusarium wilt of tomato. Crop Prot 110:275–282

Li Q, Wu Q, Zhang T, Xiang P, Bao Z, Tu W et al (2022) Phosphate mining activities affect crop rhizosphere fungal communities. Sci Total Environ 838:156196

Li X, Li D, Jiang Y, Xu J, Ren X, Zhang Y et al (2023) The effects of microbial fertilizer based *Aspergillus brunneoviolaceus* HZ23 on pakchoi growth, soil properties, rhizosphere bacterial community structure, and metabolites in newly reclaimed land. Front Microbiol 14:1091380

Liu X, Li Q, Li Y, Guan G, Chen S (2019) *Paenibacillus* strains with nitrogen fixation and multiple beneficial properties for promoting plant growth. Peer J 7:e7445

Lucini L, Colla G, Moreno MBM, Bernardo L, Cardarelli M, Terzi V et al (2019) Inoculation of *Rhizoglomus irregulare* or *Trichoderma atroviride* differentially modulates metabolite profiling of wheat root exudates. Phytochemistry 157:158–167

Ludueña LM, Anzuay MS, Angelini JG, Barros G, Luna MF, del Pilar MM et al (2017) Role of bacterial pyrroloquinoline quinone in phosphate solubilizing ability and in plant growth promotion on strain *Serratia* sp. S119. Symbiosis 72:31–43

Luo L, Zhao C, Wang E, Raza A, Yin C (2022) *Bacillus amyloliquefaciens* as an excellent agent for biofertilizer and biocontrol in agriculture: An overview for its mechanisms. Microbiol Res 259:127016

Mahmood F, Shahid M, Hussain S, Shahzad T, Tahir M, Ijaz M et al (2017) Potential plant growth-promoting strain *Bacillus* sp. SR-2-1/1 decolorized azo dyes through NADH-ubiquinone: oxidoreductase activity. Bioresour Technol 235:176–184

Maurya S, Rai D, Rai B, Dubey S (2019) Growth promotion effect of *Trichoderma* isolates on tomato seedlings. J Plant Dis Sci 14:115–118

Meena VS, Maurya BR, Bahadur I (2014) Potassium solubilization by bacterial strain in waste mica. Bangladesh J Bot 43(2):235–237

Meliani A, Bensoltane A (2016) Biofilm-mediated heavy metals bioremediation in PGPR *Pseudomonas*. J Bioremed Biodegr 7:2

Mishra J, Arora NK (2018) Secondary metabolites of fluorescent pseudomonads in biocontrol of phytopathogens for sustainable agriculture. Appl Soil Ecol 125:35–45

Mishra J, Fatima T, Arora NK (2018) Role of secondary metabolites from plant growth-promoting rhizobacteria in combating salinity stress. In: Egamberdieva D, Ahmad P (eds) Plant microbiome: stress response. Microorganisms for sustainability, vol 5. Springer, Singapore, pp 127–163

Mishra I, Fatima T, Egamberdieva D, Arora NK (2020) Novel bioformulations developed from *Pseudomonas putida* BSP9 and its biosurfactant for growth promotion of *Brassica juncea* (L.). Plan Theory 9(10):1349

Mishra P, Mishra J, Arora NK (2021) Plant growth promoting bacteria for combating salinity stress in plants–recent developments and prospects: a review. Microbiol Res 252:126861

Mishra P, Bhattacharya A, Verma P, Bharti C, Arora NK (2022) Plant growth-promoting bacteria as biostimulants of crops in saline agroecosystems. In: Arora NK, Bouizgarne B (eds) Microbial BioTechnology for sustainable agriculture, microorganisms for sustainability, vol 33. Springer, Singapore, pp 205–235

Mishra P, Mishra J, Arora NK (2023) Biofortification revisited: addressing the role of beneficial soil microbes for enhancing trace elements concentration in staple crops. Microbiol Res 275:127442

MNM (2020) Biostimulants market by active ingredient (humic substances, amino acids, seaweed extracts, microbial amendments), crop type (fruits & vegetables, cereals, turf & ornamentals), application method, form, and region—global forecast to 2025. Available online https://www.marketsandmarkets.com/Market-Reports/biostimulant-market-1081.html?gclid=CjwKCAjwvZv0BRA8EiwAD9T2Vce-jO-PrqoWBmCMQcDedgb8CtLzwE0Soc6hhdxsx-Fs8Ny5BEIwqBoC0XoQAvD_BwE

Moon SJ, Yoon MH (2019) Plant growth promotion effect of *Arthrobacter enclensis* Yangsong-1 isolated from a button mushroom bed. J Mushroom 17:12–18

Mumtaz MZ, Ahmad M, Jamil M, Hussain T (2017) Zinc solubilizing *Bacillus* spp. potential candidates for biofortification in maize. Microbiol Res 202:51–60

Murgia I, Arosio P, Tarantino D, Soave C (2012) Biofortification for combating 'hidden hunger' for iron. Trends Plant Sci 17(1):47–55

Nagrale DT, Chaurasia A, Kumar S, Gawande SP, Hiremani NS, Shankar R et al (2023) PGPR: the treasure of multifarious beneficial microorganisms for nutrient mobilization, pest biocontrol and plant growth promotion in field crops. World J Microbiol Biotechnol 39(4):100

Nestel P, Bouis HE, Meenakshi JV, Pfeiffer W (2006) Biofortification of staple food crops. J Nutr 136(4):1064–1067

Nocelli N, Bogino PC, Banchio E, Giordano W (2016) Roles of extracellular polysaccharides and biofilm formation in heavy metal resistance of rhizobia. Materials 9:418

Ojala JC, Jarrell WM, Menge JA, Johnson ELV (1983) Influence of mycorrhizal fungi on the mineral nutrition and yield of onion in saline soil 1. J Agron 75(2):255–259

Olowe OM, Akanmu AO, Asemoloye MD (2020) Exploration of microbial stimulants for induction of systemic resistance in plant disease management. Ann Appl Biol 177(3):282–293

Orhan E, Esitken A, Ercisli S, Turan M, Sahin F (2006) Effects of plant growth promoting rhizobacteria (PGPR) on yield, growth and nutrient contents in organically growing raspberry. Sci Hortic 111(1):38–43

Otieno N, Lally RD, Kiwanuka S, Lloyd A, Ryan D, Germaine KJ et al (2015) Plant growth promotion induced by phosphate solubilizing endophytic Pseudomonas isolates. Front Microbiol 6:745

Padda KP, Puri A, Zeng Q, Chanway CP, Wu X (2017) Effect of GFP-tagging on nitrogen fixation and plant growth promotion of an endophytic diazotrophic strain of *Paenibacillus polymyxa*. Botany 95:933–942

Pandey S, Ghosh PK, Ghosh S, De TK, Maiti TK (2013) Role of heavy metal resistant *Ochrobactrum* sp. and *Bacillus* spp. strains in bioremediation of a rice cultivar and their PGPR like activities. J Microbiol 51:11–17

Panhwar QA, Jusop S, Naher UA, Othman R, Razi MI (2013) Application of potential phosphate-solubilizing bacteria and organic acids on phosphate solubilization from phosphate rock in aerobic rice. Sci World J 2013:10

Patel P, Trivedi G, Saraf M (2018) Iron biofortification in mungbean using siderophore producing plant growth promoting bacteria. Environ Sustain 1(4):357–365

Piccinin GG, Braccini AL, Dan LG, Scapim CA, Ricci TT, Bazo GL (2013) Efficiency of seed inoculation with *Azospirillum brasilense* on agronomic characteristics and yield of wheat. Ind Crop Prod 43:393–397

Ping XU, Liang LZ, Xiao YD, Jing XU, Jiang PK, Ren FS (2014) Response of soil phosphorus required for maximum growth of *Asparagus officinalis* L. to inoculation of arbuscular mycorrhizal fungi. Pedosphere 24(6):776–782

Prakash J, Arora NK (2020) Development of *Bacillus safensis*-based liquid bioformulation to augment growth, stevioside content, and nutrient uptake in *Stevia rebaudiana*. World J Microbiol Biotechnol 36:1–13

Puri A, Padda KP, Chanway CP (2016) Seedling growth promotion and nitrogen fixation by a bacterial endophyte *Paenibacillus polymyxa* P2b-2R and its GFP derivative in corn in a long-term trial. Symbiosis 69:123–129

Purwaningsih O, Kusumastuti CT, Nugroho YS, Morib CY (2019) The effect of *Rhizobium japonicum* on the growth of soybean cultivars in coastal area. J Ilmu Pertanian (Agric Sci J) 4:33–39

Rana A, Joshi M, Prasanna R, Shivay YS, Nain L (2012) Biofortification of wheat through inoculation of plant growth promoting rhizobacteria and cyanobacteria. Eur J Soil Biol 50:118–126

Rawat R, Tewari L (2011) Effect of abiotic stress on phosphate solubilization by biocontrol fungus *Trichoderma* sp. Curr Microbiol 62:1521–1526

Riva V, Riva F, Vergani L, Crotti E, Borin S, Mapelli F (2020) Microbial assisted phytodepuration for water reclamation: environmental benefits and threats. Chemosphere 241:124843

Robertson GP, Vitousek PM (2009) Nitrogen in agriculture: balancing the cost of an essential resource. Annu Rev Environ Resour 34:97–125

Rouphael Y, Colla G (2020) Editorial: Biostimulants in Agriculture. Front Plant Sci 11:40

Ruzzi M, Aroca R (2015) Plant growth-promoting rhizobacteria act as biostimulants in horticulture. Sci Hortic 196:124–134

Saa S, Rio OD, Castro S, Brown PH (2015) Foliar application of microbial and plant based biostimulants increases growth and potassium uptake in almond (*Prunus dulcis* [Mill.] DA Webb). Front Plant Sci 6:87

Safdarian M, Askari H, Shariati V, Nematzadeh G (2019) Transcriptional responses of wheat roots inoculated with *Arthrobacter nitroguajacolicus* to salt stress. Sci Rep 9:1–12

Sahodaran NK, Arun AK, Ray JG (2019) Native arbuscular mycorrhizal fungal isolates (*Funneliformis mosseae* and *Glomus microcarpum*) improve plant height and nutritional status of banana plants. Exp Agric 55:924–933

Sakamoto K, Kaji T (2017) Estimation of the root colonization of soybean by an arbuscular mycorrhizal fungus, *Gigaspora rosea*, based on specific fatty acid profiles. Soil Sci Plant Nutr 63:536–542

Sandhya V, Ali SZ (2015) The production of exopolysaccharide by *Pseudomonas putida* GAP-P45 under various abiotic stress conditions and its role in soil aggregation. Microbiology 84:512–519

Sarmah R, Sarma AK (2022) Phosphate solubilizing microorganisms: a review. Commun Soil Sci Plant Anal:1–10

Sattar A, Naveed M, Ali M, Zahir ZA, Nadeem SM, Yaseen M et al (2019) Perspectives of potassium solubilizing microbes in sustainable food production system: a review. Appl Soil Ecol 133:146–159

Sellappan R, Thangavel K, Uthandi S (2023) Bioprotective potential of maize apoplastic fluid bacterium (*Bacillus amyloliquefaciens*) and arbuscular mycorrhizal fungi (*Glomus intraradices*) against *Spodoptera frugiperda* infestation in maize. Physiol Mol. Plant Pathol 127:102050

Shah DA, Sen S, Shalini A, Ghosh D, Grover M, Mohapatra S (2017) An auxin secreting *Pseudomonas putida* rhizobacterial strain that negatively impacts water-stress tolerance in *Arabidopsis thaliana*. Rhizosphere 3:16–19

Shahabivand S, Padash A, Aghaee A, Nasiri Y, Fathi Rezaei P (2018) Plant biostimulants (*Funneliformis mosseae* and humic substances) rather than chemical fertilizer improved biochemical responses in peppermint. Plant Physiol 8:2333–2344

Shahid M, Hameed S, Tariq M, Zafar M, Ali A, Ahmad N (2015) Characterization of mineral phosphate-solubilizing bacteria for enhanced sunflower growth and yield-attributing traits. Ann Microbiol 65(3):1525–1536

Shaikh S, Saraf M (2017) Biofortification of *Triticum aestivum* through the inoculation of zinc solubilizing plant growth promoting rhizobacteria in field experiment. Biocatal Agric Biotechnol 9:120–126

Silva LG, Camargo RC, Mascarin GM, Nunes PSDO, Dunlap C, Bettiol W (2022) Dual functionality of *Trichoderma*: biocontrol of *Sclerotinia sclerotiorum* and biostimulant of cotton plants. Front Plant Sci 13:983127

Sindhu G, Murali M, Thriveni M, Anupama N, Amruthesh K (2018) Research article growth promotion and disease resistance in muskmelon induced by crude proteins of *Penicillium verruculosum* against gummy stem blight disease. J Crop Sci 10:160–167

Singh A, Sarma BK, Singh HB, Upadhyay RS (2014) Chapter 40—*Trichoderma*: a silent worker of plant rhizosphere. In: Gupta VK, Schmoll M, Herrera-Estrella A, Upadhyay RS, Druzhinina I, Tuohy MG (eds) Biotechnology and biology of Trichoderma. Elsevier, Amsterdam, pp 533–542

Sirohi G, Upadhyay A, Srivastava PS, Srivastava S (2015) PGPR mediated zinc biofertilization of soil and its impact on growth and productivity of wheat. J Soil Sci Plant Nutr 15(1):202–216

Sohrabi M, Mohammadi H, Mohammadi A (2018) Influence of AM fungi, *Glomus mosseae* and *Glomus intraradices* on chickpea growth and root-rot disease caused by *Fusarium solani* f. sp. pisi under greenhouse conditions. J Agric Sci Technol 17(7):1919–1929
Speck JJ, James EK, Sugawara M, Sadowsky MJ, Gyaneshwar P (2019) An alkane sulfonate monooxygenase is required for symbiotic nitrogen fixation by *Bradyrhizobium diazoefficiens* (syn. *Bradyrhizobium japonicum*) USDA110T. Appl Environ Microbiol 85(24):e01552–e01519
Suhag M (2016) Potential of biofertilizers to replace chemical fertilizers. Int Adv Res J Sci Eng Technol 3(5):163–167
Sun Y, Liu Y, Liu X, Dang X, Dong X, Xie Z (2020) *Azorhizobium caulinodans* c-di-GMP phosphodiesterase Chp1 involved in motility, EPS production, and nodulation of the host plant. Appl Microbiol Biotechnol 104:2715–2729
Tabassum B, Khan A, Tariq M, Ramzan M, Khan MSI, Shaid N et al (2017) Bottlenecks in commercialisation and future prospects of PGPR. Appl Soil Ecol 121:102–117
Tewari S, Arora NK (2018) Role of salicylic acid from *Pseudomonas aeruginosa* PF23 EPS+ in growth promotion of sunflower in saline soils infested with phytopathogen *Macrophomina phaseolina*. Environ Sustain 1:49–59
Tian S, Lu L, Xie R, Zhang M, Jernstedt J, Hou D et al (2015) Supplemental macronutrients and microbial fermentation products improve the uptake and transport of foliar applied zinc in sunflower (*Helianthus annuus* L.) plants. Studies utilizing micro X-ray florescence. Front Plant Sci 5:808
Trenkel MET (2021) Slow-and controlled-release and stabilized fertilizers: an option for enhancing nutrient use efficiency in agriculture. International Fertilizer Industry Association (IFA)
Trinh CS, Lee H, Lee WJ, Lee SJ, Chung N, Han J et al (2018) Evaluation of the plant growth-promoting activity of *Pseudomonas nitroreducens* in *Arabidopsis thaliana* and *Lactuca sativa*. Plant Cell Rep 37:873–885
Upadhyay SK, Saxena AK, Singh JS, Singh DP (2019) Impact of native ST-PGPR (*Bacillus pumilus*; EU927414) on PGP traits, antioxidants activities, wheat plant growth and yield under salinity. Clim Chang Environ Sustain 7:157–168
Valat L, Deglène-Benbrahim L, Kendel M, Hussenet R, Le Jeune C, Schellenbaum P et al (2018) Transcriptional induction of two phosphate transporter 1 genes and enhanced root branching in grape plants inoculated with *Funneliformis mosseae*. Mycorrhiza 28:179–185
Vani MS, Hindumathi A, Reddy BN (2018) Beneficial effect of arbuscular mycorrhizal fungus, *Glomus fasciculatum* on plant growth and nutrient uptake in tomato. Indian Phytopathol 71:115–122
Vassilev N, Vassileva M, Martos V, Garcia del Moral LF, Kowalska J, Tylkowski B et al (2020) Formulation of microbial inoculants by encapsulation in natural polysaccharides: focus on beneficial properties of carrier additives and derivatives. Front Plant Sci 11:270
Velázquez E, Silva LR, Ramírez-Bahena MH, Peix A (2016) Diversity of potassium-solubilizing microorganisms and their interactions with plants. In: Meena V, Maurya B, Verma J, Meena R (eds) Potassium solubilizing microorganisms for sustainable agriculture. Springer, New Delhi, pp 99–110
Verma JP, Yadav J, Tiwari KN, Kumar A (2013) Effect of indigenous *Mesorhizobium* spp. and plant growth promoting rhizobacteria on yields and nutrients uptake of chickpea (*Cicer arietinum* L.) under sustainable agriculture. Ecol Eng 51:282–286
Verma M, Mishra J, Arora NK (2019) Plant growth-promoting rhizobacteria: diversity and applications. In: Sobti R, Arora NK, Kothari R (eds) Environmental biotechnology: for sustainable future. Springer, Singapore, pp 129–173
Verma M, Singh A, Dwivedi DH, Arora NK (2020) Zinc and phosphate solubilizing *rhizobium radiobacter* (LB2) for enhancing quality and yield of loose leaf lettuce in saline soil. Environ Sustain 3(2):209–218
Visconti D, Fiorentino N, Cozzolino E, Woo SL, Fagnano M, Rouphael Y (2020) Can *Trichoderma*-based biostimulants optimize N use efficiency and stimulate growth of leafy vegetables in greenhouse intensive cropping systems? Agronomy 10:121

Volpiano CG, Lisboa BB, São José JFB, de Oliveira AMR, Beneduzi A, Passaglia LMP et al (2018) Rhizobium strains in the biological control of the phytopathogenic fungi *Sclerotium (Athelia) rolfsii* on the common bean. Plant Soil 432:229–243

Vurukonda SSKP, Vardharajula S, Shrivastava M, SkZ A (2016) Enhancement of drought stress tolerance in crops by plant growth promoting rhizobacteria. Microbiol Res 184:13–24

Wahid F, Sharif M, Steinkellner S, Khan MA, Marwat KB, Khan SA (2016) Inoculation of arbuscular mycorrhizal fungi and phosphate solubilizing bacteria in the presence of rock phosphate improves phosphorus uptake and growth of maize. Pak J Bot 48(2):739–747

Woo SL, Hermosa R, Lorito M, Monte E (2023) *Trichoderma*: a multipurpose, plant-beneficial microorganism for eco-sustainable agriculture. Nat Rev Microbiol 21(5):312–326

Worldometers (2020). Retrieved from https://www.worldometers.info/worldpopulation/population-by-country/

Xiao JX, Hu CY, Chen YY, Yang B, Hua J (2014) Effects of low magnesium and an arbuscular mycorrhizal fungus on the growth, magnesium distribution and photosynthesis of two citrus cultivars. Sci Hortic 177:14–20

Xie X, Lin H, Peng X, Xu C, Sun Z, Jiang K et al (2016) Arbuscular mycorrhizal symbiosis requires a phosphate transceptor in the *Gigaspora margarita* fungal symbiont. Mol Plant 9:1583–1608

Yakhin OI, Lubyanov AA, Yakhin IA, Brown PH (2017) Biostimulants in plant science: a global perspective. Front Plant Sci 7:2049

Yang J, Lan L, Jin Y, Yu N, Wang D, Wang E (2022) Mechanisms underlying legume–rhizobium symbioses. J Integr Plant Biol 64(2):244–267

Yazdani M, Bahmanyar MA, Pirdashti H, Esmaili MA (2009) Effect of phosphate solubilization microorganisms (PSM) and plant growth promoting rhizobacteria (PGPR) on yield and yield components of corn (*Zea mays* L.). World Acad Eng Technol 49:90–92

Yin Z, Fan B, Roberts DP, Chen S, Shi F, Buyer JS et al (2017) Enhancement of maize growth and alteration of the rhizosphere microbial community by phosphate-solubilizing fungus *Aspergillus aculeatus* P93. J Agric Biotech 2:1

You M, Fang S, MacDonald J, Xu J, Yuan ZC (2020) Isolation and characterization of *Burkholderia cenocepacia* CR318, a phosphate solubilizing bacterium promoting corn growth. Microbiol Res 233:126395

Zaim S, Bekkar AA, Belabid L (2017) Rhizobium as a crop enhancer and biofertilizer for increased non-legume production. In: Hansen A, Choudhary D, Agrawal P, Varma A (eds) Rhizobium biology and biotechnology. Soil biology, vol 50. Springer, Cham, pp 25–37

Zhang Y, Zhuang WY (2020) *Trichoderma brevicrassum* strain TC967 with capacities of diminishing cucumber disease caused by *Rhizoctonia solani* and promoting plant growth. Biol Control 142:104151

Zhang H, Sun Y, Xie X, Kim MS, Dowd SE, Paré PW (2009) A soil bacterium regulates plant acquisition of iron via deficiency-inducible mechanisms. Plant J 58(4):568–577

Zhao J, Liu W, Liu D, Lu C, Zhang D, Wu H et al (2018) Identification and evaluation of *Aspergillus tubingensis* as a potential biocontrol agent against grey mould on tomato. J Gen Plant Pathol 84:148–159

Chapter 3
Effect of Engineered Nanoparticles on Rhizospheric Microbes

Ved Prakash, Sneha Tripathi, Padmaja Rai, Samarth Sharma, Kavita Tiwari, Shivani Mahra, Shweta Rana, Vivek Kumar, and Shivesh Sharma

Abstract The numerous applications of nanoparticles (NPs) are gaining momentum in various industrial sectors. The augmentation of NPs due to increased usage aids their entry into the terrestrial ecosystem through waste streams, treatment plants, and fertilizer applications. The excessive release of these engineered nanoparticles (ENPs) affects the plant growth-promoting rhizobacteria (PGPR), which are paramount for maintaining soil fertility. The PGPR inhabit the rhizosphere of the plant roots and derive their nutrition from the root exudates. The ENPs change the physiological and functional attributes of these useful microbes. In addition, ENPs alter the soil exoenzyme profile of these microbes. The exoenzymes are crucial in regulating the biogeochemical cycles and soil fertility. The interactions between ENPs and the PGPR may disturb the overall soil ecosystem, its fertility, thus impacting crop production. There is limited understanding on how concentration of NPs may influence the plant growth-promoting traits of microbes in synergistic as well as antagonistic ways. There are very few ecotoxicological studies on these NPs, and so far, there is no specific legislation or regulation for proper disposal of ENPs. The present book chapter elucidates the positive and negative role of ENPs over rhizospheric microbes.

Keywords Engineered nanoparticles · Rhizosphere · PGPR · Soil microorganism · Soil enzyme

---

V. Prakash · S. Tripathi · P. Rai · S. Sharma · K. Tiwari · S. Mahra · S. Sharma (✉)
Department of Biotechnology, Motilal Nehru National Institute of Technology Allahabad, Prayagraj, India
e-mail: shiveshs@mnnit.ac.in

S. Rana
Department of Physical and Natural Sciences, FLAME University, Pune, India

V. Kumar
Himalayan School of Biosciences, Swami Rama Himalayan University Jolly Grant, Dehradun, India

N. K. Arora, B. Bouizgarne (eds.), *Microbial Biotechnology for Sustainable Agriculture Volume 2*, Microorganisms for Sustainability 34,
https://doi.org/10.1007/978-981-97-2355-3_3

3.1 Introduction

The immoderate use of engineered nanoparticles (ENPs) in various industry and consumer products has raised global concerns about their fate in the biological system and hence demands the assessment of associated risks (Sharma et al. 2012; Arvidsson et al. 2020; Solano et al. 2021). Nanoparticles profoundly affect microbes and plants (Ma et al. 2010; Yu et al. 2013; Ameen et al. 2021). The terrestrial and plant systems have been inundated with ENPs, which indicate potential health concerns (Boxall et al. 2007; Crane et al. 2008; Ma et al. 2010; Ge et al. 2011; Hong et al. 2021). Addressing the present and future concerns associated with ENPs requires immediate attention (Schulte et al. 2008; Johnston et al. 2013). Studies on nanotoxicology and the associated nano-risks have increased with the spike in worldwide production of ENPs (Oberdörster et al. 2005). Rhizobacteria help provide resilience to plants to withstand various abiotic stresses, thus aiding in the increased agricultural production to attain sustainable development goals for global food security (Vishwakarma et al. 2017; Verma et al. 2018; Upadhyay et al. 2019; Prakash et al. 2020; Arora and Fatima 2022). Plant growth is supported by diverse microorganisms that inhabit the rhizospheric region where the number of microbial genes outnumbers the plant genes (Mendes et al. 2013; Alawiye and Babalola 2019); thus it is referred to as plant's second genome. Nanoparticle stress leads to alterations of microbial structure in the rhizosphere; therefore its critical assessment for the impacts on bulk and rhizospheric soil is important (Sillen et al. 2015). Various studies have confirmed the deleterious role of ENPs on the soil rhizosphere community (Mishra and Kumar 2009; Ge et al. 2011; Frenk et al. 2013; Yadav et al. 2014; Cao et al. 2016; Pallavi Mehta et al. 2016; Tripathi et al. 2017; Singh et al. 2019; Vishwakarma et al. 2020; Ajiboye et al. 2023), which ultimately affects the microbes as well as plant health. Notably, studies pertaining to ENPs exposure on microbes and its supporting community are scarce.

A deep understanding of how rhizospheric microbes and their enzyme activity pattern respond to ENPs is critical, as microbial communities being integral to soil systems respond to their alternations (Sadowsky and Schortemeyer 1997). ENPs may lead to shifts in microbial diversity and richness (Kumar et al. 2011). As microbes maintain overall soil health and restore fertility any changes in their community pattern will impact the entire soil ecosystem (Gupta et al. 2022; Arora 2023). Thus, insights to decipher microbe-nanoparticle interactions in natural systems and their transformation processes are highly required.

3.2 Current Regulation on ENPs Production and Disposal

Nanotechnology has marked its presence in almost all domains of industrial sectors including agriculture, transport, energy, healthcare, and electronics. As per the Lux Research report (2006), this sector was expected to reach a market of USD 3 trillion by 2015. The Grand View Research's Market analysis reports the global nanomaterials market size was valued at USD 10.88 billion in 2022 and is expected to register at a CAGR (compound annual growth rate) of 14.8% from 2023 to 2030 (Nanomaterials Market Size, Share & Growth 2023 Report).

There is no specific regulation set for nanotechnology in the European Union (EU). As per the *European Energy and Environmental Law Review*, the safety standards for use of ENPs cannot be applied to all systems mainly because of their undeciphered toxicity, bioaccumulation, and degradability in natural environment; the same quality standards and methodology are used for ecotoxicological evaluation (Baun et al. 2016) (Fig. 3.1).

Various countries have jointly taken efforts to devise a suitable nomenclature but a defined nomenclature system for nanomaterials is not established. An insufficient regulatory framework and legislation pose threat from ENPs to soil systems, which is of great concern (Ganzleben and Hansen 2012; Suppan 2013). Sewage sludge and biosolids are major contributors of nanoparticles (NPs) in the terrestrial systems (Brar et al. 2010). European Directive 86/278/EEC defines threshold values for heavy metals in sewage sludge that could be utilized for agricultural purposes (Council Directive 1986), but devising a set of guidelines and agreements for ENPs is still remote (Van calster 2006). The EU has introduced certain regulations on nanomaterials, namely Registration, Evaluation, Authorization and Restriction of Chemicals ((EC) 1907/2006) (REACH), Classification, Labelling and Packaging ((EC) 1272/2008) (CLP), Biocidal Products Regulation ((EU) 528/2012) (BPR), Novel Foods ((EU) 2015/2283)45, Cosmetic Products ((EC) 1223/2009) (Buzek and Ask 2009; Brinch et al. 2016, ECHA 2015. European Commission. Regulation (EC) No 1907/2006, Regulation (EC) no. 258/97; 2006).

Apart from the above-mentioned guidelines, two prominent guidelines are followed by developed countries namely: The Organization for Economic Co-operation and Development (OECD) and International Organization for Standardization (ISO), but it is still not clear if both guidelines can be adapted to nanomaterials (Hirano 2009). This ambiguous scenario for the regulation of ENPs demands the design of a framework at global scale for the use and disposal of nanomaterials in industrial setups.

3.3 Sources of ENPs Entry to Soil and Their Mode of Action Over Microbes

Increased usage of ENPs in a diverse range of commodities for various applications raises concerns regarding their associated risk to the environment. ENPs cause harm to soil systems and their persistence may lead to contamination of groundwater (da Silva Júnior et al. 2022). Both natural and anthropogenic sources release nanomaterials in the environment (Buzea et al. 2007; Buseck and Adachi 2008; Sadik 2013; Strambeanu et al. 2015; Zaveri et al. 2022). The application sector of ENPs

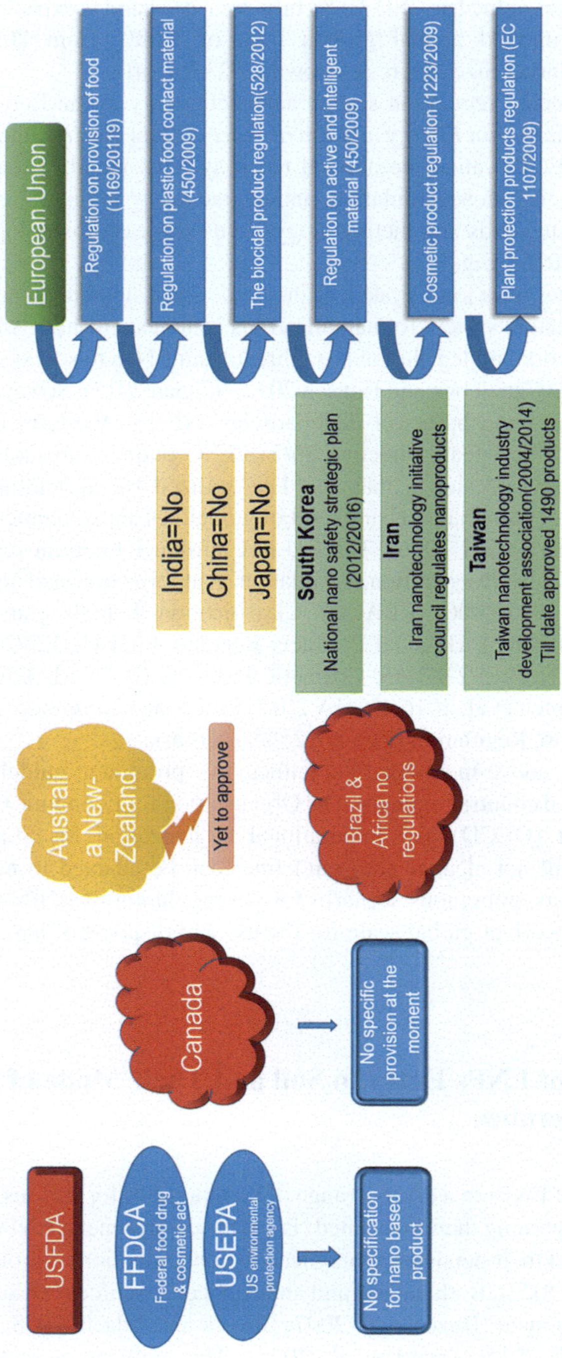

Fig. 3.1 Existing regulation of agri-nano products at global level

prominently involves antimicrobial paints, catalysts, cosmetics, electronics, fuel cells, metallurgy, textiles, wastewater streams, and many more, which pose the highest possibility to adulterate soil (Molteni et al. 2006; Ko et al. 2007; Kumar et al. 2008; Qiu et al. 2011; Yang et al. 2012; Abramova et al. 2013; Lu et al. 2015; Tan et al. 2017; Maťátková et al. 2022) (Fig. 3.2). Thus, these NPs released either at the point of production or at disposal may contaminate soil, and surface waters may prove a menace to the environment and associated microbes.

The rise in application of silver NPs (AgNPs) in antimicrobial products and consumer goods causes potential risk to the environment (Glover et al. 2011; Ahmed and Ogulata 2022) and it paves way to the soil ecosystem through diverse pathways (Anjum et al. 2013). Wastewater treatment plants act as point sources of ENPs where they get added to the environment as sludge or as treated water. Sewage sludge may contain AgNPs, ZnONPs, and TiO_2NPs, which are used in various sectors of industry (Lombi et al. 2012; Jośko and Oleszczuk 2013; McGillicuddy et al. 2017). The sludge often subjected to landfills (Kim et al. 2005) may contaminate the soil. NPs being reactive in nature interact with other waste components to form new compounds (Soni et al. 2015). Rajput et al. (2018) evaluated the ill effects of ENPs and concluded that their exposure negatively impacts microbial biomass and enzymatic activity and hence influences the soil microbial community and diversity. On the contrary, ENPs like silica have played a positive role in soil microbial population, enhanced total biomass, and enriched silica content (Rangaraj et al. 2014). Similarly, nanosilica promoted bacterial population growth by enhancing the protein level of bacterial biomass and the nutrient content of soil (Karunakaran et al. 2013a, b).

Exploration of nanotechnology in environmental remediation to reduce organic pollutants such as pesticides, solvents, polychlorinated biphenyls, aromatic hydrocarbons, personal care commodities, and pharmaceuticals from soil has opened new areas of its application (Li et al. 2016). Research found proven usage of amphiphilic polyurethane (APU) NPs in detoxification of polyaromatic hydrocarbons from soil that sorb contaminants strongly (Tungittiplakorn et al. 2004). In a similar way various studies have concluded the implication of ENPs in soil remediation, for instance iron oxide ENPs in arsenic, chromium, and ibuprofen decontamination (Shipley

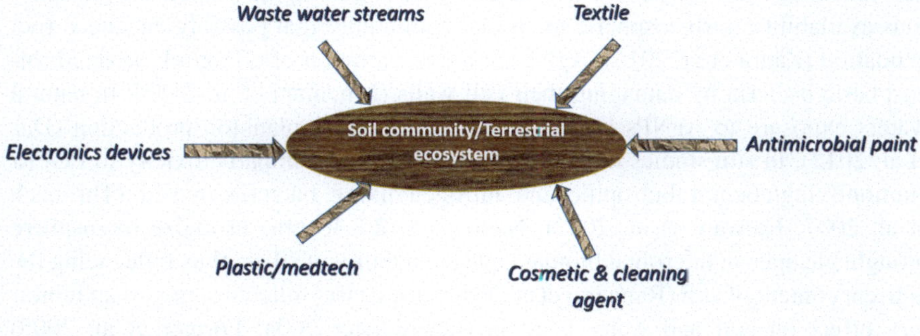

Fig. 3.2 Different sources of nanoparticles entry to the soil

et al. 2011; Machado et al. 2013; Su et al. 2016). However, various applications of ENPs lead to their gradual accumulation in the soil. The nano-fertilizers further boost the concentration of ENPs in the soil (Liu and Lal 2015; Servin et al. 2015; Solanki et al. 2015; Zahra et al. 2022). Thus, all factors collectively contribute to augmenting NPs in the terrestrial ecosystems. The problem of an increased number of NPs in the terrestrial ecosystems requires a systemic approach for sustainable development.

3.3.1 Nanoparticles Effect on Rhizospheric Bacterial Community

ENPs infiltrate in to the soil system through agrochemicals or the application of biosolids, which could potentially hamper soil bacterial community. Nano-CeO_2 and nano-ZnO at concentrations of 0.1 and 0.5 g kg^{-1}, respectively, showed to change soil bacterial communities (Ge et al. 2014). Silver NPs adversely affected rhizospheric bacteria compared to bulk soil. For maize grown in soil and rhizospheric communities unaffected by the ENPs, an increase in biomass content was observed, clearly indicating the importance of the rhizospheric microbial community in augmenting overall plant health (Sillen et al. 2015). AgNPs at concentrations of 0–75 ppm on wheat, cowpea, and brassica species changed the rhizospheric bacterial diversity, which was further affected by the grown plant species (Pallavi Mehta et al. 2016). Cao et al. (2016) demonstrated the role of arbuscular mycorrhizal fungi (AMF) in reducing the negative impact of FeONPs on rhizospheric bacterial diversity. Even lower dose of titanium oxide NPs was found to affect the soil microbial communities (Burke et al. 2014). But another study involving TiO_2 nanoparticles demonstrated that they improved soil microbial population by 21.4–48.1% along with enhancing soil enzyme activity (Raliya et al. 2015). Molybdenum NPs in colloidal solution at a concentration of 8 mg/L stimulated the growth of agronomically important microflora (Taran et al. 2014). The impact of ZnO nanoparticles at doses ranging from 0 to 100 mg/kg significantly altered the soil bacterial community without greatly affecting alpha diversity (Xu et al. 2018). Soil amended with TiO_2 and Fe_3O_4 at a dose of 0–250 mg kg^{-1} modified phosphorous availability in rhizospheric microbial community that possibly enhanced root exudation (Zahra et al. 2015). AgNPs at a concentration of 60 μg/mL affected rhizospheric bacteria by damaging their cell walls (Mirzajani et al. 2013). In natural waters exposure to AgNPs led to reduction in bacterioplankton production (Das et al. 2012). In situ studies have shown that nano-silver imparts toxicity to PGPR/ammonifying/chemolithotrophic and nitrogen fixing bacteria in soil (Throbäck et al. 2007; Kalwani et al. 2022). Nanosilica (20–40 nm) in maize rhizosphere brought changes in microbial biomass and availability of silica, thus influencing the nutrient content of soil (Rangaraj et al. 2014). Likewise, fullerenes impart an inhibitory effect on soil and water bacteria (Oberdörster 2004; Fortner et al. 2005)

(Table 3.1). But Nyberg et al. (2008) found that C60 fullerene did not have a negative impact on anaerobic microbes (Nyberg et al. 2008). The effect of gold NPs on soils showed to alter microbial community as well as soil enzyme activities subsequently having an impact on the nutrient recycling potential of the soils (Asadishad et al. 2017). Impact of zinc and copper oxide NPs showed altering microbial community structure in culture dependent and independent setups (Collins et al. 2012). In two different soil types, the effect of copper oxide and iron oxide NPs showed to alter bacterial community composition and conferred oxidative and hydrolytic activity on bacteria. It was also found that clay fraction and organic content significantly reduced NPs toxicity (Frenk et al. 2013). Graphene oxide NPs negatively impact soil microbes, specifically *Bacillus* species in a dose- and time-dependent manner. Results demonstrated to affect cell viability of *Bacillus cereus*, *Bacillus mycoides*, *Bacillus megaterium*, *Bacillus marisflavi*, and *Bacillus subtilis* (Gurunathan 2015). Overall, various findings demonstrated metal NPs confer ecotoxicity affecting plant stimulatory soil bacteria (Mishra and Kumar 2009). The studies mentioned above highlighted the negative impacts of ENPs on soil microbial diversity but Timmusk et al. (2018) reported that nano-titiana significantly promoted the performance of plant growth-promoting bacteria, suggesting that ENPs can also have beneficial effects on rhizospheric microbes (Table 3.1).

3.3.2 ENPs Role on Soil Enzyme Activity

Soil enzyme activities are the indicators of microbial diversity and hence one of the approaches to monitor alterations in soil chemistry (Alkorta et al. 2003; Kaur et al. 2022; Wang et al. 2022). The impact of ENPs on soil enzyme activity was examined and evaluated by Asadishad et al. (2018). Silver and zinc NPs altered enzyme activity while copper oxide (nCuO) or titanium dioxide (nTiO_2) NPs did not affect the activity. The effect of surface coating and size of NPs on enzyme activity were evaluated. In another study, the effect of Mo-ENPs was assessed on catalase, invertase, phosphatase, and urease activity. Compared to nFe_3O_4, nTiO_2, and nCeO_2, nZnO strongly altered enzyme activity as well as the microbial community (You et al. 2018). Another study showed the effect of TiO_2 and ZnO on soil protease, catalase, and peroxidase activities that were reduced by the effect of ENPs (Du et al. 2011). The ZnNPs or Zn^{2+} were applied to *C. sativus* for a duration of 8 weeks following that soil enzyme activities such as dehydrogenase, β-glucosidase, and acid phosphatase were evaluated; the ENPs decreased their activity and a large decrease was observed in dehydrogenase activity (Kim et al. 2011). ZnONPs also showed to alter phosphorous and nitrogen cycles in the soil apart from impairing soil enzyme activity (Kwak et al. 2016). The rhizosphere of *Iris wilsonii*, *Arundo donax*, and *Typha orientalis* when treated with AgNPs negatively impacted soil exoenzyme activity but it was also dependent on plant species too (Shin et al. 2012; Cao et al. 2017). Likewise, CuONPs reduced β-glucosidase activity along with altering the activity of dehydrogenase, acid phosphatase, and β-glucosidases (Kim et al. 2013).

Table 3.1 Influence of certain ENPs on soil rhizospheric microbes

Nanoparticles	Strain	Effect	Parameter evaluated	References
Silver nanoparticles (AgNPs)	*Bacillus thuringiensis*	Accumulation of envelope protein precursors	Analysis of proteins pattern	Mirzajani et al. (2014)
AgNPs	*Bacillus thuringiensis* SBURR1	"pit" formation	Membrane leakage	Mirzajani et al. (2013)
AgNPs	*Pseudomonas putida* KT2440	Cytotoxicity	Antibacterial activity	Gupta et al. (2015)
Iron oxide nanoparticles (FeONPs)	Proteobacteria, Actinobacteria, Acidobacteria, Bacteroidetes, Chloroflexi, Planctomycetes, Gemmatimonadetes, Firmicutes, and Nitrospirae	Decreased soil dissolved organic carbon	Microcalorimetric measurement	Cao et al. (2016)
Titanium oxide (TiO_2) and zinc oxide (ZnO) nanoparticles	Soil microcosm	Reduced both microbial biomass and diversity	Soil microbial biomass	Ge et al. (2011)
Copper oxide (CuO) and zinc oxide (ZnO) nanoparticles	*Pseudomonas chlororaphis* O6	Accumulation of intracellular reactive oxygen species	Membrane damage	Dimkpa et al. (2011)
CuONPs	*Phragmites australis*	Increased microbial abundance	Community structure	Fernandes et al. (2017)
ZnONPs	Rhizosphere microbes	Enhanced microbial populations	P-solubilizing enzymes	Raliya et al. (2016)
ZnONPs	*Sinorhizobium meliloti, Serendipita indica*			
CuO, ZnO, and Fe_2O_3 nanoparticles	*Rhizobium* sp. OS1	Affect the physiological expression of rhizobial species	Indole acetic acids, siderophores, and cyanogenic compounds	Oves et al. (2014)
Fullerene	*Escherichia coli* and *Bacillus subtilis*	Inhibited growth	Respiratory activity	Fortner et al. (2005)
Nano-CuO and nano-ZnO	Soil bacterial community	Inhibition of bacterial growth	Growth	Rousk et al. (2012)

(continued)

Table 3.1 (continued)

Nanoparticles	Strain	Effect	Parameter evaluated	References
Cobalt, iron, nickel, and silver nanoparticles	*Flavobacterium* and *Niastella*	Affect soil bacterial diversity	Pyrosequence analysis	Shah et al. (2014)
TiO_2 and CuONPs	Microcosm	Imparted toxicity to microbes	Microbial biomass, enzyme activities	Xu et al. (2015)
TiO_2NPs	*Nitrosomonas europaea*	Increased cell permeability	Cell wall/ membrane	Fang et al. (2010)
CuONPs	Rhizobiales and Sphingobacteriaceae	Negatively affected	Bacterial hydrolytic activity, oxidative potential, community composition	Frenk et al. (2013)
ZnONPs	*Pseudomonas chlororaphis* O6	Disrupt cell signaling	Cell signaling molecules	Goodman et al. (2016)
AgNPs	Acidobacteria, Actinobacteria, Cyanobacteria, and Nitrospirae	Bacteria number showed inverse relation with concentration of NPs used	Survival rate	Juan et al. (2017)
ZnONPs	*B. subtilis* and *Pseudomonas aeruginosa*	Hampered enzymes activity	Soil enzyme activity	Santimano and Kowshik (2013)
CeO_2NPs	Soil microorganism	β-Glucosidase and urease activity were inhibited	Soil enzyme activity	Li et al. (2017)
Fe_3O_4NPs	Actinobacteria, Duganella, Streptomycetaceae, or Nocardioides	Positive impact on specific set of bacteria	Molecular fingerprinting and sequencing analysis	He et al. (2011)
CeO_2 and ZnONPs	*Sinorhizobium meliloti, S. melba*	Imparted toxicity	Minimum inhibitory concentration, disk diffusion tests, and dynamic growth	Bandyopadhyay et al. (2012)
TiO_2	PGPRs, *B. subtilis*, and *Pseudomonas fluorescens*	Growth inhibition and toxicity	Microbial susceptibility	Karunakaran et al. (2013b)
Tunsten (WO_3) NPs (tungsten)	*Azotobacter vinelandii*	Negative impact on growth	Growth	Allard et al. (2013)

(continued)

Table 3.1 (continued)

Nanoparticles	Strain	Effect	Parameter evaluated	References
Nanosilica alumina (Al_2O_3)	*B. megaterium*, *B. brevis*, *P. fluorescens*, and *Azotobacter vinelandii*	Up to 1000 mg L − 1 nanosilica were nontoxic to PGPRs while Al2O3 NPs imparted toxicity	Ecotoxicological study	Karunakaran et al. (2014)
TiO_2NPs	*Rhizobium leguminosarum* bv. *viciae* 3841	Altered cell morphology and polysaccharide composition of the cell wall	Morphological study	Fan et al. (2014)
Graphene oxide (GO) NPs	*B. megaterium*, *B. cereus*, *B. subtilis*, *B. mycoides*, and *B. marisflavi*	Cytotoxicity	Cell viability	Gurunathan (2015)
AgNPs	Proteobacteria and Planctomycetes	Increase in microbe population	Sensitivity of bacteria	Juan et al. (2017)
ZnONPs	Actinobacteria Acidobacteria Chloroflexi Proteobacteria Bacteroidetes Firmicutes Cyanobacteria	Alteration in bacterial community structure	Community composition	Xu et al. (2018)
AgNPs ZnONPs	Proteobacteria, Actinobacteria, and Firmicutes	Decreased relative abundance	Community composition	Chavan and Nadanathangam (2019)
n-ZnO and n-TiO2	*B. subtilis*	Showed impact on biofilm growth	Biofilm growth Proline assay Sporulation assay	Eymard-Vernain et al. (2018a)
Ag-NPs	*B. subtilis*	Affected secretome	Morphology, agglomeration state, structure	Eymard-Vernain et al. (2018b)
Ag-NPs	Rhizosphere bacteria	Inhibited growth of soil bacteria	PCR-DGGE Bacterial diversity	Karimi and Sadeghi (2019)
Ag-NPs	*Azotobacter nigricans*, *Azorhizophilus pasali*	Lethal effect on bacteria	Phylogenetic analysis	Rezayatmand and Doudi (2018)
Ag-NPs	*Proteobacteria Actinobacteria Bacteroidetes*	Decreased bacterial diversity	Microbial community composition	Vitali et al. (2019)

(continued)

Table 3.1 (continued)

Nanoparticles	Strain	Effect	Parameter evaluated	References
Fe0/PVP nanoparticles	*Alternaria* sp. 4D *P. viride*	Growth suppression	Inhibition activity	Sidorenko et al. 2020
nCr2O3 and nCeO2	*Proteobacteria* and *Bacteroidia*	Alterations in taxonomic composition	Community composition	Luo et al. (2020)

Higher AgNP concentration influenced dehydrogenase activity in the rhizosphere of *Typha orientalis* Presl, showing significant inhibition (Cao et al. 2016). AgNPs can be further utilized for decontamination of soil infested with heavy metals (Gaddam et al. 2015; Rahmatpour et al. 2017).

3.4 Conclusion and Future Prospects

Nanomaterials are emerging as a boon and a bane at the same time. ENPs have found multiple uses in industry and consumer products. Their market has increased in the past few years. While they have properties that help the development and growth of plants, which is critical to addressing the global food security problem, they also possess functional aspects that are detrimental to the propagation of the bacterial soil community and their enzymatic activities during the plant-ENPs interaction in the rhizosphere (Khan et al. 2021; Khanna et al. 2021; Kumar et al. 2021). Various routes of exposure determine the bioavailability of ENPs in soil, causing bioaccumulation and affecting soil microbes and plant systems. The regular deposition of ENPs in the soil system directly or indirectly may affect soil health. No legitimate guidelines exist for the production, usage, and disposal of ENPs along with the regulation of ENPs released in soil systems. Determination of toxicity caused by ENPs need to be streamlined and made coherent across countries. The clarity and parity in regulating the harmful effects of ENPs is critical to combat nanotoxicity (Foulkes et al. 2020). The field of nanotechnology is still at a budding stage and so judicious usage of NPs and impact assessment of nanotoxicity is important to safeguard environmental health (Mishra et al. 2022). There is a need to enhance our understanding of the interactions between plants and ENPs, as well as between microbial communities and ENPs, with a focus on the overall interaction between plants, ENPs, and rhizospheric bacteria. Future studies can focus on interaction between these three stakeholders. The toxicity of ENPs on microbial and plant system is well identified; however the interaction of ENPs and soil compositions as well as organisms leads to transformation of ENPs reducing their toxic effects to a significant level. Nonetheless, accumulation of ENPs in soils may represent a source of metals but its impact on microbe-plant system needs to be studied. More studies are required to fully understand the mechanisms by which NPs reduce the bacterial count and enzymatic activity in the rhizosphere.

Acknowledgments Authors are thankful to Director, MNNIT Allahabad, Prayagraj, for providing necessary facilities.

References

Abramova A, Gedanken A, Popov V, Ooi EH, Mason TJ, Joyce EM et al (2013) A sonochemical technology for coating of textiles with antibacterial nanoparticles and equipment for its implementation. Mater Lett 96:121–124

Ahmed T, Ogulata RT (2022) A review on silver nanoparticles-green synthesis, antimicrobial action and application in textiles. J Natur Fibers 19(14):8463–8484

Ajiboye TT, Ajiboye TO, Babalola OO (2023) Impacts of binary oxide nanoparticles on the soybean plant and its rhizosphere, associated phytohormones, and enzymes. Molecules 28(3):1326

Alawiye TT, Babalola OO (2019) Bacterial diversity and community structure in typical plant rhizosphere. Diversity 11(10):179

Alkorta I, Aizpurua A, Riga P, Albizu I, Amézaga I, Garbisu C (2003) Soil enzyme activities as biological indicators of soil health. Rev Environ Health 18(1):65–73

Allard P, Darnajoux R, Phalyvong K, Bellenger JP (2013) Effects of tungsten and titanium oxide nanoparticles on the diazotrophic growth and metals acquisition by *Azotobacter vinelandii* under molybdenum limiting condition. Environ Sci Technol 47(4):2061–2068

Ameen F, Alsamhary K, Alabdullatif JA, ALNadhari S (2021) A review on metal-based nanoparticles and their toxicity to beneficial soil bacteria and fungi. Ecotoxicol Environ Saf 213:112027

Anjum NA, Gill SS, Duarte AC, Pereira E, Ahmad I (2013) Silver nanoparticles in soil–plant systems. J Nanopart Res 15(9):1896

Arora P (2023) Life below land: the need for a new sustainable development goal. Front Environ Sci 11:1215282. https://doi.org/10.3389/fenvs.2023.1215282

Arora NK, Fatima T (2022) Role of plant growth promoting microbes in managing sustainability of stressed agroecosystems. Environ Sustain 5:1–3

Arvidsson R, Hansen SF, Baun A (2020) Influence of natural organic matter on the aquatic ecotoxicity of engineered nanoparticles: recommendations for environmental risk assessment. NanoImpact 20:100263

Asadishad B, Chahal S, Cianciarelli V, Zhou K, Tufenkji N (2017) Effect of gold nanoparticles on extracellular nutrient-cycling enzyme activity and bacterial community in soil slurries: role of nanoparticle size and surface coating. Environ Sci Nano 4(4):907–918

Asadishad B, Chahal S, Akbari A, Cianciarelli V, Azodi M, Ghoshal S et al (2018) Amendment of agricultural soil with metal nanoparticles: effects on soil enzyme activity and microbial community composition. Environ Sci Technol 52(4):1908–1918

Bandyopadhyay S, Peralta-Videa JR, Plascencia-Villa G, José-Yacamán M, Gardea-Torresdey JL (2012) Comparative toxicity assessment of CeO_2 and ZnO nanoparticles towards *Sinorhizobium meliloti*, a symbiotic alfalfa associated bacterium: use of advanced microscopic and spectroscopic techniques. J Hazard Mater 241:379–386

Baun A, Hartmann NB, Grieger KD, Hansen SF (2016) Setting the limits for engineered nanoparticles in European surface waters–are current approaches appropriate? J Environ Monit 11(10):1774–1781

Boxall AB, Tiede K, Chaudhry Q (2007) Engineered nanomaterials in soils and water: how do they behave and could they pose a risk to human health? Future Med:919–927

Brar SK, Verma M, Tyagi RD, Surampalli RY (2010) Engineered nanoparticles in wastewater and wastewater sludge–evidence and impacts. Waste Manag 30:504–520

Brinch A, Hansen SF, Hartmann NB, Baun A (2016) EU regulation of nanobiocides: challenges in implementing the biocidal product regulation (BPR). Nano 6(2):33

Burke DJ, Zhu S, Pablico-Lansigan MP, Hewins CR, Samia ACS (2014) Titanium oxide nanoparticle effects on composition of soil microbial communities and plant performance. Biol Fertil Soils 50(7):1169–1173

Buseck PR, Adachi K (2008) Nanoparticles in the atmosphere. Elements 4(6):389–394

Buzea C, Pacheco II, Robbie K (2007) Nanomaterials and nanoparticles: sources and toxicity. Biointerphases 2(4):MR17-71

Buzek J, Ask B (2009) Regulation (EC) No 1223/2009 of the European Parliament and of the Council of 30 November 2009 on cosmetic products. Off J Eur Union L. 342

Cao J, Feng Y, Lin X, Wang J (2016) Arbuscular mycorrhizal fungi alleviate the negative effects of iron oxide nanoparticles on bacterial community in rhizospheric soils. Front Environ Sci 18:4–10

Cao C, Huang J, Cai WS, Yan CN, Liu JL, Jiang YD (2017) Effects of silver nanoparticles on soil enzyme activity of different wetland plant soil systems. Soil Sediment Contam 26(5):558–567

Chavan S, Nadanathangam V (2019) Effects of nanoparticles on plant growth-promoting bacteria in Indian agricultural soil. Agronomy 9:140

Collins D, Luxton T, Kumar N, Shah S, Walker VK, Shah V (2012) Assessing the impact of copper and zinc oxide nanoparticles on soil: a field study. PLoS One 7(8):e42663

Council Directive (1986) Council directive 86/278/EEC of 12 June 1986 on the protection of the environment, and in particular of the soil, when sewage sludge is used in agriculture. Official J L 181(04–07):0006–0012

Crane M, Handy RD, Garrod J, Owen R (2008) Ecotoxicity test methods and environmental hazard assessment for engineered nanoparticles. Ecotoxicology 17(5):421

da Silva Júnior AH, Mulinari J, de Oliveira PV, de Oliveira CRS, Júnior FWR (2022) Impacts of metallic nanoparticles application on the agricultural soils microbiota. J Hazard Mater Adv 7:100103

Das P, Williams CJ, Fulthorpe RR, Hoque ME, Metcalfe CD, Xenopoulos MA (2012) Changes in bacterial community structure after exposure to silver nanoparticles in natural waters. Environ Sci Technol 46(16):9120–9128

Dimkpa CO, Calder A, Britt DW, McLean JE, Anderson AJ (2011) Responses of a soil bacterium, *Pseudomonas chlororaphis* O6 to commercial metal oxide nanoparticles compared with responses to metal ions. Environ Pollut 159(7):1749–1756

Du W, Sun Y, Ji R, Zhu J, Wu J, Guo H (2011) TiO_2 and ZnO nanoparticles negatively affect wheat growth and soil enzyme activities in agricultural soil. J Environ Monit 13(4):822–828

ECHA (2015) Guidance on the application of the CLP criteria; guidance to Regulation (EC) no 1272/2008 on classification, labelling and packaging (CLP) of substances and mixtures

European Commission (2006) Regulation (EC) No 1907/2006 of the European Parliament and of the Council of 18 December 2006 concerning the Registration, Evaluation, Authorisation and Restriction of Chemicals (REACH), establishing a European Chemicals Agency, amending Directive 1999/45/EC and repealing Council Regulation (EEC) No 793/93 and Commission Regulation (EC) No 1488/94 as well as Council Directive 76/769/EEC and Commission Directives 91/155/EEC, 93/67/EEC, 93/105/EC and 2000/21/EC. Off J Eur Union 396:1–849

Eymard-Vernain E, Luche S, Rabilloud T, Lelong C (2018a) Impact of nanoparticles on the *Bacillus subtilis* (3610) competence. Sci Rep 8:1–8

Eymard-Vernain E, Lelong C, Pradas del Real AE, Soulas R, Bureau S, Tardillo Suarez V et al (2018b) Impact of a model soil microorganism and of its secretome on the fate of silver nanoparticles. Environ Sci Technol 52(1):71–78

Fan R, Huang YC, Grusak MA, Huang CP, Sherrier DJ (2014) Effects of nano-TiO_2 on the agronomically-relevant Rhizobium–legume symbiosis. Sci Total Environ 466:503–512

Fang X, Yu R, Li B, Somasundaran P, Chandran K (2010) Stresses exerted by ZnO, CeO_2 and anatase TiO_2 nanoparticles on the Nitrosomonas europaea. J Colloid Interface Sci 348(2):329–334

Fernandes JP, Almeida CM, Andreotti FF, Barros L, Almeida T, Mucha AP (2017) Response of microbial communities colonizing salt marsh plants rhizosphere to copper oxide nanopar-

ticles contamination and its implications for phytoremediation processes. Sci Total Environ 581:801–810
Fortner JD, Lyon DY, Sayes CM, Boyd AM, Falkner JC, Hotze EM (2005) C60 in water: nanocrystal formation and microbial response. Environ Sci Technol 39(11):4307–4316
Foulkes R, Man E, Thind J, Yeung S, Joy A, Hoskins C (2020) The regulation of nanomaterials and nanomedicines for clinical application: current and future perspectives. Biomater Sci 8(17):4653–4664
Frenk S, Ben-Moshe T, Dror I, Berkowitz B, Minz D (2013) Effect of metal oxide nanoparticles on microbial community structure and function in two different soil types. PLoS One 8(12):e84441
Gaddam DP, Devamma N, Prasad TN (2015) Evaluation of the effect of indigenous mycogenic silver nanoparticles on soil exo-enzymes in barite mine contaminated soils. Appl Nanosci 5(4):505–513
Ganzleben C, Hansen SF (2012) Environmental exposure to nanomaterials–data scoping study: final report, Milieu, Brussels, Belgium
Ge Y, Schimel JP, Holden PA (2011) Evidence for negative effects of TiO_2 and ZnO nanoparticles on soil bacterial communities. Environ Sci Technol 45:1659–1664
Ge Y, Priester JH, Van De Werfhorst LC, Walker SL, Nisbet RM, An YJ et al (2014) Soybean plants modify metal oxide nanoparticle effects on soil bacterial communities. Environ Sci Technol 48(22):13489–13496
Glover RD, Miller JM, Hutchison JE (2011) Generation of metal nanoparticles from silver and copper objects: nanoparticle dynamics on surfaces and potential sources of nanoparticles in the environment. ACS Nano 5(11):8950–8957
Goodman J, Mclean JE, Britt DW, Anderson AJ (2016) Sublethal doses of ZnO nanoparticles remodel production of cell signaling metabolites in the root colonizer *Pseudomonas chlororaphis* O6. Environ Sci-Nano 3(5):1103–1113
Gupta IR, Anderson AJ, Rai M (2015) Toxicity of fungal-generated silver nanoparticles to soil-inhabiting *Pseudomonas putida* KT2440, a rhizospheric bacterium responsible for plant protection and bioremediation. J Hazard Mater 286:48–54
Gupta G, Dhar S, Kumar A, Choudhary AK, Dass A, Sharma VK et al (2022) Microbes-mediated integrated nutrient management for improved rhizo-modulation, pigeonpea productivity, and soil bio-fertility in a semi-arid agro-ecology. Front Microbiol 13:924407
Gurunathan S (2015) Cytotoxicity of graphene oxide nanoparticles on plant growth promoting rhizobacteria. J Ind Eng Chem 32:282–391
He S, Feng Y, Ren H, Zhang Y, Gu N, Lin X et al (2011) The impact of iron oxide magnetic nanoparticles on the soil bacterial community. J Soils Sediments 11(8):1408–1417
Hirano S (2009) A current overview of health effect research on nanoparticles. Environ Health Prev Med 14(4):223
Hong J, Wang C, Wagner DC, Gardea-Torresdey JL, He F, Rico CM (2021) Foliar application of nanoparticles: mechanisms of absorption, transfer, and multiple impacts. Environ Sci Nano 8(5):1196–1210
Johnston H, Pojana G, Zuin S, Jacobsen NR, Møller P, Loft S et al (2013) Engineered nanomaterial risk. Lessons learnt from completed nanotoxicology studies: potential solutions to current and future challenges. Crit Rev Toxicol 43(1):1–20
Jośko I, Oleszczuk P (2013) The influence of ZnO and TiO_2 nanoparticles on the toxicity of sewage sludges. Environ Sci-Proc Imp 15(1):296–306
Juan WANG, Kunhui SHU, Zhang LI, Youbin SI (2017) Effects of silver nanoparticles on soil microbial communities and bacterial nitrification in suburban vegetable soils. Pedosphere 27(3):482–490
Kalwani M, Chakdar H, Srivastava A, Pabbi S, Shukla P (2022) Effects of nanofertilizers on soil and plant-associated microbial communities: emerging trends and perspectives. Chemosphere 287:132107

Karimi E, Sadeghi A (2019) Impact of silver nanoparticles on bacterial flora of the cotton rhizosphere and bulk soil. 72–14:63 فصلنامهعلمیژنتیکنوین

Karunakaran G, Suriyaprabha R, Manivasakan P, Yuvakkumar R, Rajendran V, Prabu P et al (2013a) Effect of nanosilica and silicon sources on plant growth promoting rhizobacteria, soil nutrients and maize seed germination. IET Nanobiotechnol 7(3):70–77

Karunakaran G, Suriyaprabha R, Manivasakan P, Yuvakkumar R, Rajendran V, Kannan N et al (2013b) Impact of nano and bulk ZrO_2, TiO_2 particles on soil nutrient contents and PGPR. J Nanosci Nanotechnol 13(1):678–685

Karunakaran G, Suriyaprabha R, Manivasakan P, Rajendran V, Kannan N (2014) Influence of nano and bulk SiO_2 and Al_2O_3 particles on PGPR and soil nutrient contents. Curr Nanosci 10(4):604–612

Kaur H, Kalia A, Sandhu JS, Dheri GS, Kaur G, Pathania S (2022) Interaction of TiO2 nanoparticles with soil: effect on microbiological and chemical traits. Chemosphere 301:134629

Khan ST, Adil SF, Shaik MR, Alkhathlan HZ, Khan M, Khan M (2021) Engineered nanomaterials in soil: their impact on soil microbiome and plant health. Plan Theory 11(1):109

Khanna K, Kohli SK, Handa N, Kaur H, Ohri P, Bhardwaj R et al (2021) Enthralling the impact of engineered nanoparticles on soil microbiome: a concentric approach towards environmental risks and cogitation. Ecotoxicol Environ Saf 222:112459

Kim EH, Cho JK, Yim S (2005) Digested sewage sludge solidification by converter slag for landfill cover. Chemosphere 59(3):387–395

Kim S, Kim J, Lee I (2011) Effects of Zn and ZnO nanoparticles and Zn2+ on soil enzyme activity and bioaccumulation of Zn in *Cucumis sativus*. Chem Ecol 27(1):49–55

Kim S, Sin H, Lee S, Lee I (2013) Influence of metal oxide particles on soil enzyme activity and bioaccumulation of two plants. J Microbiol Biotechnol 23(9):1279–1286

Ko SH, Park I, Pan H, Grigoropoulos CP, Pisano AP, Luscombeet CK et al (2007) Direct nanoimprinting of metal nanoparticles for nanoscale electronics fabrication Nano Lett 7(7):1869–1877

Kumar A, Vemula PK, Ajayan PM, Verma JP (2008) Silver-nanoparticle-embedded antimicrobial paints based on vegetable oil. Nat Mater 7(3):236

Kumar N, Shah V, Walker VK (2011) Perturbation of an arctic soil microbial community by metal nanoparticles. J Hazard Mater 190:816–822

Kumar A, Sharma PK, Singh S, Verma JP (2021) Impact of engineered nanoparticles on microbial communities, soil health and plants. In: Singh P, Singh R, Verma P, Bhadouria R, Kumar A, Kaushik M (eds) Plant-microbes-engineered Nano-particles (PM-ENPs) nexus in agroecosystems. Advances in science, technology & innovation. Springer, Cham, pp 201–215

Kwak JI, Yoon SJ, An YJ (2016) Long-term effects of ZnO nanoparticles on exoenzyme activities in planted soils. Environ Eng Res 22(2):224–229

Li Q, Chen X, Zhuang J, Chen X (2016) Decontaminating soil organic pollutants with manufactured nanoparticles. Environ Sci Pollut R 23(12):11533–11548

Li B, Chen Y, Liang WZ, Mu L, Bridges WC, Jacobson AR et al (2017) Influence of cerium oxide nanoparticles on the soil enzyme activities in a soil-grass microcosm system. Geoderma 299:54–62

Liu R, Lal R (2015) Potentials of engineered nanoparticles as fertilizers for increasing agronomic productions. Sci Total Environ 514:131–139

Lombi E, Donner E, Tavakkoli E, Turney TW, Naidu R. Miller BW et al (2012) Fate of zinc oxide nanoparticles during anaerobic digestion of wastewater and post-treatment processing of sewage sludge. Environ Sci Technol 46(16):9089–9096

Lu PJ, Huang SC, Chen YP, Chiueh LC, Shih DYC (2015) Analysis of titanium dioxide and zinc oxide nanoparticles in cosmetics. J Food Drug Anal 23(3):587–594

Luo J, Song Y, Liang J, Li J, Islam E, Li T (2020) Elevated CO_2 mitigates the negative effect of CeO_2 and Cr_2O_3 nanoparticles on soil bacterial communities by alteration of microbial carbon use. Environ Pollut 114456

Lux Research (2006) The nanotech report 4th edition. 2011 February; accessed from the website: https://portal.luxresearchinc.com/research/document excerpt/2778

Ma X, Geiser-Lee J, Deng Y, Kolmakov A (2010) Interactions between engineered nanoparticles (ENPs) and plants: phytotoxicity, uptake and accumulation. Sci Total Environ 408(16):3053–3061

Machado S, Stawiński W, Slonina P, Pinto AR, Grosso JP, Nouws HPA et al (2013) Application of green zero-valent iron nanoparticles to the remediation of soils contaminated with ibuprofen. Sci Total Environ 461:323–329

Maťátková O, Michailidu J, Miškovská A, Kolouchová I, Masák J, Čejková A (2022) Antimicrobial properties and applications of metal nanoparticles biosynthesized by green methods. Biotechnol Adv 58:107905

McGillicuddy E, Murray I, Kavanagh S, Fogarty A, Cormican M et al (2017) Silver nanoparticles in the environment: sources, detection and ecotoxicology. Sci Total Environ 575:231–246

Mendes R, Garbeva P, Raaijmakers JM (2013) The rhizosphere microbiome: significance of plant beneficial, plant pathogenic, and human pathogenic microorganisms. FEMS Microbiol Rev 37:634–663

Mirzajani F, Askari H, Hamzelou S, Farzaneh M, Ghassempour A (2013) Effect of silver nanoparticles on *Oryza sativa* L. and its rhizosphere bacteria. Ecotoxicol Environ Safe 88:48–54

Mirzajani F, Askari H, Hamzelou S, Schober Y, Römpp A, Ghassempour A et al (2014) Proteomics study of silver nanoparticles toxicity on *Bacillus thuringiensis*. Ecotoxicol Environ Safe 100:122–130

Mishra VK, Kumar A (2009) Impact of metal nanoparticles on the plant growth promoting rhizobacteria. Dig J Nanomater Biostruct 4:587–592

Mishra AK, Das R, Sahoo S, Biswal B (2022) Global regulations and legislations on nanoparticles usage and application in diverse horizons. Compre Anal Chem:261–290

Molteni G, Bianchi CL, Marinoni G, Santo N, Ponti A (2006) Cu/Cu-oxide nanoparticles as catalyst in the "click" azide–alkyne cycloaddition. New J Chem 30(8):1137–1139

Nyberg L, Turco RF, Nies L (2008) Assessing the impact of nanomaterials on anaerobic microbial communities. Environ Sci Technol 42(6):1938–1943

Oberdörster E (2004) Manufactured nanomaterials (fullerenes, C60) induce oxidative stress in the brain of juvenile largemouth bass. Environ Health Perspect 112(10):1058–1062

Oberdörster G, Oberdörster E, Oberdörster J (2005) Nanotoxicology: an emerging discipline evolving from studies of ultrafine particles. Environ Health Perspect 113(7):823–839

Oves M, Khan MS, Zaidi A, Ahmed AS, Azam A (2014) Production of plant-growth promoting substances by nodule forming symbiotic bacterium rhizobium sp. os1 is influenced by Cuo, Zno and Fe_2O_3 nanoparticles. IIOAB J 5(4):1–11

Pallavi Mehta CM, Srivastava R et al (2016) Impact assessment of silver nanoparticles on plant growth and soil bacterial diversity. 3 Biotech 6:254

Prakash V, Khan MY, Rai P, Prasad R, Tripathi DK, Sharma S (2020) Exploring plant rhizobacteria synergy to mitigate abiotic stress: a new dimension toward sustainable agriculture. In: Tripathi DK, Chauhan DK, Prasad SM, Ramawat N, Singh VP, Sharma S, Dubey NK (eds) Plant life under changing environment. Academic Press, Elsevier, pp 861–882

Qiu JD, Wang GC, Liang RP, Xia XH, Yu HW (2011) Controllable deposition of platinum nanoparticles on graphene as an electrocatalyst for direct methanol fuel cells. J Phys Chem C 115(31):15639–15645

Rahmatpour S, Shirvani M, Mosaddeghi MR, Nourbakhsh F, Bazarganipour M (2017) Dose–response effects of silver nanoparticles and silver nitrate on microbial and enzyme activities in calcareous soils. Geoderma 285:313–322

Rajput VD, Minkina T, Sushkova S, Tsitsuashvili V, Mandzhieva S, Gorovtsov A et al (2018) Effect of nanoparticles on crops and soil microbial communities. J Soils Sediments 18(6):2179–2187

Raliya R, Biswas P, Tarafdar JC (2015) TiO2 nanoparticle biosynthesis and its physiological effect on mung bean (*Vigna radiata* L.). Biotechnol Rep 5:22–26

Raliya R, Tarafdar JC, Biswas P (2016) Enhancing the mobilization of native phosphorus in the mung bean rhizosphere using ZnO nanoparticles synthesized by soil fungi. J Agric Food Chem 64(16):3111–3118

Rangaraj S, Gopalu K, Rathinam Y, Periasamy P, Venkatachalam R, Narayanasamy K (2014) Effect of silica nanoparticles on microbial biomass and silica availability in maize rhizosphere. Biotechnol Appl Biochem 61(6):668–675

Rezayatmand Z, Doudi M (2018) The effect of silver nanoparticles on the growth of nitrogen fixing and acc deaminase producing bacteria isolated from sunflower rhizosphere. Int J Mol Clin Microbiol 8:912–921

Rousk J, Ackermann K, Curling SF, Jones DL (2012) Comparative toxicity of nanoparticulate CuO and ZnO to soil bacterial communities. PLoS One 7(3):e34197

Sadik OA (2013) Anthropogenic nanoparticles in the environment. Environ Sci-Proc Imp 15(1):19–20

Sadowsky M, Schortemeyer M (1997) Soil microbial responses to increased concentrations of atmospheric CO_2. Glob Change Biol 3:217–224

Santimano MC, Kowshik M (2013) Altered growth and enzyme expression profile of ZnO nanoparticles exposed non-target environmentally beneficial bacteria. Environ Monit Assess 185(9):7205–7214

Schulte P, Geraci C, Zumwalde R, Hoover M, Kuempel E (2008) Occupational risk management of engineered nanoparticles. J Occup Environ Hyg 5(4):239–249

Servin A, Elmer W, Mukherjee A, De la Torre-Roche R, Hamdi H, White JC et al (2015) A review of the use of engineered nanomaterials to suppress plant disease and enhance crop yield. J Nanopart Res 17:1–21

Shah V, Collins D, Walker VK, Shah S (2014) The impact of engineered cobalt, iron, nickel and silver nanoparticles on soil bacterial diversity under field conditions. Environ Res Lett 9(2):024001

Sharma V, Kumar A, Dhawan A (2012) Nanomaterials: exposure, effects and toxicity assessment. Proc Natl Sci India B: Biol Sci 82:3–11

Shin YJ, Kwak JI, An YJ (2012) Evidence for the inhibitory effects of silver nanoparticles on the activities of soil exoenzymes. Chemosphere 88(4):524–529

Shipley HJ, Engates KE, Guettner AM (2011) Study of iron oxide nanoparticles in soil for remediation of arsenic. Sci Total Environ 13(6):2387–2397

Sidorenko A, Rastimesina I, Postolachi O, Fedorov V, Gutul T, Vaseashta A (2020) The toxic effect of trifluralin on soil microorganisms in the presence of Fe 0/PVP nanoparticles. In: Sidorenko A, Hahn H (eds) Functional nanostructures and sensors for CBRN defence and environmental safety and security. NATO Science for Peace and Security Series C: Environmental Security, pp 113–123

Sillen WM, Thijs S, Abbamondi GR, Janssen J, Weyens N, White JC et al (2015) Effects of silver nanoparticles on soil microorganisms and maize biomass are linked in the rhizosphere. Soil Biol Biochem 91:14–22

Singh J, Vishwakarma K, Ramawat N, Rai P, Singh VK, Mishra RK et al (2019) Nanomaterials and microbes' interactions: a contemporary overview. 3 Biotech 9:1–14

Solanki P, Bhargava A, Chhipa H, Jain N, Panwar J (2015) Nano-fertilizers and their smart delivery system. In: Rai M, Ribeiro C, Mattoso L, Duran N (eds) Nanotechnologies in food and agriculture. Springer, Cham, pp 81–101

Solano R, Patiño-Ruiz D, Tejeda-Benitez L, Herrera A (2021) Metal-and metal/oxide-based engineered nanoparticles and nanostructures: a review on the applications, nanotoxicological effects, and risk control strategies. Environ Sci Pollut Res 28:16962–16981

Soni D, Naoghare PK, Saravanadevi S, Pandey RA (2015) Release, transport and toxicity of engineered nanoparticles. Rev Environ Contam Toxicol 234., Springer, Cham:1–47

Strambeanu N, Demetrovici L, Dragos D (2015) Natural sources of nanoparticles. In: Lungu M, Neculae A, Bunoiu M, Biris C (eds) Nanoparticles promises and risks. Springer, Cham, pp 9–19

Su H, Fang Z, Tsang PE, Zheng L, Cheng W, Fang J et al (2016) Remediation of hexavalent chromium contaminated soil by biochar-supported zero-valent iron nanoparticles. J Hazard Mater 318:533–540

Suppan S (2013) Nanomaterials in soil: our future food chain? Institute for Agriculture and Trade Policy (IATP), Minneapolis

Tan A, Teng J, Zeng X, Fu D, Zhang H (2017) Fabrication of aluminium matrix hybrid composites reinforced with SiC microparticles and TiB_2 nanoparticles by powder metallurgy. Powder Metall 60(1):66–72

Taran NY, Gonchar OM, Lopatko KG, Batsmanova LM, Patyka MV, Volkogon MV (2014) The effect of colloidal solution of molybdenum nanoparticles on the microbial composition in rhizosphere of *Cicer arietinum* L. Nanoscale Res Lett 9:1–8

Throbäck IN, Johansson M, Rosenquist M, Pell M, Hansson M, Hallin S (2007) Silver (Ag+) reduces denitrification and induces enrichment of novel nirK genotypes in soil. FEMS Microbiol Lett 270(2):189–194

Timmusk S, Seisenbaeva G, Behers L (2018) Titania (TiO_2) nanoparticles enhance the performance of growth-promoting rhizobacteria. Sci Rep-UK 8(1):617

Tripathi DK, Tripathi A, Shweta SS, Singh Y, Vishwakarma K et al (2017) Uptake, accumulation and toxicity of silver nanoparticle in autotrophic plants, and heterotrophic microbes: a concentric review. Front Microbiol 8:07

Tungittiplakorn W, Lion LW, Cohen C, Kim JY (2004) Engineered polymeric nanoparticles for soil remediation. Environ Sci Technol 38(5):1605–1610

Upadhyay N, Vishwakarma K, Singh J, Verma RK, Prakash V, Jain S (2019) Plant-microbe-soil interactions for reclamation of degraded soils: potential and challenges. In: Phyto and rhizo remediation. Springer, Singapore, pp 147–173

Van Calster G (2006) Regulating nanotechnology in the European Union. Eur Energy Environ Law Rev 15(8):238–247

Verma RK, Sachan M, Vishwakarma K et al (2018) Role of PGPR in sustainable agriculture: molecular approach toward disease suppression and growth promotion. In: Role of rhizospheric microbes in soil. Springer, Singapore, pp 259–290

Vishwakarma K, Mishra M, Jain S, Singh J, Upadhyay N, Verma RK et al (2017) Exploring the role of plant-microbe interactions in improving soil structure and function through root exudation: a key to sustainable agriculture. In: Singh D, Singh H, Prabha R (eds) Plant-microbe interactions in agro-ecological perspectives. Springer, Singapore, pp 467–487

Vishwakarma K, Singh VP, Prasad SM, Chauhan DK, Tripathi DK, Sharma S (2020) Silicon and plant growth promoting rhizobacteria differentially regulate AgNP-induced toxicity in *Brassica juncea*: implication of nitric oxide. J Hazard Mater 390:121806

Vitali F, Raio A, Sebastiani F, Cherubini P, Cavalieri D, Cocozza C (2019) Environmental pollution effects on plant microbiota: the case study of poplar bacterial-fungal response to silver nanoparticles. Appl Microbiol Biotechnol 103:8215–8227

Wang G, Gao Q, Yang Y, Hobbie SE, Reich PB, Zhou J (2022) Soil enzymes as indicators of soil function: a step toward greater realism in microbial ecological modeling. Glob Chang Biol 28(5):1935–1950

Xu C, Peng C, Sun L, Zhang S, Huang H, Chen Y et al (2015) Distinctive effects of TiO_2 and CuO nanoparticles on soil microbes and their community structures in flooded paddy soil. Soil Biol Biochem 86:24–33

Xu J, Luo X, Wang Y, Feng Y (2018) Evaluation of zinc oxide nanoparticles on lettuce (*Lactuca sativa* L.) growth and soil bacterial community. Environ Sci Pollut Res 25:6026–6035

Yadav RC, Patra AK, Purakayastha TJ, Singh R, Kumar C (2014) Effect of engineered nanoparticles of Fe and Zn oxides on enzyme activity and bacterial abundance in soil at ambient and elevated atmospheric CO_2. Proc Natl Acad Sci India Sect B Biol Sci 84:649–656

Yang Y, Zhang C, Hu Z (2012) Impact of metallic and metal oxide nanoparticles on wastewater treatment and anaerobic digestion. Environ Sci-Proc Imp 15(1):39–48

You T, Liu D, Chen J et al (2018) Effects of metal oxide nanoparticles on soil enzyme activities and bacterial communities in two different soil types. J Soils Sediments 18(1):211–221

Yu SJ, Yina YG, Liu JU (2013) Silver nanoparticles in the environment. Environ Sci 15:78–92

Zahra Z, Arshad M, Rafique R et al (2015) Metallic nanoparticle (TiO_2 and Fe_3O_4) application modifies rhizosphere phosphorus availability and uptake by *Lactuca sativa*. J Agric Food Chem 63(31):6876–6882

Zahra Z, Habib Z, Hyun H, Shahzad HMA (2022) Overview on recent developments in the design, application, and impacts of nanofertilizers in agriculture. Sustainability 14(15):9397

Zaveri RA, Wang J, Fan J et al (2022) Rapid growth of anthropogenic organic nanoparticles greatly alters cloud life cycle in the Amazon rainforest. Sci Adv 8(2):eabj0329

Chapter 4
Nanobiofungicides: Effective Tools for Control of Fungal Pathogens

Juhi Goyal, Preet Jain, Priti Yadav, Priyank Upadhyay, and Nitish Rai

Abstract Nanobiofungicides are durable and highly efficient formulations for the control of phytopathogenic fungi, even at lower dosages of active ingredients when used in the fields. The synthesis of nanobiofungicides involves microorganisms such as bacteria, algae, yeast, fungi, actinomycetes, and viruses. Several types of nanobiofungicide formulation processes exist, such as nanoemulsions, inorganic nanoparticles such as titanium dioxide and silica, and polymer-based nanoformulations. Nanobiofungicides may exhibit antifungal activity depending on the employed microorganism and could be due to antibiosis, parasitism, modification in plant growth, activating the plant's defense mechanism.

The advancement of nanotechnology has increased the number of nanoparticles (NPs) entering the environment. To protect plants from the potentially harmful effects of nanomaterials, an increasing amount of research has focused on determining the toxicity of the NPs commonly used in industry. Therefore, this chapter highlights several aspects of nanobiofungicides, which include their preparation strategies, mechanism of action, delivery, agricultural application, and environmental impact.

Keywords Nanoparticles · Biofungicides · Nanoformulations · Phytopathogenic fungi · Plant protection

J. Goyal · P. Jain · P. Yadav · P. Upadhyay
Department of Biotechnology, Mohanlal Sukhadia University, Udaipur, Rajasthan, India

N. Rai (✉)
Department of Biotechnology, Mohanlal Sukhadia University, Udaipur, Rajasthan, India

Department of Zoology, University of Lucknow, Lucknow, Uttar Pradesh, India
e-mail: nitish.rai@mlsu.ac.in

N. K. Arora, B. Bouizgarne (eds.), *Microbial Biotechnology for Sustainable Agriculture Volume 2*, Microorganisms for Sustainability 34,
https://doi.org/10.1007/978-981-97-2355-3_4

4.1 Introduction

Fungi are a group of spore-producing, filamentous eukaryotes that are omnipresent and are endophytic, saprophytic, or pathogenic (Dayarathne et al. 2020; Bhunjun et al. 2022; El-Baky and Amara 2021). About 15,000 species of fungi are reported to be pathogenic to plants and are a predominant cause for over 70% of chief crop diseases (Patel et al. 2014). Leaf spot, scab, damping off, gall, canker, powdery mildew, dieback, rust, anthracnose, root rot, wilt, and blight are some of the common diseases caused by fungal plant pathogens (Iqbal et al. 2018; Jain et al. 2019; Hussain and Usman 2019; Hariharan and Prasannath 2021). Fungal diseases cause a major reduction in the quantity and quality of crops. It has been reported that *Pyricularia oryzae*, responsible for rice blast disease, and *Puccinia recondita*, which causes leaf rust in wheat, lead to a 21% and 30% decrease in rice and wheat yield, respectively (Patel et al. 2014). Decreased crop yield not only results in huge economic losses but also questions the food security of the ever-growing population (O'Brien 2017; Thambugala et al. 2020).

Fungicides act by killing or inhibiting the growth of fungal cells or their spores. There is no denying that since the introduction of chemical fungicides, fungal diseases in plants could be significantly controlled, thereby drastically increasing crop productivity and associated monetary gains (Gianessi and Reigner 2006; Shuping and Eloff 2017). Low cost, rapid availability, and easy to use features have led to the misuse and overuse of chemical fungicides (Youssef et al. 2019; El-Baky and Amara 2021). Extensive use of such fungicides has serious damaging and toxic effects on humans, environment, and non-target organisms such as earthworms and other beneficial soil microbes, which ultimately creates an imbalance in the ecosystem (Patel et al. 2014; Shuping and Eloff 2017). Chemical fungicides are reported to be associated with cancer, hormonal dysregulation, and respiratory diseases (Hoppin et al. 2017; Piel et al. 2019; Werder et al. 2020). Another crucial challenge is the emergence of fungicide resistant fungal pathogens which arises due to over usage of chemical fungicides. Additionally, these fungicides are hard to remove, their degradation is very gradual, and they leach downward resulting in the contamination of water sources (Possiede et al. 2009; Stamatis et al. 2010). Therefore, alternate ways of controlling fungal pathogens are needed that are eco-friendly, non-toxic, affordable, and effective. One such alternative is use of biofungicides (Nega 2014; Porteous-Álvarez et al. 2021).

Biofungicides are specific formulations derived from living organisms, particularly microbes such as bacteria, fungi, actinomycetes, and yeast or metabolites produced by them that work against phytopathogenic fungi (Liu et al. 2021; Wash et al. 2022). Some of the microorganisms that are employed as biofungicides are *Trichoderma* spp., *Pseudomonas syringae*, *Candida oleophila*, *Epicoccum nigrum*, *Pantoea agglomerans*, *Penicillium frequentans*, *Bacillus safensis*, etc. (De Cal et al. 2012; Moraes Bazioli et al. 2019; Meher et al. 2020; Prakash et al. 2022). Microbial metabolites that are commonly used as biofungicides are pyrrolnitrin, blasticidin S, validamycins, kasugamycin, polyoxins, phellinsin A, 2,4-diacetylphloroglucinol, soraphen A, fenpiclonil, strobilurin A, azoxystrobin, tubercidin, gliotoxin, gliovirin, etc. (Bartlett et al. 2002; Lee et al. 2004; Kim et al. 2006; Mishra et al. 2022;

Muhammad et al. 2022). Although biofungicides are natural compounds and play a prominent role in the development of sustainable agriculture, they still have certain drawbacks. These are sensitive to temperature changes, have slow response against fungal phytopathogens, and loose efficiency on repeated applications (Khakimov et al. 2020). To overcome such shortcomings, nanotechnology is being explored in conjunction with biofungicides to develop superior antifungal agents (Sekhon 2014).

Nanotechnology is a technology based on nanoscale materials with a size of 0.1–100 nm, with great potential applications in diverse areas (Morones et al. 2005). Nanomaterials can be synthesized using chemicals, which results in toxicity and elevated price (Kaushal 2018). Thus, plant-based (green synthesis) or microbe-based nanoparticle synthesis is preferred. The development of nanostructures to curb plant fungal infections involves a reconsideration of the biological effects of prevailing antimicrobial agents by modifying their size to escalate their efficiency. Such nanobioformulations that inhibit pathogenic fungi are termed nanobiofungicides (Nuruzzaman et al. 2016; Bhattacharyya et al. 2016; Kaushal 2018).

The common preferences for synthesizing nanobiofungicides are microorganisms. This is because of their extensive diversity, tiny size, ability to be genetically manipulated, good biosorption capability, metal ion adsorption in a selective manner, and relatively hassle-free downstreaming procedure (Kaushal 2018). Nanobiofungicides have several advantages over biofungicides such as the higher surface area to volume ratio, the self-regulating ability which enables them to release fungicide in appropriate amounts and in a controlled way, thereby reducing the excessive use of fungicides (Sharon et al. 2010; Bhattacharyya et al. 2016) (Fig. 4.1). Additionally, the chance of interaction with a pathogen and its destruction is elevated due to the exclusive optical properties of nanostructures. However, the production of nanobiofungicides is still in its initial stage (Prasad et al. 2016; Kaushal 2018).

This chapter provides an overview about nanobiofungicides along with their preparation strategies. Further, the proposed mechanism of action and delivery method of nanobiofungicide is presented. This chapter also discusses utility of

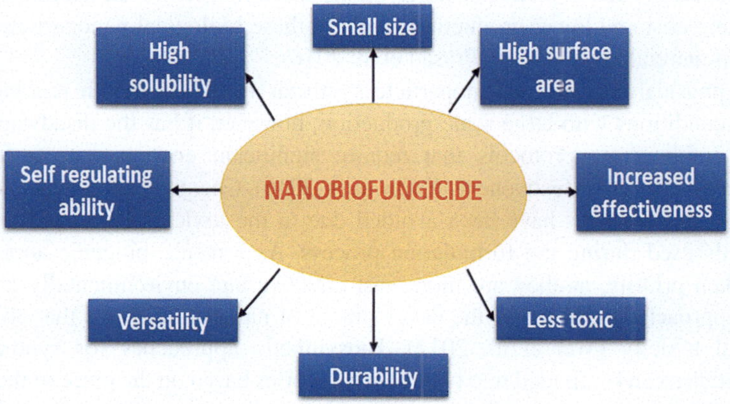

Fig. 4.1 Advantages of nanobiofungicides

nanobiofungicides in agriculture against several phytopathogens and their impact on environmental sustainability is also highlighted.

4.2 Nanobiofungicides: Preparation Strategies

The synthesis of nanoparticles can be achieved through two approaches, namely the top-down approach and the bottom-up approach. The top-down approach involves reducing the size of a material of interest using chemical and physical methods to produce nanoparticles in the nanoscale range. Conversely, the bottom-up approach involves the self-organization of atoms, molecules, or smaller particles to synthesize nanoparticles. Compared to the top-down approach, the bottom-up approach is more cost-effective and flexible and has been extensively used over the past two decades (Prasad 2014). Traditionally, physical and chemical methods have been employed for the synthesis of nanoparticles. These methods include the sol-gel technique, solvothermal synthesis, chemical reduction, laser ablation, and inert gas condensation. However, these techniques are often expensive and environmentally hazardous. Furthermore, the use of certain chemicals during the synthesis process can limit the nanoparticles' application in clinical and pharmaceutical settings (Aziz et al. 2015).

Diverse microorganisms like yeast, algae, fungi, bacteria, actinomycetes, and virus are involved in nanobiofungicide synthesis (Prasad 2014; Prasad et al. 2014, 2016; Aziz et al. 2015; Ahmad et al. 2002) (Table 4.1). Several types of nanobiofungicide formulation processes exist like nanoemulsions, inorganic nanoparticles such as silica and titanium dioxide, polymer-based nanoformulations, etc. There are several rapidly advancing techniques to synthesize metal nanoparticles with precise shapes and sizes. The production of nanomaterial using "metal-microorganism" is gaining popularity on a global level. Green technology-based nanoparticle production is promoted with characteristics of cost-effectiveness, particle amendment, and wide application in diverse fields. Biological nanoparticles can be obtained from various sources such as plants, bacteria, algae, and fungi. However, the downstream processing cost and lower production levels of these biological nanoparticles limit their commercial applications (Prasad et al. 2016).

The physical process of nanoparticle synthesis involves smooth regulation of desired conditions with large scale production; however, it has the disadvantage of having cumbersome protocols that require significant energy, non-eco-friendly nature, and less cost-effectiveness (Yadav et al. 2016). Chemical approaches for synthesizing nanoparticles have been avoided due to the toxicity associated with the chemicals used during the formulation process. As a result, biogenic approaches have taken priority, as they are more cost-effective and environmentally friendly. These approaches have led to the development of nanoformulations that eliminate chemical toxicity (Wei et al. 2012). Biosynthetic approaches for synthesizing nanoparticles can be divided into two main categories based on the place of the nanoformulation: intra and extracellular synthesis. Intracellular biosynthesis involves proteins, enzymes, and cofactors, which reduce metal ions to produce the nanoparticle intracellularly. Extracellular biosynthesis of nanoparticles occurs through the

Table 4.1 Microbes employed in the preparation of nanobioformulations acting against phytopathogenic fungi

S. no.	Microorganism	Nano-formulation	Target fungus	Disease/condition	References
1.	*Serratia* sp.	AgNPs	*Bipolaris sorokiniana*	Spot blotch in wheat	Lamsal et al. (2011b)
2.	*Bacillus* species	AgNPs	*Fusarium oxysporum*	Wilt disease of tobacco, legumes, cucurbits, banana, tomato, and sweet potatoes	Gopinath and Velusamy (2013)
3.	*Macrophomina phaseolina* and *Rhizoctonia*	Cu-chitosan NPs	*A. alternata*	Leaf spot disease	Saharan et al. (2013)
4.	*Cucumber* and pansy plant	Silica-silver NPs	*Sphaerotheca pannosa var Rosae*	Cucurbit powdery mildew	Banik and Sharma (2011)
5.	*Oomycete* pythium aphanidermatum	β-D-glucan NPs	*Curcuma longa (syn C. domestica)*	Rhizome rot disease	Anusuya and Sathiyabama (2015)
6.	*Gayadomonas* sp. TNPM15	Tellurium NPs	*Fusarium oxysporum* and *Alternaria alternata*	Panama disease, Black rot	Abd El-Ghany et al. (2023)
7.	*Trichoderma harzianum*	AgNPs	*Rhizoctonia solani* (RS9), *Fusarium fujikuroi* (FF10), and *Macrophomina phaseolina* (MP4)	Cotton seedling diseases	Zaki et al. (2022)
8.	*Monascus purpureus*	Pegylated zinc nanoparticles (PEGylated ZnNPs)	Including *Fusarium solani, Fusarium sporotrichioides, Fusarium oxysporum, Rhizoctonia solani,* and *macrophomina phaseolina*	Food spoilage fungi	Ammar et al. (2021)
9.	*Solanum rostrata*	Myco AgNPs	*Fusarium graminearum, Fusarium udum, Rhizoctonia solani, Aspergillus niger*	*Fusarium* head blight in wheat and barley, wilt disease on pigeon pea, root rot, Black mold	Akther and Hemalatha (2019)
10.	*Alternaria* sp.	AgNPs	*Fusarium* spp.	Rots, wilts, and damping off	Win et al. (2020)

(continued)

Table 4.1 (continued)

S. no.	Microorganism	Nano-formulation	Target fungus	Disease/ condition	References
11.	*Aspergillus niger*	Fungal chitosan nanocomposite	*Botrytis cinerea*, *Rhizopus stolonifera*, *Aspergillus niger*	Gray mold of strawberry, black mold	Melo et al. (2020)
12.	*Curvularia affinis*	AgNPs	*Alternaria solani*	Early blight of tomato	Halder et al. (2018)

secretion of biomolecules and the interaction with functional groups present on the cell wall (Salunke et al. 2016). Several studies have investigated the potential of using biosynthesized nanoparticles to control plant diseases (Senthil-Nathan 2015). The extracellular biosynthesis approach has been used to prepare silver nanoparticles from *Bacillus* sp. strain GP23, which showed antimicrobial properties and suppressed wilt disease in legumes, banana, tobacco, cucurbits, sweet potato, and tomato, caused by *Fusarium oxysporum* (Gopinath and Velusamy 2013). The extracellular biosynthesis of silver nanoparticles using the SBT8 strain of *Bacillus* has been reported to have antimicrobial properties and acted as a biocatalyst (Yurtluk et al. 2018). The nanoparticles synthesized from silver nitrate using *Bacillus subtilis* and *Bacillus amyloliquefaciens* also exhibited antibacterial and antifungal properties against *Candida albicans* (Ghiuță et al. 2018). The nanoscale *Bacillus thuringiensis* chitinases are effective in managing *Caenorhabditis elegans*, which is a nematode species commonly found in soil and is considered a model organism for studying various biological processes. The use of nanoscale *B. thuringiensis* chitinases as biopesticidal delivery system has the potential to extend its use in farming applications for managing nematode populations that can cause crop damage (Qin et al. 2020). The extracellular synthesis of nanoparticles by the fungi *Pestalotiopsis pauciseta* and the endophytic fungus *Pestalotiopsis* sp. can be harnessed as antimicrobial agents, either alone or in combination with antibiotics. Studies have shown that the bacterium *Serratia* sp. BHU-S4 can combat the fungus *Bipolaris sorokiniana*, which causes spot blotch disease in wheat, through the extracellular synthesis of silver nanoparticles (Lamsal et al. 2011a). Future research can enhance the techniques employed for the management of microbial infections (Abd-Elsalam 2013; Kah et al. 2013; Vardhana and Kathiravan 2015). The use of metal nanoparticles to control plant diseases has also been explored. For controlling diseases of grape trees and other fruit trees, copper nanoparticles that were dissolved in water, can be used as fungicide (Lee et al. 2013; Bhattacharyya et al. 2015). Silver nanoparticle coated seeds have shown enhanced germination potential (Parveen and Rao 2015). Cu-chitosan and chitosan nanoparticles have shown uniform size and stability, which could account for their increased antifungal activity against *Aalternaria alternata*, *Macrophomina phaseolina*, and *Rhizoctonia* in in vitro studies (Saharan et al. 2013). Cu-chitosan and Cu-chitosan nanoparticles have shown uniform size and stability, which could account for their increased antifungal activity against *Rhizoctonia*, *M. phaseolina*, and *A. alternata* in in vitro studies (Saharan et al. 2013).

So, various types of nanoparticles have been developed in colloidal suspensions or powder forms with nanoscale dimensions to be used as pesticide formulations.

These formulations have demonstrated efficacy in improving the stability of the active organic compound (such as UV, thermal, and hydrolysis stability), as well as in promoting foliar settling, reducing foliar leaching, and enabling systemic action and synergism, among other benefits. These mechanisms collectively result in a reduced dose of pesticide application.

4.3 Underlying Mechanisms

The studies on the precise mechanism for the antifungal activity of nanobiofungicides are quite limited. The antifungal mechanism of nanobiofungicides could be an amalgamation of various modes of biocidal activity of microbes, their targeted release from nanoformulations, and the interaction of nanoparticles at the molecular level (Bhattacharyya et al. 2016; Kaushal 2018) (Fig. 4.2).

Depending on the employed microbe, antifungal activity depicted by nanobiofungicides could be due to antibiosis, parasitism, modification in plant growth, activating the plant's defense mechanism, etc. (Shrivastava et al. 2014). Antibiosis is a mechanism by which chemical substances or metabolites are produced by microbes that inhibit the growth of several other pathogenic microbes (Bhar et al. 2022). The most reported cases of fungi producing antimicrobial substances include isolates of *Gliocladium* and *Trichoderma*. It has been reported that *Gliocladium virens* secretes gliotoxin, which plays a dominant role in controlling the activity of pathogenic fungi *Pythium ultimum* and *Rhizoctonia solani* (Pankhurst and Lynch 2005). Specific chemical and physical properties of nanoformulations have the potential to alter the physiological and biochemical processes of plants, thereby affecting its growth and promotion. It has been found that on the application of nanoparticles, there was an induction in the growth of *Dracocephalum moldavica* (Gohari et al. 2020). Many *B. subtilis* strains coated in silver, gold, or aluminum can greatly boost plant growth and development, which helps in controlling phytopathogen infection hence making them suitable nanobiofungicides (Gouda et al. 2018).

The primary step in the action of nanobiofungicides could be the uptake of nanoparticles by the fungi. There are three ways to internalize nanoparticles. First is the direct internalization of nanomaterial in the cell wall, the second is the adsorption through specific receptors followed by internalization, and the third is internalization mediated by ion transport protein (Kalia et al. 2020). The fungal cell wall consists of chitin, lipids, and polysaccharides with β-1,3-D-glucan and β-1,6-D-glucan proteins being the dominant mannoproteins (Patil and Chandrasekaran 2020). The activity of the enzyme β-glucan synthase is inhibited by nanoparticles, resulting in alteration in the formation of N-acetylglucosamine in the cell wall. Enzyme inhibition leads to irregularities such as increased cell wall thickening, cell membrane liquefaction, disarrangement of cytoplasmic contents, and disassociation of cell wall from cytoplasmic contents, which ultimately results in fungal death (Arciniegas-Grijalba et al. 2017; Alghuthaymi et al. 2021).

It has been reported that negatively charged nanoparticles can attach to fungal membranes due to electrostatic interactions. This attachment causes membrane

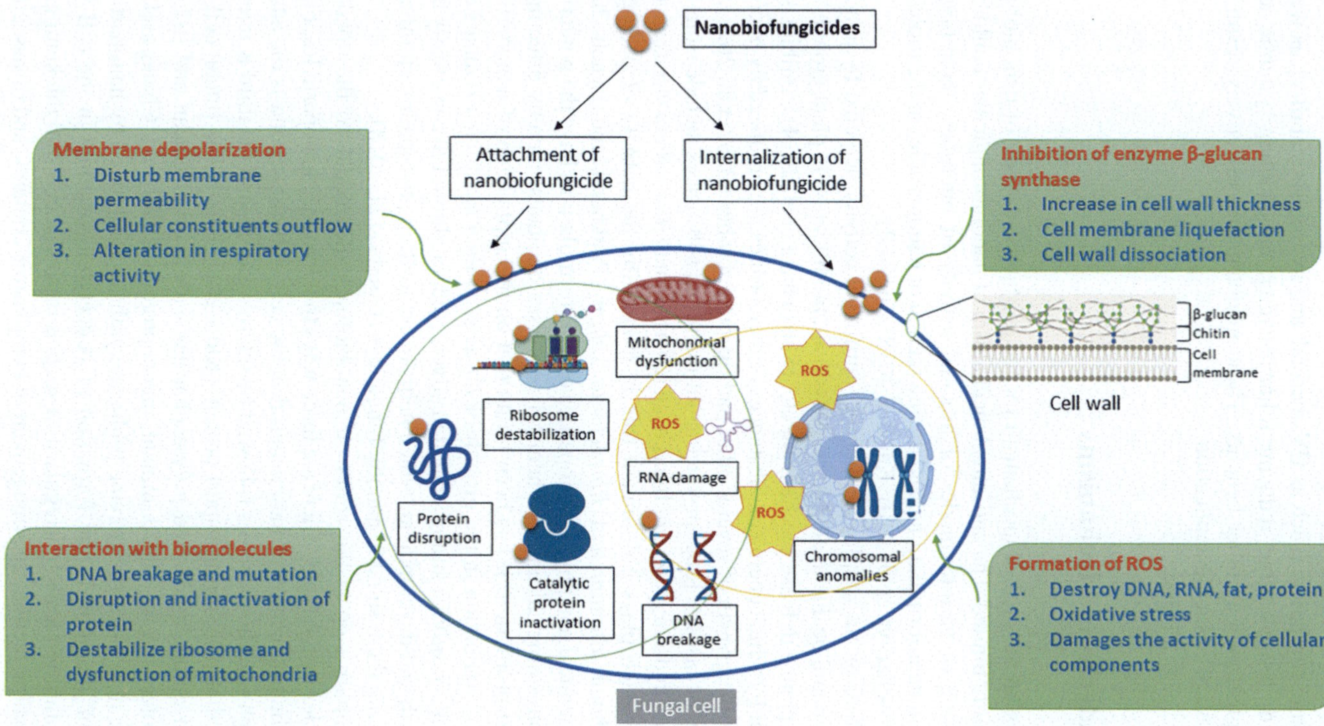

Fig. 4.2 Mechanisms of antifungal activity of nanobiofungicides

depolarization, which disrupts the permeability of the membrane resulting in the discharge of cellular components like DNA, RNA, proteins, metabolites, and enzymes (Alghuthaymi et al. 2015). Damage in the cell membrane is not only responsible for perforations but also leads to inflammation of the membrane. Such kind of cell damage has been observed through transmission electron microscopy. Additionally, membrane depolarization also alters other physical properties such as it disturbs respiratory activity which kills pathogens (Ibrahim et al. 2020).

At the molecular level, the nanobiofungicides can form complexes with various biomolecules distorting their structures, and nucleic acid anomalies like breakage of DNA, chromosomal aberrations, and catalytic protein inactivation (Patra et al. 2012). It has also been reported that nanoparticles act against fungal pathogens by disrupting their proteins, destabilizing the ribosomes, triggering mitochondrial dysfunction, elevating the reactive oxygen species (ROS) formation, and making them deprived of antioxidants. ROS formation severely damages many cellular components such as DNA, RNA, fats, and proteins, which leads to cell death (Shoeb et al. 2013). To validate the action of ROS, Lipovsky et al. (2011) identified the enhanced expression of lipid peroxidation in the cells, which is a marked indicator of ROS production. The alterations in the levels of enzymes such as glutathione dismutase, ascorbate peroxidase, and superoxide dismutase also signal ROS formation (Yu et al. 2020).

A wide range of microbes could be utilized for the production of nanobiofungicides. Bacteria are frequently used as insecticides; they can also be used to protect plants against harmful fungi. A lot of attention is given to members of the Bacillaceae, particularly *Bacillus* spp., as microbial control agents (Kumar et al. 2021). Bacteria can reduce metal ions through active adsorption and are hence employed for the production of nanoformulations that act against fungal pathogens (Kumari et al. 2019). It has been reported that members of *Bacillus* species secrete secondary metabolites with antagonistic properties (Ashwini and Srividya 2014). A major number of molecules such as siderophores, lipopeptides, and bacteriocins show antifungal activity and are produced by *Bacillus* species through non-ribosomal means (Fira et al. 2018). Additionally, chitinase is reported to be secreted by some of the *Bacillus* species like *B. licheniformis*, *B. stearothermophilus*, *B. amyloliquefaciens*, *B. thuringiensis*, *B. megaterium*, and *B. subtilis*. The chitinolytic action, i.e., degradation of chitin and inhibition of fungal growth is best shown by the chitinase secreted by *B. subtilis* (Karunya et al. 2011). Widnyana and Javandira (2016) reported the inhibition of the fungal pathogens from the soybean seeds by *B. subtilis*. AP3 and PRBS1 are the two *B. subtilis* strains that have shown antagonistic effects against fungal species such as *R. solani*, *Sclerotinia sclerotiorum*, *Colletotrichum truncatum*, and *Phomopsis*. *Bacillus methylotrophicus* NKG1, isolated from *Pinus koraiensis* root area, is reported to show antifungal activity and also acts as a fertilizer (Ge et al. 2016). The lipopeptides secreted by *B. subtilis* are reported to control the pathogenic fungal species such as *Penicillium*, *Botrytis*, and *Aspergillus* (Meena and Kanwar 2015; de Andrade et al. 2019).

Similar to bacteria, certain fungal species also show antagonistic behavior toward other phytopathogenic fungi. *Aspergillus*, *Trichoderma*, and *Fusarium* have been reported to be utilized in the formation of silver nanoparticles. A study was conducted to investigate the extracellular biosynthesis of silver nanoparticles using an endophytic fungus, *Epicoccum nigrum*, which was isolated from *Phellodendron amurense*. Such nanoparticles inhibited the growth of pathogenic fungi such as *Fusarium solani*, *Cryptococcus neoformans*, *Aspergillus fumigatus*, *C. albicans*, *Aspergillus flavus*, and *Sporothrix schenckii* (Qian et al. 2013). Abd El-Aziz et al. (2015) reported that silver nanoparticles biosynthesis was attained by *F. solani*, and was found to be active against several fungal pathogens that commonly infect barley, wheat, and maize seeds. Balakumaran et al. (2015) employed *Guignardia mangiferae* fungus for the synthesis of silver nanoparticles and observed their activity against fungi *R. solani*, *Curvularia lunata*, and *Colletotrichum* sp. In another study, fungi, *Aspergillus versicolor* was used for producing silver nanoparticles which displayed a concentration-dependent effect against two fungi, *Sclerotinia sclerotiorum* and *Botrytis cinerea*, responsible for strawberry plant infection (Elgorban et al. 2016). The mechanism for such antifungal activity could be due to the combined effect of silver ions and the fungi used for the biosynthesis of nanoparticles. The silver ion disrupts DNA replication, affects the enzymes involved in the respiratory chain, and affects adenosine triphosphate (ATP) production (Feng et al. 2000; Yamanaka et al. 2005; Elgorban et al. 2016).

4.4 Methods of Delivery

The conventional method of fungicide application to crops includes spraying and dusting, which involves the use of hand-held and energy-operated equipment. The conventional dispersal system has some drawbacks. It requires fungicide application on plants in large amounts, which in turn elevates the production of fungicides, increases the user's exposure to the fungicide's active ingredient during dispersal, negatively affects the non-target organisms, and leads to the formation of resistant fungal strains (Oliveira et al. 2020). After the dispersal of fungicides, several environmental factors such as temperature, humidity, light, radiation, surrounding microbes, soil properties, etc. alter the physical and chemical properties of fungicide, thereby affecting its efficiency and causing premature degradation (Oliveira et al. 2020). Nanobiofungicides help overcome these issues through their unique delivery mechanisms. They may be applied to the plants by spraying and dusting but are required in very fewer amounts compared to the chemical fungicides, i.e., they include low-volume, high-value applications (Manjunatha et al. 2016).

Nanobiofungicides can be formulated in different forms such as nanospheres, nanocapsules, nanocrystals, nanotubes, nanoneedles, nanocomposites, etc. These nanostructured systems act as effective carriers of active compounds and help to maintain their stability (Ghormade et al. 2011). The nanocarriers act as a protective container for active components, preventing their breakdown, preserving their activity, and shielding them from the environmental factors mentioned earlier (Maruyama et al. 2016).

The form of the nanocarrier is crucial in defining the release and protection profile of the active ingredient since the geometry of the nanostructures is connected to the surface area per unit of volume. The shape, size, and material from which nanostructures are prepared play a dominant role in the protection, release, and dissolution of the active compounds (Cooper 2010; Flores-céspedes et al. 2015). Nanobiofungicides can be designed depending on the speediness required for the release of the active compound such as quick release, selective release, moisture release, slow-release, magnetic release, and release by heat, pH, etc. Different materials are used for the preparation of nanobiofungicides such as polymers, proteins/peptides, lipids/waxes, clay minerals, etc. (Rani et al. 2017; Kumar et al. 2018; Abdollahdokht et al. 2022). Nanoformulations preferred in fungicide delivery could be polymer-based, lipid-based, inorganic nanoparticles, nanofibers, nanogels, etc. (Oliveira et al. 2020). Polymeric substances particularly biodegradable polymers like chitosan, gelatin, collagen, etc. are mostly used. Based on these polymers, nanoformulations, like polymer micelle, polymersome, nanospheres, nanocapsules, are formulated and used for fungicide delivery. For lipid-based delivery, liposome, nanoemulsion, and micelle could be employed. In the case of formulations made of inorganic nanostructures, porous, gold, quantum, and iron-oxide nanoparticles could be used (Oliveira et al. 2020; Shekhar et al. 2021).

Nanobiofungicides are made up of a shell or membrane structure that encloses and holds the active substance in its core (Irfan et al. 2018). There are different mechanisms for the delivery of fungicidal compounds or active agents from the central part of the nanostructure; such as diffusion, erosion, dissolution, swelling, and fragmentation. These mechanisms may act together or alone (Peppas et al. 2000; Liechty et al. 2010; Oliveira et al. 2020). In the diffusion process, the release rate of the active component is governed by various factors. It is affected by the chemical properties of the active compound such as the volatility, polarity, molecular weight, etc., and the physicochemical properties of the matrix (support), such as density, cross-linking, physical state, etc. The release rate also depends on the physical properties of particles like shape, crystallinity, and size (Peppas et al. 2000; Liechty et al. 2010). In the process of erosion, the active compounds are released only when the carrier matrix erodes, i.e., chemical degradation of the matrix occurs. Some specific environmental conditions are responsible for triggering erosion. Physical factors like low/high temperature, chemicals like the presence of strong acids/bases, pH, and biological factors like enzyme action, play a dominant role in the erosion process. Dissolution of nanostructure occurs through specific external conditions, releasing the active agent only when the carrier matrix gets dissolved (Liechty et al. 2010). In the case of swelling, the matrix gets swelled due to the solvents or the environmental conditions they are exposed to, causing a release of the bioactive agents. The expansion of the interior pore size is what causes the particles to swell. When the pore size equals the molecular size of the active substance, it serves as a pathway for its release to the outside environment. In this method, the swelling rate and diffusion time of the active agent through the matrix will have the biggest impacts on the fungicide's release rate (Peppas et al. 2000). In the case of fragmentation, disturbances like mechanical, climatic, physical, chemical, and enzymatic cause cracks and breakage in the nanostructure. The fragmentation rate of the carrier matrix greatly regulates the release of the fungicide. Through these aforementioned mechanisms, the fungicides are released by the mass transfer process in a controlled and continuous way, which assists them to reach their target pathogen in an optimized manner. This release is often affected by several factors such as the thickness, biodegradability, and mechanical properties of the coating, the physicochemical properties of the soil and its water content, and the density of the bioactive agent (Irfan et al. 2018).

4.4.1 Nanomaterials in Agriculture

One of the biggest challenges for the sustainable development of agriculture is to reduce the use of toxic chemicals, which include fertilizers, insecticides, pesticides, which may end up in the food chain and responsible for various diseases in both animals and humans. The excess use of fungicides and pesticides has proven to be hazardous since there is too much runoff (Arora 2018). Several fungicides and pesticides are being banned due to their hazardous effects on the health of animals and humans due to bioaccumulation. Currently, there is a need to give more emphasis on the use and development of such products and technologies which are more environment friendly. Nanotechnology has introduced various options, which are being implicated for providing a new solution to the problem. Nanoparticles are highly efficient for delivery as they have a very small size and larger size to surface area ratio. A high surface area ratio ensures maximum utilization and a small size is good for permeability. Nanoparticles are being used in agriculture to enhance the growth of the plant, protect the plant from stress, diseases, pests, and fungus infections, and protect the postharvest product (Mittal et al. 2020).

There are various types of nanoparticles which are currently being developed, such as metal and metal oxide based, carbon based nanofibres and nanotubes, and organic nanoparticles; out of these, one of the most widely used and tested nanoparticles are metal based nanoparticles. Metal based nanoparticles have metal as the active agent, which can be in either oxidized or non-oxidized form. Such nanoparticles are easy to produce by green methods, physical methods, or chemical methods (Dikshit et al. 2021). Organic nanoparticles have advantages that they are biodegradable and can be used as a carrier for any suitable chemical.

Carbon nanomaterial research studies have revealed diverse and promising agri-applications (Alghuthaymi et al. 2021). Carbon nanoparticles occur in a variety of forms like nanotubes, fullerenes, graphene, etc. They are highly reactive in nature and they have high adsorption capacity, which is responsible for both positive and negative effects on the plant health and environment (Mukherjee et al. 2016). Carbon nanoparticles have been shown to inhibit the growth of pathogenic fungi. In an in vitro study, surface modified multi-walled carbon nanotubes have shown the inhibition of spore germination and elongation in *Fusarium graminearum*, through water depletion by aggregates formed around the spores (Wang et al. 2017). A study conducted by Hao et al. (2020) using various carbon nanoparticles indicated toxic effects on soil micro-biome and rice plants.

Metal toxicity in microbes occurs due to exposure to trace amounts of metals, such as copper, lead, iron, or zinc (Robinson et al. 2021). Various kinds of metal nanoparticles are developed as antifungal agents for plant protection. Several studies have evaluated the antifungal properties of silver nanoparticles (Das and Ahmed 2012; Hawar et al.

2022). Silver ions deprive the fungal cells from the potassium ion by increasing the efflux rate of potassium ions resulting in changing the membrane balance (Vagabov et al. 2008). Silver is known to be toxic to inactivate thiol groups of the proteins, disrupt fungal cell wall, resulting in cell lysis and DNA mutation, thereby disrupting membrane processes and dissociating enzyme complexes and ultimately blocking the respiratory chain in the fungus. Silver nanoparticles have shown inhibitory action against *Bipolaris sorokiniana* responsible for spot blotch disease in wheat (Mishra et al. 2014), *Phytophthora* spp., *M. phaseolina* (Mohamed and Elshahawy 2022), and others.

Zinc oxide nanoparticles are tested for the protection of the postharvest product from diseases caused by various fungi like *F. oxysporum*, *Aspergillus niger*, *Penicillium expansum*, and *Botrytis cinerea* (Narendhran and Sivaraj 2016; Yehia and Ahmed 2013). ZnO NPs reduce ergosterol biosynthesis, change membrane permeability, reduce cell wall integrity, increase ROS production, reduce hyphal growth, increase hyphal branching, change hyphal morphology and interrupt conidia and conidiophore development and limit the reproduction capacity (Galván Márquez et al. 2018; Lanfranco et al. 2002; Malandrakis et al. 2022).

Nanomaterials including copper are effective against *Fusarium* (Borgatta et al. 2018). Pathogenic fungi like *C. albicans*, *C. neoformans*, *A. niger*, and *F. oxysporum* are being targeted using biogenic nanoparticles such as gold (Au), silver (Ag), zinc oxide (ZnO), and copper (Cu) nanoparticles (Ashajyothi et al. 2016). Sawai and Yoshikawa (2004) showed antifungal activities of MgO, CaO, and ZnO nanoparticles against fungi *A. niger*, *Rhizopus stolonifer*, and antimicrobial activity against several microbes. TiO_2 has been used for photocatalytic treatment of the water and the significant photocatalytic inhibition of *F. oxysporum* by titanium dioxide nanoparticles was shown by Anbeaki et al. (2009) compared to culture not exposed to light. Although silicon is considered non-essential element for the plant, its interaction with the plant has shown beneficial effects in stress acclimatization and enhancing growth and biomass of plant, seed germination, tissue differentiation, and plant defense (Rastogi et al. 2019). Mesoporous silica gel nanoparticles loaded with the fungicide are effective in the slow and prolonged release of the fungicide, overcoming the problem of rapid loss of fungicide (Chen et al. 2011).

Different organic materials are being tried and tested in agriculture as they give additional benefits over others in terms of control over the environment and product damage. Organic nanoparticles are used as carriers encapsulating active substances, which help in reduced toxicity, improved release of active substances, reduced amount required and enabling the precise targeting of active substances. Many products are developed from the plant byproducts like cellulose-based nanoparticles. Chitosan based organic nanoparticles are being developed as antifungal nanomaterial as well as they are being used to encapsulate various types of materials like fungicides or metal ions for the control of plant fungal diseases (Saharan et al. 2015; Karaman et al. 2016). Chitosan can penetrate the fungal cell wall and affect important cell processes like mRNA synthesis and essential proteins and enzyme production (Kong et al. 2010). Other organic nanoparticles such as glucan nanoparticles (GNPs) from β glucan and (Anusuya and Sathiyabama 2015) polylactic-co-glycolic acid (PLGA) nanoparticles (Fukamachi et al. 2019) are also known to show their potential in plant fungal diseases management (Table 4.2).

Table 4.2 Nanomaterials utilized against agriculturally relevant phytopathogens and their effect

S.no	Type of nanomaterial	Nanomaterial	Synthesis	Size	Phytopathogen	Plant	Effect	References
1	Metal	Silver nanoparticle (NP)	Biosynthesis by Serratia sp. BHU-S4	10–20 nm	*Bipolaris sorokiniana*	*Triticum aestivum*	Showed antifungal activity in in vitro as well in vivo studies	Mishra et al. (2014)
2	Miniemulsion	Lignin-based nanocarriers encapsulating fungicide	Miniemulsion polymerization with fungicide encapsulation	50–500 nm	*Phaeomoniella chlamydospora* and *Phaeoacremonium minimum*	*Vitis vinifera*	Inhibition of fungal growth after 96 h in vitro and boscalid-loaded lignin NCs exhibited long-term activity in *in planta* studies	Machado et al. (2020)
3	Metal oxide	Nano-CeO_2	CeO_2 were received from Meliorum Technologies (USA)	8 ± 1 nm, aggregate to 231 ± 16 nm in deionized (DI) water	Fusarium oxysporum	*Solanum lycopersicum*	Suppression of Fusarium wilt and improved chlorophyll in tomato plants	Adisa et al. (2018)
4	Metal nanoparticle	Ag NP	Greenhouse-raised *A. absinthium* plants were dried at room temperature and used for AgNPs synthesis	–	*Phytophthora* spp.	Solanum tuberosum and solanum lycopersicum	Mycelium growth inhibition and germination inhibition, *in planta* and in vitro inhibition of *Phytophthora*	Ali et al. (2015)

5	Metal nanoparticle	CuO NP; $Cu_3(PO_4)_2 \cdot 3H_2O$	Commercially obtained CuO nanoparticles; $Cu_3(PO_4)_2 \cdot 3H_2O$ nanosheets using a polyol method	30 nm; 151 nm thickness 13 nm	*Fusarium*	*Citrullus lanatus*	Significant effects by CuO NP on disease at 1000 mg/L and significant inhibition of fungal disease by $Cu_3(PO_4)_2 \cdot 3H_2O$	Borgatta et al. (2018)
6	Metal nanoparticle	Cu NP	Chemical synthesis	50–300 nm.	*A. alternata*, *F. oxysporum*, *C. lunata*, and *P. destructiva*	In vitro	Inhibited all the plant pathogenic fungi used in study	Brahmanwade et al. (2016)
7	Silica coated nanoparticle	Si NP	Mesoporous silica nanoparticles and chitosan-coated mesoporous silica nanoparticle	35–40 nm	*Fusarium oxysporum*	*Citrullus lanatus*	Significant reduction in disease severity and increased fruit yield	Buchman et al. (2019)
8	Biopolymer based NP	Cu Chitosan NP	Chitosan gelated TPP with addition of $CuSO_4$ followed by ultrasonification	150 nm	*Alternaria solani* and *Fusarium oxysporum*	*Solanum lycopersicum*	Effective disease control in tomato plants with early blight and Fusarium wilt	Saharan et al. (2015)
9	Metal NP	Zn NP	Fungal synthesis using *Aspergillus niger*	96.43 nm	*Alternaria solani*	*Solanum tuberosum* L.	MIC of 130.568 mg/L was determined	Singh et al. (2022)

(continued)

Table 4.2 (continued)

S.no	Type of nanomaterial	Nanomaterial	Synthesis	Size	Phytopathogen	Plant	Effect	References
10	Biodegradable polymer NP	Poly lactic-co-glycolic acid (PLGA) nanoparticles (NPs) encapsulating cyazofamid	Precipitation solvent evaporation technique	126 nm	Phytophthora infestans	Solanum lycopersicum	In vitro antimicrobial activity assay inhibited zoospores release and in vivo showed suppression of infection in plant leaf	Fukamachi et al. (2019)
11	Metal nanoparticle	Cu NP	Commercially obtained	25 nm	Botrytis cinerea	Malus sylvestris	Study showed use of Cu-NPs in combination with conventional fungicides as an environmentally safe, and sustainable strategy to control resistant *B. cinerea*	Malandrakis et al. (2020)
12	Metal nanoparticle	Zn NP in combination with fungicides	Commercially obtained	<100 nm	Boscalid-resistant *Alternaria alternata*	In vitro	Potential synergistic role of ZnO in ROS generation effectively inhibited *A. alternata* alone or in combination with boscalid	Malandrakis et al. (2022)

13	Metal nanoparticle	Ag NP	Green synthesis using AgNs *Trichoderma harzianum*	31.13 nm	*Sclerotinia sclerotiorum* and *Sclerotium rolfsii*	Phaseolus vulgaris and Helianthus annuus	Inhibition of mycelial linear growth, sclerotia formation, dry weight, and reduction in disease severity in plant	El-Ashmony et al. (2022)
14	Metal nanoparticle	Ag NP	Green synthesis using Citrus sinensis L. osbeck peel extracts	32–47 nm	*M. phaseolina.*	Faba bean	Pre-treatment of Faba bean seeds reduced charcoal rot diseases and raised the survival of plants under greenhouse conditions and under M. phaseolina infested soil	Mohamed and Elshahawy (2022)
15	Silica nanoparticle	Mesoporous silica loaded with Pyoluteorin	Chemical synthesis	2.4 nm	*Phytophthora capsici*	In vitro	Prolonged released of fungicide for 28 days	Chen et al. (2011)

4.5 Environmental Sustainability of Nanobiofungicides

Various advancements in food production have occurred with the introduction of new technologies to produce food to feed the world's increasing population. One of the technologies that have had a significant impact on food productivity is the use of pesticides and fertilizers, which has dramatically increased food productivity. Rising global food production using sustainable approaches to feed a growing population is one of the world's current fundamental challenges (Hazarika et al. 2022). The toxicity of metal nanoparticles is determined by various factors, including solubility and binding specificity to biological sites among others. The electrostatic interaction of nanoparticles with membranes and their accumulation in the cytoplasm could be responsible for nanotoxicity (Rana and Kalaichelvan 2013; Aziz et al. 2015, 2016).

The rapid development of nanotechnologies has led to the release of large quantities of manufactured nanoparticles into the environment, posing potential risks to human health and plant life. Consequently, a growing body of research has focused on assessing the toxicity of commonly used nanomaterials (Yang and Watts 2005; Rana and Kalaichelvan 2013; Du et al. 2017; Tripathi et al. 2017a, b, c). Factors such as solubility and binding specificity to biological sites, determine the toxicity of metal nanoparticles, which have demonstrated antibacterial, anticandidal, and antifungal properties (Aziz et al. 2016; Patra and Baek 2017). The cytotoxicity of metal nanoparticles is influenced by the charge at the membrane surface, but the effectiveness of nano-toxic effects is undeniably linked to the structure of the targeted cell wall, with mold being the most sensitive. The electrostatic interaction of membranes with nanoparticles and nanoparticle accumulation in the cytosol are possible mechanisms of nanotoxicity (Aziz et al. 2015, 2016; Rana and Kalaichelvan 2013).

Photoactive nanoparticles, such as TiO2, ZnO, SiO2, and fullerenes, can generate excited electrons upon exposure to light, leading to the formation of superoxide radicals when in the presence of oxygen via direct electron transfer. When organisms are exposed to both nanoparticles and UV light, their toxicity can be observed as visible light has a lower energy than UV light. In response to oxidative stress caused by the production of reactive oxygen species (ROS), cells increase the expression of protective enzymes or genes, which can be measured as a means of determining the context of ecotoxicity and toxicity (Kovochich et al. 2005; Vannini et al. 2014). In vitro studies have demonstrated the production of ROS by TiO2 and fullerenes (Sayes et al. 2004), whereas other studies have suggested that nanoparticles (silicon nanoparticles and fullerenes) may rescue against oxidative stress (Tripathi et al. 2016b, 2017d; Daroczi et al. 2006; Venkatachalam et al. 2017). To further elucidate this dichotomy, additional research is needed to better understand cell–nanoparticles interactions, as well as the mechanisms of nanoparticles metabolism in organisms and specific cells.

In addition, various polymers have been used as drug and agrochemical delivery agents, such as gelatin, cellulose acetate phthalate, gum, chitosan, polylactic acid, polybutadiene, polylactic-glycolic acid, polyvinyl alcohol (PVA), polyhydroxyalkanoates, polyacrylamide, and polystyrene (Spadola et al. 2020; Ashfaq et al. 2019).

Ecotoxicological research is concerned with studying the environmental impact of materials and the complexities of natural systems. Extensive research is needed to understand the long-term effects of nanoparticles exposure in the environment and identify potential adaptive mechanisms (Cox et al. 2017; Singh et al. 2017). Additionally, more research is needed to investigate nanoparticles bioaccumulation in the food chain and their interactions with other pollutants in the environment. Nanoparticles enter the cellular system, translocate to the shoot, and accumulate in various aerial parts, which increases their potential to cycle through the ecosystem at various trophic levels. The accumulation of nanoparticles affects the rate of transpiration, respiration, photosynthesis, and food material translocation (Tripathi et al. 2016b; Du et al. 2017). The toxicity grade is related to the surface and surface properties of nanoparticles. Therefore, ecotoxicity is critical because it establishes a direct link between the negative effects of nanoparticles and organisms, such as microorganisms, plants, and humans, at different trophic levels (Rana and Kalaichelvan 2013; Tripathi et al. 2016a).

4.5.1 Future Prospects

As thoroughly discussed in the present work, the use of nanobiofungicides has shown great potential for controlling fungal pathogens. Several studies have demonstrated the efficacy of nanobiofungicides against various fungal diseases, including powdery mildew, gray mold, and Fusarium wilt. The advantages of nanobiofungicides, such as their high selectivity, biodegradability, and low toxicity, make them a promising alternative to conventional fungicides (Mubeen et al. 2023) In addition, the use of nanobiofungicides can help reduce the development of resistance in fungal pathogens, as nanoparticles can target multiple cellular pathways and interfere with fungal growth and development.

Moreover, nanobiofungicides can also be used as delivery vehicles for other bioactive molecules, such as plant growth regulators, micronutrients, and herbicides, to improve plant health and productivity. Nanoparticles can protect these bioactive molecules from degradation and enhance their uptake and translocation in plants (Kutawa et al. 2021). This approach can help reduce the amount of agrochemicals used in agriculture and improve the sustainability of crop production.

However, there are still several bottlenecks that need to be addressed before nanobiofungicides can be widely adopted in agriculture (Pan et al. 2023; Mubeen et al. 2023; Kutawa et al. 2021). One of the main challenges is the lack of standardized methods for the synthesis and characterization of nanoparticles, which can affect their stability, efficacy, and safety. Moreover, the potential risks associated with the release of nanoparticles into the environment and their impact on non-target organisms need to be thoroughly evaluated.

To overcome these bottlenecks and further advance the development of nanobiofungicides, future research should focus on the following areas:

- Synthesis and Characterization of Nanoparticles: Develop standardized protocols for the synthesis, characterization, and functionalization of nanoparticles to ensure their stability, efficacy, and safety.
- Mode of Action: Investigate the mode of action of nanobiofungicides against fungal pathogens to optimize their efficacy and reduce the risk of resistance.
- Formulation and Delivery: Develop efficient and sustainable methods for the formulation and delivery of nanobiofungicides, including encapsulation, coating, and spraying, to enhance their stability and efficacy.
- Environmental and Health Risk Assessment: Conduct thorough risk assessments of nanobiofungicides to evaluate their potential impact on the environment and human health and develop mitigation strategies to minimize the risks.

4.6 Conclusion

Nanobiofungicides have emerged as effective tools for the control of fungal pathogens. The use of nanotechnology in agriculture has gained considerable attention in recent years, and nanobiofungicides are one of the most promising areas of research in this field. The combination of nanotechnology and biologically derived compounds allows for targeted and effective treatments that can overcome many of the limitations of traditional fungicides, including the development of resistance and negative impacts on the environment and human health. Studies have demonstrated that nanobiofungicides can effectively inhibit the growth and reproduction of a range of fungal pathogens, including those that are resistant to traditional fungicides. Furthermore, these formulations can be easily applied to crops using existing spraying equipment, making them a practical and cost-effective option for farmers. The requirement for environmental sustainability to maintain a balance between future needs and current resources is a critical issue on a global scale. Several examples of nanotechnology and microbes mediated management of fungal control show the way toward sustainable and effective solution. Potential bacteria with nanoparticles play an important role in environmental stability and conservation in an environmentally friendly and efficient manner.

In conclusion, nanobiofungicides represent a major breakthrough in the development of sustainable and effective tools for controlling fungal pathogens. With continued research and development, these formulations have the potential to revolutionize the way we manage plant diseases and improve crop yields, thereby contributing to global food security. However, more research is needed to fully understand the potential impacts of these formulations on the environment and human health and to develop appropriate regulations and guidelines for their use.

Acknowledgements The authors are grateful to the authorities of Mohanlal Sukhadia University, Udaipur, for supporting this work.

Conflict of InterestThe authors declare no conflict of interest.

References

Abd El-Aziz ARM, Al-Othman MR, Mahmoud MA, Metwaly HA (2015) Biosynthesis of silver nanoparticles using *Fusarium solani* and its impact on grain borne fungi. Dig J Nanomater Bios 10:655–662

Abd El-Ghany MN, Hamdi SA, Korany SM, Elbaz RM, Farahat MG (2023) Biosynthesis of novel tellurium nanorods by *Gayadomonas* sp. TNPM15 isolated from mangrove sediments and assessment of their impact on spore germination and ultrastructure of phytopathogenic fungi. Microorganisms 11(3):558

Abd-Elsalam KA (2013) Nano platforms for plant pathogenic fungi management. App Phys A 100:829–834

Abdollahdokht D, Gao Y, Faramarz S, Poustforoosh A, Abbasi M, Asadikaram G et al (2022) Conventional agrochemicals towards nano-biopesticides: an overview on recent advances. Chem Biol Technol Agric 9(1):1–9

Adisa IO, Reddy Pullagurala VL, Rawat S, Hernandez-Viezcas JA, Dimkpa CO, Elmer WH et al (2018) Role of cerium compounds in *Fusarium* wilt suppression and growth enhancement in tomato (*Solanum lycopersicum*). J Agric Food Chem 66(24):5959–5970

Ahmad A, Mukherjee P, Mandal D, Senapati S, Khan MI, Kumar R et al (2002) Enzyme mediated extracellular synthesis of CdS nanoparticles by the fungus, *Fusarium oxysporum*. J Am Chem Soc 124(41):12108–12109

Akther T, Hemalatha S (2019) Mycosilver nanoparticles: synthesis, characterization and its efficacy against plant pathogenic fungi. Bio Nano Sci 9:296–301

Alghuthaymi MA, Almoammar H, Rai M, Said-Galiev E, Abd-Elsalam KA (2015) Myconanoparticles: synthesis and their role in phytopathogens management. Biotechnol Biotechnol Equip 29(2):221–236

Alghuthaymi MA, Kalia A, Bhardwaj K, Bhardwaj P, Abd-Elsalam KA, Valis M et al (2021) Nanohybrid antifungals for control of plant diseases: current status and future perspectives. J Fungi 7(1):48

Ali M, Kim B, Belfield KD, Norman D, Brennan M, Ali GS (2015) Inhibition of *Phytophthora parasitica* and *P. Capsici* by silver nanoparticles synthesized using aqueous extract of *Artemisia absinthium*. Phytopathology 105(9):1183–1190

Ammar HA, Alghazaly MS, Assem Y, Zeid AAA (2021) Bioengineering and optimization of PEGylated zinc nanoparticles by simple green method using *Monascus purpureus*, and their powerful antifungal activity against the most famous plant pathogenic fungi, causing food spoilage. Environ Nanotechnol Monit Manag 16:100543

Anbeaki RAA, Hameed FR, Gassim FAG (2009) Antifungal activity of titanium dioxide photocatalysis against *Fusarium oxysporum* f. sp. *lycopersici*. Euphrates J Agric Sci 1(3):14–26

Anusuya S, Sathiyabama M (2015) Foliar application of β–D-glucan nanoparticles to control rhizome rot disease of turmeric. Int J Biol Macromol 72:1205–1212

Arciniegas-Grijalba PA, Patiño-Portela MC, Mosquera-Sánchez LP, Guerrero-Vargas JA, Rodríguez-Páez JE (2017) ZnO nanoparticles (ZnO-NPs) and their antifungal activity against coffee fungus *Erythricium salmonicolor*. Appl Nanosci 7:225–241

Arora NK (2018) Agricultural sustainability and food security. Environ Sustain 1(3):217–219

Ashajyothi C, Prabhurajeshwar C, Handral HK (2016) Investigation of antifungal and anti-mycelium activities using biogenic nanoparticles: an eco-friendly approach. Environ Nanotechnol Monit Manag 5:81–87

Ashfaq M, Talreja N, Chuahan D, Srituravanich W (2019) Polymeric nanocomposite-based agriculture delivery system: emerging technology for agriculture. In: Jamal F (ed) Genetic engineering—a glimpse of techniques and applications. IntechOpen, UK, p 77

Ashwini N, Srividya S (2014) Potentiality of *Bacillus subtilis* as biocontrol agent for management of anthracnose disease of chilli caused by *Colletotrichum gloeosporioides* OGC1. 3 Biotech 4(2):127–136

Aziz N, Faraz M, Pandey R, Sakir M, Fatma T, Varma A et al (2015) Facile algae derived route to biogenic silver nanoparticles: synthesis, antibacterial and photocatalytic properties. Langmuir 31:11605–11612

Aziz N, Pandey R, Barman I, Prasad R (2016) Leveraging the attributes of Mucor hiemalis-derived silver nanoparticles for a synergistic broad-spectrum antimicrobial platform. Front Microbiol 7:1984

Balakumaran MD, Ramachandran R, Kalaichelvan PT (2015) Exploitation of endophytic fungus, *Guignardia mangiferae* for extracellular synthesis of silver nanoparticles and their in vitro biological activities. Microbiol Res 178:9–17

Banik S, Sharma P (2011) Plant pathology in the era of nanotechnology. Indian Phytopathol 64(2):120–127

Bartlett DW, Clough JM, Godwin JR, Hall AA, Hamer M, Parr-Dobrzanski B (2002) The Strobilurin fungicides. Pest Manag Sci 58(7):649–662

Bhar A, Jain A, Das S (2022) Development and regulation of microbial pesticides in the post-genomic era. In: Rakshit A, Meena VS, Abhilash PC, Sarma BK, Singh HB, Leonardo F, Parihar M, Singh AK (eds) Biopesticides, vol 2. Elsevier, Netherlands, pp 285–299

Bhattacharyya A, Antoney PU, Raja Naika H, Reddy SJ, Adeyemi MM, Omkar (2015) Nanotechnology and butterflies: a mini review. J Appl Biosci 41:27–32

Bhattacharyya A, Duraisamy P, Govindarajan M, Buhroo AA, Prasad R (2016) Nano-biofungicides: emerging trend in insect Pest control. In: Prasad R (ed) Advances and applications through fungal nanobiotechnology. Fungal biology. Springer, Cham, pp 307–319

Bhunjun CS, Niskanen T, Suwannarach N, Wannathes N, Chen YJ, McKenzie EH et al (2022) The numbers of fungi: are the most speciose genera truly diverse? Fungal Divers 114(1):387–462

Borgatta J, Ma C, Hudson-Smith N, Elmer W, Plaza Perez CD, De La Torre-Roche R et al (2018) Copper based nanomaterials suppress root fungal disease in watermelon (*Citrullus lanatus*): role of particle morphology, composition and dissolution behavior. ACS Sustain Chem Eng 6(11):14847–14856

Brahmanwade K, Shende S, Bonde S, Gade A, Rai M (2016) Fungicidal activity of Cu nanoparticles against Fusarium causing crop disease. Environ Chem Lett 14:229–235

Buchman JT, Elmer WH, Ma C, Landy KM, White JC, Haynes CL (2019) Chitosan-coated mesoporous silica nanoparticle treatment of *Citrullus lanatus* (watermelon): enhanced fungal disease suppression and modulated expression of stress-related genes. ACS Sustain Chem Eng 7(24):19649–19659

Chen J, Wang W, Xu Y, Zhang X (2011) Slow-release formulation of a new biological pesticide, pyoluteorin, with mesoporous silica. J Agric Food Chem 59(1):307–311

Cooper ER (2010) Nanoparticles: a personal experience for formulating poorly water-soluble drugs. J Control Release 141(3):300–302

Cox A, Venkatachalam P, Sahi S, Sharma N (2017) Reprint of: silver and titanium dioxide nanoparticle toxicity in plants: a review of current research. Plant Physiol Biochem 110:33–49

Daroczi B, Kari G, McAleer MF, Wolf JC, Rodeck U, Dicker AP (2006) In vivo radioprotection by the fullerene nanoparticle DF-1 as assessed in a zebrafish model. Clin Cancer Res 12(23):7086–7091

Das D, Ahmed G (2012) Silver nanoparticles damage yeast cell wall. J Biotechnol 3:36–39

Dayarathne MC, Mridha AU, Wang Y (2020) Diagnosis of fungal plant pathogens using conventional and molecular approaches. In: Kurouski D (ed) Diagnostics of plant diseases. IntechOpen, UK, pp 19–42

De Andrade FM, de Assis PT, Souza TP, Guimarães PHS, Martins AD, Schwan RF et al (2019) Beneficial effects of inoculation of growth promoting bacteria in strawberry. Microbiol Res 223:120–128

De Cal A, Larena I, Guijarro B, Melgarejo P (2012) Use of biofungicides for controlling plant diseases to improve food availability. Agriculture 2(2):109–124

Dikshit PK, Kumar J, Das AK, Sadhu S, Sharma S, Singh S et al (2021) Green synthesis of metallic nanoparticles: applications and limitations. Catalysts 11(8):902

Du W, Tan W, Peralta-Videa JR, Gardea-Torresdey JL, Ji R, Yin Y et al (2017) Interaction of metal oxide nanoparticles with higher terrestrial plants: physiological and biochemical aspects. Plant Physiol Biochem 110:210–225

El-Ashmony RM, Zaghloul NS, Milošević M, Mohany M, Al-Rejaie SS, Abdallah Y et al (2022) The biogenically efficient synthesis of silver nanoparticles using the fungus *Trichoderma harzianum* and their antifungal efficacy against *Sclerotinia sclerotiorum* and *Sclerotium rolfsii*. J Fungi 8(6):597

El-Baky NA, Amara AAAF (2021) Recent approaches towards control of fungal diseases in plants: an updated review. J Fungi 7(11):900

Elgorban AM, Aref SM, Seham SM, Elhindi KM, Bahkali AH, Sayed SR et al (2016) Extracellular synthesis of silver nanoparticles using *aspergillus versicolor* and evaluation of their activity on plant pathogenic fungi. Mycosphere 7(6):844–852

Feng QL, Wu J, Chen GQ, Cui FZ, Kim TN, Kim JO (2000) A mechanistic study of the antibacterial effect of silver ions on *Escherichia coli* and *Staphylococcus aureus*. J Biomed Mater Res Part B: Appl Biomater 52:662–668

Fira D, Dimkić I, Berić T, Lozo J, Stanković S (2018) Biological control of plant pathogens by bacillus species. J Biotechnol 285:44–55

Flores-Céspedes F, Martínez-Domínguez GP, Villafranca-Sánchez M, Fernández-Pérez M (2015) Preparation and characterization of azadirachtin alginate biosorbent based formulations: water release kinetics and photodegradation study. J Agric Food Chem 63(38):8391–8398

Fukamachi K, Konishi Y, Nomura T (2019) Disease control of Phytophthora infestans using cyazofamid encapsulated in poly lactic-co-glycolic acid (PLGA) nanoparticles. Colloids Surf A Physicochem Eng Asp 557:315–322. Elsevier

Galván Márquez I, Ghiyasvand M, Massarsky A, Babu M, Samanfar B, Omidi K et al (2018) Zinc oxide and silver nanoparticles toxicity in the baker's yeast, Saccharomyces cerevisiae. PloS one 13(3):e0193111

Ge B, Liu B, Nwet TT, Zhao W, Shi L, Zhang K (2016) *Bacillus methylotrophicus* strain NKG-1, isolated from Changbai Mountain, China, has potential applications as a biofertilizer or biocontrol agent. PLoS One 11(11):e0166079

Ghiuță I, Cristea D, Croitoru C, Kost J, Wenkert R, Vyrides I et al (2018) Characterization and antimicrobial activity of silver nanoparticles, biosynthesized using *Bacillus* species. Appl Surf Sci 438:266–273

Ghormade V, Deshpande MV, Paknikar KM (2011) Perspectives for nanobiotechnology enabled protection and nutrition of plants. Biotechnol Adv 29(6):792–803

Gianessi LP, Reigner N (2006) The importance of fungicides in U.S. crop production. Outlooks Pest Manag 17:209–213

Gohari G, Mohammadi A, Akbari A, Panahirad S, Dadpour MR, Fotopoulos V et al (2020) Titanium dioxide nanoparticles (TiO2 NPs) promote growth and ameliorate salinity stress effects on essential oil profile and biochemical attributes of *Dracocephalum moldavica*. Sci Rep 10(1):1–14

Gopinath V, Velusamy P (2013) Extracellular biosynthesis of silver nanoparticles using *Bacillus* sp. GP-23 and evaluation of their antifungal activity towards *Fusarium* oxysporum. Spectrochim Acta A Mol Biomol Spectrosc 106:170–174

Gouda S, Kerry RG, Das G, Paramithiotis S, Shin HS, Patra JK (2018) Revitalization of plant growth promoting rhizobacteria for sustainable development in agriculture. Microbiol Res 206:131–140
Halder MA, Mondal SO, Ray SA, Kundu SU (2018) Myco-synthesized silver nanoparticles from Curvularia affinis showing inhibitory activity against phyto-pathogenic fungus Alternaria solani. J Mycopathol Res 56(1):1–4
Hao Y, Ma C, White JC (2020) Carbon-based nanomaterials alter the composition of the fungal endophyte community in rice (*Oryza sativa* L.). Environ Sci Nano 7:2047–2060
Hariharan G, Prasannath K (2021) Recent advances in molecular diagnostics of fungal plant pathogens: a mini review. Front Cell Infect Microbiol 10:1–14
Hawar SN, Al-Shmgani HS, Al-Kubaisi ZA, Sulaiman GM, Dewir YH, Rikisahedew JJ (2022) Green synthesis of silver nanoparticles from Alhagi graecorum leaf extract and evaluation of their cytotoxicity and antifungal activity. J Nanomater 2022:1–8
Hazarika A, Yadav M, Yadav DK, Yadav HS (2022) An overview of the role of nanoparticles in sustainable agriculture. Biocatal Agric Biotechnol 43:102399
Hoppin JA, Umbach DM, Long S, London SJ, Henneberger PK, Blair A et al (2017) Pesticides are associated with allergic and non-allergic wheeze among male farmers. Environ Health Perspect 125:535–543
Hussain F, Usman F (2019) Fungal biotic stresses in plants and its control strategy. In: de Oliveira AB (ed) Abiotic and biotic stress in plants. IntechOpen, UK, pp 61–68
Ibrahim IM, Abdelmalek DH, Elshahat ME, Elfky AA (2020) COVID-19 spike-host cell receptor GRP78 binding site prediction. J Infect 80(5):554–562
Iqbal Z, Khan MA, Sharif M, Shah JH, ur Rehman MH, Javed K (2018) An automated detection and classification of citrus plant diseases using image processing techniques: a review. Comput Electron Agric 153:12–32
Irfan SA, Razali R, KuShaari K, Mansor N, Azeem B, Ford Versypt AN (2018) A review of mathematical modeling and simulation of controlled-release fertilizers. J Control Release 271:45–54
Jain A, Sarsaiya S, Wu Q, Lu Y, Shi J (2019) A review of plant leaf fungal diseases and its environment speciation. Bioengineered 10(1):409–424
Kah M, Beulke S, Tiede K (2013) Nano-pesticides: state of knowledge, environmental fate and exposure modelling. Crit Rev Environ Sci Technol 43(16):1823–1867
Kalia A, Elsalam Abd KA, Kuca K (2020) Zinc-based nanomaterials for diagnosis and management of plant diseases: ecological safety and future prospects. J Fungi 6(4):222
Karaman DŞ, Sarwar S, Desai D, Björk EM, Odén M, Chakrabarti P et al (2016) Shape engineering boosts antibacterial activity of chitosan coated mesoporous silica nanoparticle doped with silver: a mechanistic investigation. J Mater Chem B 4:3292–3304
Karunya SK, Reetha D, Saranraj P, Milton DJ (2011) Optimization and purification of chitinase produced by *Bacillus subtilis* and its antifungal activity against plant pathogens. Int J Pharm Biol Arch 2(6):1680–1685
Kaushal M (2018) Role of microbes in plant protection using intersection of nanotechnology and biology. In: Abd-Elsalam K, Prasad R (eds) Nanobiotechnology applications in plant protection. Nanotechnology in the life sciences. Springer, Cham, pp 111–135
Khakimov AA, Omonlikov AU, Utaganov SBU (2020) Current status and prospects of the use of biofungicides against plant diseases. GSC Biol Pharm Sci 13(03):119–126
Kim HJ, Hwang IS, Kim BS, Hwang BK (2006) Isolation and in vitro and in vivo antifungal activity of phenylacetic acid produced by *Micromonospora aurantiaca* strain JK-1. Plant Pathol J 22(1):75
Kong M, Chen XG, Xing K, Park HJ (2010) Antimicrobial properties of chitosan and mode of action: a state of the art review. Int J Food Microbiol 144:51–63
Kovochich M, Xia T, Xu J, Yeh J, Nel A (2005) Principles and procedures to assess nanoparticles. Environ Sci Technol 39(5):1250–1256
Kumar S, Nehra M, Dilbaghi N, Tankeshwar K, Kim KH (2018) Recent advances and remaining challenges for polymeric nanocomposites in healthcare applications. Prog Polym Sci 80:1–38

Kumar P, Pandhi S, Mahato DK, Kamle M, Mishra A (2021) *Bacillus*-based nano-bioformulations for phytopathogens and insect–pest management. Egypt J Biol Pest Control 31:1–1

Kumari B, Mallick M, Solanki MK, Hora A, Mani M (2019) Applying nanotechnology to bacteria: an emerging technology for sustainable agriculture. In: Kumar A, Singh AK, Choudhary KK (eds) Role of plant growth promoting microorganisms in sustainable agriculture and nanotechnology. Elsevier, Netherlands, pp 121–143

Kutawa AB, Ahmad K, Ali A, Hussein MZ, Abdul Wahab MA, Adamu A et al (2021) Trends in nanotechnology and its potentialities to control plant pathogenic fungi: a review. Biology 10(9):881

Lamsal K, Kim SW, Jung JH, Kim YS, Kim KS, Lee YS (2011a) Application of silver nanoparticles for the control of *Colletotrichum* species in vitro and pepper anthracnose disease in field. Mycobiology 39:194–199

Lamsal K, Kim SW, Jung JH, Kim YS, Kim KS, Lee YS (2011b) Inhibition effects of silver nanoparticles against powdery mildew on cucumber and pumpkin. Mycobiology 39:26–32

Lanfranco L, Balsamo R, Martino E, Perotto S, Bonfante P (2002) Zinc ions alter morphology and chitin deposition in an ericoid fungus. Eur J Histochem 46:341–350

Lee JY, Moon SS, Yun BS, Yoo ID, Hwang BK (2004) Thiobutacin, a novel antifungal and anti-oomycete antibiotic from *Lechevalieria aerocolonigenes*. J Nat Prod 67(12):2076–2078

Lee KJ, Park SH, Govarthanan M (2013) Synthesis of silver nanoparticles using cow milk and their antifungal activity against phytopathogens. Mater Lett 105:128–131

Liechty WB, Kryscio DR, Slaughter BV, Peppas NA (2010) Polymers for drug delivery systems. Annu Rev Chem Biomol Eng 1:149–173

Lipovsky A, Nitzan Y, Gedanken A, Lubart R (2011) Antifungal activity of ZnO nanoparticles-the role of ROS mediated cell injury. Nanotechnology 22(10):105101

Liu X, Cao A, Yan D, Ouyang C, Wang Q, Li Y (2021) Overview of mechanisms and uses of biopesticides. Int J Pest Manag 67(1):65–72

Machado TO, Beckers SJ, Fischer J, Müller B, Sayer C, de Araújo PH et al (2020) Bio-based lignin nanocarriers loaded with fungicides as a versatile platform for drug delivery in plants. Biomacromolecules 21(7):2755–2763

Malandrakis AA, Kavroulakis N, Chrysikopoulos CV (2020) Synergy between cu-NPs and fungicides against *Botrytis cinerea*. Sci Total Environ 703:135557

Malandrakis AA, Kavroulakis N, Chrysikopoulos CV (2022) Zinc nanoparticles: mode of action and efficacy against boscalid-resistant *Alternaria alternata* isolates. Sci Total Environ 829:154638

Manjunatha SB, Biradar DP, Aladakatti YR (2016) Nanotechnology and its applications in agriculture: a review. J Agric Sci 29(1):1–13

Maruyama CR, Guilger M, Pascoli M, Bileshy-José N, Abhilash PC, Fraceto LF et al (2016) Nanoparticles based on chitosan as carriers for the combined herbicides imazapic and imazapyr. Sci Rep 6(1):1–5

Meena KR, Kanwar SS (2015) Lipopeptides as the antifungal and antibacterial agents: applications in food safety and therapeutics. Biomed Res Int 2015:473050

Meher J, Rajput RS, Bajpai R, Teli B, Sarma BK (2020) Trichoderma: a globally dominant commercial biofungicide. In: Manoharachary C, Singh HB, Varma A (eds) Trichoderma: agricultural applications and beyond. Soil biology, vol 61. Springer, Cham, pp 195–208

Melo NF, de Lima MA, Stamford TL, Galembeck A, Flores MA, de Campos Takaki GM et al (2020) In vivo and in vitro antifungal effect of fungal chitosan nanocomposite edible coating against strawberry phytopathogenic fungi. Int J Food Sci Technol 55(11):3381–3391

Mishra S, Singh BR, Singh A, Keswani C, Naqvi AH, Singh HB (2014) Biofabricated silver nanoparticles act as a strong fungicide against *Bipolaris sorokiniana* causing spot blotch disease in wheat. PLoS One 9(5):e97881

Mishra J, Mishra I, Arora NK (2022) 2,4-Diacetylphloroglucinol producing Pseudomonas fluorescens JM-1 for management of ear rot disease caused by *Fusarium moniliforme* in *Zea mays* L. 3 Biotech 12:138

Mittal D, Kaur G, Singh P, Yadav K, Azmal AS (2020) Nanoparticle-based sustainable agriculture and food science: recent advances and future outlook. Front Nanotechnol 2:579954

Mohamed YMA, Elshahawy IE (2022) Antifungal activity of photo-biosynthesized silver nanoparticles (AgNPs) from organic constituents in orange peel extract against phytopathogenic *Macrophomina phaseolina*. Eur J Plant Pathol 162(3):725–738

Moraes Bazioli J, Belinato JR, Costa JH, Akiyama DY, Pontes JG, Kupper KC et al (2019) Biological control of citrus postharvest phytopathogens. Toxins 11(8):460

Morones JR, Elechiguerra JL, Camacho A, Holt K, Kouri JB, RamÃrez JT et al (2005) The bactericidal effect of silver nanoparticles. Nanotechnology 16:2346–2353

Mubeen I, Mfarrej MF, Razaq Z, Iqbal S, Naqvi SA, Hakim F et al (2023) Nanopesticides in comparison with agrochemicals: outlook and future prospects for sustainable agriculture. Plant Physiol Biochem 30:107670

Muhammad M, Wahab RA, Huyop F, Rusli MH, Yaacob SNS, Teo HL (2022) An overview of the potential role of microbial metabolites as greener fungicides for future sustainable plant diseases management. J Crop Prot 11(1):1–27

Mukherjee A, Majumdar S, Servin AD, Pagano L, Dhankher OP, White JC (2016) Carbon nanomaterials in agriculture: a critical review. Front Plant Sci 7:172

Narendhran S, Sivaraj R (2016) Biogenic ZnO nanoparticles synthesized using *L. aculeata* leaf extract and their antifungal activity against plant fungal pathogens. Bull Mater Sci 39:1–5

Nega A (2014) Review on concepts in biological control of plant pathogens. J Biol Agric Healthc 4:33–35

Nuruzzaman M, Rahman MM, Liu Y, Naidu R (2016) Nanoencapsulation, nano-guard for pesticides: a new window for safe application. J Agric Food Chem 64:1447–1448

O'Brien PA (2017) Biological control of plant diseases. Australas Plant Pathol 46(4):293–304

Oliveira MS, Cordova LG, Peres NA (2020) Efficacy and baseline sensitivity of succinate-dehydrogenase-inhibitor fungicides for management of colletotrichum crown rot of strawberry. Plant Dis 104(11):860–2865

Pan X, Guo X, Zhai T, Zhang D, Rao W, Cao F et al (2023) Nanobiopesticides in sustainable agriculture: developments, challenges, and perspectives. Environ Sci Nano 2023

Pankhurst CE, Lynch JM (2005) Biocontrol of soilborne plant diseases. Encyclopedia of soils in the environment. Elsevier, Netherlands, pp 129–136

Parveen A, Rao S (2015) Effect of nanosilver on seed germination and seedling growth in *Pennisetum glaucum*. J Clust Sci 26:693–701

Patel N, Desai P, Patel N, Jha A, Gautam HK (2014) Agronanotechnology for plant fungal disease management: a review. Int J Curr Microbiol App Sci 3(10):71–84

Patil S, Chandrasekaran R (2020) Biogenic nanoparticles: a comprehensive perspective in synthesis, characterization, application and its challenges. J Genet Eng Biotechnol 18(1):67

Patra JK, Baek KH (2017) Antibacterial activity and synergistic antibacterial potential of biosynthesized silver nanoparticles against foodborne pathogenic bacteria along with its anticandidal and antioxidant effects. Front Microbiol 8:167

Patra P, Mitra S, Debnath N, Goswami A (2012) Biochemical-, biophysical-, and microarray-based antifungal evaluation of the buffer-mediated synthesized nano zinc oxide: an in vivo and in vitro toxicity study. Langmuir 28:16966–16978

Peppas NA, Bures P, Leobandung W, Ichikawa H (2000) Hydrogels in pharmaceutical formulations. Eur J Pharm Biopharm 50(1):27–46

Piel C, Pouchieu C, Carles C, Béziat B, Boulanger M, Bureau M et al (2019) Agricultural exposures to carbamate herbicides and fungicides and central nervous system tumour incidence in the cohort AGRICAN. Environ Int 130:104876

Porteous-Álvarez AJ, Maldonado-González MM, Mayo-Prieto S, Lorenzana A, Paniagua-García AI, Casquero PA (2021) Green strategies of powdery mildew control in hop: from organic products to nanoscale carriers. J Fungi 7(6):490

Possiede YM, Gabardo J, Kava-Cordeiro V, Galli-Terasawa LV, Azevedo JL, Glienke C (2009) Fungicide resistance and genetic variability in plant pathogenic strains of Guignardia citricarpa. Braz J Microbiol 40:308–313

Prakash J, Egamberdieva D, Arora NK (2022) A novel *Bacillus safensis*-based formulation along with mycorrhiza inoculation for controlling *Alternaria alternata* and simultaneously improving growth, nutrient uptake, and steviol glycosides in *Stevia rebaudiana* under field conditions. Plants 11:1857

Prasad R (2014) Synthesis of silver nanoparticles in photosynthetic plants. J Nanopart 10 https://doi.org/10.1155/2014/963961

Prasad R, Kumar V, Prasad KS (2014) Nanotechnology in sustainable agriculture: present concerns and future aspects. Afr J Biotechnol 13:705–713

Prasad R, Pandey R, Barman I (2016) Engineering tailored nanoparticles with microbes: quo vadis? Wiley Interdiscip Rev Nanomed Nanobiotechnol 8(2):316–330

Qian Y, Yu H, He D, Yang H, Wang W, Wan X et al (2013) Biosynthesis of silver nanoparticles by the endophytic fungus *Epicoccum nigrum* and their activity against pathogenic fungi. Bioprocess Biosyst Eng 36:1613–1619

Qin J, Tong Z, Zhan Y, Buisson C, Song F, He K et al (2020) A *Bacillus thuringiensis* chitin-binding protein is involved in insect peritrophic matrix adhesion and takes part in the infection process. Toxins 12(4):252

Rana S, Kalaichelvan P (2013) Ecotoxicity of nanoparticles. Int Sch Res Notices 2013

Rani R, Dahiya S, Dhingra D, Dilbaghi N, Kim KH, Kumar S (2017) Evaluation of anti-diabetic activity of glycyrrhizin-loaded nanoparticles in nicotinamide-streptozotocin-induced diabetic rats. Eur J Pharm Sci 106:220–230

Rastogi A, Tripathi DK, Yadav S, Chauhan DK, Živčák M, Ghorbanpour M et al (2019) Application of silicon nanoparticles in agriculture. 3 Biotech 9(3):1–11

Robinson JR, Isikhuemhen OS, Anike FN (2021) Fungal–metal interactions: a review of toxicity and homeostasis. J Fungi 7(3):225

Saharan V, Mehrotraa A, Khatik R, Rawal P, Sharma SS, Pal A (2013) Synthesis of chitosan-based nanoparticles and their in vitro evaluation against phytopathogenic fungi. Int J Biol Macromol 62:677–683

Saharan V, Sharma G, Yadav M, Choudhary MK, Sharma SS, Pal A et al (2015) Synthesis and in vitro antifungal efficacy of Cu–chitosan nanoparticles against pathogenic fungi of tomato. Int J Biol Macromol 75:346–353

Salunke BK, Sawant SS, Lee SI, Kim BS (2016) Microorganisms as efficient biosystem for the synthesis of metal nanoparticles: current scenario and future possibilities. World J Microbiol Biotechnol 32(5):1–16

Sawai J, Yoshikawa T (2004) Quantitative evaluation of antifungal activity of metallic oxide powders (MgO, CaO and ZnO) by an indirect conductimetric assay. J Appl Microbiol 96(4):803–809

Sayes CM, Fortner JD, Guo W, Lyon D, Boyd AM, Ausman KD et al (2004) The differential cytotoxicity of water-soluble fullerenes. Nano Lett 4(10):1881–1887

Sekhon BS (2014) Nanotechnology in Agri-food production: an overview. Nanotechnol Sci Appl 7:31–53

Senthil-Nathan S (2015) A review of biopesticides and their mode of action against insect pests. In: Thangavel P, Sridevi G (eds) Environmental Sustainability. Springer, New Delhi, pp 49–63

Sharon M, Choudhary AK, Kumar R (2010) Nanotechnology in agricultural diseases and food safety. J Phytology 2(4):83–92

Shekhar S, Sharma S, Kumar A, Taneja A, Sharma B (2021) The framework of nanopesticides: a paradigm in biodiversity. Mater Adv 2:6569–6588

Shoeb M, Singh BR, Khan JA, Khan W, Singh BN, Singh BH et al (2013) ROS-dependent anticandidal activity of zinc oxide nanoparticles synthesized by using egg albumin as a biotemplate. Adv Nat Sci Nanosci Nanotechnol 4:3

Shrivastava D, Singhai AK, Yadav RK (2014) Effect of lime and rice husk ash on engineering properties of black cotton soil. Int J Eng Res Sci Technol 3(2):292–296

Shuping DSS, Eloff JN (2017) The use of plants to protect plants and food against fungal pathogens : a review. Afr J Tradit Complement Altern Med 14(4):120–127

Singh S, Vishwakarma K, Singh S, Sharma S, Dubey NK, Singh VK et al (2017) Understanding the plant and nanoparticle interface at transcriptomic and proteomic level: a concentric overview. Plant Gene 11:265–272

Singh A, Gaurav SS, Shukla G, Rani P (2022) Assessment of mycogenic zinc nano-fungicides against pathogenic early blight (*Alternaria solani*) of potato (*Solanum tuberosum* L.). Mater Today Proc 49:3528–3537

Spadola G, Sanna V, Bartoli J, Carcelli M, Pelosi G, Bisceglie F et al (2020) Thiosemicarbazone nano-formulation for the control of Aspergillus flavus. Environ Sci Pollut Res 27:20125–20135

Stamatis N, Helac D, Konstantinou I (2010) Occurrence and removal of fungicides in municipal sewage treatment plant. J Hazard Mater 175:829–835

Thambugala KM, Daranagama DA, Phillips AJL, Kannangara SD, Promputtha I (2020) Fungi vs. fungi in biocontrol: an overview of fungal antagonists applied against fungal plant pathogens. Front Cell Infect Microbiol 10: 604923

Tripathi DK, Singh S, Singh S, Dubey NK, Chauhan DK (2016a) Impact of nanoparticles on photosynthesis: challenges and opportunities. Mater Focus 5(5):405–411

Tripathi DK, Singh S, Singh VP, Prasad SM, Chauhan DK, Dubey NK (2016b) Silicon nanoparticles more efficiently alleviate arsenate toxicity than silicon in maize cultiver and hybrid differing in arsenate tolerance. Front Environ Sci 4: 46

Tripathi DK, Singh S, Singh S, Pandey R, Singh VP, Sharma NC et al (2017a) An overview on manufactured nanoparticles in plants: uptake, translocation, accumulation and phytotoxicity. Plant Physiol Biochem 110:2–12

Tripathi DK, Singh S, Singh S, Srivastava PK, Singh VP, Singh S et al (2017b) Nitric oxide alleviates silver nanoparticles (AgNps)-induced phytotoxicity in Pisum sativum seedlings. Plant Physiol Biochem 110:167–177

Tripathi DK, Singh S, Singh VP, Prasad SM, Dubey NK, Chauhan DK (2017c) Silicon nanoparticles more effectively alleviated UV-B stress than silicon in wheat (*Triticum aestivum*) seedlings. Plant Physiol Biochem 110:70–81

Tripathi DK, Tripathi A, Singh S, Singh Y, Vishwakarma K, Yadav G et al (2017d) Uptake, accumulation and toxicity of silver nanoparticle in autotrophic plants, and heterotrophic microbes: a concentric review. Front Microbiol 8:07

Vagabov V, Yu Ivanov A, Kulakovskaya T, Kulakovskaya E, Petrov V, Kulaev I (2008) Efflux of potassium ions from cells and spheroplasts of *Saccharomyces cerevisiae* yeast treated with silver and copper ions. Biochemistry 73:1224–1227

Vannini C, Domingo G, Onelli E, De Mattia F, Bruni I, Marsoni M et al (2014) Phytotoxic and genotoxic effects of silver nanoparticles exposure on germinating wheat seedlings. J Plant Physiol 171(13):1142–1148

Vardhana J, Kathiravan G (2015) Biosynthesis of silver nanoparticles by endophytic fungi *Pestalotiopsis pauciseta* isolated from the leaves of *Psidium guajava* Linn. Int J Pharm Sci Rev Res 31(1):29–31

Venkatachalam P, Jayaraj M, Manikandan R, Geetha N, Rene ER, Sharma NC et al (2017) Zinc oxide nanoparticles (ZnONPs) alleviate heavy metal-induced toxicity in Leucaena leucocephala seedlings: a physiochemical analysis. Plant Physiol Biochem 110:59–69

Wang X, Zhou Z, Chen F (2017) Surface modification of carbon nanotubes with an enhanced antifungal activity for the control of plant fungal pathogen. Materials 10(12):1375

Wash P, Batool A, Mulk S, Nazir S, Yasmin H, Mumtaz S et al (2022) Prevalence of antimicrobial resistance and respective genes among *Bacillus* spp., a versatile bio-fungicide. Int J Environ Health Res 19:22

Wei X, Luo M, Li W, Yang L, Liang X, Xu L et al (2012) Synthesis of silver nanoparticles by solar irradiation of cell-free *Bacillus* amyloliquefaciens extracts and $AgNO_3$. Bioresour Technol 103(1):273–278

Werder EJ, Engel LS, Satagopan J, Blair A, Koutros S, Lerro CC et al (2020) Herbicide, fumigant, and fungicide use and breast cancer risk among farmers' wives. Environ Epidemiol 4(3):e097
Widnyana IK, Javandira C (2016) Activities *Pseudomonas* spp. and *Bacillus* sp. to stimulate germination and seedling growth of tomato plants. Agric Agric Sci Procedia 9:419–423
Win TT, Khan S, Fu P (2020) Fungus (*Alternaria* sp.) mediated silver nanoparticles synthesis, characterization, and screening of antifungal activity against some phytopathogens. J Nanotechnol 2020:1–9
Yadav A, Verma A, Yadav K (2016) Fungal nanoparticles: an emerging tool in medical biology. In: Prasad R (ed) Advances and applications through fungal nanobiotechnology. Fungal biology. Springer, Cham, pp 213–240
Yamanaka M, Hara K, Kudo J (2005) Bactericidal actions of a silver ion solution on *Escherichia coli*, studied by energy-filtering transmission electron microscopy and proteomic analysis. Appl Environ Microbiol 71:7589–7593
Yang L, Watts DJ (2005) Particle surface characteristics may play an important role in phytotoxicity of alumina nanoparticles. Toxicol Lett 158(2):122–132
Yehia R, Ahmed OF (2013) In vitro study of the antifungal efficacy of zinc oxide nanoparticles against *Fusarium oxysporum* and *Penicillium expansum*. Afr J Microbiol Res 7:1917–1923
Youssef K, de Oliveira AG, Tischer CA, Hussain I, Roberto SR (2019) Synergistic effect of a novel chitosan/silica nanocomposites-based formulation against gray mold of table grapes and its possible mode of action. Int J Biol Macromol 141:247–258
Yu Z, Li Q, Wang J, Yu Y, Wang Y, Zhou Q et al (2020) Reactive oxygen species-related nanoparticle toxicity in the biomedical field. Nanoscale Res Lett 15(1):115
Yurtluk T, Akçay FA, Avcı A (2018) Biosynthesis of silver nanoparticles using novel *Bacillus* sp SBT8. Prep Biochem Biotechnol 48(2):151–159
Zaki SA, Ouf SA, Abd-Elsalam KA, Asran AA, Hassan MM, Kalia A et al (2022) Trichogenic silver-based nanoparticles for suppression of fungi involved in damping-off of cotton seedlings. Microorganisms 10(2):344

Chapter 5
Biotechnological Applications of Phosphate Solubilizing Microorganisms: Biological Alternative to Improve Phosphorus Availability

Fatima Zahra Aliyat, Mohammed El Guilli, Laila Nassiri, and Jamal Ibijbijen

Abstract Phosphorus (P) is abundant in nature and, along with nitrogen and potassium, is regarded as a crucial component of plants. Plants acquire P in the form of monovalent (H_2PO_4) and divalent (HPO_4) orthophosphate anions from the soil solution. Although the amount of P in the soil is generally important, the forms available to plants are minimal. The inorganic P and organic P are the forms that are unavailable to plants; the inorganic forms are associated with calcium in alkaline soils and aluminum and iron compounds in acidic soils, whereas organic P forms are associated with organic matter. Furthermore, P is quickly immobilized and blocked in the soil; therefore, a large amount of the phosphate input during chemical fertilization is immobilized. P-solubilizing microorganisms, through rhizospheric and symbiotic relationships, considerably modify the quantity of P that the plant needs to acquire for its growth. The number of species identified as P solubilizers has increased recently as a result of numerous studies on phosphate solubilization by microorganisms; a variety of these microorganisms, including bacteria, fungi, and actinomycetes, are capable of P solubilization. Moreover, several studies have investigated the P-solubilizing microorganisms' performance as biofertilizers with the goal of

F. Z. Aliyat
Environment and Valorization of Microbial and Plant Resources Unit, Faculty of Sciences, Moulay Ismail University, Meknes, Morocco

African Genome Center, Mohammed VI Polytechnic University (UM6P), Ben Guerir, Morocco

M. El Guilli
National Institute of Agricultural Research, Regional Center for Agricultural Research, Kenitra, Morocco

L. Nassiri · J. Ibijbijen (✉)
Environment and Valorization of Microbial and Plant Resources Unit, Faculty of Sciences, Moulay Ismail University, Meknes, Morocco
e-mail: j.ibijbijen@fs.umi.ac.ma

N. K. Arora, B. Bouizgarne (eds.), *Microbial Biotechnology for Sustainable Agriculture Volume 2*, Microorganisms for Sustainability 34,
https://doi.org/10.1007/978-981-97-2355-3_5

enhancing crop productivity while preserving soil health. Moreover, the biotechnological applications of these microorganisms as a biological alternative to chemicals have now become a focus of attention for several researchers and agricultural decision-makers. This chapter focuses on the variety of microorganisms that can solubilize phosphate, and how they can be used in agriculture as eco-friendly approaches to enhance the availability of phosphorus in soil.

Keywords Phosphate-solubilizing microorganisms · Phosphorus bioavailability · Biofertilizers · Agriculture · Biotechnology

5.1 Introduction

Phosphorus (P) is one of the main macronutrients for plants (Lu et al. 2023; Chen et al. 2006). Due to the limited reserves of degradable natural P, a targeted and environmentally friendly supply is necessary for agriculture. This is particularly crucial in intensive agriculture, where a supply of P as fertilizer is essential to obtain good yields (Aliyat et al. 2022).

P is found in soils in varied proportions of organic and inorganic forms (Li et al. 2022). In soil, inorganic P occurs principally in insoluble mineral complexes, some of these traits may manifest following frequent usage of chemical fertilizers. Under natural conditions in the soil, P rapidly precipitates into sparingly soluble complexes composed of various types of phosphates (Bashan et al. 2013; Richardson 2001).

Soil microorganisms like bacteria, fungus, actinomycetes are involved in a variety of processes that result in the transformation of soil P and are thus an integral part of the soil P cycle. Soil microorganisms, in particular, are effective at solubilizing and mineralizing inorganic and organic P from the total soil P, thereby releasing and unblocking it (Beltran-Medina et al. 2023; Rodríguez and Fraga 1999). At present, the main purpose in managing soil P is to optimize crop production and minimize P loss from soils. Recently, P-solubilizing microorganisms have attracted attention. The ability of P-solubilizing microorganisms to transform insoluble forms of P into a bioavailable form is a crucial attribute of plant growth-promoting bacteria that helps to enhance plant yields (Alaylar et al. 2020). Consequently, the utilization of P-solubilizing microorganisms as inoculants has become increasingly appealing for boosting P uptake by plants (Beltran-Medina et al. 2023; Chen et al. 2006). Moreover, according to Yazdani et al. (2009), the application of P-solubilizing microorganisms as a biofertilizer might minimize the application of phosphorus fertilizer by 50% without significant crop yield decreases, this procedure includes the P-solubilizing microorganisms application to plants, plant surfaces, and soils. These bacteria colonize the rhizosphere or inside of the plant and promote growth by providing phosphorus to developing plants (Khan et al. 2009).

In recent years, various studies have presented a large number of new P-solubilizing microorganisms (Chawngthu et al. 2020; Kaur and Kaur 2020; Matter et al. 2020; Rfaki et al. 2019; Manzoor et al. 2017; Behera et al. 2017). A significant portion of these studies were initial explorations, limited to in vitro experiments and examining the application of plant-based solutions in the field. The majority of these investigations assumed that the capacity to solubilize P would result in the availability of this nutrient for plant absorption in soil.

The excessive and continuous application of chemical fertilizers has contributed to environmental impacts, and agronomists are interested in finding alternate solutions to ensure sustainable yields and protect soil health. A modern approach to farming, also known as sustainable agriculture, includes environmentally responsible agricultural practices that preserve the soil ecosystem's long-term ecological equilibrium. The application of microbial inoculants as biofertilizers, like PSM, is an environmentally safe solution to further mineral fertilizer applications.

In this context, this chapter provides a thorough overview of phosphate-solubilizing microorganisms and highlights the biodiversity and biotechnological applications of phosphate-solubilizing microorganisms as a biological alternative to enhancing P availability to promote plant growth.

5.2 Phosphorus in Soil and its Availability

Soil is a natural ecosystem and an ecological environment of persistent biological processes, primarily determined by the chemical composition of its original material and the growth of the plant it maintains (Zhao et al. 2022; Behera et al. 2017). P is an indispensable nutrient for the growth of plants, as it performs critical functions such as energy metabolism, signal transduction, structural support, and genetic transmission. Its significance lies in aiding plant development, including cell division and the creation of fresh tissues (Dissanayaka et al. 2018; Lobo et al. 2019). Moreover, its key function within the plant is related to the conservation and conversion of energy (ATP), the production of structural compounds (nucleotides, nucleic acids, coenzymes, phospholipids, etc.) involved in root development, the formation of grains, fruits, and fibers, the ripening of fruits, and the germination of seeds. Likewise, plants can only absorb P from orthophosphate anions (H_2PO^{4-} and HPO_4^{2-}) (Lobo et al. 2019; Sims and Sharpley 2005). According to Zou et al. (1992), just ~0.1% of the total P is in a soluble form available for plant uptake.

P is very dynamic in soil and is subject to a variety of chemical processes in soil. Besides, P soil forms can usually be classified as P solution soil, insoluble organic P, and inorganic P (Behera et al. 2017; Panhwar et al. 2009) (Fig. 5.1). P is frequently present in apatite, a category of phosphate minerals. Apatite wear generates low concentrations of P anion, approximately one percent of overall soil phosphorus. P anions take part in processes that restrict their availability to plants, whether in the form of crystalline salts, salts in solution, or salts adsorbed by soil colloids (Barea and Richardson 2015; Kafkafi et al. 1988). In addition, P ions may be

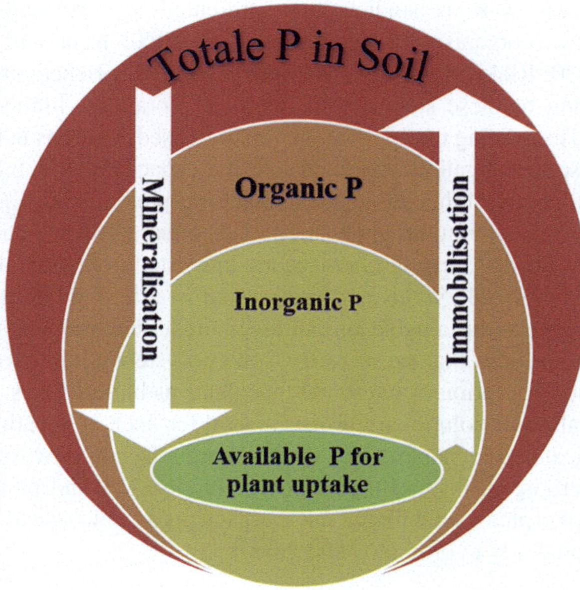

Fig. 5.1 Dynamic of P in Soil

immediately adsorbed by soil colloids or form highly stable complexes with iron, aluminum, or manganese hydroxides that form part of such colloids (McGechan 2002).

In soil, the immobilization and fixation of P transform it into an unavailable form (Vanderdeelen 1995). More than 70–90% of the P fertilizers used are immobilized in the soil, making the availability of P nutrients inaccessible for plant uptake (Holford 1997; Kalayu 2019; Kumar et al. 2023); in addition, because of its rapid immobilization in the soil, the efficiency of chemical P fertilizers applied rarely reaches 30%, and immobilized in aluminum and iron phosphate forms in acid soils (Ducousso-Détrez et al. 2022; Norrish and Rosser 1983; Sims and Sharpley 2005), or in calcium phosphate form in neutral to alkaline soils (Shen et al. 2011; Wu et al. 2020). Khan et al. (2009) suggested that cumulative P due to continuous chemical fertilization in agricultural soils will be enough to sustain maximum agricultural productivity for approximately 100 years if available (Fig. 5.2). Due to its high significance, numerous experiments and studies have been conducted to develop biotechnological and ecological remedies for providing plants with P.

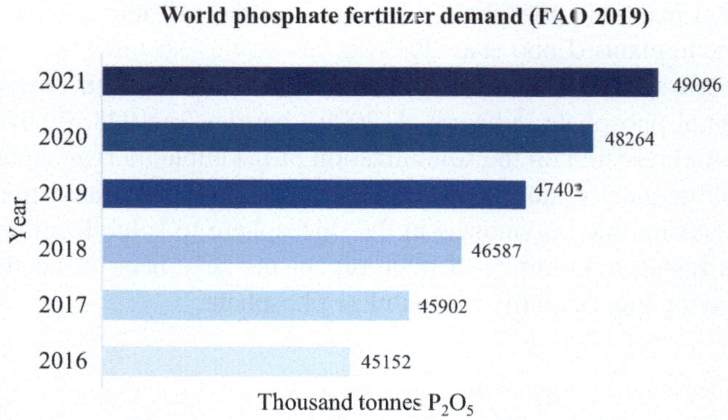

Fig. 5.2 World phosphate fertilizer demand (modified from Rawat et al. 2020)

5.3 Importance of Phosphate-Solubilizing Microorganisms in Plant Phosphate Nutrition

While the population of the planet continues to rise, one of the biggest obstacles will be the demands put on agriculture for potential food production. The application of fertilizers has greatly improved the production of crops. Since 1950, for example, the profitability of growing wheat has quadrupled. Despite fertilizer application improves growth of plants, from an ecological perspective, the soil's physical and chemical characteristics can be altered, leading to potential pollution.

Over the years, there has been an increase in demand for and usage of phosphate fertilizers to ensure a constant supply of P in plants (Rawat et al. 2020). Consequently, the manufacture of chemical phosphatic fertilizers is a high-energy operation that demands USD 4 billion per year of energy to fulfill the global demand (Goldstein et al. 1993; Khan et al. 2009). The possibility that almost 70–90% of additional phosphatic fertilizer is precipitated by metal cation complexes found in the soils further complicates the situation (Holford 1997; Kalayu 2019). The utilization of fossil energy sources in the production of fertilizers and the possibility of a reduction in supply have brought the issue of mineral phosphate solubilization to the forefront of discussions regarding the future of fertilizer development costs. As a result, the solubilization of phosphates has emerged as a topic of extensive research due to its significance in the field of agriculture.

In soil, the majority of living species are microorganisms that are an essential part of the planet's biological diversity. These microorganisms play a crucial role in several processes in diverse environments, including the acquisition of plant nutrients.

Phosphate solubilizing microorganisms were found to be the most successful ecological candidates for phosphoric plant nutrition (Rawat et al. 2020; Sharma et al. 2013). These species have the ability for phosphate solubilization and may also transform insoluble phosphate compounds in the soil into soluble forms that are accessible to plants (Lobo et al. 2019).

Rhizospheric species were recognized as early as 1903 for their function in solubilizing mineral phosphate (Khan et al. 2009); besides, in 1948, Pikovskaya published the initial research on the solubilization of insoluble inorganic phosphate by microbes. Subsequently, numerous experiments have investigated the ability of naturally occurring microorganisms in the rhizosphere to solubilize mineral phosphates. As a result, a multitude of microorganisms have been isolated and documented as having the capability to solubilize phosphate.

5.4 Phosphate-Solubilizing Microorganisms: Type, Diversity, and Functions

Soil is a hot spot for a diversity of microorganisms, which are an important component of the soil and may affect its fertility through their beneficial activities (Khan et al. 2009). Rhizospheric microorganisms play a role in promoting various soil processes, including but not limited to decomposition, nutrient mobilization and mineralization, nutrient release, and solubilization of phosphate (Rawat et al. 2020). Phosphate solubilizing microorganisms are species that have the potential to convert insoluble P into a plant-available form. Phosphate solubilizing microorganisms are diverse in nature. According to the literature, there are three major categories, bacteria belonging to the genera *Serratia* and *Pantoea* (Aliyat et al. 2020, 2022; Chen et al. 2006; Rfaki et al. 2019), *Pseudomonas*, *Enterobacter*, *Rhizobium*, and *Bacillus* (Joshi et al. 2019; Mendoza-Arroyo et al. 2020; Paul and Sinha 2017; Prakash and Arora 2019; Wang et al. 2020a, b), and fungi belonging to the genera *Penicillium*, *Aspergillus*, *Trichoderma*, *Acremonium*, *Hymenella*, and *Glomus* (Antunes et al. 2007; Busato et al. 2021; Gaind and Nain 2015; Rawat et al. 2020), actinomycetes like *Streptomyces* and *Micromonospora* (Hamdali et al. 2008; Sadeghi et al. 2012) (Table 5.1). Chen et al. (2006) reported that phosphate solubilizing bacteria make up 1–50% of the total population in soil, while fungi account for 0.1–0.5% (Fig. 5.3).

Conversely, apart from their capacity to solubilize phosphates, P-solubilizing microorganisms have diversified functions. They are known to boost plant growth through the production of growth-promoting metabolites. Multiple studies have explored the growth-promoting effects of phosphate solubilizing microorganisms in plants under various conditions. These microorganisms can synthesize growth-promoting hormones like auxins, cytokinins, and gibberellins that facilitate cell differentiation, shoot growth, root production, and germination (Brijesh Singh et al. 2022; Gul et al. 2023; Rawat et al. 2020). Commonly, many soil microorganisms may secrete auxins, particularly indole acetic acid (IAA), indole butyric acid, or other similar compounds, through tryptophan metabolism. According to Astriani et al. (2020), the novel bacterium *Pseudomonas plecoglossicida* isolated from soybean rhizosphere solubilized P (75.39 mg/L) and likewise produced plant growth

Table 5.1 Phosphate-solubilizing microorganisms isolated and reported recently

Microorganisms	Type	Origin	Culture medium used	Organic acids	References
Pseudomonas fluorescens, P. putida, and P. taiwanensis, Stenotrophomonas maltophilia, and *Pantoea agglomerans*	Bacteria	Soil samples from five Tunisian regions	Pikovskaya (PVK) and NBRIP with TCP sole source of phosphorus	–	Amri et al. (2023)
Bacillus subtilis	Bacteria	Rhizosphere of Eragrostis cynosuroides	PVK with TCP sole source of phosphorus	–	Kumari et al. (2023)
Pseudomonas migulae, pseudomonas frederiksbergensis, Aeromonas media, and *Pseudomonas vancouverensis*	Bacteria	The Yellow River wetland nature reserve in Luoyang city, China	NBRIP with TCP sole source of phosphorus	–	Sun et al. (2023)
Burkholderia, Pseudomonas, Bacillus, Enterobacter, Pantoea, Paraburkholderia, Cronobacter, Ralstonia, Curtobacterium, and Massilia	Bacteria	Soil samples collected from the dry regions of Tharaka Nith, Embu counties in Eastern Kenya	PVK	–	Kirui et al. (2022)
A. hydei, Gongronella hydei, Penicillium soli, Talaromyces yunnanensis	Fungi	Air and soil in Yunnan, China	PVK with TCP sole source of phosphorus	–	Doilom et al. (2020)
Pseudomonas sp., *Serratia* sp.	Bacteria	Soil, Colombia	SMRS1 agar modified with phosphate rock	–	Blanco-Vargas et al. (2020)

(continued)

Table 5.1 (continued)

Microorganisms	Type	Origin	Culture medium used	Organic acids	References
Pantoea agglomerans, Pseudomonas brassicacearum subsp. neoaurantiaca, Pantoea stewartii subsp. indologenes, Serratia rubidaea, Pseudomonas lactis	Bacteria	Phosphate solid sludge Morocco	NBRIP with TCP as sole source of phosphorus	–	Aliyat et al. (2020)
Trichoderma sp.	Fungi	Amazonian soils, Brazil	TSM medium	Citric acid, phytic acid, lactic acid, fumaric acid, ascorbic acid, isocitric acid, malic acid	Bononi et al. (2020)
Burkholderia, Klebsiella, Pseudomonas, Achromobacter, Sphingobacterium	Bacteria	Rhizosphere soil of Jerusalem artichoke, Thailand	Pikovskaya's agar (PKV) medium containing 5% tricalcium phosphate	Tartaric acid, oxalic acid, gluconic acid, lactic acid, acetic acid, malic acid	Nacoon et al. (2020)
Penicillium oxalicum, Aspergillus niger	Fungi			Oxalic acid, formic acid, tartaric acid, malic acid, citric acid, acetic acid	
Penicillium oxalicum FJG21, Penicillium oxalicum FJQ5	Fungi	Soil of corn farm, Shanghai, China	Pikovskaya's agar (PKV) medium containing 0.5% tricalcium phosphate	–	Wang et al. (2020b)
Bacillus subtilis BPM12	Bacteria	Soil of corn farm, Shanghai, China	Pikovskaya's agar (PKV) medium containing 0.5% tricalcium phosphate	–	Wang et al. (2020b)

(continued)

Table 5.1 (continued)

Microorganisms	Type	Origin	Culture medium used	Organic acids	References
Enterobacter cloacae	Bacteria	Soil and rhizosphere of roots of wheat, cowpea, maize, and trifolium plants growing in fields, Sharkia, Daqahlia, and Giza districts, Egypt	NBRIP with phosphate rock as sole source of phosphorus	–	El-Saied et al. (2020)
Bacillus licheniformis QA1 and Enterobacter asburiae QF11	Bacteria	Rhizospheric soil of *Chenopodium quinoa* Willd, Benguerir, Morocco	NBRIP with phosphate rock as sole source of phosphorus	–	Mahdi et al. (2020)
Pseudomonas sp. *PHR6*	Bacteria	Soil rhizosphere of the native plant Peganum Harmala, Tunisia	Pikovskaya's agar (PKV) medium containing 5% tricalcium phosphate	–	Daly et al. (2020)
Serratia plymuthica BMA1	Bacteria	Rock phosphate, Tunisia	NBRIP with phosphate rock as sole source of phosphorus	Gluconic acid and lactic acid	Borgi et al. (2020)
Bacillus spp.	Bacteria	Rhizosphere of *Tamarix ramosissima*, Mexico	Pikovskaya's agar (PKV) medium containing 5% tricalcium phosphate	–	Abdelmoteleb and Gonzalez-Mendoza (2020)

(continued)

Table 5.1 (continued)

Microorganisms	Type	Origin	Culture medium used	Organic acids	References
Pantoea agglomerans	Bacteria	Rhizospheric soil of Araucaria growing in Qingshan River Beach, Wuhan, Hubei, China	NBRIP with phosphate rock as sole source of phosphorus	–	Li et al. (2020)
Bacillus, Burkholderia, Paenibacillus sp.	Bacteria	Soil, Mizoram, India	Pikovskaya's agar (PKV) medium containing 5% tricalcium phosphate	Gluconic oxalic, citric, tartaric, succinic, formic, and acetic acid	Chawngthu et al. (2020)
Pseudomonas Sp. *DN 13-01, Sphingobacterium suaedae T47, Bacillus pimilus X22, and Bacillus cereus 263AG5*	Bacteria	Phosphate sludge, Merah Lahrach, Morocco	NBRIP with tricalcium phosphate as sole source of phosphorus	–	(Benbrik et al. 2020)
Pantoea agglomerans NCTC 9381, *Pantoea vagans* LMG 24199, *Pseudomonas azotoformans NBRC 12693, Enterobacter ludwigii* EN-119, *Serratia quinivorans* 4364	Bacteria	Agriculture soil Morocco	NBRIP with phosphate rock as sole source of phosphorus	Citric acid, fumaric acid, oxalic acid, acetic acid, succinic acid, gluconic acid	Rfaki et al. (2019)
Bacillus megaterium YLYP1, Pseudomonas prosekii YLYP6, and Pseudomonas sp. *YLYP29*	Bacteria	Sewage sludge, China	NBRIP and Pikovskaya's medium	dl-2-aminocaprylic acid, benzenecarboxylic acid, 2-acetylbenzoic acid, N-methylpyrrole-2-carboxylic acid, and gluconic acid	Yu et al. (2019)

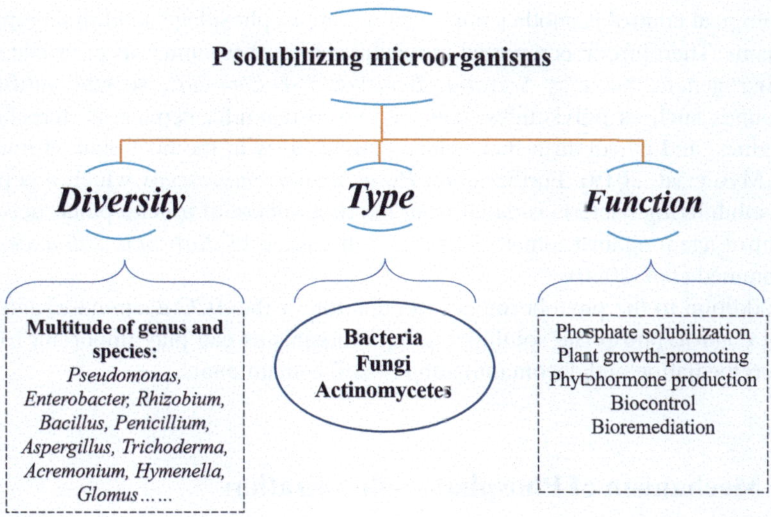

Fig. 5.3 Phosphate solubilizing microorganisms: type, diversity, and functions

hormones like IAA with 38.89 ppm. As well, Bader et al. (2020) reported in their study that *Trichoderma harzianum* solubilizes P up to 288.18 μg/ml. Moreover, a quantity of 21.14 μg/mL of IAA was generated, and upon evaluation as a biofertilizer for *Solanum lycopersicum* L., an enhancement in shoot and root biomass, leaf count, and leaf area was observed. Furthermore, *Aspergillus niger* CSR3, an endophytic fungus isolated from *Cannabis sativa*, was evaluated for its growth-promoting capabilities including phosphate solubilization and the generation of IAA and gibberellins. Additionally, its capacity to promote the growth of mutant Waito-C rice was assessed. The results indicated that treatment with *A. niger* CSR3 significantly enhanced almost all plant growth traits examined, including root–shoot length, biomass, and chlorophyll content (Lubna et al. 2018). *Enterobacter ludwigii* GAK2 is also reported to solubilize P and produce IAA and gibberellic acid (GA1, GA3). The strain inoculation promoted seed germination in a gibberellin deficient rice mutant Waito-C and rice cultivar "Hwayoungbyeo" and increased shoot length, root length, biomass, and chlorophyll content of rice plants (Lee et al. 2019). Numerous microorganisms have also been documented that can also solubilize P and produce phytohormones.

Phosphate solubilizing microorganisms are characterized by their ability to produce the ACC deaminase enzyme, which plays a crucial role in reducing the levels of ethylene in plants during stress conditions. When exposed to environmental stress, plants tend to produce higher concentrations of ethylene, which can hinder their growth. However, the ACC deaminase enzyme works to break down ACC (a precursor to ethylene) into α-ketobutyrate and ammonia, thereby reducing the levels of ethylene in plants. This process promotes the growth and development of plants even in stressful conditions (Ali and Kim 2018; Ji et al. 2020; Misra and Chauhan 2020; Rawat et al. 2020).

Biological control is another important feature of phosphate solubilizing microorganisms. Therefore, it is reported in several studies that numerous phosphate solubilizing genera, such as *Serratia*, *Bacillus*, *Pseudomonas*, secrete antifungal compounds such as polyketides, daucans, viscosinamide, peptaibols, terpenoids, phenazines, and pyrrolnitrin that reduce pathogenesis in the host plant (Liu et al. 2017; Myo et al. 2019). Furthermore, *Pseudomonas fluorescens* which is a phosphate solubilizing bacteria isolated from the rhizosphere of tomato plants acts as a biocontrol agent against tomato bacterial wilt caused by *Ralstonia solanacearum* (Mohammed et al. 2020).

In addition to the phytohormones, production of the ACC deaminase, and biological control, phosphate solubilizing microorganisms can play important role in the bioremediation and decontamination of soil contaminants.

5.5 Mechanism of Phosphate Solubilization

Phosphate solubilization microorganisms apply numerous approaches to make P available for plants to absorb (Fig. 5.4). The primary mechanisms employed by soil microorganisms for phosphate solubilization include the generation of complex or mineral-dissolving compounds, such as organic acid anions, siderophores, protons, and extracellular enzymes, as well as the liberation of P via substrate degradation.

5.5.1 *Organic Phosphate Solubilization*

Organic phosphate form may constitute 30–50% of the total P in most soils (Pradhan et al. 2017). Organic phosphate solubilization is often referred to as organic phosphorus mineralization. The process of soil organic phosphate mineralization serves

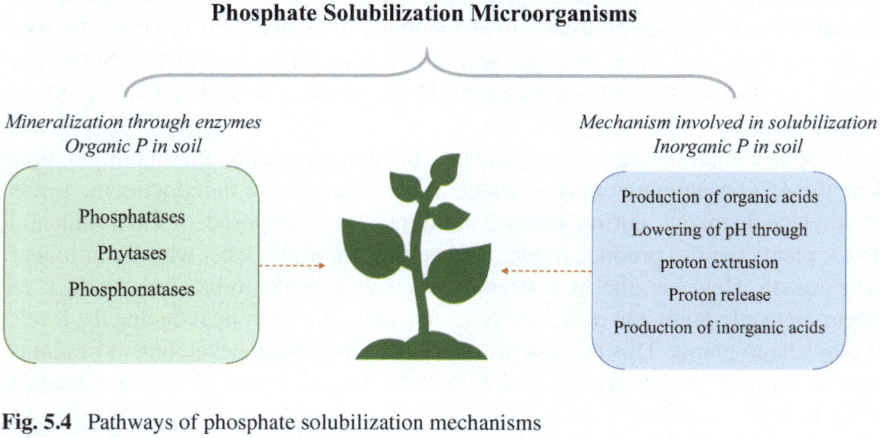

Fig. 5.4 Pathways of phosphate solubilization mechanisms

a crucial role in the agricultural P cycle, and it can account for anywhere from 4% to 90% of the total soil P (Xie et al. 2023).

Various types of soil organic phosphate compounds include nucleic acids, phospholipids, phytic acids, phosphonates, polyphosphonates, and sugar phosphates. The process of organic phosphate solubilization occurs in soil via the action of microbial enzymes such as phytase, phosphatase, phosphohydrolase, phosphonatase, and C-P lyase (Prabhu et al. 2019).

By hydrolyzing ester phosphate bonds and releasing phosphate ions, phosphatase enzymes change high molecular weight organic phosphate into low molecular weight molecules. These enzymes are classified as acidic, neutral, and alkaline phosphatases based on their maximum pH (Prabhu et al. 2019). According to Maougal et al. (2014) and Yadav and Tarafdar (2003), the mineralization process is mediated especially by the phytase and phosphatase enzymes released by the soil microorganisms. Moreover, Ahmed and Shahab (2011) suggested that the main activity is evidently the function of acid phosphatases and phytases, since their substrates are primarily found in the soil.

Phytic acid or myo-inositol phosphate compounds are hydrolyzed by the enzyme phytase. Meanwhile, phosphonatase and C-P lyase break down the ester bonds of phosphonates, such as phosphoenolpyruvate and phosphonoacetate, transforming them into hydrocarbons and phosphate ions that can be assimilated (Wanner 1996).

5.5.2 *Inorganic Phosphate Solubilization*

Microorganisms primarily solubilize inorganic phosphate by producing organic acids. This results in several outcomes, including a decrease in pH, an increase in chelation of cations bound to P, competition with P for soil adsorption sites, and the creation of soluble complexes with metal ions (such as Ca, Al, and Fe) associated with insoluble P. Consequently, this leads to the release of P (Sharma et al. 2013). Various microorganisms have been found to have the ability to solubilize insoluble inorganic phosphate compounds, such as tricalcium phosphate, dicalcium phosphate, hydroxyapatite, and rock phosphate. This has been demonstrated in several studies (Kumar et al. 2023; Song et al. 2008; Tawaraya et al. 2006).

According to Chen et al. (2006), phosphate-solubilizing microorganisms signal the production of organic acids via the direct oxidation pathway, which occurs on the outer surface of the cytoplasmic membrane, when the pH of the medium decreases. The acids produced by microbial metabolism are primarily the result of oxidative respiration or the fermentation of organic carbon sources (Alam et al. 2023; Trolove et al. 2003). Thus, these organic acids may directly dissolve the mineral phosphate as a result of an exchange of an anion of phosphate by an anion of the acid or by chelation of the Ca, Fe, Al and ions associated with the phosphate (de Oliveira Mendes et al. 2020; Rashid et al. 2004). Different organic acids such as gluconic acid, 2-ketogluconic acid, lactic acid, isovaleric acid, isobutyric acid, acetic acid, oxalic acid, and citric acid have been reported to be produced by soil microorganisms (de Oliveira Mendes et al. 2020; Patel and Goswami 2020; Rashid et al. 2004) and known for their activity of solubilizing insoluble phosphate.

5.6 Phosphate-Solubilizing Enzymes

5.6.1 Phytase

Most of the phytases (myo-inositol hexakisphosphate phosphohydrolases) are high molecular weight phosphatases. In its fundamental form, phytate serves as the main source of inositol and is the most common form in which P is stored in plant seeds and pollen (Wodzinski and Ullah 1996). Phytate degradation can also facilitate the solubilization of organic phosphate in soil. However, this application of the enzymes is not currently being utilized. It is worth noting that phytate is the primary component of soil organic P (Hayes et al. 2000; Silva et al. 2023).

A variety of soil microorganisms are known to generate phytase enzymes similar to those in the plant root system. Nowadays fungus, yeast, and some of the representatives of *Bacillus* and *Enterobacter* are considered to generate the most dynamic extracellular phytases (Kaur 2020). Phytases developed by *Bacillus* and *Enterobacter* have a strong substrate specific activity, a pH optimum in the range 3.5–7.5, and an optimum temperature of 37–70 °C (Gontia et al. 2012; Lei and Stahl 2001). Some alkaline *Bacillus* phytases are limited only to phosphomonoester bonds in phytate molecules due to their small substrate specificity (Kaur 2020).

The plants have very little potential to get P from phytate directly. However, according to Richardson (2001) when converted with the gene phytase (*phy*A) from *A. niger*, the growth and P nutrition of *Arabidopsis* plants supplied with phytate are substantially improved. This contributed to better P nutrition to the degree that plant growth and P content were equal to those of the regulated plants with inorganic phosphate.

Numerous reports exist regarding the thermally stable phytase genes (phy) produced by *Bacillus* sp. DS11 and other microorganisms capable of producing phytase (Kerovuo et al. 1998; Kim et al. 1998).

5.6.2 Phosphatases

Phosphatases function as scavengers of organic phosphoesters, which are crucial nutrients for the cell, by removing inorganic phosphates from nucleotides and sugar phosphates. Interest in these enzymes has increased over the last decade due to their potential biotechnological applications (Lu et al. 2023; Rodriguez-Viciana et al. 2006). Gram-negative bacteria have been reported for various acid phosphatase genes that facilitate the hydrolysis of various phosphate systems such as primary, secondary, cyclic, sugar, and phenols. These enzymes act as catalysts in the hydrolysis process (Rossolini et al. 1998).

5.7 Phosphate Solubilization Microorganisms Isolation: Methods and Approaches

Isolation of phosphate solubilizing microorganisms from various matrixes, typically the rhizosphere of plants, is carried out in vitro on the basis of a culture media without a supply of soluble phosphorus but supplied in its insoluble form.

Many media have been proposed for the solubilization of inorganic phosphate since Pikovskaya's initial study in 1948. These include Pikovskaya's agar, as well as other media like National Botanical Research Institute Phosphate (NBRIP) medium and Basal Sperber Medium (Nautiyal 1999; Pikovskaya 1948; Sperber 1958). Various sources of insoluble inorganic P, such as tricalcium phosphate, aluminum phosphate, iron phosphate, and zinc phosphate, are utilized (Aliyat et al. 2022; Bashan et al. 2013; Gadagi and Sa 2002; Di Simine et al. 1998). Isolation of phosphate solubilizing microorganisms is achieved by a plate screening process by visual identification of clear zones around microorganisms in media with insoluble mineral phosphate.

Microorganisms that are capable of solubilizing phosphate can be obtained through the use of enrichment culture techniques or serial dilutions from a variety of sources, including non-rhizosphere and rhizosphere soils, the rhizoplane, as well as other environments like soil from rock phosphate deposit areas and marine environments (Chawngthu et al. 2020; Gaur and Rana 1990). The first step in identifying effective phosphate-solubilizing species is to verify their ability to solubilize insoluble phosphate in a liquid culture media as a quantitative test. Once this is established, the most potent microorganisms can be selected for the development of inoculants, and their efficacy can be evaluated under greenhouse and field conditions across a range of crops.

5.8 Biotechnological Applications of Phosphate-Solubilizing Microorganisms: Biofertilizers and Sustainable Agriculture

5.8.1 Biofertilizers

The agricultural industry is tasked with the responsibility of providing sufficient amounts of healthy food to meet the needs of a growing population while safeguarding and conserving environmental resources, such as water, soil, animal, and plant biodiversity. To achieve this, farmers must acquire new knowledge, methods, and tools that allow them to produce high-quality food in large quantities while minimizing the use of chemical inputs like fertilizers and harmful phytosanitary products that endanger the environment and human health. One way to tackle this challenge is by designing cropping systems that enable the transition from agrochemicals to agroecology. Biofertilizers, which are derived from natural processes and products, offer a promising alternative for producers who must reconnect with innovative traditional agroecological practices and techniques.

Biofertilizers are microorganisms that improve the soil's nutritional quality (Ahmed and Shahab 2011; Arora et al. 2020; Iqbal et al. 2023). Bacteria, fungi, and actinomycetes are the dominant sources of biofertilizers (Pradhan et al. 2017). Plants have several associations with fungi, bacteria, and actinomycetes. Furthermore, biofertilizers can also be called "biostimulants," "growth and/or development stimulators," "soil activators," or even "phytostimulants." According to the definition adopted by the EBIC (European Biostimulants Industry Council), a biofertilizer is "a material that contains a substance and/or microorganisms whose function, when applied to plants or the rhizosphere, is to stimulate natural processes to improve and benefit nutrient uptake, nutrient efficiency, abiotic stress tolerance, and crop quality, regardless of the nutrient content of the biostimulant" (European Biostimulants Industry Council). This approach includes a wide range of microorganisms such as, for example, rhizobacteria, cyanobacteria, and fungi. These are useful for growth, yield, assimilation of nutrients, and defense of plants against various stresses.

Farmers have been satisfied with increased agricultural yields since the development of chemical fertilizers in the last century. But eventually chemical fertilizers begin to exhibit their ill-effects, such as leaching out, pollute the soil and rivers, damaging biodiversity, rendering the crop more vulnerable to disease assaults, decreasing soil productivity, and thereby creating irreparable harm to the ecosystems (Rodríguez and Fraga 1999). Biofertilizers are microbiologically active, low-cost chemicals added to soils, promoting the development of plants (Pathak and Kumar 2016). Several field studies have confirmed the performance of phosphate solubilizing microorganisms as biofertilizers (Maharana and Dhal 2022; Sharma et al. 2023).

The usage of microorganisms as biofertilizers has many benefits over synthetic chemicals, they are estimated cleaner than various of the chemicals already in use; they do not persist in the food chain; the target species rarely establish tolerance like they do when chemical agents are used; and biofertilizers are not considered detrimental to biological systems or the ecosystems (Pathak and Kumar 2016; Rodríguez and Fraga 1999).

The phosphate solubilizing microorganism's application in agronomic activities is promoted for a variety of purposes. For example, they enhance soil productivity through their continuous soil operations, improve crop productivity by improved availability of nutrients, do not trigger environmental contamination, protect plants against certain soil-borne pathogens, boost soil heath and fertility, and require low-cost technologies for their processing with a strong cost–benefit ratio (Khan et al. 2009).

The term "sustainable agriculture" refers to agricultural systems that are able to maintain their productivity and social benefits over the long term while conserving resources, providing economic support, being financially efficient, and environmentally sound. The study of rhizosphere microbial organisms has garnered considerable attention in organic farming and biosafety. A major area of focus in the future will be identifying safe and eco-friendly methods for utilizing beneficial microorganisms in sustainable agriculture.

Therefore, sustainable agriculture aims to minimize the adverse impacts on both direct and off-farm environments while maintaining sustainable production and profit levels. In developing countries around the world, and especially among poor farmers, soil fertility remains the primary limitation on crop yields. Consequently, in such situations, farmers are unlikely to adopt improved crop varieties or implement more sophisticated agricultural practices until soil fertility has been adequately restored in those regions (Ricci et al. 2019). In general, crop yields may be improved by the use of phosphate fertilizers. Nevertheless, international oil scarcity has boosted the prices of chemical fertilizers. Rising food production without losing the cost–benefit ratio thus poses scientists around the globe with a considerable task. Agronomists are already aggressively exploring alternative phosphatic fertilizer sources in order to augment and in certain cases eliminate conventional fertilizers with the goal of maintaining competitive crop yields. The utilization of multiple microbial processes occurring at the interface between the soil and plant roots, known as the rhizosphere, presents an alternative to chemical phosphatic fertilizers. There are several reasons for the usage of agricultural practices involving phosphate-solubilizing organisms. For example, for improving the efficiency of the soil through the continued microbial activity, improving plant growth and crop yield by better nutrient supply, not causing environmental pollution, boosting soil heat and cooling, protecting plants from pathogens, and low price-benefits technologies.

5.8.2 Formulation

Over time, the formulation of biofertilizers has evolved. Initially, they were developed as inoculants with inert materials (such as vermiculite, perlite, rock phosphate, and calcium sulfate) or as powders (including clay, peat coal, and soil), and from plant waste materials (such as farmyard manure, compost, soybean meal, agricultural waste, byproducts of sugar industry, and spent mushroom compost), as well as plain lyophilized bacterial cultures. The carrier-based formulation is the most extensively studied type of biofertilizer formulation (Kumar et al. 2022; Prabhu et al. 2019).

A variety of industrial formulations of biofertilizers are currently available and to ensure the optimal viability of microorganisms used in biofertilizer formulations, various techniques have been employed. These include optimizing the biofertilizer formulation, utilizing stress-tolerant and genetically engineered strains, and employing liquid biofertilizers for ease of use. Carriers are utilized are used in formulations for exploiting microorganisms as biofertilizers, and such materials could additionally contribute to microorganisms' viability prior to release into the field. Furthermore, after their release, carrier can provide an adequate micro-environment for the rapid growth of organisms (Brar et al. 2012).

Liquid biofertilizer formulations could be viewed as feasible strategy for increasing biofertilizer longevity. When compared to solid carrier-based biofertilizers, liquid formulations allow the manufacturer to add sufficient amounts of nutrients, cell defenders, and stimulators responsible for cell, spore, or cyst production to ensure long shelf-life. Traditional carrier-based biofertilizers have a storage life of around 6 months; however, liquid formulations can have an expiration date of up to 2 years (Brar et al. 2012; Kumar et al. 2022; Mahdi et al. 2010).

Furthermore, solid carrier-based biofertilizers are less thermo-tolerant, while liquid formulations can withstand temperatures of up to 55 °C. Excellent shelf-life may then be accomplished by improving the composition of a liquid biofertilizer. However, the processing expense of a liquid biofertilizer is considerably greater than that of a solid formulation.

5.9 Future Prospects

Phosphate-solubilizing microorganisms have a promising future in agriculture due to their ability to solubilize phosphate, and they can make a significant impact on sustainable agriculture. They can be used to develop environmentally friendly fertilizers that can increase crop yields while reducing the use of chemical fertilizers. Additionally, phosphate-solubilizing microorganisms can be used for crop protection that are less harmful to the environment. Also, phosphate-solubilizing microorganisms can improve soil health by promoting the growth of beneficial microorganisms that can help fix nitrogen and other nutrients in the soil. This can improve soil fertility and reduce the need for chemical fertilizers. Besides bioremediation, they can be used to clean up contaminated soils by solubilizing phosphate and other nutrients that can help remediate heavy metals and other pollutants. Overall, phosphate-solubilizing microorganisms have the potential to play an important role in sustainable agriculture and help address many of the challenges the industry is facing.

Each year hundreds of research papers are published on phosphate-solubilizing microorganisms, and recently several researchers are working on the formulations, which is one of the difficult steps in the production of biofertilizer (Faridian et al. 2023; Mayadunna et al. 2023), Also, the combination of different microorganisms in a single product is a challenge (Ain and Noraini 2023; Smirnova et al. 2023).

There are still several bottlenecks that need to be addressed for the successful implementation of phosphate-solubilizing microorganisms in agriculture. One major challenge is the lack of effective and affordable production technologies for large-scale application of phosphate-solubilizing microorganisms. Another challenge is the need for better understanding of the interactions between phosphate-solubilizing microorganisms and plants, as well as the mechanisms underlying the solubilization of phosphorus in soil.

To overcome these challenges, future research on phosphate-solubilizing microorganisms should focus on developing cost-effective and scalable production technologies, and improving our understanding of the complex interactions between phosphate-solubilizing microorganisms, plants, and soil. Additionally, there is a need for more field studies to evaluate the effectiveness of phosphate-solubilizing microorganisms in different soil and environmental conditions and to optimize their use in different agricultural systems. With continued research and development, phosphate-solubilizing microorganisms hold great promise as a sustainable solution for improving soil fertility and promoting food security.

Moreover, the use of phosphate-solubilizing microorganisms in the biofertilizer industry has gained increasing attention in recent years. The biofertilizer industry has started incorporating phosphate-solubilizing microorganisms into their products, and several commercial biofertilizers containing phosphate-solubilizing microorganisms are now available in the market (Joshi and Gauraha 2022). However, the development of efficient and cost-effective production technologies for phosphate-solubilizing microorganisms-based biofertilizers remains a challenge. The use of phosphate-solubilizing microorganisms in the biofertilizer industry has significant potential to promote sustainable agriculture and reduce the environmental impact of conventional fertilizers, but further research and development are needed to fully realize this potential.

5.10 Conclusion

To summarize, the role of phosphate solubilizing microorganisms in plant nutrition is significant as they improve the availability of phosphate by solubilizing insoluble inorganic forms. Inoculating soil with these microorganisms is an effective way of converting insoluble P into available P for plant nutrition, leading to improved plant growth, crop yield, and quality. These microorganisms promote plant growth by supplying readily absorbable P and generating phytohormones, and also act as biocontrol agents against plant pathogens by producing antifungal metabolites. Although phosphate solubilizing microorganisms have great potential as biostimulants for sustainable agriculture, their widespread use is constrained by limited understanding of the biological ecology and population dynamics in soil. Additionally, achieving food security requires the implementation of sustainable agricultural production structures that promote grain yields and income generation. In this context, novel biotechnology techniques must be developed, tested, and transferred to farmers for their eventual adoption in the fields.

Acknowledgments The authors would like to acknowledge the support through the R&D Initiative—Appel à projets autour des phosphates APPHOS—sponsored by OCP (OCP Foundation, R&D OCP, Mohammed VI Polytechnic University, National Center of Scientific and Technical

Research CNRST, Ministry of Higher Education, Scientific Research and Professional Training of Morocco MESRSFC) under the project entitled * Valorisation des boues solides des phosphates en arboriculture fruitière et foresterie *, project ID *BIO-ELG 01/2017*.

References

Abdelmoteleb A, Gonzalez-Mendoza D (2020) Isolation and identification of phosphate solubilizing *Bacillus* spp. from Tamarix ramosissima rhizosphere and their effect on growth of *Phaseolus vulgaris* under salinity stress. Geomicrobiol J 37(10):901–908. https://doi.org/10.1080/01490451.2020.1795321

Ahmed N, Shahab S (2011) Phosphate solubilization: their mechanism genetics and application. Int J Microbiol 9:4408–4412

Ain AA, Noraini M (2023) Effects of rice husk biochar (RHB) with combined inoculation of arbuscular mycorrhizal fungi (AMF) and phosphate solubilizing bacteria (PSB) on growth of maize (*Zea mays*). IOP Conf Ser Earth Environ Sci 1131. IOP Publishing:012007

Alam K, Barman M, Datta SP, Annapurna K, Shukla L (2023) Modification of inorganic fractions of phosphorus by phosphate-Solubilising microorganisms in conjunction with phosphorus fertilisation in a tropical Inceptisol. J Soil Sci Plant Nutr 23(2):2488–2497. https://doi.org/10.1007/s42729-023-01206-6

Alaylar B, Egamberdieva D, Gulluce M, Karadayi M, Arora NK (2020) Integration of molecular tools in microbial phosphate solubilization research in agriculture perspective. World J Microbiol Biotechnol 36(7):93. https://doi.org/10.1007/s11274-020-02870-x

Ali S, Kim W-C (2018) Plant growth promotion under water: decrease of waterlogging-induced ACC and ethylene levels by ACC deaminase-producing bacteria. Front Microbiol 9. https://doi.org/10.3389/fmicb.2018.01096

Aliyat FZ, Maldani M, El Guilli M, Nassiri L, Ibijbijen J (2020) Isolation and characterization of phosphate solubilizing bacteria from phosphate solid sludge of the Moroccan phosphate mines. Open Agric J 14(1):16–24

Aliyat FZ, Maldani M, El Guilli M, Nassiri L, Ibijbijen J (2022) Phosphate-solubilizing bacteria isolated from phosphate solid sludge and their ability to solubilize three inorganic phosphate forms: calcium, iron, and aluminum phosphates. Microorganisms 10(5):980

Amri M, Rjeibi MR, Gatrouni M et al (2023) Isolation, identification, and characterization of phosphate-solubilizing bacteria from Tunisian soils. Microorganisms 11(3):783. https://www.mdpi.com/2076-2607/11/3/783

Antunes PM, Schneider K, Hillis D, Klironomos JN (2007) Can the arbuscular mycorrhizal fungus Glomus intraradices actively mobilize P from rock phosphates? Pedobiologia 51(4):281–286. https://doi.org/10.1016/j.pedobi.2007.04.007

Arora NK, Fatima T, Mishra I, Verma S (2020) Microbe-based inoculants: role in next green revolution. In Environmental Concerns and Sustainable Development 2.: Biodiversity, Soil and Waste Management (Eds Shukla V and Kumar N). Springer, Singapore pp 191–246

Astriani M, Zubaidah S, Abadi AL, Suarsini E (2020) Pseudomonas plecoglossicida as a novel bacterium for phosphate solubilizing and indole-3-acetic acid-producing from soybean rhizospheric soils of East Java, Indonesia. Biodivers J Biol Divers 21:2

Bader AN, Salerno GL, Covacevich F, Consolo VF (2020) Native Trichoderma harzianum strains from Argentina produce indole-3 acetic acid and phosphorus solubilization, promote growth and control wilt disease on tomato (*Solanum lycopersicum* L.). J King Saud Univ Sci 32(1):867–873

Barea J-M, Richardson AE (2015) Phosphate mobilisation by soil microorganisms. In: Lugtenberg B (ed) Principles of plant-microbe interactions: microbes for sustainable agriculture. Springer International Publishing, Cham, pp 225–234. https://doi.org/10.1007/978-3-319-08575-3_24

Bashan Y, Kamnev AA, de-Bashan LE (2013) Tricalcium phosphate is inappropriate as a universal selection factor for isolating and testing phosphate-solubilizing bacteria that enhance plant growth: a proposal for an alternative procedure. Biol Fertil Soils 49(4):465–479. https://doi.org/10.1007/s00374-012-0737-7

Behera BC, Yadav H, Singh SK et al (2017) Phosphate solubilization and acid phosphatase activity of Serratia sp. isolated from mangrove soil of Mahanadi river delta, Odisha, India. J Genet Eng Biotechnol 15(1):169–178. https://doi.org/10.1016/j.jgeb.2017.01.003

Beltran-Medina I, Romero-Perdomo F, Molano-Chavez L, Gutiérrez AY, Silva AMM, Estrada-Bonilla G (2023) Inoculation of phosphate-solubilizing bacteria improves soil phosphorus mobilization and maize productivity. Nutr Cycl Agroecosyst 126(1):21–34. https://doi.org/10.1007/s10705-023-10268-y

Benbrik B, Elabed A, El Modafar C, Douira A, Amir S, Filali-Maltouf A, ... & Koraichi SI (2020) Reusing phosphate sludge enriched by phosphate solubilizing bacteria as biofertilizer: Growth promotion of *Zea Mays*. Biocatalysis Agricultural Biotechnol 30, 101825

Blanco-Vargas A, Rodríguez-Gacha LM, Sánchez-Castro N et al (2020) Phosphate-solubilizing *pseudomonas* sp., and *Serratia* sp., co-culture for *Allium cepa* L. growth promotion. Heliyon 6(10):e05218

Bononi L, Chiaramonte JB, Pansa CC, Moitinho MA, Melo IS (2020) Phosphorus-solubilizing *Trichoderma* spp. from Amazon soils improve soybean plant growth. Sci Rep 10(1):2858. https://doi.org/10.1038/s41598-020-59793-8

Borgi MA, Saidi I, Moula A, Rhimi S, Rhimi M (2020) The attractive *Serratia plymuthica* BMA1 strain with high rock phosphate-solubilizing activity and its effect on the growth and phosphorus uptake by *Vicia faba* L. Plants. Geomicrobiol J 37(5):437–445. https://doi.org/10.1080/01490451.2020.1716892

Brar SK, Sarma SJ, Chaabouni E (2012) Shelf-life of biofertilizers: an accord between formulations and genetics. J Biofertil Biopestici 3(5):1–2

Brijesh Singh S, Murali M, Gowtham HG et al (2022) Plant growth-promoting Rhizobacteria for sustainable agriculture. In: Nayak SK, Baliyarsingh B, Singh A, Mannazzu I, Mishra BB (eds) Advances in agricultural and industrial microbiology: Volume-2: applications of microbes for sustainable agriculture and in-silico strategies. Springer Nature Singapore, Singapore, pp 1–27. https://doi.org/10.1007/978-981-16-9682-4_1

Busato JG, Ferrari LH, Chagas Junior AF, da Silva DB, dos Santos PT, de Paula AM (2021) *Trichoderma* strains accelerate maturation and increase available phosphorus during vermicomposting enriched with rock phosphate. J Appl Microbiol 130(4):1208–1216. https://doi.org/10.1111/jam.14847

Chawngthu L, Hnamte R, Lalfakzuala R (2020) Isolation and characterization of Rhizospheric phosphate solubilizing bacteria from wetland Paddy field of Mizoram, India. Geomicrobiol J 37(4):366–375. https://doi.org/10.1080/01490451.2019.1709108

Chen YP, Rekha PD, Arun AB, Shen FT, Lai WA, Young CC (2006) Phosphate solubilizing bacteria from subtropical soil and their tricalcium phosphate solubilizing abilities. Appl Soil Ecol 34(1):33–41. https://doi.org/10.1016/j.apsoil.2005.12.002

Daly I, Jellali S, Mehri I et al (2020) Phosphorus and ammonium removal characteristics from aqueous solutions by a newly isolated plant growth-promoting bacterium. Environ Technol 41(20):2603–2617. https://doi.org/10.1080/09593330.2019.1575917

de Oliveira Mendes G et al (2020) Oxalic acid is more efficient than sulfuric acid for rock phosphate Solubilization. Miner Eng 155:106458

Di Simine CD, Sayer JA, Gadd GM (1998) Solubilization of zinc phosphate by a strain of *Pseudomonas fluorescens* isolated from a forest soil. Biol Fertil Soils 28:87–94

Dissanayaka DMSB, Plaxton WC, Lambers H, Siebers M, Marambe B, Wasaki J (2018) Molecular mechanisms underpinning phosphorus-use efficiency in rice. Plant Cell Environ 41(7):1483–1496. https://doi.org/10.1111/pce.13191

Doilom DM, Guo J-W, Phookamsak R et al (2020) Screening of phosphate-solubilizing fungi from air and soil in Yunnan, China: four novel species in Aspergillus, Gongronella, Penicillium, and Talaromyces. Front Microbiol 11. https://doi.org/10.3389/fmicb.2020.585215

Ducousso-Détrez A, Fontaine J, Lounès-Hadj Sahraoui A, Hijri M (2022) Diversity of phosphate chemical forms in soils and their contributions on soil microbial community structure changes. Microorganisms 10(3):609. https://www.mdpi.com/2076-2607/10/3/609

El-Saied R, Abd Elhamed RS, Hassanein W (2020) Possibility of decreasing consumption of chemical fertilizers with using phosphorous and potassium solubilizing bacteria inoculation on fennel. Plant Archives 20(2):315–3169

Faridian L, Baharlouei J, Fallah Nosratabad A, Kari Dolat Abad H (2023) An exploratory research on the adoption of different phosphate-solubilizing fungi for production of phosphate biofertilizers. Geomicrobiol J 40(5):493–500. https://doi.org/10.1080/01490451.2023.2196288

Gadagi RS, Sa T (2002) New isolation method for microorganisms solubilizing iron and aluminum phosphates using dyes. Soil Sci Plant Nutr 48(4):615–618. https://doi.org/10.1080/00380768.2002.10409246

Gaind S, Nain L (2015) Soil–phosphorus mobilization potential of phytate mineralizing fungi. J Plant Nutr 38(14):2159–2175. https://doi.org/10.1080/01904167.2015.1014561

Gaur A, Rana J (1990) Effect of insecticides on wheat crop inoculated with phosphate solubilizing bacteria (PSB) and VAM fungi. In: Trends in mycorrhizal research. Proceedings of the National Conference on Mycorrhiza, held at Haryana Agricultural University, Hisar, India, Feb 14-16, 1990.

Goldstein A, Rogers R, Mead G (1993) Separating phosphate from ores via bioprocessing. Biotechnology 11:1250–1254

Gontia I, Tantwai K, Singh Rajput LP, Tiwari S (2012) Transgenic plants expressing phytase gene of microbial origin and their prospective application as feed. Food Technol Biotechnol 50(1):3–10

Gul N, Wani IA, Mir RA et al (2023) Plant growth promoting microorganisms mediated abiotic stress tolerance in crop plants: a critical appraisal. Plant Growth Regul 100(1):7–24. https://doi.org/10.1007/s10725-022-00951-5

Hamdali H, Hafidi M, Virolle MJ, Ouhdouch Y (2008) Rock phosphate-solubilizing actinomycetes: screening for plant growth-promoting activities. World J Microbiol Biotechnol 24(11):2565–2575

Hayes J, Richardson A, Simpson R (2000) Components of organic phosphorus in soil extracts that are hydrolysed by phytase and acid phosphatase. Biol Fertil Soils 32:279–286

Holford HICR (1997) Soil phosphorus: its measurement, and its uptake by plants. Soil Res 35(2):227–240. https://doi.org/10.1071/S96047

Iqbal A, Shafi MI, Rafique M, et al. (2023) Biofertilizers to improve soil health and crop yields. In Sustainable Agriculture Reviews 61: biochar to improve crop production and decrease plant stress under a changing climate (Eds Fahad S et al.). Springer, Switzerland, pp 61:247-272

Ji J, Yuan D, Jin C, Wang G, Li X, Guan C (2020) Enhancement of growth and salt tolerance of rice seedlings (Oryza sativa L.) by regulating ethylene production with a novel halotolerant PGPR strain Glutamicibacter sp. YD01 containing ACC deaminase activity. Acta Physiol Plant 42(4):42. https://doi.org/10.1007/s11738-020-3034-3

Joshi SK, Gauraha AK (2022) 24—Global biofertilizer market: emerging trends and opportunities. In: Soni R, Suyal DC, Yadav AN, Goel R (eds) Trends of applied microbiology for sustainable economy. Academic Press, pp 689–697. https://doi.org/10.1016/B978-0-323-91595-3.00024-0

Joshi E, Iyer B, Rajkumar S (2019) Glucose and arabinose dependent mineral phosphate solubilization and its succinate-mediated catabolite repression in rhizobium sp. RM and RS. J Biosci Bioeng 128(5):551–557. https://doi.org/10.1016/j.jbiosc.2019.04.020

Kafkafi U, Bar-Yosef B, Rosenberg R, Sposito G (1988) Phosphorus adsorption by kaolinite and montmorillonite: II. Organic anion competition. Soil Sci Soc Am J 52(6):1585–1589. https://doi.org/10.2136/sssaj1988.03615995005200060012x

Kalayu G (2019) Phosphate solubilizing microorganisms: promising approach as biofertilizers. Int J Agron 2019:4917256. https://doi.org/10.1155/2019/4917256

Kaur G (2020) Microbial phytases in plant minerals acquisition. In: Sharma V, Salwan R, Al-Ani LKT (eds) Molecular aspects of plant beneficial microbes in agriculture. Academic Press, pp 185–194 https://doi.org/10.1016/B978-0-12-818469-1.00016-X

Kaur R, Kaur S (2020) Variation in the phosphate solubilizing bacteria from virgin and the agricultural soils of Punjab. Curr Microbiol 77(9):2118–2127. https://doi.org/10.1007/s00284-020-02080-6

Kerovuo J, Lauraeus M, Nurminen P, Kalkkinen N, Apajalahti J (1998) Isolation, characterization, molecular gene cloning, and sequencing of a novel phytase from *Bacillus subtilis*. Am Soc Microbiol 64(6):2079–2085. https://doi.org/10.1128/AEM.64.6.2079-2085.1998

Khan MS, Zaidi A, Wani PA (2009) Role of phosphate solubilizing microorganisms in sustainable agriculture—a review. In: Lichtfouse E, Navarrete M, Debaeke P, Véronique S, Alberola C (eds) Sustainable agriculture. Springer Netherlands, Dordrecht, pp 551–570. https://doi.org/10.1007/978-90-481-2666-8_34

Kim Y-O, Kim H-K, Bae K-S, Yu J-H, Oh T-K (1998) Purification and properties of a thermostable phytase from *Bacillus* sp. DS11. Enzym Microb Technol 22(1):2–7. https://doi.org/10.1016/S0141-0229(97)00096-3

Kirui CK, Njeru EM, Runo S (2022) Diversity and phosphate solubilization efficiency of phosphate solubilizing bacteria isolated from semi-arid agroecosystems of eastern Kenya. Microbiol Insights 15:11786361221088991. https://doi.org/10.1177/11786361221088991

Kumar S, Diksha, Sindhu SS, Kumar R (2022) Biofertilizers: an ecofriendly technology for nutrient recycling and environmental sustainability. Curr Res Microb Sci 3:100094. https://doi.org/10.1016/j.crmicr.2021.100094

Kumar M, Shankar A, Chaudhary S, Prasad V (2023) Phosphate solubilizing microorganisms: multifarious applications. In: Chhabra S, Prasad R, Maddela NR, Tuteja N (eds) Plant microbiome for plant productivity and sustainable agriculture. Springer Nature Singapore, Singapore, pp 245–262. https://doi.org/10.1007/978-981-19-5029-2_10

Kumari S, Kumar P, Kiran S, Kumari S, Singh A (2023) Characterization of culture condition dependent, growth responses of phosphate solubilizing bacteria (*Bacillus subtilis* DR2) on plant growth promotion of *Hordeum vulgare*. Vegetos. https://doi.org/10.1007/s42535-023-00589-2

Lee K-E, Adhikari A, Kang S-M et al (2019) Isolation and characterization of the high silicate and phosphate solubilizing novel strain Enterobacter ludwigii GAK2 that promotes growth in rice plants. Agronomy 9(3):144. https://www.mdpi.com/2073-4395/9/3/144

Lei X, Stahl C (2001) Biotechnological development of effective phytases for mineral nutrition and environmental protection. Appl Microbiol Biotechnol 57(4):474–481. https://doi.org/10.1007/s002530100795

Li L, Chen R, Zuo Z et al (2020) Evaluation and improvement of phosphate solubilization by an isolated bacterium Pantoea agglomerans ZB. World J Microbiol Biotechnol 36(2):27. https://doi.org/10.1007/s11274-019-2744-4

Li X, Wen Q, Zhang S et al (2022) Long-term changes in organic and inorganic phosphorus compounds as affected by long-term synthetic fertilisers and pig manure in arable soils. Plant Soil 472(1):239–255. https://doi.org/10.1007/s11104-021-05225-z

Liu K, Newman M, McInroy JA, Hu C-H, Kloepper JW (2017) Selection and assessment of plant growth-promoting rhizobacteria for biological control of multiple plant diseases. Phytopathology 107(8):928–936. https://doi.org/10.1094/phyto-02-17-0051-r

Lobo CB, Juárez Tomás MS, Viruel E, Ferrero MA, Lucca ME (2019) Development of low-cost formulations of plant growth-promoting bacteria to be used as inoculants in beneficial agricultural technologies. Microbiol Res 219:12–25. https://doi.org/10.1016/j.micres.2018.10.012

Lu H, Wang F, Wang Y, Lin R, Wang Z, Mao C (2023) Molecular mechanisms and genetic improvement of low-phosphorus tolerance in rice. Plant Cell Environ 46(4):1104–1119. https://doi.org/10.1111/pce.14457

Lubna AS, Hamayun M, Gul H, Lee I-J, Hussain A (2018) Aspergillus Niger CSR3 regulates plant endogenous hormones and secondary metabolites by producing gibberellins and indoleacetic acid. J Plant Interact 13(1):100–111. https://doi.org/10.1080/17429145.2018.1436199

Maharana R, Dhal NK (2022) Solubilization of rock phosphate by phosphate solubilizing bacteria isolated from effluent treatment plant sludge of a fertilizer plant. Folia Microbiol 67(4):605–615. https://doi.org/10.1007/s12223-022-00953-w

Mahdi SS, Hassan G, Samoon S, Rather H, Dar SA, Zehra B (2010) Bio-fertilizers in organic agriculture. J Phytology 2(10):42–54

Mahdi I, Fahsi N, Hafidi M, Allaoui A, Biskri L (2020) Plant growth enhancement using rhizospheric halotolerant phosphate solubilizing bacterium bacillus licheniformis QA1 and Enterobacter asburiae QF11 isolated from Chenopodium quinoa Willd. Microorganisms 8(6):948. https://www.mdpi.com/2076-2607/8/6/948

Manzoor M, Abbasi MK, Sultan T (2017) Isolation of phosphate solubilizing bacteria from maize rhizosphere and their potential for rock phosphate solubilization–mineralization and plant growth promotion. Geomicrobiol J 34(1):81–95. https://doi.org/10.1080/01490451.2016.1146373

Maougal R, Brauman A, Plassard C, Abadie J, Djekoun A, Drevon J-J (2014) Bacterial capacities to mineralize phytate increase in the rhizosphere of nodulated common bean (Phaseolus vulgaris) under P deficiency. Eur J Soil Biol 62:8–14

Matter JM, Sampaio SC, Rosa DM, Remor MB, Model KJ, Sbizzaro M (2020) Isolation of phosphate-solubilizing bacteria from subtropical soils with different fertilization histories. Biosci J 36:4

Mayadunna N et al (2023) Isolation of phosphate-solubilizing microorganisms and the formulation of biofertilizer for sustainable processing of phosphate rock. Life 13(3):782

McGechan MB (2002) SW—soil and water: sorption of phosphorus by soil, part 2: measurement methods, results and model parameter values. Biosyst Eng 82(2):115–130

Mendoza-Arroyo GE et al (2020) Inorganic phosphate Solubilization by a novel isolated bacterial strain Enterobacter Sp. ITCB-09 and its application potential as biofertilizer. Agriculture 10(9):383

Misra S, Chauhan PS (2020) ACC deaminase-producing rhizosphere competent Bacillus Spp. mitigate salt stress and promote zea mays growth by modulating ethylene metabolism. 3 Biotech 10(3):1–14

Mohammed AF, Oloyede AR, Odeseye AO (2020) Biological control of bacterial wilt of tomato caused by Ralstonia Solanacearum using pseudomonas species isolated from the rhizosphere of tomato plants. Arch Phytopathol Plant Protect 53(1–2):1–16

Myo EM et al (2019) Evaluation of Bacillus Velezensis NKG-2 for bio-control activities against fungal diseases and potential plant growth promotion. Biol Control 134:23–31

Nacoon S et al (2020) Interaction between phosphate solubilizing bacteria and arbuscular mycorrhizal fungi on growth promotion and tuber inulin content of helianthus Tuberosus L. Sci Rep 10(1):1–10

Nautiyal CS (1999) An Efficient microbiological growth medium for screening phosphate solubilizing microorganisms. FEMS Microbiol Lett 170(436):265–270

Norrish K, Rosser H (1983) Soils: an Australian viewpoint. Melbourne, CSIRO/London, UK, Australia: Academic Press; 1983. Mineral phosphate, pp 335–361

Panhwar QA, Radziah O, Sariah M, Ismail MR (2009) Solubilization of different phosphate forms by phosphate solubilizing bacteria isolated from aerobic rice. Int J Agric Biol 11(6):667–673

Patel D, Goswami D (2020) Phosphorus solubilization and mobilization: mechanisms, current developments, and future challenge. In: Advances in plant microbiome and sustainable agriculture. Springer, pp 1–20

Pathak DV, Kumar M (2016) Microbial inoculants as biofertilizers and biopesticides. In: Microbial inoculants in sustainable agricultural productivity. Springer, pp 197–209

Paul D, Sinha SN (2017) Isolation and characterization of phosphate solubilizing bacterium pseudomonas aeruginosa KUPSB12 with antibacterial potential from river ganga, India. Ann Agrar Sci 15(1):130–136
Pikovskaya RI (1948) Mobilization of phosphorus in soil in connection with vital activity of some microbial species. Mikrobiologiya 17:362–370
Prabhu N, Borkar S, Garg S (2019) Phosphate solubilization by microorganisms: overview, mechanisms, applications and advances. In: Advances in Biological Science Research. Elsevier, pp 161–176
Pradhan A, Pahari A, Mohapatra S, Mishra BB (2017) Phosphate-solubilizing microorganisms in sustainable agriculture: genetic mechanism and application. In: Advances in soil microbiology: recent trends and future prospects. Springer, pp 81–97
Prakash J, Arora NK (2019) Phosphate-solubilizing Bacillus Sp. enhances growth, phosphorus uptake and oil yield of mentha arvensis L. 3 Biotech 9(4):126
Rashid M et al (2004) Organic acids production and phosphate solubilization by phosphate solubilizing microorganisms (PSM) under in vitro conditions. Pak J Biol Sci 7(2):187–196
Rawat P, Das S, Shankhdhar D, Shankhdhar SC (2020) Phosphate-solubilizing microorganisms: mechanism and their role in phosphate solubilization and uptake. J Soil Sci Plant Nutr 21: 49-68
Rfaki A, Zennouhi O, Aliyat FZ, Nassiri L, Ibijbijen J(2019) Isolation, selection and characterization of root-associated rock phosphate solubilizing bacteria in Moroccan wheat (*Triticum aestivum* L.). Geomicrobiol 37(3), 230-241.
Ricci M, Tilbury L, Daridon B, Sukalac K (2019) General principles to justify plant biostimulant claims. Front Plant Sci 10:494
Richardson AE (2001) Prospects for using soil microorganisms to improve the acquisition of phosphorus by plants. Funct Plant Biol 28(9):897–906
Rodríguez H, Fraga R (1999) Phosphate solubilizing bacteria and their role in plant growth promotion. Biotechnol Adv 17(4–5):319–339
Rodriguez-Viciana P et al (2006) A phosphatase holoenzyme comprised of Shoc2/Sur8 and the catalytic subunit of PP1 functions as an M-Ras effector to modulate Raf activity. Mol Cell 22(2):217–230
Rossolini GM et al (1998) Bacterial nonspecific acid phosphohydrolases: physiology, evolution and use as tools in microbial biotechnology. Cell Mol Life Sci CMLS 54(8):833–850
Sadeghi A, Karimi E, Dahaji PA, Javid MG, Dalvand Y, Askari H (2012) Plant growth promoting activity of an auxin and siderophore producing isolate of Streptomyces under saline soil conditions. World J Microbiol Biotechnol 28:1503–1509
Sharma SB, Sayyed RZ, Trivedi MH, Gobi TA (2013) Phosphate solubilizing microbes: sustainable approach for managing phosphorus deficiency in agricultural soils. Springerplus 2:1–14
Sharma S, Kumari N, Prasad B (2023) Phosphate-solubilising microorganisms as potential biofertilizer: a review. Agric Rev 44(1):84–88
Shen J, Yuan L, Zhang J, Li H, Bai Z, Chen X et al (2011) Phosphorus dynamics: from soil to plant. Plant Physiol 156(3):997–1005
Silva LID, Pereira MC, Carvalho AMXD, Buttrós VH, Pasqual M, Dória J (2023) Phosphorus-solubilizing microorganisms: a key to sustainable agriculture. Agriculture 13(2):462
Sims JT, Sharpley AN (2005) Phosphorus: agriculture and the environment. Agronomy Monograph vol 46. (Eds Sims JT and Sharpley AN). American Society of Agronomy.
Smirnova I, Sadanov A, Baimakhanova G, Faizulina E, Tatarkina L (2023) Use of a consortium of agronomically important microorganisms for growing soybean ((L.) Merr.). Open Agric J 17:1
Song OR, Lee SJ, Lee YS, Lee SC, Kim KK, Choi YL (2008) Solubilization of insoluble inorganic phosphate by *Burkholderia cepacia* DA23 isolated from cultivated soil. Braz J Microbiol 39:151–156
Sperber JI (1958) The incidence of apatite-solubilizing organisms in the rhizosphere and soil. Aust J Agric Res 9(6):778–781

Sun X, Niu Y, Du Y, Geng C, Guo C, Zhao L (2023) Isolation and identification of new soil strains with phosphate-solubilizing and exopolysaccharide-producing abilities in the yellow river wetland nature reserve of Luoyang City, China. Sustainability 15(4):3607

Tawaraya K, Naito M, Wagatsuma T (2006) Solubilization of insoluble inorganic phosphate by hyphal exudates of arbuscular mycorrhizal fungi. J Plant Nutr 29(4):657–665

Trolove SN, Hedley MJ, Kirk GJD, Bolan NS, Loganathan P (2003) Progress in selected areas of rhizosphere research on P acquisition. Soil Res 41(3):471–499

Vanderdeelen J (1995) Phosphate immobilisation in an uncropped field experiment on a calcareous soil. Plant Soil 171:209–215

Wang Q, Xiao C, Feng B, Chi R (2020a) Phosphate rock solubilization and the potential for lead immobilization by a phosphate-solubilizing bacterium (Pseudomonas sp.). J Environ Sci Health Part A 55(4):411–420

Wang YY, Li PS, Zhang BX, Wang YP, Meng J, Gao YF et al (2020b) Identification of phosphate-solubilizing microorganisms and determination of their phosphate-solubilizing activity and growth-promoting capability. Bioresources 15(2):2560–2578

Wanner BL (1996) Phosphorus assimilation and control of the phosphate regulon. Escherichia coli and Salmonella: cellular and molecular biology, 1, 1357-1381

Wodzinski RJ, Ullah AHJ (1996) Phytase. In: Advances in applied microbiology, vol 42. Academic Press, pp 263–302

Wu C, Wu X, Chen S, Wu D (2020) A newly discovered humic-reducing bacterium, *Pseudomonas geniculata* PQ01, isolated from paddy soil promotes paraquat anaerobic transformation. Front Microbiol 11:2003

Xie S, Tran HT, Pu M, Zhang T (2023) Transformation characteristics of organic matter and phosphorus in composting processes of agricultural organic waste: research trends, vol 6. Mater Sci Energy Technol, p 331

Yadav RS, Tarafdar JC (2003) Phytase and phosphatase producing fungi in arid and semi-arid soils and their efficiency in hydrolyzing different organic P compounds. Soil Biol Biochem 35(6):745–751

Yazdani M, Bahmanyar MA, Pirdashti H, Esmaili MA (2009) Effect of phosphate solubilization microorganisms (PSM) and plant growth promoting rhizobacteria (PGPR) on yield and yield components of corn (*Zea mays* L.). World Acad Sci. Eng Technol 49(1):90–92

Yu LY, Huang HB, Wang XH, Li S, Feng NX, Zhao HM et al (2019) Novel phosphate-solubilising bacteria isolated from sewage sludge and the mechanism of phosphate solubilisation. Sci Total Environ 658:474–484

Zhao K, Yu H, Xue R, Stirling E, Wang Y, Xu J et al (2022) The only constant is change: endogenous circadian rhythms of soil microbial activities. Soil Biol Biochem 173:108805

Zou X, Binkley D, Doxtader KG (1992) A new method for estimating gross phosphorus mineralization and immobilization rates in soils. Plant Soil 147:243–250

Chapter 6
Diazotrophs for Enhancement of Productivity of Major Cereal Crops Under Abiotic Stresses

Priyanka Verma, Ankita Bhattacharya, Chanda Bharti, and Naveen Kumar Arora

Abstract Population explosion has created tremendous pressure on agricultural productivity of major cereal crops. Due to anthropogenic activities and improper management of natural resources, instances of abiotic stresses have increased many folds. Abiotic stresses include salinity, drought, waterlogging, extreme temperatures, and heavy metal stresses that hamper plant growth at its various growth and developmental stages by producing stress ethylene, reactive oxygen species (ROS), osmotic stresses, water stresses, cell toxicity, reducing nutrient mineralization and phytohormonal activities that enhance growth. Nitrogen is abundantly required by plants especially in stressed conditions, but usage of chemical fertilizers results in problems like eutrophication and long-term excessive use can lower soil quality. Diazotrophic bacteria are plant growth promoting bacteria (PGPB) that provide one of the most ecofriendly and sustainable ways to combat abiotic stresses along with fixing atmospheric nitrogen. There are many mechanisms applied by diazotrophic PGPB to protect plants from stresses such as by production of osmolytes, exopolysaccharides (EPS), ROS scavenging enzymes, 1-aminocyclopropane-1-carboxylate (ACC) deaminase, regulate transcription of plant innate genes for stress tolerance, regulate phytohormones, solubilize minerals like phosphorus, zinc, and potassium, and produce siderophores. Here in this chapter emphasis is given on the role of diazotrophs in abiotic stress management in major cereal crops including wheat, rice, and maize.

Keywords Diazotrophs · Abiotic stress · Salinity · Cereal crops · Drought

---

P. Verma · A. Bhattacharya · C. Bharti · N. K. Arora (✉)
Department of Environmental Science, School of Earth and Environmental Sciences, Babasaheb Bhimrao Ambedkar University, Lucknow, Uttar Pradesh, India

N. K. Arora, B. Bouizgarne (eds.), *Microbial Biotechnology for Sustainable Agriculture Volume 2*, Microorganisms for Sustainability 34,
https://doi.org/10.1007/978-981-97-2355-3_6

6.1 Introduction

With the increase in human population in last few decades, food security has become a challenge demanding more agricultural production from limited land. It is estimated by Food and Agricultural Organization (FAO) that by 2050 world population will touch 9.1 billion and a boost of up to 70% will be needed in food production (Bodirsky et al. 2015; Payumo et al. 2018). Population increase has resulted in further limitation of land area for growing crops; humans are even encroaching undisturbed habitats for agriculture, threatening the existence of several species and causing imbalances in ecosystems. One of the problems due to unsustainable and unplanned exploitation of natural resources is that mankind is now facing issues such as climate change. Effects of climate change enhance the occurrence of natural hazards like floods, droughts, landslides, storms, and cyclones, additionally they are also linked with long-term land degradation events like soil salinization, waterlogging, and fluctuations in regular synchronicities of seasons and precipitation systems on which normal crop patterns depend, exposing crops to face various abiotic stresses (Arora 2019; Arora and Mishra 2022). It is worth mentioning that out of the total arable landmasses left on Earth, one-fifth are arid or semiarid and two-third have turned saline. It is estimated that abiotic stresses caused due to climate change are responsible for around 50% losses in agricultural crop productivities (Maiti and Satya 2014; Minhas et al. 2017; Arora 2019; Verma et al. 2020).

Major cereal crops including wheat, rice, and maize are important food components for humans and livestock throughout the world but are highly sensitive to abiotic changes in their environment as they are heavily climate dependent. The effect of abiotic stresses on major cereal crops can be seen in every growth stage including reduction in germination rates, flowering, chlorophyll content, and grain yield. Abiotic stresses also affect plant metabolic pathways including photosynthesis, respiration, mineral uptake, and activities of phytoenzymes, ultimately resulting in yield loss. Stresses like salinity, drought, waterlogging cause osmotic stress, while salinity and heavy metals stres also result in ionic toxicities. Reactive oxygen species (ROS) and stress ethylene (C_2H_4) production is engaged with abiotic stresses. Wheat is among the staple crops and 30% of the total global cereal production. By 2050 wheat production has to be enhanced by 60% of the current production values so as to meet the requirement of population (FAOSTAT 2018). Many international organizations like Intergovernmental Panel on Climate Change (IPCC), International Maize and Wheat Improvement Center (CIMMYT) and United Nations Food and Agriculture Organisation (FAO) have predicted if current scenario of extreme abiotic stresses continues, it may reduce the production of wheat by 20–30% in coming years (Hossain et al. 2021). Rice is another important staple crop on which 50% of world population depends for calorie source with an estimated enhancement of 28% in its demand by 2050 (Zhu et al. 2018; Hussain et al. 2020).

Nitrogen (N_2) is one of the most abundantly required macronutrient in plants, which is present mostly in soil in forms that are unavailable to plants. That is why productivity enhancement requires large amount of N_2 and that too in a sustainable way. It is required by plants as a key component of chlorophylls, nucleic acids, and amino acids/proteins. Symbiotic and non-symbiotic biological fixation, rainfall, organic matter, and additives through fertilizers are the processes that add N_2 to soil. There are various physical, chemical and biological methods available to convert either the degraded lands to suitable arable agricultural fields or making modification in plant system for combating abiotic stresses. For meeting the gap between crop productivity and demand of ever-increasing population, farmers are leaning toward the usage of chemical fertilizers and pesticides, which are further causing land degradation and problems like land and water pollution. Eutrophication is among the severe problems that the water bodies face due to run off nitrogen and phosphorus fertilizers. Moreover, these chemicals get integrated to food chains so easily causing biomagnification and other physiological abnormalities, ultimately resulting in problems at various trophic levels. Usage of transgenic crops for tolerating array of environmental extremities does not provide sure results in the actual field conditions, also they have unforeseen impacts on the ecosystem as whole (Arora et al. 2020; Chan et al. 2020). Microbe based inoculants can be used to replace chemicals that cause pollution of natural resources. Microbe based inoculants can help with the next green revolution without any ill effect to ecosystems (Arora et al. 2020). In condition of abiotic stresses, microbes help plant by producing various low molecular weight osmolytes to maintain osmotic balance, upregulate many channels and pumps by producing signaling molecules, ROS scavenging molecules, thermoprotectant molecules, exopolysaccharides (EPS), selective uptake of minerals and water, mineral solubilization, phytohormone production, and siderophore production, helping with overall growth and development of the plant. Usage of plant growth promoting bacteria (PGPB) is emerging as a holistic and sustainable approach to combat the abiotic stresses and enhance crop productivity in increasingly hostile environmental conditions.

6.2 Effect of Abiotic Stresses on Important Cereal Crops

Environmental degradation and climate change triggered due to numerous human activities are the major driving factors behind the reduction in crop productivity. Changing climatic and ecological conditions impose stresses on plants, affecting their growth and developmental process (Calanca 2017; Yadav et al. 2020). Crops productivity is known to be affected by various biotic (pathogens, nematodes, parasites, invasive plant species, etc.) and abiotic factors (water availability, temperature, soil salinization, nutrient imbalances, light, radiations, etc.) affecting different growth parameters (Bray 2000; Koyro et al. 2012; Fatima and Arora 2023). Abiotic stresses are the most prominent reason for the declining productivity of major cereal crops (wheat, rice, and maize) and are responsible for productivity loss of main crops by more than 50% in the last few years (Calanca 2017). Due to adverse climatic conditions and overutilization of chemical fertilizers and pesticides, agricultural land is degrading faster, making crops more vulnerable toward abiotic stresses like desertification and salinization of soil (Koyro et al. 2012; Mishra et al. 2021). Salinization of soil has increased over the past few years due to irrigation with saline water, use of agro-chemicals and climate change, resulting in the degradation of fertile agricultural fields (Abbas et al. 2013; Arora et al. 2020). Due to irrigation with saline water, salt gets accumulated in the rhizosphere, increasing the concentration of sodium (Na^+) and chloride (Cl^-) ions, which reduce potassium (K^+) concentration. Salinity affects the crops by causing ions toxicity resulting in an imbalance of ionic and water homeostasis due to the generation of ROS, which hampers cell signaling, cell metabolism, N_2 metabolism, and various other physiological parameters in the plants (Grattan 2002; Zörb et al. 2019) (Fig. 6.1). Numerous micro- and macro-nutrients are essential for the growth and development of plants; among them, N_2 is one of the vital nutrients, which plays role in various developmental processes. N_2 metabolism is affected in the soil under saline conditions as salt ions show inhibitory effects against nitrate (NO_3^-) and ammonium (NH_4^+). N_2 assimilation is significant in the osmotic adjustment and the elimination of ROS. That is why disturbance in N_2 assimilation has adverse effects on various physiological traits of the affected plant (Ashraf et al. 2018). It has been observed that the saline environment is impacting the yields of the major cereal crops, which is imposing pressure on the farmers for using more chemical fertilizers in their agricultural lands for enhancing productivity.

Due to ionic disturbance, salinity is the primary cause behind yield loss in the wheat (*Triticum aestivum* L.) crop. Abdul-Kadir and Paulsen (1982) observed the effects of salinity on N_2 metabolism in wheat crops. Salt stress was found to reduce the N_2 content in the root and shoot system in the wheat plant, which was due to retardation in the activity of NO_3^- reductase enzyme and N_2 uptake. In the study conducted by Abbas et al. (2013), it was observed that salinity affects the growth, quality, and yield in the wheat crop. The wheat quality stagnated as grain protein, fiber, and fat contents declined under the saline conditions. Regulation of genes responsible for nitrogen fixation and assimilation is the primary factor that is mis-driven under saline conditions. A study showed that a decrease in NO_3^- content lowered the regulation of *OsNR1* responsible for NH_4^+ assimilation in rice (*Oryza sativa* L.). Various other crucial genes were also observed to be downregulated, affecting the synthesis of N_2-based compounds like amino acids and betaines, necessary for repairing cellular damage caused by ROS (Wang et al. 2012). It has also been studied that the soil salinity disturbs the physicochemical parameters in rice as it does not support the root establishment, which further affects the nutrient mineralization and availability to the whole plant system (Razzaq et al. 2020). In saline conditions, NO_3^- is replaced by Cl^-, which affected the growth of leaves and roots in maize (*Zea mays* L.), resulting in reduced productivity. N_2 deficiency in maize affects plant growth and degrades its nutritive value (Hessini et al. 2019). Maize cultivated under saline-alkaline stress showed retarded growth due to decreased NO_3^- content in roots and leaves. NH_4^+ content was found to be increased in saline conditions, which induced toxicity of NH_4^+ in plants, limiting the biomass. These conditions also were reported to be responsible for downregulation of important enzymatic activities, viz. nitrate reductase, glutamate pyruvate transaminase (GPT), glutamic-oxaloacetic transaminase (GOT), glutamine synthetase (GS), and

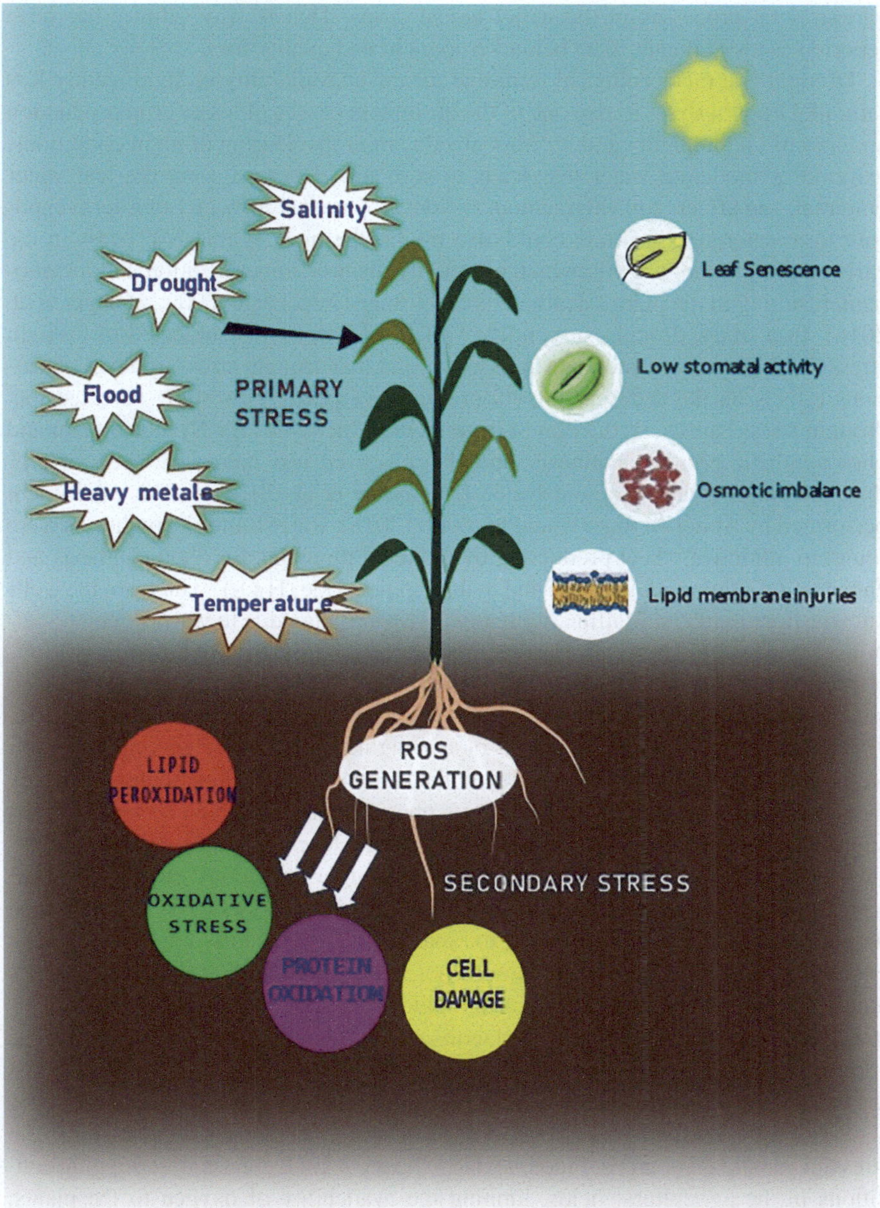

Fig. 6.1 Effects of abiotic stresses on the cereal crops

glutamate synthase (GOGAT). These enzymes are involved in growth-promoting activities, like GPT and GOT are linked with transamination; GS and GOGAT are associated with NH_4^+ assimilation. Hence, it has been noticed that salinity obstructs various vital processes in the major cereal crops. That is why salinity has been reported as a significant issue behind crop yield and quality loss.

Drought is a meteorological term that means unavailability or significantly less rainfall. Drought stress is also one of the prominent causes of lesser crop production and loss of yield, quality and nutritional value of crops. During drought conditions, extreme or moderate water loss leads to stomatal closure, lowers the leaf water potential, and affects cell enlargement and development (Fig. 6.1). Long-term exposure to a water-stressed state could also inhibit the photosynthesis process in the early stages of plant growth, which eventually disturbs mineralization and metabolism, resulting in the plant's death in its early stage (Jaleel et al. 2009; Ahanger et al. 2016). In a study, drought susceptible variety of wheat was subjected to drought stress. With the help of proteomics, it was found that drought stress imposes deleterious effects on the different growth-promoting aspects. This study revealed that drought stress causes a reduction in the enzymatic activities for N_2 metabolism and photosynthetic proteins synthesis, and also observed was increased levels of proline, methionine, arginine, and lysine (Michaletti et al. 2018). In the experiment performed by Abdel-Motagally and El-Zohri (2018), water limiting conditions were found to minimize wheat production by reducing pigments for photosynthesis and enhancing the amount of proline and hydrogen peroxide (H_2O_2),, which are the well-known stress markers. Malfunctioning of N_2 assimilation due to water deficit conditions in rice affects grain quality and productivity by interdicting various physiological and developmental processes like cell growth, cell elongation, and some critical enzymatic and antioxidant activities (Upadhyaya and Panda 2019). *ZmAS3* and *ZmAS4* are the two genes of asparagine synthetase regulating the mobilization of N_2 in leaves and other parts of plants reported to be downregulated during the drought stress during the reproductive phase of maize. These genes also play a significant role in regulating the carbon (C) and N_2 levels in maize during drought-stressed conditions (Yang et al. 2019). Under drought conditions, chlorophyll content, stomatal conductance, transpiration rate, and photosynthetic rate were reduced due to the low availability of N_2 in the maize. NO_3^- reductase activity and NO_3^- content decreased, but the plant showed positive responses after the N_2 supply, by altering various biochemical and physiological activities (Song et al. 2019). This indicated that enhancement of N_2 metabolism and mineralization could intensify the various enzymatic activities, which could be helpful in developing tolerance against the drought stress in the cereal crops.

Nowadays, floods are more frequent due to changing climatic conditions, inflicting stressful situations for crops. Flooding due to heavy rainfall creates waterlogging conditions in the agricultural fields, limiting the availability of oxygen to the plants; hence, it can also be called low oxygen stress. Lack of oxygen availability affects N_2 assimilation, NO_3^- and nitrite (NO_2^-) reduction, nitric oxide (NO) synthesis, which could adversely affect the N_2 metabolism in plants. Under the oxygen deficit conditions, plant cells are unable to function properly because of impact on aerobic respiration, affecting the cell's energy availability, resulting in cell and plant death (Limami et al. 2014; Diab and Limami 2016) (Fig. 6.1). In low oxygen stress, there is a reduction in the translation of some critical proteins because of reduced N_2 metabolism.

During heavy precipitation, the root system in rice plants faces anoxic stress condition, which disables the plant cell to assimilate the nutrients, and cells are damaged due to peroxidation (Xiong et al. 2020). In a field experiment of maize conducted by Ren et al. (2017), it has been found that waterlogging declined the crucial enzymatic activities of nitrate reductase (NR), GS, and glutamate dehydrogenase (GDH), which play a potential role in N_2 mineralization and uptake. Translocation of N_2 from stem to leaves is marked to be obstructed in a stressed environment, which affects the further transportation of N_2 to the grains thus reducing the crop yields.

Heavy metal toxicity is emerging as a concerning issue regarding food security, crop production, environmental sustainability, and soil microbial diversity. Heavy metals possess metallic characteristics, specific gravity >5.0 g cm^{-3} and atomic density more than five times greater than water (Hawkes 1997; Rascio and Navari-Izzo 2011). These metals are one of the major reasons behind environmental pollution, whether soil or water. Human activities such as urbanization, industrialization, extensive use of chemical fertilizers and pesticides, etc., are mainly responsible for prompting the heavy metal toxicity in soil (Shahid et al. 2015). Heavy metal toxicity/stress is meant by the presence of heavy metals in more than desirable amounts, which imposes stress on the microbiota of the soil, affecting crop productivity and soil quality (Gill 2014; Vanisree et al. 2022; Riyazuddin et al. 2022) (Fig. 6.1). Heavy metal stress interferes with various developmental traits of plants like photosynthesis, respiration, stomatal mechanisms, nutrient assimilation, N_2 metabolism, and enzymatic activities (Gill 2014; Doni et al 2014; Seneviratne et al. 2019). The study based on the satellite images obtained for multi-temporal sentinel-2 showed the difference between rice crops growing under cadmium (Cd) stress and unstressed conditions. Results of the study revealed the low chlorophyll level in the rice-growing under the stressed condition, which was due to malfunctioning of metabolic activities (Liu et al. 2018). Heavy metal contamination by deposition of nickel (Ni) and lead (Pb) in the root and leaves of the wheat plant was reported to have damaging effects on wheat productivity and grains quality. There was loss of crop yield as consequence of reduced antioxidant activities and other metabolic activities necessary for N_2 and C assimilation (Rizvi et al. 2019). In the experiment conducted by AbdElgawad et al. (2020), maize seedlings were subjected to an excess amount of four heavy metals, viz., zinc (Zn), copper (Cu), Ni, Cd. Different parts of plants did not show the same amount of heavy metals as roots, and older leaves exhibited higher heavy metal concentration. Older leaves and root systems were affected by the inhibition of growth-enhancing activities, generation of ROS, reduction in antioxidant activities, and nutrient mineralization in the root zone.

6.3 Nitrogen Fixation in Major Cereal Crops

As a key component of amino acids, nucleic acids, proteins, adenosine triphosphate (ATP), and chlorophyll, N_2 is the most important nutrient for crops (Wang et al. 2017). Although it is abundant in the atmosphere (78%), its availability to plants is limited because plants can only use it in NO_3^- or NH_4^+ forms (Bøckman 1997; Huang et al. 2000). Chemical fertilizers, composts, atmospheric nitrogen fixation (lightning), and biological nitrogen fixation (BNF) are all ways for crop plants to obtain N_2 (Vance et al. 2015). Nitrogen fertilizer usage and synthesis for cereal crops production have surged ninefold since the 1980s and are predicted to expand by 40–60% over the next 40 years (Bloch et al. 2020). The production and usage of synthetic N_2 fertilizers made from ammonia (NH_3) is projected to use 2% of global energy production (Glendining et al. 2011), which is costly in terms of both money and the environment (the process emits greenhouse gases and promotes eutrophication in water bodies) (Rockström et al. 2009; Geddes et al. 2015). Hence utilizing microorganisms that interact with cereals whether in the soil, on the root surface (epiphytes), or residing inside the roots (endophytes) as N_2 fixers is one method to reduce synthetic fertilizers use (Geddes et al. 2015).

Diazotrophs are prokaryotes that transform atmospheric N_2 into NH_3 (Fig. 6.2). They include cyanobacteria, free-living soil bacteria, associative bacteria (*Azospirillum*, *Enterobacter*), and symbiotic bacteria (*Rhizobium* and *Bradyrhizobium*) (Postgate 1982; Rosenblueth et al. 2018a, b). They infiltrate plants through root colonization, seed inoculation, and the cortex (Patel and Archana 2017; Pfeiffer et al. 2013). Diazotrophs are a subgroup of plant growth promoting rhizobacteria (PGPR) that can improve plant growth and productivity by supplying nitrogen and establishing a healthier plant that is better equipped to fight diseases and infections and survive under stress conditions. Some nitrogen fixers are also known to be able to dissolve insoluble phosphates, zinc, potassium, produce phytohormones, chelate iron, and suppress phytopathogens (Arora et al. 2013; Nag and Das 2022). BNF through diazotrophs is an alternative source of N_2 in cereal crops that can reduce the use of artificial fertilizers (Bhardwaj et al. 2014). BNF provides 12–70% of total N_2 uptake in field crops (70% N_2 uptake in maize and 50–90% of N_2 in wheat grains), via diazotrophic bacteria, making it the most important source of N_2 in the agricultural system (Montañez et al. 2012; Soumare et al. 2020; Rosenblueth et al. 2018a, b; Rogers and Oldroyd 2014).

Although N_2 fixation is a high-energy process that requires 16 moles of ATP to fix 1 mole of N_2, it is a necessary step in the N_2 cycle (Sun et al. 2022). As a result, diazotrophic bacteria have evolved a mechanism that uses enzyme nitrogenase to catalyze the conversion of dinitrogen to NH_4^+ (Wen et al. 2022). Nitrogenase is a two-metal enzyme complex that consists of dinitrogenase protein (MoFe protein), which contains a molybdenum-iron cofactor and is the site of N_2 reduction, and dinitrogenase reductase protein (Fe protein), which transfers electrons from an electron donor to the nitrogenase protein. In diazotrophs, several genes are involved in the synthesis and assembly of functional nitrogenase. nifH (dinitrogenase reductase), nifD and K (subunits of dinitrogenase), nifEN (scaffold for iron-molybdenum cofactor), nifB (P-cluster), and the regulatory gene nifA are some of the crucial nif genes that are conserved in the majority of the diazotrophic bacteria (Venkateshwaran 2014; Brusamarello-Santos et al. 2017; Curatti and Rubio 2014; Poza-Carrión et al. 2015; Kumar et al. 2018). Among other N_2-fixing genes, nifH has been used to

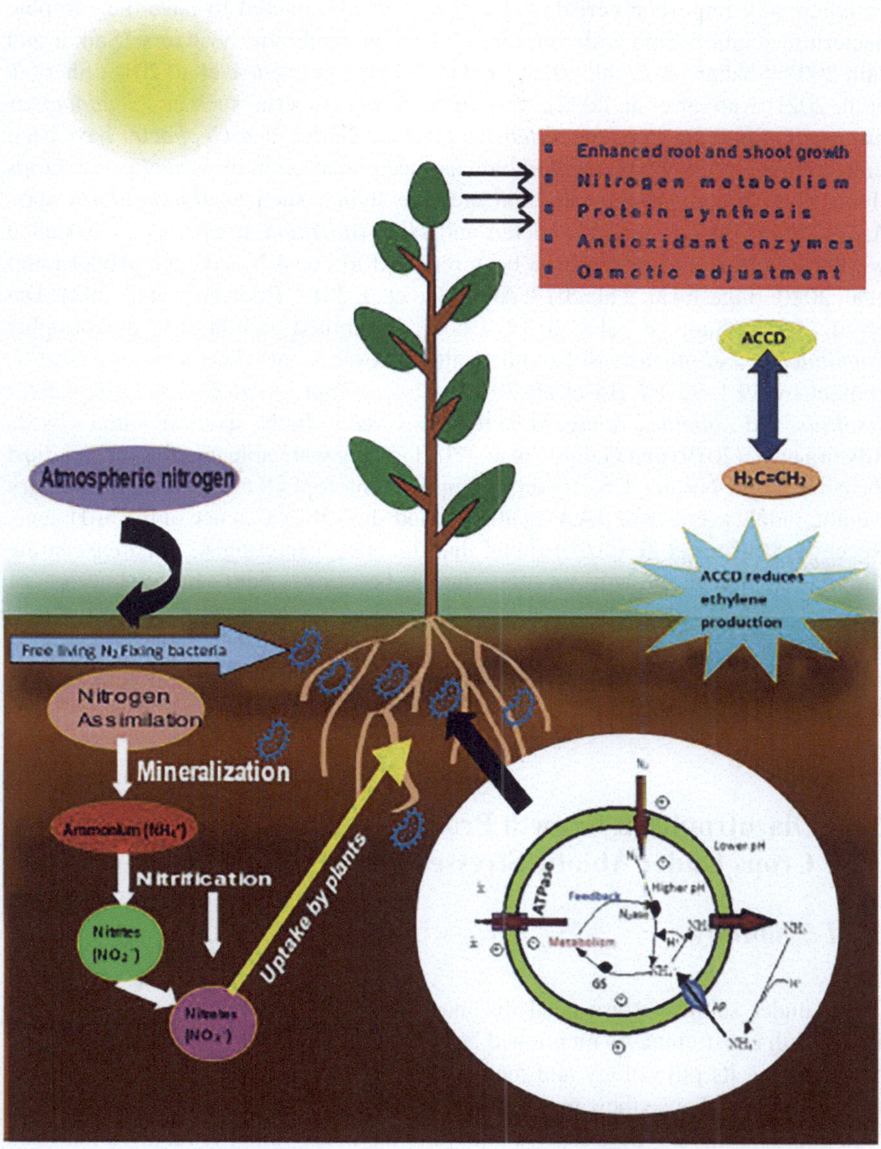

Fig. 6.2 Nitrogen fixation by diazotrophs in cereal crops

identify diazotrophic communities in major cereal crops (Wang et al. 2018a, b; Li et al. 2020).

Many non-leguminous plants have been found to be able to fix N_2 associatively or endophytically with N_2-fixing bacteria. Rice, wheat, and maize, among other commercially important cereal crops, have been discovered to have diazotrophic bacterium relationships with symbiotic and non-symbiotic N_2 fixers (Saikia and Jain 2007; Shabanamo et al. 2018; Li et al. 2020; Van Deynze et al. 2018; Sheoran et al. 2021; Katiyar et al. 2022). Several N_2-fixing bacteria, such as *Pseudomonas protegens*, *Gluconacetobacter*, *Klebsiella*, *Paenibacillus*, and *Geobacter* have been linked to increased N_2 content and biomass accumulation in important cereal crops (Fox et al. 2016). Diazotrophs that are free living, such as *Burkholderia* spp., *Azospirillum lipoferum*, *Azotobacter*, and *Herbaspirillum seropedicae* associated with rice, wheat, and maize have been reported to boost N_2 content (Abdel Latef et al. 2020; Bageshwar et al. 2017; Ambrosio et al. 2017; Rodrigues et al. 2021; Dai et al. 2021; Santi et al. 2013). Under N_2-limited conditions, diazotrophic *Paenibacillus beijingensis* isolated from the rhizosphere of wheat increased total N_2 content by 49.1–92.3% (Li et al. 2019). *Azospirillum brasilense*, *Achromobacter insolitus*, and *Zoogloea ramigera* were discovered to be N_2 fixers in studies by da Silveira et al. (2019) and Galindo et al. (2021). They were able to specifically adjust the activities of NR and GS, as well as improve chlorophyll content, root, shoot dry weight, indole acetic acid (IAA) synthesis, and show the existence of the nifH gene. Recently, Pedrosa et al. (2020) found that the NH_4^+-excreting *A. brasilense* strain stimulated maize growth in the field, boosting yield by 4.7–29%, or 460.5–1769.3 kg ha^{-1}. Non-leguminous plants in association with the diazotrophs deal with the N_2 deficient conditions and play significant role in N_2 fixation (Soumare et al. 2020). Numerous associations of N_2-fixing microbes are emerging as greener and sustainable replacement of N_2 fertilizers for the cereal crops.

6.4 Diazotrophs in Growth Promotion of Major Cereal Crops Under Abiotic Stresses

6.4.1 Salinity

Plants under salinity stress need the macronutrient N_2 for combating the stress and growth constraints. As mentioned before salinity reduces the ability of the crop by impacting its physiology and metabolism. Cereal crops generally do not have ability to fix N_2, hence they are dependent on some external source of usable form of nitrogen. Although these crops do not have an innate ability to fix N_2, it has been established that they are both internally and externally occupied with an array of microbes that can fix N_2 along with tolerating the ionic extremes of Na^+ salt. Salt buildup inflicts ionic, oxidative, and water imbalance stresses in plant system. Halo-tolerant plant growth promoting bacteria (HT-PGPB) through a number of physiological biochemical and molecular responses help plant to cope with these tensions. These salt tolerance mechanisms employed by PGPB involve maintenance of Na^+ to other ionic ratios in plant cells by making modifications in plant cell envelops through several ionic porters, enhancing expression of stress-responsive genes in plant cell, like salt overly sensitive (SOS) genes, sodium hydrogen exchanger (NHX), high-affinity K^+ transporter (HKT) transporters, etc. (Mishra et al. 2021). Arabidopsis K^+ Transporter (AKT) enhanced transcription levels of genes that encode antioxidant enzymes like superoxide dismutase (SOD), peroxidase (POD), catalase (CAT), NR, glutathione reductase (GR), polyphenol oxidase (PPO), guaiacol peroxidase (GP), monohydrate dehydrogenase (MDHAR), and dehydroascorbate reductase (DHAR) in plants. Accumulation of low molecular weight osmolytes inside plant cells, enhancing water uptake by plant cells by increasing expressions of aquaporin's genes, producing EPS molecules that make water and nutrient selectively available and act as barrier between plant cell and ionic salts, regulation of stress C_2H_4 by 1-aminocyclopropane-1-carboxylate (ACC) deaminase (ACCD) enzyme and other phytohormones are some important mechanisms of diazotrophic HT-PGPB (Arora et al. 2020) (Table 6.1).

Salt tolerant strains of *Bacillus halotolerans* and *Lelliottia amnigena* isolated from wheat rhizosphere in Egypt were found to be excellent fixer of N_2. Their application on wheat improved N_2 and protein content in grains along with showing other plant growth promoting (PGP) and salt tolerating traits in both pot and field studies (El-Akhdar et al. 2020). A strain of *Alcaligenes* sp. was found to enhance growth in rice by showing different PGP characters like mineral solubilization, siderophore production, and phytohormone production along with fixation of N_2 even in different levels of salinity. EPS production, which is considered as one of the most promising approach to prevent osmotic stress around root zone was found to be enhanced by up to 263% in salinity level of 300 mM NaCl concentration as against non-saline control conditions. It was also found that the growth parameter levels were kicked up when the secondary metabolite EPS was exogenously applied along with microbial culture subsequently boosting stress tolerant metabolites like proline and antioxidant enzymes (Fatima et al. 2020a). Five bacterial strains, *Bacillus safensis* HL1HP11, *Bacillus pumilus* HL3RS14, *Kocuria rosea* HL1RP8, *Enterobacter aerogenes* AT1HP4, and *Aeromonas veronii* AT1RP10 were utilized as N_2 fixers in enhancing the productivity of maize under saline conditions. These strains showed various PGP traits like IAA production, phosphate solubilization, N_2 fixation, and production of ACCD (Mukhtar et al. 2020).

Microbes isolated from some salt tolerant plants that are naturally thriving in harsh saline soils are also found to be effective in showing plant growth promotion in major cereal crops by various mechanisms along with fixing N_2 as they have an innate ability to do so. For instance, it was found that *Curtobacterium albidum* SRV4 strain isolated from rhizosphere of salt tolerant plants growing in saline soils was able to show growth promotion in paddy majorly by fixing N_2, showing ACCD activity, producing EPS and other PGP activities (Vimal et al. 2019). Diazotrophic *Bacillus* strains were found to enhance growth of Durum wheat under salinity stress by producing ACCD, enhancing seed germination rate, N_2 and protein content in grains, and improving other mineral contents in plant. Along with providing growth promotion under salinity stress, the PGPB was also found to control Fusarium head blight in plant (Brahim et al. 2022). Endophytic diazotroph identified as *Proteus mirabilis* was found to show PGP traits under saline conditions

Table 6.1 Mechanism of diazotrophic bacteria in amelioration of abiotic stresses in major cereal crops

Abiotic stresses	Diazotrophic bacteria	Effect/mechanism for plant growth promotion	Major cereal crops	References
Salinity	*Azotobacter*	Increased total nitrogen, biomass, and grain yield	Wheat (*Triticum aestivum* L.)	Chaudhary et al. (2013)
	Bacillus aryabhattai, *Achromobacter denitrificans*, and *Ochrobactrum intermedium*	Showed phosphate solubilization, and IAA production at 1.16% salt stress, enhanced expression of salt-responsive plant genes: BZ8, SOS1, GIG, and NHX1	Rice (*Oryza sativa* L.)	Sultana et al. (2020)
	Azotobacter and *Azospirillum*	Exhibited the highest IAA production, showed nitrogenase activity, produced gibberellins, and tolerate at 5% NaCl stress	Wheat (*Triticum aestivum* L.)	Omer (2017)
	Azospirillum and *Pseudomonas* spp.	Solubilized inorganic tricalcium phosphate, produced IAA, exhibited ACC deaminase activity, and tolerate salt stress up to 6.5% NaCl	Wheat (*Triticum aestivum* L.)	Rajput et al. (2018)
	Bacillus, *Pantoea*, *Marinobacterium*, *Acinetobacter*, *Enterobacter*, *Pseudomonas*, *Rhizobium*, and *Sinorhizobium*	Enhanced seed germination, IAA production	Wheat (*Triticum aestivum* L.)	Sorty et al. (2016)
	Bacillus tequilensis and *Bacillus aryabhattai*	Enhanced osmoprotectant properties, such as proline and total soluble sugar (TSS), antioxidant enzymes like superoxide dismutase (SOD), peroxidase (POD), and catalase (CAT)	Rice (*Oryza sativa* L.)	Shultana et al. (2021)
	Flavobacterium johnsoniae, *Pseudomonas putida*, *Achromobacter xylosoxidans*, and *Azotobacter chroococcum*	Grown in NaCl concentrations (0–3.5%) in N-free broth	Wheat (*Triticum aestivum* L.)	Abdelwahab et al. (2018)
	Azospirillum	Produced IAA, solubilized phosphate, showed production of HCN and ammonia	Maize (*Zea mays* v.)	Hamoum et al. (2015)
	Klebsiella variicola	Exhibited 1 aminocyclopropane-1-carboxylate deaminase (ACCD), IAA, EPS, and osmolytes—such as proline, sugars, proteins, and amino acids	Wheat and maize	Kusale et al. (2021)
	Azospirillum brasilense	Showed increase in photosynthetic pigments, enzymatic activities, decreased proline content, improved growth and yield components	Maize (*Zea mays* v.)	Abo-Kora (2016)
	Paenibacillus castaneae VPB1 and *Paenibacillus stellifer*	Tolerate salinity stress up to 15% NaCl, showed phosphorus solubilization, siderophore, and IAA production	Rice (*Oryza sativa* L.)	Gopal and Anandham (2020)
	Bacillus halotolerans and *Lelliottia amnigena*	Produced EPS and IAA, possess ability for P solubilization, increased K^+ in root-shoot	Wheat (*Triticum aestivum* L.)	El-Akhdar et al. (2020)
	Alcaligenes faecalis	Retained ionic imbalance by regulating Na^+ and K^+ ions, proficient to indole acetic acid production, positive results for CAT, gelatinase, and NH_3, solubilizing phosphate	Wheat (*Triticum aestivum* L.)	Babar et al. (2021)
	Bacillus, *Halobacillus*, and *Pseudomonas*	P solubilization activity and IAA production, produce ACC deaminase, siderophores, and HCN, showed maximum increase in dry weights of root and shoot	Maize (*Zea mays* v.)	Mukhtar et al. (2020)
	Enterobacter cloacae HG1	Produced iron chelators and ACCD, tolerate salt stress up to 10% NaCl, increases root length, plant height, fresh weight, and dry weight	Wheat (*Triticum aestivum* L.)	Ji et al. (2020)

Drought	*Azotobacter chroococcum*	Enhanced chlorophyll contents, SOD, CAT, and nifH gene increased, reduced proline content	Rice (*Oryza sativa* L.)	Kumar et al. (2021)
	Leclercia adecarboxylata and *Agrobacterium fabrum* were newly reported and Bacillus amyloliquefaciens and *Pseudomonas aeruginosa w*	Improved seedling growth under osmotic stress, showed highest amount of ACC deaminase and IAA production	Wheat (*Triticum aestivum* L.)	Danish and Zafar-ul-Hye (2019)
	Achromobacter xylosoxidans and *Enterobacter cloacae*	Had potential for higher "1-aminocyclopropane-1-carboxylate" (ACC)-deaminase synthesis, better nutrient solubilization and IAA production	Maize (*Zea mays v.*)	Danish et al. (2020)
	Gluconacetobacter diazotrophicus Pal5	Stimulate "induced systemic tolerance" (IST), increased plant biomass, and osmoprotectant solutes in shoots	Red rice (*Oryza sativa* L.)	Filgueiras et al. (2020)
	Gluconacetobacter diazotrophicus Pal5	Enhanced the richness of genes involved in N fixation (nifH), increased plant biomass, chlorophyll content, plant nitrogen uptake, and water use efficiency	Maize (*Zea mays* L.)	Tufail et al. (2021)
	Azotobacter spp.	Enhanced yield and yield components, as well as nitrogen N, P, and K contents, reduced proline accumulations in shoots	Wheat (*Triticum aestivum* L.)	Saad et al. (2020)
Extreme temperature	*Alcaligenes faecalis*, *Arthrobacter* sp., *Bacillus siamensis*, *Bacillus subtilis*, *Delftia acidovorans*, *Methylobacterium mesophilicum*, *Methylobacterium* sp., *Pseudomonas poae*, *Pseudomonas putida*, and *Pseudomonas stutzeri*	Produced 1-aminocyclopropane-1-carboxylate deaminase siderophore, ammonia and hydrogen cyanide, solubilize Zn and phosphate	Wheat (*Triticum aestivum* L.)	Verma et al. (2019)
	Herbaspirillum seropedicae; Azospirillum brasilense	Increased N accumulation in shoots	Maize (*Zea mays v.*)	Breda et al. (2020)
Metal toxicity	*Pseudomonas* sp. and *Pseudomonas fluorescens*	Exploit 1-aminocyclopropane-1-carboxylate (ACC) as sole nitrogen source and enhanced root elongation	Wheat (*Triticum aestivum* L.)	Govindasamy et al. (2015)
	Agrobacterium, *Pseudomonas*, and *Brevundimonas*	Showed the expression of NifH gene, and surfaced positive result for ammonia production, IAA production, cellulase and protease activity	Wheat (*Triticum aestivum* L.)	Patel et al. (2020)
	Azotobacter chroococcum	Persisted up to 1400 and 2000 μg mL^{-1} of Cu and Pb, secreted 65.5 and 60.8 μg mL^{-1} IAA at 400 μg mL^{-1} each of Cu and Pb, lowered the levels of proline, malondialdehyde (MDA), and antioxidant enzymes	Maize (*Zea mays* L.)	Rizvi and Khan (2018)
	Azotobacter	Chelate cadmium, improve shoot and root length, enhance chlorophyll content, carotenoids, and protein	Wheat (*Triticum aestivum* L.)	Hassan et al. (2016)
	Herbaspirillum sp. *GW103*	Produced IAA, siderophores, and ACCD, significantly increased Pb and Zn accumulation in shoots and roots, increased protein content, catalase, and superoxide dismutase levels	Maize (*Zea mays* L.)	Praburaman et al. (2017)
	Enterobacter sp.	Tolerate (Cd^{2+}, Pb^{2+} As^{3+}, Ni^{2+}, Hg^{2+}) up to certain levels, exhibit P solubilization, IAA production, ACCD activity improves seed germination, root–shoot length, root–shoot biomass, seedling vigor index, and chlorophyll content	Rice (*Oryza sativa* L.)	Pramanik et al. (2018b)
	Enterobacter aerogenes MCC 3092	Resistance to Cd^{2+}, Pb^{2+}, and As^{3+}, possess IAA production, nitrogen fixation, phosphate solubilization, ACCD activity, reduced oxidative stress (through antioxidants)	Rice (*Oryza sativa* L.)	Pramanik et al. (2018a)
	Enterobacter bugandensis	Subsided the uptake of Cd by providing essential elements (N and C sources)	Wheat (*Triticum aestivum* L.)	Han et al. (2021)
	Rhodopseudomonas palustris and Rubrivivax gelatinosus A	Express nifH, vnfG, and anfG, which are the Mo, V, and Fe nitrogenase genes, eradicated Cd and Zn, produced 5-aminolevulinic acid (ALA) and IAA	Rice (*Oryza sativa* L.)	Sakpirom et al. (2019)

(Verma et al. 2021). Bacterium isolated from rhizospheric saline soil identified to be diazotrophic genus *Azospirillum*, alleviated salinity stress and promoted growth of maize (Hamoum et al. 2015). Diazotrophic strains *A. brasilense* and *Rhizobium tropici* mitigated salinity stress in maize crop by showing various mechanisms like upregulation of ROS detoxifying enzyme's genes *APX1*, *CAT1*, *SOD2*, and *SOD4* in leaves and *APX2* in roots (Fukami et al. 2017).

6.4.2 Drought

Drought stress is one of the most severe abiotic stresses, which limits crop productivity and results in yield losses all over the world (Kour et al. 2019). Drought stress is reported to lower cereal crop production by 9–10%, and it is anticipated to cause problems for crop plants on more than half of the world's arable land by 2050 (Lesk and Anderson 2021; Vinocur and Altman 2005). Drought stress lowers nutrient intake, which can be countered by introducing N_2-fixing bacteria (Saikia et al. 2018). To overcome the drought effect on plants, these bacteria produce ACCD, EPS, volatile organic compounds, induce osmolyte accumulation, scavenge ROS, upregulate or downregulate stress-responsive genes and antioxidants, expressions of genes for aquaporins (AQPs), and upregulate or downregulate stress-responsive genes and antioxidants, apart from providing fixed N_2 (Prasad et al. 2019).

C_2H_4 is a plant hormone; however, it has a negative effect on plants when it is present in large concentrations. ACCD producing bacteria convert ACC (an immediate precursor of C_2H_4 biosynthesis in higher plants via methionine) to α-ketobutyrate and NH_3, reducing C_2H_4 levels during stressful situations (Glick 2014; Glick et al. 1998; Glick 2004). Many drought-tolerant diazotrophic bacteria have been found associated with wheat and maize crops, including *Pseudomonas aeruginosa*, *Bacillus amyloliquefaciens*, *Azospirillum* sp., *Achromobacter xylosoxidans*, and *Leclercia adecarboxylata*, as well as *Enterobacter cloacae* exhibiting ACCD activity (Arzanesh et al. 2011; Danish and Zafar-ul-Hye 2019; Zarei et al. 2020). Water status, osmolyte accumulation, chlorophyll content, and carotenoid content in wheat leaves (*T. aestivum*) were enhanced by ACCD and EPS generating *Variovorax paradoxus*, *Achromobacter* spp., *Ochrobactrum anthropic*, *Pseudomonas* sp., and *Serratia marcescens*. Siderophore, ammonia, IAA, and solubilisation of Zn and phosphate were also reported to be done by these bacteria, promoting plant growth (Khan and Singh 2021; Chandra et al. 2020).

EPS synthesis is another key method for diazotrophs to cope with drought stress (Ilyas et al. 2021). EPS are a mixture of sugar polymers that create biofilms and aid attachment to plant root surfaces (Mian et al. 1978). EPS protects microorganisms from water stress by improving water retention, protecting bacterial colonies from desiccation, increasing the soil's water holding capacity, and controlling the diffusion of organic carbon sources (Roberson and Firestone 1992; Chenu and Roberson 1996; Gulez et al. 2014; Fatima and Arora 2023). Drought stress in wheat and maize was eased by EPS-producing N_2-fixing *Bacillus subtilis* and *A. brasilense*, which produced a significant amount of sucrose, protein, uronic acid, and osmolytes while also increasing the synthesis of antioxidant enzymes (SOD) (Vurukonda et al. 2016). Another study by Bano et al. (2013) found that inoculation of N_2 fixing *A. lipoferum* strain enhanced shoot and root fresh weight, shoot and root dry weight, shoot length and root length of maize under drought stress.

6.4.3 Waterlogging

BNF in non-leguminous crops is induced by the non-symbiotic association of N_2-fixers, as there is no nodule formation. The waterlogging situation converts rhizospheric conditions from aerobic to anaerobic, restraining the needful processes in the plant system. Diazotrophic communities can fix N_2 in oxygen deficit or low oxygen levels. In rice cultivation, there is a waterlogged condition favoring the N_2 fixation mechanism driven by cyanobacteria and photosynthetic bacteria found free living in the upper soil (Ladha and Reddy 2003; Pittol et al. 2016). Endophytic diazotrophs have been reported in N_2 fixation under waterlogged conditions because of their capability to penetrate deep inside the tissue (Verma et al. 2001). Stress-induced C_2H_4 production under the flooding conditions had adverse impacts on leaves, root growth, and flowering in the cereal crops, which can be resolved by synthesis of ACCD by diazotrophic bacteria. ACCD gene (acdS) is noteworthy in the breakdown of ACC, which lowers the C_2H_4 production, making the plant resistant against the flooding stress (Selvakumar et al. 2012; Sapre et al. 2019; Saghafi et al. 2020). Waterlogging priming can ameliorate stress and develop tolerance in the crops. In this method, priming is done by sodium azide to induce stress tolerance in plants, which is helpful in maintaining the growth-promoting traits and yield qualities (Vwioko et al. 2019).

6.4.4 Temperature Extremes

Drastically changing climatic conditions due to global warming are imposing threat of heat stress to the food security and crop productivity. Studies have shown that earth surface temperature has increased at the rate of approximately 0.2 °C per decade in last 30 years, and with the ongoing trend of greenhouse gas emission it is estimated to rise by 2.5–4.5 °C by the end of twenty-first century (Solomon 2007; Bernstein et al. 2008; Wang et al. 2018a; Arora and Mishra 2023; Arora 2024). Asseng et al. (2015) predicted that there will be a 6% decline in wheat production with every 1 °C rise of temperature, which is estimated equivalent to about 42 million tons of wheat. Winter wheat that is grown in many cold countries is affected by rise in temperature due to heat stress caused by global warming (Rezaei et al. 2015). It is reported that climate change could show a reduction of 10–15% in paddy yield, which can cause rise in market prices by 32–37% as estimated by International

Food Policy Research Institute (IFPRI) (Li 2018). Heat stress elicit a cascade of physiological, molecular, and biochemical changes at molecular levels that ultimately result in reduced crop yield. In plants, symptoms of heat stress include wilting, leaf damage, fruit drop, and bolting (Shaffique et al. 2022). Interaction between plants and heat tolerant microbes enhances accumulation of heat tolerant osmolytes such as proline, glycine betaine, soluble sugar, and trehalose. Endophytic and epiphytic diazotrophs belonging to N_2 fixing *Azospirillum* genus were able to produce a high molecular weight glycol-protein called lectins that can induce cellular responses to produce stress metabolites and signaling pathways in plant cells conferring plant to fight heat stress (Alen'kina and Nikitina 2021). Two N_2 fixing bacterial strains, *B. amyloliquefaciens* UCMB5113 and *A. brasilense* NO40, were tested under greenhouse for management of short term heat stress in wheat seedlings and showed lower levels of accumulated ROS thus resulted in downregulation of many antioxidant enzymes related genes in plants. After bacterial inoculation, the plants no longer needed to elevate their defense mechanisms by ascorbate peroxidase (*APX1*), *S*-adenosylmethionine synthetase (*SAMS1*), and heat shock proteins (*HSP17.8*), the ascorbate-glutathione pathway antioxidant enzymes, and other common stress metabolites (Abd El-Daim et al. 2014). The bacterial strains isolated from *Solanum lycopersicum* L., *Cajanus cajan*, *Glycine max*, and *Triticum* spp. that included *Bacillus cereus*, *Pseudomonas* spp., *Serratia liquefaciens*, *P. fluorescens*, and *Pseudomonas putida* helped in the production of phytohormones, antioxidant enzymes, ACCD, and fixed N_2, consequently mitigating heat stress in these plants (Kang et al. 2019). *Burkholderia phytofirmans* PsJN and *Curvularia protuberata* Cp4666D, which were reported in *T. aestivum* and *S. lycopersicum,* respectively, caused remarkable thermotolerance in these plants under heat stress by suppressing production of free radicals. These microbes enhanced N_2 content in plants and additionally increased phytohormone production (Rana et al. 2021; Shaffique et al. 2022).

Low temperatures also create stressful conditions for the growth, development, and survival of the plants, resulting in restricting the production of cereal crops in the colder regions. Low temperature stress is further divided into chilling temperature stress (0–15 °C) and freezing temperature stress (<0 °C). Stressful conditions under low temperatures adversely affect the morphological, biochemical, and molecular aspects of the plants. Cold stress have deleterious impact on the major developmental process resulting in poor germination rate, wilting, reduced growth, necrosis, etc. (Bhandari and Nayyar 2013). During the exposure of cold stress, plants promote the generation of ROS, which causes cellular damage to plants. PGPR have caliber to scavenge the ROS generated during stressed conditions and promote the mineralization and solubilization of nutrients along with N_2 fixation (Mitra et al. 2021). Rhizospheric bacterial species *Kosakonia* sp. CIR2 and *Staphylococcus* sp. CSR1T2 were found to enhance the survival rates of paddy under cold stress by many folds when compared to non-inoculated control (de Souza et al. 2021). Cereal crop maize is susceptible to even slight lowering in temperature. It has been found that if temperature falls below 15 °C, it negatively affects root development, nutrient uptake, and hormonal imbalances in maize. It was found that integrated application of micronutrients like Zn and Mn, seaweed extracts, along with PGPB *Pseudomonas* sp., *B. amyloliquefaciens,* and *Bacillus simplex* was able to protect maize crop from chilling injuries (Bradáčová et al. 2016). Cold-tolerant strains *Lysinibacillus fusiformis* YJ4 and *Lysinibacillus sphaericus* YJ5 have potential to alleviate the adverse effect of cold stress on the maize crop by boosting up the N_2 and other mineral contents along with inducing systemic resistance against oxidative and osmotic stresses (Jha and Mohamed 2023). Psychrotolerant bacterium *B. amyloliquefaciens* IARI-HHS2-30 isolated from Indian Himalayas was found to be capable in N_2 fixation and ACCD activities in the wheat crop as well as promoting other growth attributes like solubilization of zinc, potassium, and phosphate, production of IAA, siderophores, and biocontrol against *Rhizoctonia solani* and *Macrophomina phaseolina* even at 4 °C (Verma et al. 2015).

6.4.5 Heavy Metal Toxicity

Problem of heavy metals contamination has increased due to various anthropogenic activities. It is evident from recent researches that PGPR can increase the growth in plants without accumulating toxic metals. Heavy metals inhibit enzyme functions in plants by disrupting nucleic acid structure and displacing essential nutrients from plant systems. Heavy metal tolerant PGPR reduce bioavailability of these toxic ions by mechanisms such as creating permeability barrier, intracellular and extracellular chelation and sequestration of metals, detoxification to lesser toxic forms of metal ions, acidification and precipitation of metals, and excretion of toxic metals by pumps (Mishra et al. 2017; Etesami 2018; Arora and Chauhan 2021). Metal biosorption is another method applied by bacteria in which they produce EPS (Rizvi et al. 2020). Many diazotrophs belonging to genus *Azotobacter*, *Bacillus*, *Achromobacter,* and *Pseudomonas* are able to show toxic metal resistance, for example, a PGPR, *P. aeruginosa,* which also improved N_2 and P content in wheat, was able to save wheat plants from oxidative stress posed by Zn toxicity (Islam et al. 2014). In another example, strains of *Pseudomonas moraviensis*, *B. halotolerans*, *Enterobacter hormaechei*, and *Pseudomonas frederiksbergensis* isolated from Jujube (*Ziziphus lotus*) showed tolerance against heavy metals like Cu, Cd, and Ni and also improved growth by increasing uptake of N_2 by 17% in wheat (Fahsi et al. 2021). N_2 fixing strains of fluorescent pseudomonads, *Ralstonia eutropha*, *Exiguobacterium aurantiacum*, *Enterobacter ludwigii,* and *Klebsiella pneumoniae* were found to enhance growth in wheat by tolerating different heavy metal ions like Cd, mercury (Hg), chromium (Cr), arsenic (As), and Cu along with showing many other growth promoting attributes in wheat (Verma et al. 2015; Gontia-Mishra et al. 2016; Wang et al. 2018b; Rizvi and Khan 2017). Consortium of Cd resistant *Ochrobactrum* sp., Pb resistant *Bacillus* sp., and As resistant *Bacillus* sp. was used to treat rice crop, which enhanced seed germination, root elongation, siderophore production, ACCD activity and reducing antioxidant stress along with enhancing uptake of N_2 (Pandey et al. 2013). *Halomonas* species isolated from rhizosphere of *Avicennia marina* a mangrove plant, which effectively showed production of EPS and tolerated heavy metal stress along with salinity in rice also fixed N_2 (Mukherjee et al. 2019). A diazotrophic PGP *Herbaspirillum* sp. has been found to improve growth of maize crop by reducing the ill effects of heavy metals Pb and Zn along with exhibiting PGP traits like siderophore, ACCD and phytohormone production (Praburaman et al. 2017). Amelioration of heavy metal stress imposed on productivity of wheat due to higher concentration of Cd was done in a study conducted by Hassan et al. (2016). Enhancement in plants growth and crop productivity under stressed conditions was achieved by bioinoculation of crop with *Azotobacter* and other N_2 fixing strains.

6.5 Future Prospects

Challenge of food productivity for increasing population in upcoming years along with threats like global warming and climate change would be the major issues in the future. Abiotic stresses are the emerging factors due to changing climate and global heating prevailing across the world. Disturbance in the precipitation patterns gives rise to various abiotic stresses, including salinity, drought, cold stress, heat stress, heavy metal toxicity, and waterlogging, destroying the quality of crops along with deterioration of arable land quality. Salinity alone has caused degradation of approximately 1125 million hectares out of which human induced salinization and solidification accounts for 76 million hectares of surface area of world and it is increasing currently at the rate of 1.5 million hectares per year (Hossain 2019). Mitigating the quality of available degraded land mass instead of disturbing undisturbed habitats should be the major focus for future productivity demands. Major cereal crops including wheat, rice, and maize provide approximately half of the human population with much required carbohydrates, proteins, and fats and are much preferred for consumption (Fatima et al. 2020b). Nitrogen being the top required nutrient by plants is also the most applied agro-chemical in fields resulting in various environmental consequences. Application of resistant diazotrophic microbe-based inoculants can provide solution against various abiotic stresses in combination with fulfilling nitrogen requirement of crops. Recent research findings have shown that incorporating diazotrophs under abiotic stress is the most beneficial and sustainable way to enhance crop productivity instead of employing chemical fertilizers, transgenic crops or making other soil and plant modifications. Some research works found that diazotrophs can enhance cell growth, cell metabolism, nutrient mineralization, antioxidant and enzymatic activities in plant systems, making them tolerant against several abiotic stresses. However, being living cells there are many constraints that need to be overcome at the industrial level production and industry to field application of these beneficial diazotrophic microbes as biofertilizers. Hence research needs to be focused on these shortcomings. It is also proven that metabolites produced by the microbes, for example, EPS, phytohormones, and osmolytes, under stress conditions if included in bioinoculants, can successfully help in abiotic stress ameliorations

(Mishra and Arora 2018). Hence, the future research on biofertilizers using abiotic stress tolerant diazotrophs should emphasize on removing the gap between laboratory, industry, and field applications.

6.6 Conclusion

Crop production is facing challenges due to increasing population and pollution from various anthropogenic activities. Climate change and global warming at alarming rate are triggering the drastic increase in temperature, which is one of the major reasons behind emerging abiotic stresses. To achieve sustainable development goals (SDGs), there is urgent need to employ ecofriendly means for improving soil health and crop productivity. Cereal crops like rice, maize, and wheat are most important food grains but affected due to various abiotic stresses. Several research studies have been done in recent years on the role of PGPR in amelioration of salinity, drought, heat stress, and metal toxicity. PGPR help the plants to develop tolerance against stress by various mechanisms and N_2 fixation is one of the very important characteristics. Diazotrophic PGPR are important microbes to improve food productivity under stress conditions, ensure food security and must be the focus of research in coming time.

References

Abbas G, Saqib M, Rafique Q, Rahman AU, Akhtar J, Haq MAU (2013) Effect of salinity on grain yield and grain quality of wheat (*Triticum aestivum* L.). Pak J Bot 50:185–189

Abd El-Daim IA, Bejai S, Meijer J (2014) Improved heat stress tolerance of wheat seedlings by bacterial seed treatment. Plant Soil 379:337–350

Abdel Latef AAH, Abu Alhmad MF, Kordrostami M, Abo-Baker ABAE, Zakir A (2020) Inoculation with *Azospirillum lipoferum* or *Azotobacter chroococcum* reinforces maize growth by improving physiological activities under saline conditions. J Plant Growth Regul 39:1293–1306

AbdElgawad H, Zinta G, Hamed BA, Selim S, Beemster G, Hozzein WN (2020) Maize roots and shoots show distinct profiles of oxidative stress and antioxidant defense under heavy metal toxicity. Environ Pollut 258:113705

Abdel-Motagally FMF, El-Zohri M (2018) Improvement of wheat yield grown under drought stress by boron foliar application at different growth stages. J Saudi Soc Agric Sci 17(2):178–185

Abdelwahab RAI, Cherif A, Cristina CRUZ, Nabti E (2018) Extracts from marine macroalgae and Opuntia *ficus indica* cladodes enhance halotolerance and enzymatic potential of diazotrophic rhizobacteria and their impact on wheat germination under salt stress. Pedosphere 28(2):241–254

Abdul-Kadir SM, Paulsen GM (1982) Effect of salinity on nitrogen metabolism in wheat. J Plant Nutr 5(9):1141–1151

Abo-Kora HA (2016) Endophytic colonization of maize (*Zea mays* v.) root plants by PGPRs under salinity stress. Nat Sci 14(7):34–51

Ahanger MA, Morad-Talab N, Abd-Allah EF, Ahmad P, Hajiboland R (2016) Plant growth under drought stress: significance of mineral nutrients. In: Ahmad P (ed) Water stress and crop plants: a sustainable approach, vol 2. Wiley, New York, pp 649–668
Alen'kina SA, Nikitina VE (2021) Stimulating effect from lectins of associative bacteria of the genus Azospirillum on the germination and morphometric characteristics of spring wheat sprouts in simulated abiotic stress. Russ J Plant Physiol 68(2):315–321
Ambrosio R, Ortiz-Marquez JCF, Curatti L (2017) Metabolic engineering of a diazotrophic bacterium improves ammonium release and biofertilization of plants and microalgae. Metab Eng 40:59–68
Arora N (2019) Earth: 50 years challenge. Environ Sustain 2:1–3
Arora NK, Chauhan R (2021) Heavy metal toxicity and sustainable interventions for their decontamination. Environ Sustain 4:1–3
Arora NK, Mishra I (2022) Current scenario and future directions for sustainable development goal 2: a roadmap to zero hunger. Environ Sustain 5:129–133
Arora NK, Mishra I (2023) Sustainable development goal 13: recent progress and challenges to climate action. Environ Sustain 6:297–301
Arora NK, Tewari S, Singh R (2013) Multifaceted plant-associated microbes and their mechanisms diminish the concept of direct and indirect PGPRs. In: Arora N (ed) Plant microbe symbiosis: fundamentals and advances. Springer, New Delhi, pp 411–449
Arora NK, Fatima T, Mishra J, Mishra I, Verma S, Verma R et al (2020) Halo-tolerant plant growth promoting rhizobacteria for improving productivity and remediation of saline soils. J Adv Res 26:69–82
Arora P (2024) COP28: ambitions realities and future. Environ Sust 7(1):107–113. https://doi.org/10.1007/s42398-024-00304-0
Arzanesh MH, Alikhani HA, Khavazi K, Rahimian HA, Miransari M (2011) Wheat (*Triticum aestivum* L.) growth enhancement by Azospirillum sp. under drought stress. World J Microbiol Biotechnol 27:197–205
Ashraf M, Shahzad SM, Imtiaz M, Rizwan MS (2018) Salinity effects on nitrogen metabolism in plants–focusing on the activities of nitrogen metabolizing enzymes: a review. J Plant Nutr 41(8):1065–1081
Asseng S, Ewert F, Martre P, Rötter RP, Lobell DB, Cammarano D et al (2015) Rising temperatures reduce global wheat production. Nat Clim Chang 5(2):143–147
Babar M, Rasul S, Aslam K, Abbas R, Manzoor I, Hanif MK (2021) Mining of halo-tolerant plant growth promoting rhizobacteria and their impact on wheat (*Triticum aestivum* L.) under saline conditions. J King Saud Univ Sci 33(3):101372
Bageshwar UK, Srivastava M, Pardha-Saradhi P, Paul S, Gothandapani S, Jaat RS et al (2017) An environmentally friendly engineered Azotobacter strain that replaces a substantial amount of urea fertilizer while sustaining the same wheat yield. Appl Environ Microbiol 83(15):E00590-17
Bano Q, Ilyas N, Bano A, Zafar N, Akram A, Hassan F (2013) Effect of Azospirillum inoculation on maize (*Zea mays* L.) under drought stress. Pak J Bot 45(S1):13–20
Bernstein L, Bosch P, Canziani O, Chen Z, Christ R, and Riahi K (2008) IPCC, 2007: Climate Change 2007: Synthesis Report. Geneva: IPCC. ISBN 2-9169-122-4
Bhandari K, Nayyar H (2013) Low temperature stress in plants: an overview of roles of cryoprotectants in defense. In: Ahmad P, Wani M (eds) Physiological mechanisms and adaptation strategies in plants under changing environment, vol 1. Springer, New York, pp 193–265
Bhardwaj D, Ansari MW, Sahoo RK, Tuteja N (2014) Biofertilizers function as key player in sustainable agriculture by improving soil fertility, plant tolerance and crop productivity. Microb Cell Factories 13:1–10
Bloch SE, Ryu MH, Ozaydin B, Broglie R (2020) Harnessing atmospheric nitrogen for cereal crop production. Curr Opin Biotechnol 62:181–188
Bøckman OC (1997) Fertilizers and biological nitrogen fixation as sources of plant nutrients: perspectives for future agriculture. Plant Soil 194:11–14

Bodirsky BL, Rolinski S, Biewald A, Weindl I, Popp A, Lotze-Campen H (2015) Global food demand scenarios for the 21st century. PLoS One 10(11):e0139201

Bradáčová K, Weber NF, Morad-Talab N, Asim M, Imran M, Weinmann M (2016) Micronutrients (Zn/Mn), seaweed extracts, and plant growth-promoting bacteria as cold-stress protectants in maize. Chem Biol Technol Agric 3(1):1–10

Brahim AH, Ben Ali M, Daoud L, Jlidi M, Akremi I, Hmani H, Feto NA, Ben Ali M. (2022) Biopriming of Durum wheat seeds with endophytic diazotrophic bacteria enhances tolerance to fusarium head blight and salinity. Microorganisms. 5;10(5):970.

Bray EA (2000) Response to abiotic stress. In: Biochemistry and molecular biology of plants. American Society of Plant Physiologists, Rockville, pp 1158–1203

Breda FADF, Alves GC, Lopez BDO, de Aragão AR, Araújo AP, Reis VM (2020) Inoculation of diazotrophic bacteria modifies the growth rate and grain yield of maize at different levels of nitrogen supply. Arch Agron Soil Sci 66(14):1948–1962

Brusamarello-Santos LC, Gilard F, Brulé L, Quilleré I, Gourion B, Ratet P et al (2017) Metabolic profiling of two maize (*Zea mays* L.) inbred lines inoculated with the nitrogen-fixing plant-interacting bacteria *Herbaspirillum seropedicae* and *Azospirillum brasilense*. PLoS One 12(3):e0174576

Calanca PP (2017) Effects of abiotic stress in crop production. In: Ahmed M, Stockle C (eds) Quantification of climate variability, adaptation, and mitigation for agricultural sustainability. Springer, Cham, pp 165–180

Chan RL, Trucco F, Otegui ME (2020) Why are second-generation transgenic crops not yet available in the market? J Exp Bot 71(22):6876–6880

Chandra D, Srivastava R, Glick BR, Sharma AK (2020) Rhizobacteria producing ACC deaminase mitigate water-stress response in finger millet (*Eleusine coracana* (L.) Gaertn.). 3 Biotech 10:1–15

Chaudhary D, Narula N, Sindhu SS, Behl RK (2013) Plant growth stimulation of wheat (*Triticum aestivum* L.) by inoculation of salinity tolerant azotobacter strains. Physiol Mol Biol Plant 19:515–519

Chenu C, Roberson EB (1996) Diffusion of glucose in microbial extracellular polysaccharide as affected by water potential. Soil Biol Biochem 28(7):877–884

Curatti L, Rubio LM (2014) Challenges to develop nitrogen-fixing cereals by direct nif-gene transfer. Plant Sci 225:130–137

da Silveira APD, Iório RDPF, Marcos FCC, Fernandes AO, de Souza SACD, Kuramae EE et al (2019) Exploitation of new endophytic bacteria and their ability to promote sugarcane growth and nitrogen nutrition. Antonie Van Leeuwenhoek 112:283–295

Dai X, Song D, Zhou W, Liu G, Liang G, He P et al (2021) Partial substitution of chemical nitrogen with organic nitrogen improves rice yield, soil biochemical indictors and microbial composition in a double rice cropping system in south China. Soil Tillage Res 205:104753

Danish S, Zafar-ul-Hye M (2019) Co-application of ACC-deaminase producing PGPR and timber-waste biochar improves pigments formation, growth and yield of wheat under drought stress. Sci Rep 9(1):5999

Danish S, Zafar-Ul-Hye M, Hussain S, Riaz M, Qayyum MF (2020) Mitigation of drought stress in maize through inoculation with drought tolerant ACC deaminase containing PGPR under axenic conditions. Pak J Bot 52(1):49–60

de Souza EM, Lamb TI, Lamb TA, dos Santos Silva A, da Fré de Carvalho S, Nyland V (2021) Rhizospheric soil from rice paddy presents isolable bacteria able to induce cold tolerance in rice plants. J Soil Sci Plant Nutr 21(3):1993–2006

Diab H, Limami AM (2016) Reconfiguration of N metabolism upon hypoxia stress and recovery: roles of alanine aminotransferase (AlaAT) and glutamate dehydrogenase (GDH). Plants 5(2):25

Doni F, Isahak A, Che Mohd Zain CR, Wan Yusoff WM (2014) Physiological and growth response of rice plants (*Oryza sativa* L.) to Trichoderma spp. inoculants. AMB Exp 4(1):1–7

El-Akhdar I, Elsakhawy T, Abo-Koura HA (2020) Alleviation of salt stress on wheat (*Triticum aestivum* L.) by plant growth promoting bacteria strains *Bacillus halotolerans* MSR-H4 and *Lelliottia amnigena* MSR-M49. J Adv Microbiol 20:44–58

Etesami H (2018) Bacterial mediated alleviation of heavy metal stress and decreased accumulation of metals in plant tissues: mechanisms and future prospects. Ecotoxicol Environ Saf 147:175–1917

Fahsi N, Mahdi I, Mesfioui A, Biskri L, Allaoui A (2021) Plant growth-promoting rhizobacteria isolated from the Jujube (*Ziziphus lotus*) plant enhance wheat growth, Zn uptake, and heavy metal tolerance. Agriculture 11(4):316

FAOSTAT (2018) Wheat production statistics. 2018. http://www.fao.org/faostat/en/#data/QC

Fatima T, Arora NK (2023) Structural, compositional characterization of novel exopolysaccharide from halotolerant strain of *Pseudomonas entomophila* PE3 and their role in growth promotion of sunflower under saline condition. Environ Exp Bot 213:105418

Fatima T, Mishra I, Verma R, Arora NK (2020a) Mechanisms of halotolerant plant growth promoting Alcaligenes sp. involved in salt tolerance and enhancement of the growth of rice under salinity stress. 3 Biotech 10(8):1–12

Fatima Z, Ahmed M, Hussain M, Abbas G, Ul-Allah S, Ahmad S (2020b) The fingerprints of climate warming on cereal crops phenology and adaptation options. Sci Rep 10(1):1–21

Filgueiras L, Silva R, Almeida I, Vidal M, Baldani JI, Meneses CHSG (2020) *Gluconacetobacter diazotrophicus* mitigates drought stress in *Oryza sativa* L. Plant Soil 451:57–73

Fox AR, Soto G, Valverde C, Russo D, Lagares A Jr, Zorreguieta Á et al (2016) Major cereal crops benefit from biological nitrogen fixation when inoculated with the nitrogen-fixing bacterium *Pseudomonas protegens* Pf-5 X940. Environ Microbiol 18(10):3522–3534

Fukami J, de la Osa C, Ollero FJ, Megías M, Hungria M (2017) Co-inoculation of maize with *Azospirillum brasilense* and *Rhizobium tropici* as a strategy to mitigate salinity stress. Funct Plant Biol 45(3):328–339

Galindo FS, Strock JS, Pagliari PH (2021) Nutrient accumulation affected by corn stover management associated with nitrogen and phosphorus fertilization. Agriculture 11(11):1118

Geddes BA, Ryu MH, Mus F, Costas AG, Peters JW, Voigt CA et al (2015) Use of plant colonizing bacteria as chassis for transfer of N2-fixation to cereals. Curr Opin Biotechnol 32:216–222

Gill M (2014) Heavy metal stress in plants: a review. Int J Adv Res 2(6):1043–1055

Glendining MJ, Dailey AG, Powlson DS, Richter GM, Catt JA, Whitmore AP (2011) Pedotransfer functions for estimating total soil nitrogen up to the global scale. Eur J Soil Sci 62(1):13–22

Glick BR (2004) Bacterial ACC deaminase and the alleviation of plant stress. Adv Appl Microbiol 56:291–312

Glick BR (2014) Bacteria with ACC deaminase can promote plant growth and help to feed the world. Microbiol Res 169(1):30–39

Glick BR, Penrose DM, Li J (1998) A model for the lowering of plant ethylene concentrations by plant growth-promoting bacteria. J Theor Biol 190(1):63–68

Gontia-Mishra I, Sapre S, Sharma A, Tiwari S (2016) Alleviation of mercury toxicity in wheat by the interaction of mercury-tolerant plant growth-promoting rhizobacteria. J Plant Growth Regul 35(4):1000–1012

Gopal NO, Anandham R (2020) Plant growth promotion and alleviation of salts in rice by *Paenibacillus* spp. J Pharmacogn Phytochem 9(4):1485–1489

Govindasamy V, Senthilkumar M, Annapurna K (2015) Effect of mustard rhizobacteria on wheat growth promotion under cadmium stress: characterization of acdS gene coding ACC deaminase. Ann Microbiol 65:1679–1687

Grattan S (2002) Irrigation water salinity and crop production, vol 9. UCANR Publications, Oakland

Gulez G, Altıntaş A, Fazli M, Dechesne A, Workman CT, Tolker-Nielsen T, Smets BF (2014) Colony morphology and transcriptome profiling of *Pseudomonas putida* KT 2440 and its mutants deficient in alginate or all EPS synthesis under controlled matric potentials. Microbiol Open 3(4):457–469

Hamoum H, Bouznad A, Mekhaldi A, Djibaoui R (2015) Screening for PGP activities of diazotrophic bacteria isolated from saline soil and their effect on maize growth under saline stress. J Pure App Microbiol 9(Spl. Edn. 2):93–101

Han H, Zhang H, Qin S, Zhang J, Yao L, Chen Z et al (2021) Mechanisms of *Enterobacter bugandensis* TJ6 immobilization of heavy metals and inhibition of Cd and Pb uptake by wheat based on metabolomics and proteomics. Chemosphere 276:130157

Hassan W, Bashir S, Ali F, Ijaz M, Hussain M, David J (2016) Role of ACC-deaminase and/or nitrogen fixing rhizobacteria in growth promotion of wheat (*Triticum aestivum* L.) under cadmium pollution. Environ Earth Sci 75:1–14

Hawkes JS (1997) Heavy metals. J Chem Edu 74:1369–1374

Hessini K, Issaoui K, Ferchichi S, Saif T, Abdelly C, Siddique KH et al (2019) Interactive effects of salinity and nitrogen forms on plant growth, photosynthesis and osmotic adjustment in maize. Plant Physiol Biochem 139:171–178

Hossain MS (2019) Present scenario of global salt affected soils, its management and importance of salinity research. Int Res J Biol Sci 1:1–3

Hossain A, Skalicky M, Brestic M, Maitra S, Ashraful Alam M, Syed MA et al (2021) Consequences and mitigation strategies of abiotic stresses in wheat (*Triticum aestivum* L.) under the changing climate. Agronomy 11(2):241

Huang J, Reneau RB Jr, Hagedorn C (2000) Nitrogen removal in constructed wetlands employed to treat domestic wastewater. Water Res 34(9):2582–2588

Hussain S, Huang J, Huang J, Ahmad S, Nanda S, Anwar S et al (2020) Rice production under climate change: adaptations and mitigating strategies. In: Fahad S et al (eds) Environment, climate, plant and vegetation growth. Springer, Cham, pp 659–686

Ilyas M, Nisar M, Khan N, Hazrat A, Khan AH, Hayat K et al (2021) Drought tolerance strategies in plants: a mechanistic approach. J Plant Growth Regul 40:926–944

Islam F, Yasmeen T, Ali Q, Ali S, Arif MS, Hussain S et al (2014) Influence of *Pseudomonas aeruginosa* as PGPR on oxidative stress tolerance in wheat under Zn stress. Ecotoxicol Environ Saf 104:285–293

Jaleel CA, Manivannan P, Wahid A, Farooq M, Al-Juburi HJ, Somasundaram R et al (2009) Drought stress in plants: a review on morphological characteristics and pigments composition. Int J Agric Biol 11(1):100–105

Jha Y, Mohamed HI (2023) Inoculation with *Lysinibacillus fusiformis* strain YJ4 and *Lysinibacillus sphaericus* strain YJ5 alleviates the effects of cold stress in maize plants. Gesunde Pflanz 75(1):77–95

Ji C, Liu Z, Hao L, Song X, Wang C, Liu Y et al (2020) Effects of Enterobacter cloacae HG-1 on the nitrogen-fixing community structure of wheat rhizosphere soil and on salt tolerance. Front Plant Sci 11:1094

Kang D, Park J, Shim K (2019) Heat treatment at an early age has effects on the resistance to chronic heat stress on broilers. Animals 9(12):1022

Katiyar P, Kumar S, Arora NK (2022) Interactions of nitrogen-fixing bacteria and cereal crops: an important dimension. In: Maheshwari DK, Dobhal R, Dheeman S (eds) Nitrogen fixing bacteria: sustainable growth of non-legumes, vol 36. Springer Nature, Singapore, pp 169–194

Khan A, Singh AV (2021) Multifarious effect of ACC deaminase and EPS producing Pseudomonas sp. and *Serratia marcescens* to augment drought stress tolerance and nutrient status of wheat. World J Microbiol Biotechnol 37:1–17

Kour D, Rana KL, Yadav AN, Yadav N, Kumar V, Kumar A et al (2019) Drought-tolerant phosphorus-solubilizing microbes: biodiversity and biotechnological applications for alleviation of drought stress in plants. In: Sayyed R, Arora N, Reddy M (eds) Plant growth promoting rhizobacteria for sustainable stress management. Microorganisms for sustainability, vol 12. Springer, Singapore, pp 255–308

Koyro HW, Ahmad P, Geissler N (2012) Abiotic stress responses in plants: an overview. In: Environmental adaptations and stress tolerance of plants in the era of climate change. Springer, Berlin, pp 1–28

Kumar S, Suyal DC, Bhoriyal M, Goel R (2018) Plant growth promoting potential of psychrotolerant Dyadobacter sp. for pulses and finger millet and impact of inoculation on soil chemical properties and diazotrophic abundance. J Plant Nutr 41(8):1035–1046

Kumar U, Kaviraj M, Rout S, Chakraborty K, Swain P et al (2021) Combined application of ascorbic acid and endophytic N-fixing *Azotobacter chroococcum* Avi2 modulates photosynthetic efficacy, antioxidants and growth-promotion in rice under moisture deficit stress. Microbiol Res 250:126808

Kusale SP, Attar YC, Sayyed RZ, El Enshasy H, Hanapi SZ, Ilyas N et al (2021) Inoculation of *Klebsiella variicola* alleviated salt stress and improved growth and nutrients in wheat and maize. Agronomy 11(5):927

Ladha JK, Reddy PM (2003) Nitrogen fixation in rice systems: state of knowledge and future prospects. Plant Soil 252:151–167

Lesk C, Anderson W (2021) Decadal variability modulates trends in concurrent heat and drought over global croplands. Environ Res Lett 16(5):055024

Li M (2018) Climate change to adversely impact grain production in China by 2030. IFPRI, Washington, DC

Li Z, Tian D, Wang B, Wang J, Wang S, Chen HY et al (2019) Microbes drive global soil nitrogen mineralization and availability. Glob Chang Biol 25(3):1078–1088

Li X, Sun P, Zhang Y, Jin C, Guan C (2020) A novel PGPR strain *Kocuria rhizophila* Y1 enhances salt stress tolerance in maize by regulating phytohormone levels, nutrient acquisition, redox potential, ion homeostasis, photosynthetic capacity and stress-responsive genes expression. Environ Exp Bot 174:104023

Limami AM, Diab H, Lothier J (2014) Nitrogen metabolism in plants under low oxygen stress. Planta 239(3):531–541

Liu M, Wang T, Skidmore AK, Liu X (2018) Heavy metal-induced stress in rice crops detected using multi-temporal Sentinel-2 satellite images. Sci Total Environ 637:18–29

Maiti RK, Satya P (2014) Research advances in major cereal crops for adaptation to abiotic stresses. GM Crops Food 5(4):259–279

Mian FA, Jarman TR, Righelato RC (1978) Biosynthesis of exopolysaccharide by *Pseudomonas aeruginosa*. J Bacteriol 134(2):418–422

Michaletti A, Naghavi MR, Toorchi M et al (2018) Metabolomics and proteomics reveal drought-stress responses of leaf tissues from spring-wheat. Sci Rep 8:5710

Minhas PS, Rane J, Pasala RK (2017) Abiotic stresses in agriculture: an overview. In: Minhas P, Rane J, Pasala R (eds) Abiotic stress management for resilient agriculture. Springer, New York, pp 3–8

Mishra J, Arora NK (2018) Secondary metabolites of fluorescent pseudomonads in biocontrol of phytopathogens for sustainable agriculture. Appl Soil Ecol 125:35–45

Mishra J, Singh R, Arora NK (2017) Alleviation of heavy metal stress in plants and remediation of soil by rhizosphere microorganisms. Front Microbiol 8:1706

Mishra P, Mishra J, Arora NK (2021) Plant growth promoting bacteria for combating salinity stress in plants—recent developments and prospects: a review. Microbiol Res 252:126861

Mitra D, Rodriguez AMD, Cota FIP, Khoshru B, Panneerselvam P, Moradi S et al (2021) Amelioration of thermal stress in crops by plant growth-promoting rhizobacteria. Physiol Mol Plant Pathol 115:101679

Montañez A, Blanco AR, Barlocco C, Beracochea M, Sicardi M (2012) Characterization of cultivable putative endophytic plant growth promoting bacteria associated with maize cultivars (*Zea mays* L.) and their inoculation effects in vitro. Appl Soil Ecol 58:21–28

Mukherjee P, Mitra A, Roy M (2019) Halomonas rhizobacteria of *Avicennia marina* of Indian Sundarbans promote rice growth under saline and heavy metal stresses through exopolysaccharide production. Front Microbiol 10:1207

Mukhtar S, Zareen M, Khaliq Z, Mehnaz S, Malik KA (2020) Phylogenetic analysis of halophyte-associated rhizobacteria and effect of halotolerant and halophilic phosphate-solubilizing biofertilizers on maize growth under salinity stress conditions. J Appl Microbiol 128(2):556–573

Nag P, Das S (2022) Microbiome to the rescue: nitrogen cycling and fixation in non-legumes. In: Maheshwari DK, Dobhal R, Dheeman S (eds) Nitrogen fixing bacteria: sustainable growth of non-legumes, vol 36. Springer Nature, Singapore, pp 195–214

Omer AM (2017) Using diazotrophic endophytes in improving some cereal production under saline desert condition. Egypt J Desert Res 67(1):210–229

Pandey S, Ghosh PK, Ghosh S, De TK, Maiti TK (2013) Role of heavy metal resistant Ochrobactrum sp. and Bacillus spp. strains in bioremediation of a rice cultivar and their PGPR like activities. J Microbiol 51(1):11–17

Patel JK, Archana G (2017) Diverse culturable diazotrophic endophytic bacteria from Poaceae plants show cross-colonization and plant growth promotion in wheat. Plant Soil 417:99–116

Patel JK, Agrawal R, Sidhdhapara R (2020) Root associated bacterial endophytes from Poaceae plants: identification, characterization and plant growth promotion. J Microbiol Biotechnol Food Sci 10(3):478–483

Payumo JG, Assem S, Bhooshan N, Galhena H, Mbabazi R, Maredia K (2018) Managing agricultural research for prosperity and food security in 2050: comparison of performance, innovation models and prospects. Open Agric J 12(1):20

Pedrosa FO, Oliveira ALM, Guimarães VF, Etto RM, Souza EM, Furmam FG et al (2020) The ammonium excreting *Azospirillum brasilense* strain HM053: a new alternative inoculant for maize. Plant Soil 451:45–56

Pfeiffer TŽ, Mihaljević M, Stević F, Špoljarić D (2013) Periphytic algae colonization driven by variable environmental components in a temperate floodplain lake. Int J Limnol 49(3):179–190

Pittol M, Durso L, Valiati VH, Fiuza LM (2016) Agronomic and environmental aspects of diazotrophic bacteria in rice fields. Ann Microbiol 66(2):511–527

Postgate JR (1982) Biological nitrogen fixation: fundamentals. Philos Trans R Soc Lond B Biol Sci 296(1082):75–385

Poza-Carrión C, Echavarri-Erasun C, Rubio LM (2015) Regulation of nif gene expression in *Azotobacter vinelandii*. In: Biological nitrogen fixation. Wiley, Hoboken, pp 99–108

Praburaman L, Park SH, Cho M, Lee KJ, Ko JA, Han SS et al (2017) Significance of diazotrophic plant growth-promoting Herbaspirillum sp. GW103 on phytoextraction of Pb and Zn by *Zea mays* L. Environ Sci Pollut Res 24:3172–3180

Pramanik K, Mitra S, Sarkar A, Maiti TK (2018a) Alleviation of phytotoxic effects of cadmium on rice seedlings by cadmium resistant PGPR strain *Enterobacter aerogenes* MCC 3092. J Hazard Mater 351:317–329

Pramanik K, Mitra S, Sarkar A, Soren T, Maiti TK (2018b) Characterization of a Cd2+-resistant plant growth promoting rhizobacterium (Enterobacter sp.) and its effects on rice seedling growth promotion under Cd2+-stress in vitro. Agric Nat Resour 52(3):215–221

Prasad PVV, Djanaguiraman M, Jagadish SVK, Ciampitti IA (2019) Drought and high temperature stress and traits associated with tolerance. In: Ciampitti IA, Vara Prasad PV (eds) Sorghum: a state of the art and future perspectives, Agronomy monographs, vol 58. Wiley, New York, pp 241–265

Rajput L, Imran A, Mubeen F, Hafeez FY (2018) Wheat (*Triticum aestivum* L.) growth promotion by halo-tolerant PGPR-consortium. Soil Environ 37(2):178–189

Rana S, Chauhan R, Walia A, Sharma G, Datt N (2021) Beneficial microbes in agriculture under abiotic stress conditions: an overview. Pharma Innov 10:360–368

Rascio N, Navari-Izzo F (2011) Heavy metal hyperaccumulating plants: how and why do they do it? And what makes them so interesting? Plant Sci 180(2):169–181

Razzaq A, Ali A, Safdar LB, Zafar MM, Rui Y, Shakeel A (2020) Salt stress induces physiochemical alterations in rice grain composition and quality. J Food Sci 85(1):14–20

Ren B, Dong S, Zhao B, Liu P, Zhang J (2017) Responses of nitrogen metabolism, uptake and translocation of maize to waterlogging at different growth stages. Front Plant Sci 8:1216

Rezaei EE, Siebert S, Ewert F (2015) Intensity of heat stress in winter wheat-phenology compensates for the adverse effect of global warming. Environ Res Lett 10(2):024012

Riyazuddin R, Nisha N, Ejaz B, Khan MIR, Kumar M, Ramteke PW (2022) A comprehensive review on the heavy metal toxicity and sequestration in plants. Biomolecules 12(1):43

Rizvi A, Khan MS (2017) Biotoxic impact of heavy metals on growth, oxidative stress and morphological changes in root structure of wheat (*Triticum aestivum* L.) and stress alleviation by *Pseudomonas aeruginosa* strain CPSB1. Chemosphere 185:942–952

Rizvi A, Khan MS (2018) Heavy metal induced oxidative damage and root morphology alterations of maize (*Zea mays* L.) plants and stress mitigation by metal tolerant nitrogen fixing *Azotobacter chroococcum*. Ecotoxicol Environ Saf 157:9–20

Rizvi A, Ahmed B, Zaidi A, Khan MS (2019) Heavy metal mediated phytotoxic impact on winter wheat: oxidative stress and microbial management of toxicity by *Bacillus subtilis* BM2. RSC Adv 9(11):6125–6142

Rizvi A, Zaidi A, Ameen F, Ahmed B, AlKahtani MD, Khan MS (2020) Heavy metal induced stress on wheat: phytotoxicity and microbiological management. RSC Adv 10(63):38379–38403

Roberson EB, Firestone MK (1992) Relationship between desiccation and exopolysaccharide production in a soil Pseudomonas sp. Appl Environ Microbiol 58(4):1284–1291

Rockström J, Steffen W, Noone K, Persson Å, Chapin FS, Lambin EF et al (2009) A safe operating space for humanity. Nature 461(7263):472–475

Rodrigues MÂ, Torres LDND, Damo L, Raimundo S, Sartor L, Cassol LC et al (2021) Nitrogen use efficiency and crop yield in four successive crops following application of biochar and zeolites. J Soil Sci Plant Nutr 21:1053–1065

Rogers C, Oldroyd GE (2014) Synthetic biology approaches to engineering the nitrogen symbiosis in cereals. J Exp Bot 65(8):1939–1946

Rosenblueth M, Martínez-Romero J, Ramírez-Puebla ST, Vera-Ponce de León A, Rosas-Pérez T, Bustamante-Brito R et al (2018a) Endosymbiotic microorganisms of scale insects. TIP Revista especializada en ciencias químico-biológicas 21(1):53–69

Rosenblueth M, Ormeño-Orrillo E, López-López A, Rogel MA, Reyes-Hernández BJ, Martínez-Romero JC et al (2018b) Nitrogen fixation in cereals. Front Microbiol 9:1794

Saad M, Abokoura H, Abdellatif K, El-Latif M, Aly M (2020) Application of alginate bead encapsulated N2-fixing bacteria is improving wheat yield under drought stress. Plant Arch 20:1735–1747

Saghafi D, Lajayer BA, Ghorbanpour M (2020) Engineering bacterial ACC deaminase for improving plant productivity under stressful conditions. In: Molecular aspects of plant beneficial microbes in agriculture. Elsevier, Amsterdam, pp 259–277

Saikia SP, Jain V (2007) Biological nitrogen fixation with non-legumes: an achievable target or a dogma? Curr Sci 92(3):317–322

Saikia J, Sarma RK, Dhandia R, Yadav A, Bharali R, Gupta VK et al (2018) Alleviation of drought stress in pulse crops with ACC deaminase producing rhizobacteria isolated from acidic soil of Northeast India. Sci Rep 8(1):3560

Sakpirom J, Kantachote D, Siripattanakul-Ratpukdi S, McEvoy J, Khan E (2019) Simultaneous bioprecipitation of cadmium to cadmium sulfide nanoparticles and nitrogen fixation by *Rhodopseudomonas palustris* TN110. Chemosphere 223:455–464

Santi C, Bogusz D, Franche C (2013) Biological nitrogen fixation in non-legume plants. Ann Bot 111(5):743–767

Sapre S, Gontia-Mishra I, Tiwari S (2019) ACC deaminase-producing bacteria: a key player in alleviating abiotic stresses in plants. In: Kumar A, Meena V (eds) Plant growth promoting rhizobacteria for agricultural sustainability. Springer, Singapore, pp 267–291

Selvakumar G, Panneerselvam P, Ganeshamurthy AN (2012) Bacterial mediated alleviation of abiotic stress in crops. In: Maheshwari D (ed) Bacteria in agrobiology: stress management. Springer, Berlin, pp 205–224

Seneviratne M, Rajakaruna N, Rizwan M, Madawala HMSP, Ok YS, Vithanage M (2019) Heavy metal-induced oxidative stress on seed germination and seedling development: a critical review. Environ Geochem Health 41(4):1813–1831

Shabanamo S, Divya K, George TK, Rishad KS, Sreekumar TS, Jisha MS (2018) Characterization and in planta nitrogen fixation of plant growth promoting endophytic diazotrophic *Lysinibacillus sphaericus* isolated from rice (*Oryza sativa*). Physiol Mol Plant Pathol 102:46–54

Shaffique S, Khan MA, Wani SH, Pande A, Imran M, Kang SM et al (2022) A review on the role of endophytes and plant growth promoting rhizobacteria in mitigating heat stress in plants. Microorganisms 10(7):1286

Shahid M, Khalid S, Abbas G, Shahid N, Nadeem M, Sabir M et al (2015) Heavy metal stress and crop productivity. In: Hakeem K (ed) Crop production and global environmental issues. Springer, Cham, pp 1–25

Sheoran S, Kumar S, Kumar P, Meena RS, Rakshit S (2021) Nitrogen fixation in maize: breeding opportunities. Theor Appl Genet 134:1263–1280

Shultana R, Kee Zuan AT, Yusop MR, Saud HM, El-Shehawi AM (2021) *Bacillus tequilensis* strain 'UPMRB9' improves biochemical attributes and nutrient accumulation in different rice varieties under salinity stress. PLoS One 16(12):e0260869

Solomon S (2007) Climate change 2007—the physical science basis: working group I contribution to the fourth assessment report of the IPCC, vol 4. Cambridge University Press, Cambridge

Song Y, Li J, Liu M, Meng Z, Liu K, Sui N (2019) Nitrogen increases drought tolerance in maize seedlings. Funct Plant Biol 46(4):350–359

Sorty AM, Meena KK, Choudhary K, Bitla UM, Minhas PS, Krishnani KK (2016) Effect of plant growth promoting bacteria associated with halophytic weed (*Psoralea corylifolia* L) on germination and seedling growth of wheat under saline conditions. Appl Biochem Biotechnol 180:872–882

Soumare A, Diedhiou AG, Thuita M, Hafidi M, Ouhdouch Y, Gopalakrishnan S et al (2020) Exploiting biological nitrogen fixation: a route towards a sustainable agriculture. Plants 9(8):1011

Sultana S, Paul SC, Parveen S, Alam S, Rahman N, Jannat B et al (2020) Isolation and identification of salt-tolerant plant-growth-promoting rhizobacteria and their application for rice cultivation under salt stress. Can J Microbiol 66(2):144–160

Sun ZC, Ma TY, Xu SQ, Guo HR, Hu CC, Chen CJ et al (2022) Levels and variations of soil bioavailable nitrogen among forests under high atmospheric nitrogen deposition. Sci Total Environ 838:156405

Tufail MA, Touceda-González M, Pertot I, Ehlers RU (2021) *Gluconacetobacter diazotrophicus* Pal5 enhances plant robustness status under the combination of moderate drought and low nitrogen stress in Zea mays L. Microorganisms 9(4):870

Upadhyaya H, Panda SK (2019) Drought stress responses and its management in rice. In: Hasanuzzaman M, Fujita M, Nahar K, Biswas JK (eds) Advances in rice research for abiotic stress tolerance. Woodhead Publishing, Sawston, pp 177–200

Van Deynze A, Zamora P, Delaux PM, Heitmann C, Jayaraman D, Rajasekar S et al (2018) Nitrogen fixation in a landrace of maize is supported by a mucilage-associated diazotrophic microbiota. PLoS Biol 16(8):e2006352

Vance D, Neary L, Sueker J, Nelson D (2015) Nitrate remediation: the importance of the advanced nitrogen cycle in remedy selection. Remediat J 26(1):93–105

Vanisree CR, Sankhla MS, Singh P, Jadhav EB, Verma RK, Awasthi KK et al (2022) Heavy metal contamination of food crops: transportation via food chain, human consumption, toxicity and management strategies. In: Environmental impact and remediation of heavy metals, vol 316. IntechOpen, London

Venkateshwaran M (2014) Exploring the feasibility of transferring nitrogen fixation to cereal crops. In: Lugtenberg B (ed) Principles of plant-microbe interactions. Springer, Cham, pp 403–410

Verma SC, Ladha JK, Tripathi AK (2001) Evaluation of plant growth promoting and colonization ability of endophytic diazotrophs from deep water rice. J Biotechnol 91(2–3):127–141

Verma C, Singh P, Kumar R (2015) Isolation and characterization of heavy metal resistant PGPR and their role in enhancement of growth of wheat plant under metal (cadmium) stress condition. Arch Appl Sci Res 7(7):37–43

Verma P, Yadav AN, Khannam KS, Mishra S, Kumar S, Saxena AK et al (2019) Appraisal of diversity and functional attributes of thermotolerant wheat associated bacteria from the peninsular zone of India. Saudi J Biol Sci 26(7):1882–1895

Verma KK, Song XP, Li DM, Singh M, Rajput VD, Malviya MK et al (2020) Interactive role of silicon and plant–rhizobacteria mitigating abiotic stresses: a new approach for sustainable agriculture and climate change. Plants 9(9):1055

Verma S, Verma R, Verma P, Bharti C, Arora NK (2021) Salt tolerant endophytic and diazotrophic strain of *Proteus Mirabilis* Pd25 and its effect on the growth of wheat under saline conditions. NVEO 8(5):13172–13183

Vimal SR, Patel VK, Singh JS (2019) Plant growth promoting *Curtobacterium albidum* strain SRV4: an agriculturally important microbe to alleviate salinity stress in paddy plants. Ecol Indic 105:553–562

Vinocur B, Altman A (2005) Recent advances in engineering plant tolerance to abiotic stress: achievements and limitations. Curr Opin Biotechnol 16(2):123–132

Vurukonda SSKP, Vardharajula S, Shrivastava M, SkZ A (2016) Enhancement of drought stress tolerance in crops by plant growth promoting rhizobacteria. Microbiol Res 184:13–24

Vwioko ED, El-Esawi MA, Imoni ME, Al-Ghamdi AA, Ali HM, El-Sheekh MM (2019) Sodium azide priming enhances waterlogging stress tolerance in okra (*Abelmoschus esculentus* L.). Agronomy 9(11):679

Wang H, Wu Z, Zhou Y, Han J, Shi D (2012) Effects of salt stress on ion balance and nitrogen metabolism in rice. Plant Soil Environ 58(2):62–67

Wang Y, Lu J, Ren T, Hussain S, Guo C, Wang S et al. (2017) Effects of nitrogen and tiller type on grain yield and physiological responses in rice. AoB Plants 9(2): plx012

Wang J, Vanga SK, Saxena R, Orsat V, Raghavan V (2018a) Effect of climate change on the yield of cereal crops: a review. Climate 6(2):41

Wang XH, Luo WW, Wang Q, He LY, Sheng XF (2018b) Metal(loid)-resistant bacteria reduce wheat Cd and As uptake in metal(loid)-contaminated soil. Environ pollut 241:529–539

Wen Z, Browning TJ, Dai R, Wu W, Li W, Hu X et al (2022) The response of diazotrophs to nutrient amendment in the South China Sea and western North Pacific. Biogeosciences 19(22):5237–5250

Xiong Q, Zhong L, Du J, Zhu C, Peng X, He X et al (2020) Ribosome profiling reveals the effects of nitrogen application translational regulation of yield recovery after abrupt drought-flood alternation in rice. Plant Physiol Biochem 155:42–58

Yadav S, Modi P, Dave A, Vijapura A, Patel D, Patel M (2020) Effect of abiotic stress on crops. In: Sustainable crop production, vol 3. IntechOpen, Rijeka

Yang M, Geng M, Shen P, Chen X, Li Y, Wen X (2019) Effect of post-silking drought stress on the expression profiles of genes involved in carbon and nitrogen metabolism during leaf senescence in maize (*Zea mays* L.). Plant Physiol Biochem 135:304–309

Zarei T, Moradi A, Kazemeini SA, Akhgar A, Rahi AA (2020) The role of ACC deaminase producing bacteria in improving sweet corn (*Zea mays* L. var saccharata) productivity under limited availability of irrigation water. Sci Rep 10(1):20361

Zhu C, Kobayashi K, Loladze I, Zhu J, Jiang Q, Xu X et al (2018) Carbon dioxide (CO2) levels this century will alter the protein, micronutrients, and vitamin content of rice grains with health consequences for the poorest rice-dependent countries. Sci Adv 4(5):eaaq1012

Zörb C, Geilfus CM, Dietz KJ (2019) Salinity and crop yield. Plant Biol 21:31–38

Chapter 7
Role of Transcriptomics in Elucidating Mechanism of Abiotic Stress Tolerance in Plants

Mohd. Zahid Rizvi, Murtaza Abid, S. N. Pandey, and M. M. Abid Ali Khan

Abstract Abiotic stresses like drought, high temperature spells and salinity can seriously deteriorate the crop yield, which can worsen the food supply scenario at global level. Recently, transcriptomics approach has been used for a finer and deeper analysis of stress response mechanisms and adaptation of plants experiencing stress conditions. Comparative study of stress-tolerant and stress-sensitive plant varieties at transcriptomics level can shed further light on stress-adaptive strategies in plants. Employing these approaches, some stress-associated mechanisms have been observed and further research is being done in terms of single cell transcriptomic analysis and developmental stage-specific analysis for further refined knowledge. Recently, many refined techniques such as microarray analysis, RNA-seq analysis, genomic-scale expressed sequence tags and next generation sequencing have been developed which are now being utilized for generating stress response related data and will enhance our understanding of stress-responsive mechanisms. The information obtained through these approaches will contribute toward developing plants especially crop plants better equipped to counter and adapt to various abiotic stress conditions without compromising their yield potential. Plant growth promoting rhizobacteria are an important group of rhizospheric microorganisms which help plants in mitigation of abiotic stress by acting as biostimulants and inducing modifications at molecular, biochemical and physiological levels. In this chapter, stress-responsive and adaptive mechanisms at transcriptomics level in plants exposed to various abiotic stress conditions such as high temperature, drought and salinity have been discussed. Transcriptomic alterations induced by plant growth promoting rhizobacteria in abiotic stress tolerance in plants are also discussed.

M. Z. Rizvi
Department of Botany, Shia Post Graduate College, Lucknow, Uttar Pradesh, India

M. Abid · S. N. Pandey
Department of Botany, University of Lucknow, Lucknow, Uttar Pradesh, India

M. M. Abid Ali Khan (✉)
Department of Botany, Shia Post Graduate College, Lucknow, Uttar Pradesh, India

Department of Botany, University of Lucknow, Lucknow, Uttar Pradesh, India

N. K. Arora, B. Bouizgarne (eds.), *Microbial Biotechnology for Sustainable Agriculture Volume 2*, Microorganisms for Sustainability 34,
https://doi.org/10.1007/978-981-97-2355-3_7

Keywords Abiotic stress · Plants · Drought · Plant growth promoting rhizobacteria · Salinity · Transcriptomics

7.1 Introduction

The global human population has been increasing swiftly and this trend may intensify in coming years, which is a problem for current and future human populations as far as their food requirements are concerned (Ng et al. 2017; Sinha et al. 2021). Climate change, which was earlier perceived as local phenomenon, has actually consequences at global level, affecting agricultural productivity worldwide (Ahmad and Prasad 2011). There is change in structure and function in crops at physiological, biochemical as well as in molecular responses when exposed to stresses like enhanced temperature, drought, excess salt, etc., which later on affect their production in a negative way (Sharma et al. 2013; Yang et al. 2020). As their survival strategy, plants show adaptive responses, when faced with stress. Change in normal osmotic state by excess salt and drought stress causes disturbance of ionic homeostasis at cellular level (Zhu 2001; Wang et al. 2003). There is damage to functionality and structure of proteins when faced with oxidative stress as consequence of excess salt and enhanced temperature (Wang et al. 2003). Adaptation mechanisms in most of the cases in plants involve a core network of numerous signalling pathways (Zhu 2001), regulatory genes, induction of compounds counteracting to oxidative disturbance, stress-tackling proteins and stocking of solutes involved in counteracting the damage, which may be caused by osmotic stress (Wang et al. 2003; Kang and Khan 2009). Gaining the knowledge about adaptive strategy of plants to varied abiotic stresses at biochemical and biomolecular level can be very helpful in maintaining and even alleviating yields of plants under stress exposure (Zhu 2016; Zhu et al. 2019). Although advancements have been made in understanding stressed plants' responses and counteracting damage resulting from stress exposure and development of varieties with high yield even under stress conditions, still more research is required in this direction (Wohlbach et al. 2008).

Understanding plant's adaptive responses to a variety of stresses at transcriptomic level can be very helpful in deeper understanding of mechanism behind these types of responses (Liu et al. 2014; Rasheed et al. 2016b). In counteraction of plants to various stresses, transcription of concerned genes is controlled, which subsequently affects their transcriptome (Larkindale and Vierling 2008; Qin et al. 2008). Regulation of genes at transcriptional level affects many cellular processes at molecular level like transcription, translation, signalling and metabolism, which are core activities behind plant's stress response. Recently, many researchers have conducted studies to understand the mechanism of adaptive responses of multi-stress tolerance in plants. Latest in these studies are omics approaches, which involve study of large data at molecular level for detection of responsive genes (studied under genomics), their transcripts such as mRNA (analyzed in transcriptomics studies) and effector molecules, proteins (proteomics study) and metabolites (metabolomics analysis) (Basim et al. 2021). Recent technological advances in sophisticated technologies, such as microarrays, RNA sequencing, genomic-scale expressed sequence tags (ESTs), genomic sequencing and complementary DNA (cDNA) microarray analysis have been helpful in generating transcriptomics data by studying and identifying the genes of the 'osmome', 'xemome' or 'thermome'. 'Osmome', 'xemome' or 'thermome' are the gene complements behind adaptation against osmotic stress, drying or temperature variability, respectively (Roychoudhury and Chakraborty 2013). Microarray technology is very helpful in observing the expression of numerous genes simultaneously. DNA microarrays are one of the most commonly used microarrays, which are basically very small slides (gene chips or DNA chips) with spots having definite physical locations on slide. Each spot harbors a previously known DNA sequence or gene. The DNA molecular sequences in each slide act as probes to detect gene expression (transcriptome or the group of messenger RNA (mRNA) transcripts expressed by a set of genes). Chang (1983) for the first time put light on concept of microarrays (Fig. 7.1). ESTs are short length cDNA sequences employed to 'tag' the gene which is the source of mRNA (Adams et al. 1991). ESTs may be employed to gain knowledge of gene transcripts and gene sequences (Fig. 7.2). ESTs may be used in construction of probes for DNA microarrays. Due to the development of recent precise and rapid techniques, the expression pattern of large number of genes and their co-expression partners in response to different types of abiotic stresses can be studied in detail in short time (Atkinson et al. 2013). Another technique, meta-analysis can be very helpful in studying differential gene expression. Through this technique, genes and their products, which are behind core response to stress, can be identified and they can be employed in plant breeding programs for stress resistance (Shaar-Moshe et al. 2015; Ashrafi-Dehkordi et al. 2018). In this chapter, transcriptomics analysis studies in plants experiencing abiotic stress conditions such as high temperature, drought and salinity have been discussed.

7.2 Effect of Abiotic Factors, Drought, Salinity and Elevated Temperature on Transcriptome

Consequences of abiotic stress exposure in plants are triggering physiological adaptations like changes in osmotic potential, decreasing leaf turgor pressure and subsequently reduced plant growth adaptations at molecular and transcriptomics level (Tardieu et al. 2018; Li et al. 2021a). Roots of *Arabidopsis* are the organs which are more affected by drought stress, thus exhibiting more transcriptomic changes as compared to shoots of *Arabidopsis*. The reason may be many novel genes behind response of roots to drought (Rasheed et al. 2016a, Bashir et al. 2018). Phytohormones like abscisic acid (ABA) occupy a central position in plant reaction to salinity, drought and high temperature stress. Osmotic stress is observed frequently in cells experiencing these stresses (Laloum et al. 2018). The stress signals perceived by ABA intracellularly, and transduction of signals by ABA cause induction of transcription factors and ion channels, which then initiate and sustain adaptive changes like closing of stomata, synthesis of compounds providing protection to osmotic

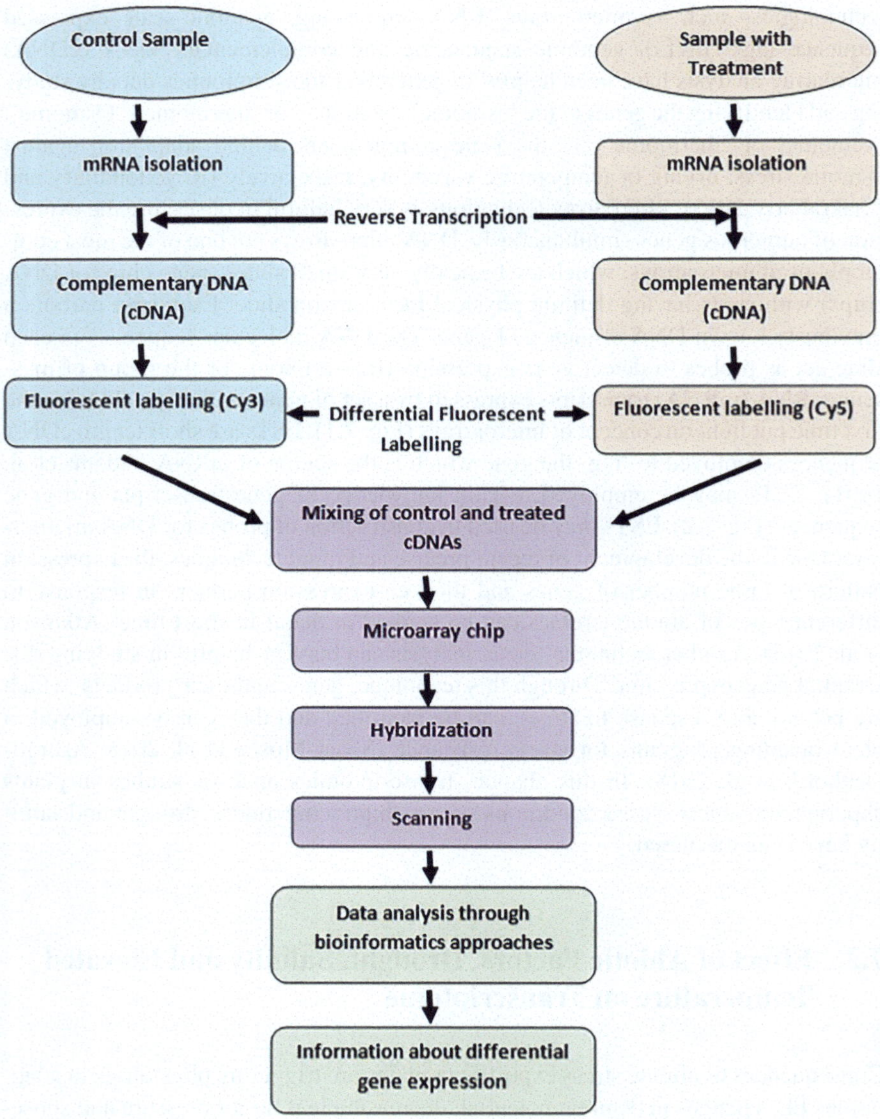

Fig. 7.1 cDNA microarray technology

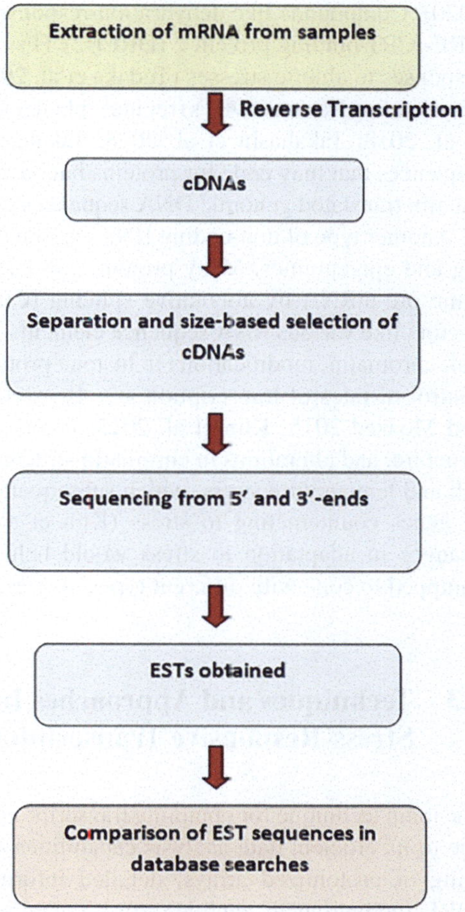

Fig. 7.2 Generation of expressed sequence tags (ESTs)

changes and the induction of many other stress-responsive genes especially osmotic stress-responsive (OR) genes.

The overall result of these responses is increased capacity of plants to tackle water stress (Laloum et al. 2018). The OR genes include those controlling storage of many solute compounds related to coping with osmotic stress (including proline, reducing sugar and polyamine compounds such as spermidine and spermine), antioxidant compounds and many types of late embryogenesis abundant (LEA) proteins. Another type of OR genes are those associated with regulation of signalling pathways like transcription factors (TFs), mostly related to the basic leucine zipper (bZIP), mitogen-activated protein kinases (MAPKs), calcium-dependent protein kinases (CDPKs), protein phosphatases and proteinases like phosphoesterases and phospholipase C (Roychoudhury et al. 2013). The changes in TFs induced by the kinases trigger their attachment to the upstream regions of target genes. These genes then initiate and sustain many responses involved in adaptation to stress (Zhao et al.

2020). Compounds like dehydration-responsive element/C-repeat (DRE/CRT) and DRE-/CRT-binding protein 2 (DREB2) TFs are associated with ABA-independent responses to abiotic stresses (Todaka et al. 2015). The peptide hormones and small open reading frames (sORFs) regulate plant response to abiotic stresses (Nakaminami et al. 2018; Takahashi et al. 2018; Takahashi and Shinozaki 2019). sORFs are sequences that may code for proteins but have shorter lengths as compared to other usually translated genomic DNA sequences (Andrews and Rothnagel 2014).

Another type of non-coding RNAs are involved in control of transcription, splicing and epigenomics. Many proteins can be synthesized from the same gene and same pre-mRNA by alternative splicing (Chekanova 2015; Laloum et al. 2018). Factors like various RNA sequence elements, changes in structure of nuclear material, chromatin, modifications in histone proteins associated with nucleic acids and control of rates of transcription are also significant for alternate splicing (Holoch and Moazed 2015; Kim et al. 2015; Pajoro et al. 2017). The changes in histone structure, and alterations in chromatin structure in plants take place due to drought, salt and temperature stress, which subsequently cause alterations in the expression of genes counteracting to stress (Kim et al. 2015). A deeper understanding of changes in adaptation to stress would help in developing plant varieties better equipped to cope with different types of stresses.

7.3 Techniques and Approaches Involved in Obtaining Stress-Responsive Transcriptomics Data

The main technique for obtaining transcriptomics data has been microarray analysis due to its efficient data analysis capabilities and cost-effectiveness. By employing tiling or customized arrays, detailed information about non-coding RNAs and sORFs coding small peptides can be obtained (Hanada et al. 2013; Bashir et al. 2014). Another technology, RNA sequencing (RNA-seq) analysis, is being preferred nowadays for transcriptomic analysis due to its cost-effectiveness, capability of very detailed data analysis and specific analysis of alterations in transcriptome and whole genome as a result of exposure to abiotic stresses (Krizek et al. 2016; Xuan et al. 2022). RNA-seq is a technology that is employed to gain knowledge about amount and sequences of RNA at any given time (Fig. 7.3). It can be used to examine the dynamic transcriptomics pattern within our cells (Wang et al. 2009). Detailed analysis of transcriptomics data can reveal new genes and promoters (Rasheed et al. 2016b, 2018; Kim et al. 2017). In this context, genes whose function is unclear could be studied to reveal their functionalities under varied stress conditions. This data could be mined to reveal genetic markers that can be potentially applied for marker-assisted breeding (Nguyen et al. 2017). The response of plants to same abiotic stress may be different depending upon developmental stage, intensity of stress, time period for which plant is exposed to stress and point of time at which data is taken. This may be due to the triggering of different set of genes during

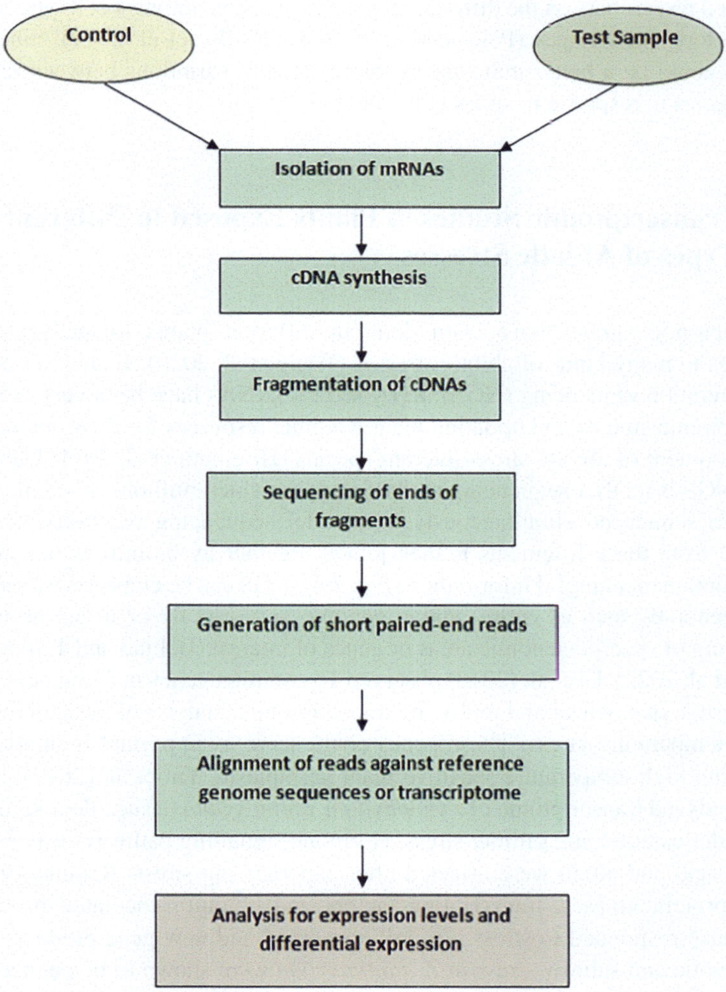

Fig. 7.3 An overview of RNA sequencing technique

different conditions (Rasheed et al. 2016b). Diurnal rhythms are another factor which control transcriptional alterations in counteraction to abiotic stress (Webb et al. 2019). Transcriptomics data may also change for different tissues and different cells within the same tissue. For these reasons, single cell transcriptomic analysis may be a better indicator of fine transcriptomic changes in response to abiotic stress (Liu and Trapnell 2016). But single cell transcriptomic studies are not easy and involve high technical expertise (Kiselev et al. 2019; Tung et al. 2023). The difference in transcriptomic data on cell to cell basis may be due to variable switching 'on' and switching 'off' of transcription of genes (Liu and Trapnell 2016). Single cell transcriptomic analysis in *Arabidopsis* roots and embryos has enlightened concerned researchers on the differential gene expression amongst cell types at various developmental stages (Palovaara et al. 2017; Shulse et al. 2019). Single cell analysis could be a better indicator of transcriptomic variations between different plant species in response to stress (Libault et al. 2017).

7.4 Transcriptomic Studies in Plants Exposed to Different Types of Abiotic Stresses

Transcriptomic studies have been done in different plants to analyze plant's responses to many kinds of abiotic stresses (Wang et al. 2020). Techniques such as next generation sequencing (NGS), RNA-seq for sRNAs have been very helpful in transcriptomic studies and updating plant genomic resources for their potential use in development of abiotic stress-tolerant species (Hiremath et al. 2011; Gaur et al. 2012). NGS is a DNA sequencing methodology in which millions of small parts of DNA are sequenced simultaneously in parallel sequencing reactions. The data obtained from these fragments is then joined together by bioinformatics applications to obtain meaningful information (Fig. 7.4). NGS can be employed to sequence whole genomes such as entire human genome very rapidly or it can be used in sequencing of specific genomic areas or genes of interest (Behjati and Tarpey 2013; Ashraf et al. 2022). Li et al. (2013) observed 16 common genes in 4 monocot plants, switch grass, rice, wheat and maize, by transcriptomic analysis of these in response to high temperature stress. These genes could serve as important biomarkers for identifying high temperature sensitive plant germplasm. Yin et al. (2019), in their study, analyzed transcriptome of *Zygophyllum xanthoxylum* (a succulent xerophyte) roots under osmotic and salinity stress. Hormone signalling pathways especially of abscisic acid and auxin were affected after experiencing stress. Regulatory genes coding protein kinases, transcription factors and ubiquitin-mediated proteolysis system also responded to stress. Overall conserved and new gene products linked with osmotic and salinity stress in *Z. xanthoxylum* were shown to be connected. In the transcriptome sequencing of *Agave sisalana* by Sarwar et al. (2019), many genes and pathways linked with response to abiotic stress such as stress-hormones, antioxidants, biosynthesis of wax and reactive oxygen species (ROS) were identified, in addition to transcripts related to numerous other drought-responsive genes. In the study of Wakasa et al. (2014), endoplasmic reticulum (ER) stress response and ER stress-related novel genes were identified in rice plants. Rasmussen et al. (2013) analyzed stress-related responses by comparing transcriptomic changes in ten *Arabidopsis thaliana* ecotypes experiencing varied kinds of single stresses and their combinations.

It was exhibited that 61% of the transcriptomic alterations in counteraction to double stress combinations were varied and unpredictable from the responses to only single stress exposure of plants. In field environments, plants frequently experience multiple stresses simultaneously instead of only single stress. Therefore, understanding of regulatory pathways and their components activated in

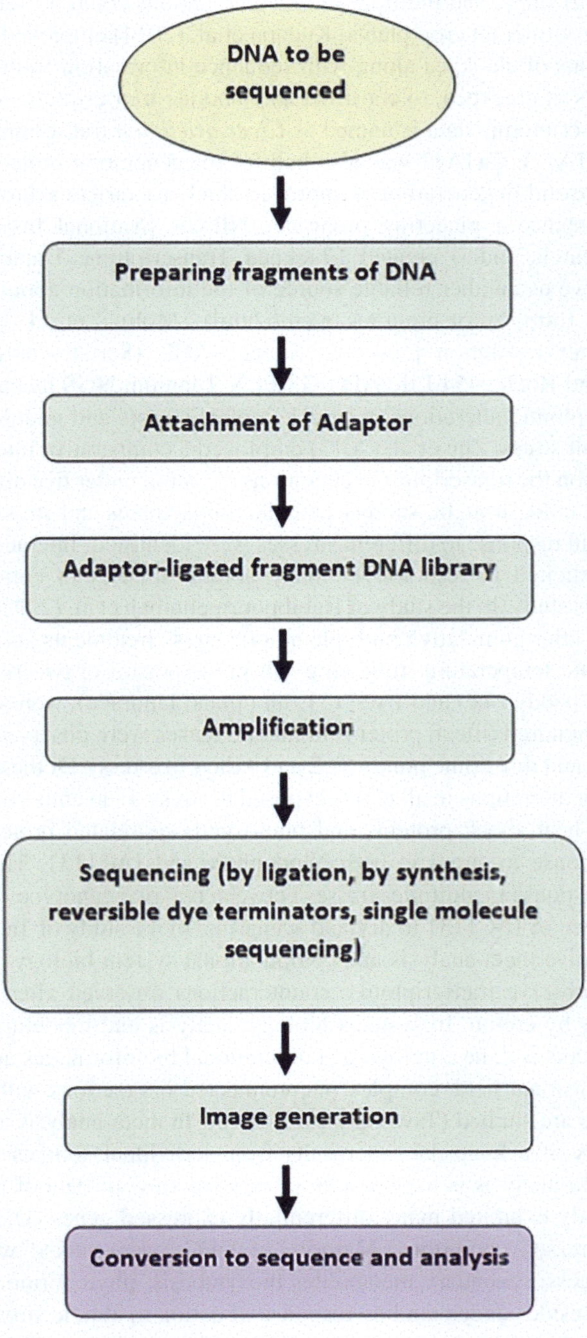

Fig. 7.4 Next generation sequencing (NGS)

counteraction to single and multiple stress combinations could be helpful in development of stress-tolerant crop plants. Kudapa et al. (2014) employed many Sanger ESTs collections of chickpea along with sequence information from two different NGS platforms of chickpea, to construct an elaborate transcriptome data of chickpea. This transcriptomic data is named as *Cicer arietinum* transcriptome assembly version 2 (CaTAv2). CaTAv2 was also helpful for generating molecular markers, which can be useful in generation of improved chickpea varieties through appropriate stress resistance engineering programs. NIPGR (National Institute of Plant Genome Research, India) created Chickpea Transcriptome Database (CTDB), which will serve as another reliable source of the information about the chickpea transcriptome (http://www.nipgr.res.in/ctdb.html). Molina et al. (2008, 2011) employed better resolution power of Super SAGE (Serial Analysis of Gene Expression) and Roche 454 Life/APG GS FLX Titanium NGS technique to study global transcriptomic alterations in chickpea plant's roots and nodules exposed to drought and salt stress. Zhu et al. (2013) employed a comparative microarray strategy to work upon the transcriptomic alterations of cotton under five different abiotic stresses: ABA, cold, drought, salinity and alkalinity. Genes and stress-linked pathways induced in response to different stresses were identified. Interactions of genes or pathways elicited in response to many abiotic stresses in cotton were also observed in this study. In the study of Habibpourmehraban et al. (2022), physiological alterations after cumulative multiple abiotic stress treatments such as drought, salt and extreme temperature, following salt pre-exposure in two rice genotypes, Nipponbare (a paddy rice) and IAC1131 (an upland landrace), were observed. Six thousand two hundred fifteen proteins in mature leaves were observed between the two genotypes and three time points (0, 2 and 4 days of stress). Of these, 41 proteins exhibited major alterations in all plants exposed to stress. Late embryogenesis abundant proteins, heat shock proteins and photosynthesis-related proteins exhibited consistent response to stress in both Nipponbare and IAC1131. The differential translational response to multiple stresses between two rice genotypes is in line with known resistance of IAC1131 to dryland scenarios. In the study of Tahmasebi et al. (2019), integrative meta-analysis and computational system biology analysis were performed to observe transcriptomic counteractions observed after experiencing abiotic stresses by cotton. In systems biology, analysis and modeling of complex biological systems is done employing computational bioinformatics approaches. In this multidimensional field, complex interconnected interactions within large biological systems are studied (Tavassoly et al. 2018). In meta-analysis approach, statistical analysis of a huge data of results from individual sources is done. The purpose of meta-analysis is to obtain an integrative view of data of results (Glass 1976). The study exhibited many differentially expressed genes (DEGs) between normal and stressed conditions. Majority of DEGs were linked with ubiquitin dependent process, secondary metabolites biosynthesis, phytohormone and signal transduction. Besides understanding response of cotton to abiotic stress, their work can also be employed in using candidate DEGs in development of molecular biomarkers and stress-resistant cotton plants. In study of Biniaz et al. (2022), a meta-analysis was done to analyze differential and shared genes, pathways and transcriptomic mechanisms associated with *Arabidopsis* response to biotic and abiotic stresses. It was observed that 21 DEGs have common control during exposure to biotic and abiotic stresses. WRKY45 and ATXTH22 were exceptions, which were, respectively, up/down- and down/upregulated upon exposure to biotic and abiotic stresses, while other common DEGs were upregulated during exposure to all biotic and abiotic stresses. The transcription factors (TFs) bHLH, MYB and WRKY were observed to be the common TFs during exposure to biotic and abiotic stresses. miRNAs which have common expression during both biotic and abiotic stresses were ath-miR414 and ath-miR5658. This study may be helpful in developing plants better adapted to biotic and/or abiotic stresses. In the study of Cohen and Leach (2019), rice transcriptome data in public domain were employed to identify genes and pathways controlled by abiotic stress, biotic stress and both types of stresses. Photosynthesis was found to be a downregulated pathway in adaptation to all kinds of stresses. Most probably downregulation of photosynthetic pathway is part of a protective response against plant photo-oxidative damage during stress (Brestic et al. 2016). In stress-tolerant varieties of rice, there is upregulation of photosynthesis and photosynthetic efficiency returned to normal, which is important for consistent yield even under stress exposure (Zhang et al. 2016). In line with these observations, overexpression of a master regulator of photosynthesis increased tolerance of rice to drought (Ambavaram et al. 2014). In the study of Cohen and Leach (2019), many stress-elicited phytohormone-responsive genes, especially genes downstream of ABA, jasmonic acid (JA) and salicylic acid (SA) were identified. These inter-related phytohormonal pathways are very important in plant's response to stress. Potential genes to study for enhancing tolerance of rice to stress and subsequently yield enhancement were also given in their work. Filichkin et al. (2018) studied leaf, root and stem xylem transcriptomes exposed to drought, salinity or temperature stress conditions. Different abiotic stress treatments changed the transcriptome of poplar through alternative splicing, differential intron retention (enhanced or decreased retention of specific introns in response to stress) and transcript isoform ratio change in a stress or tissue-specific way (or both). Findings from this study indicate toward the phenomenon that there are changes in relative ratios between fully spliced and intron-containing isoforms in response to stress, which is possibly important in adapting transcriptome to abiotic stresses. In the study of González-Schain et al. (2016), RNA-seq analysis in rice exhibited that expression of genes encoding heat shock proteins and heat shock factors during anthesis are activated and the genes encoding many transcription factor families, or genes concerned with signal transduction and metabolic pathways are repressed in a heat-tolerant rice cultivar compared with a heat-sensitive cultivar. This study pointed toward requirement of proper expression of protective chaperones in anthers prior to anthesis to cope with damage caused by stress and for fertilization. Chaperones are group of proteins which usually help in protein folding under physiological and stress scenarios. They help in preventing non-specific aggregation (Beissinger and Buchner 1998). Heat shock proteins are a type of chaperones, which are induced on high temperature exposure and help in preventing incorrect protein folding and aggregation into nonfunctional structures, which may be otherwise induced by stress (https://www.nature.com/subjects/chaperones). In many research works, transcriptional or metabolic changes (or both) were analyzed in wheat and triticale to work out the stress response mechanisms in these plants, when exposed to stress (Cagirici et al. 2017; Ullah et al. 2017; Gálvez et al. 2019). Transcriptomic studies have exhibited temporal changes in the expression of genes that are involved in adaptation to high temperature and drought in wheat (*Triticum aestivum* L.). Many wheat homeologs have shown expression partitioning when experiencing normal and abiotic stress conditions, which may be the reason behind the wide adaptability of hexaploid wheat in response to environmental conditions like stress (Liu et al. 2015). Varshney et al. (2017) have done comparative studies of the genome sequences of various pearl millet varieties, which may be helpful in improvement of stress tolerance of this important crop in arid zones. In their work, significant enrichment for wax biosynthesis genes was emphasized, which may be behind heat and drought resistance in this crop. Some important recent transcriptomic studies in plants exposed to different types of abiotic stresses are discussed below (Table 7.1).

7.5 Transcriptomics in Elucidating Role of Plant Growth Promoting Rhizobacteria in Mitigating Abiotic Stresses

The plant rhizosphere harbors various microorganisms having considerable impact on plant growth and development or adaptation and survival under stressful conditions (Genitsaris et al. 2020). Plant growth promoting rhizobacteria (PGPR) are important microorganisms amongst them. PGPR consist of microorganisms living inside plants as endophytes besides those living on exterior of plant but are differentially attached with the rhizospheric plane (Bhat et al. 2020). When plants are exposed to abiotic stress, PGPR associated with them adopt myriad of strategies which help in survival of plants. Some of these strategies are secretion of osmolytes like proline, glycine betaine and choline. These osmolytes perform role of osmoprotectants; production of 1-aminocyclopropane-1-carboxylic acid (ACC) deaminase, which reduces quantity of ethylene in plants and helps them in coping with abiotic stress (Gamalero and Glick 2015; Fatima and Arora 2023); secretion of plant growth regulators (cytokinins, gibberellins and auxins). They enhance root growth subsequently increasing water and nutrient absorption; secretion of volatile compounds, which help in mitigating stress (e.g., 2*R*,3*R*-butanediol stimulates closing of stomata); secretion of polysaccharide compounds, which enhance permeability by altering soil characteristics and maintaining appropriate water potential; altering root cell membrane elasticity, which helps in coping with stress (Conrath et al. 2002; Beckers and Conrath 2007). Antioxidants generated by PGPR have also been observed to play important role in stress adaptation by destroying reactive oxygen species (ROS) (Hasanuzzaman et al. 2021). Misra and Chauhan (2020) observed role of PGPR harboring ACC deaminase. PGPR in plants helped in controlling ethylene quantity by metabolizing ACC through ACC deaminase enzyme and thus

Table 7.1 Some important transcriptomic studies in plants during abiotic stress

Plants	Types of stress	Transcriptomic studies	References
Grape (*Vitis vinifera*)	Waterlogging	Transcriptomic investigations in grapevine revealed that waterlogging has been observed to increase expression of numerous genes associated with stress tolerance (including antioxidant system), anaerobic fermentation and glycolysis-associated genes, however, many genes related to chlorophyll synthesis and photosynthesis were downregulated. Genes associated with amino acid metabolic process and some hormones were also functionalized. These responses helped in decreasing damage to grapes due to waterlogging stress. Ethylene and ABA play a major role in plant's response to waterlogging, genes encoding class-2 nonsymbiotic hemoglobin were found to be important in waterlogging acclimation response	Zhu et al. (2018)
Eucalyptus globulus	Cold	In cold acclimated *E. globulus* plants, RNA-seq analysis identified 51 EglWRKY genes and expression of many EglWRKY genes was activated in leaf and root under cold stress scenarios	Aguayo et al. (2019)
Cotton (*Gossypium* spp.)	Drought, cold, salt, alkalinity, chilling and waterlogging	Integrative meta-analysis and computational system biology analysis were performed to elucidate transcriptomic responses in cotton during different types of abiotic stresses. The meta-analysis exhibited 1465 DEGs between normal and stressed scenarios. The DEGs were linked with ubiquitin dependent process, secondary metabolites biosynthesis, phytohormone and signal transduction	Tahmasebi et al. (2019)

(continued)

Table 7.1 (continued)

Plants	Types of stress	Transcriptomic studies	References
Sweet potato (*Ipomoea batatas* L.)	Drought	Transcriptomic analysis revealed a total of 11,359 DEGs in sweet potato after exposure to drought stress (PEG 6000). Amongst these DEGs, 7666 were upregulated and 3693 were downregulated. Abscisic acid, ethylene and jasmonic acid signalling pathways were found to be significantly involved in the drought resistance of sweet potato	Zhu et al. (2019)
	Oxidative and drought	Overexpression of *Arabidopsis* YUCCA6 gene in transgenic sweet potato plants under control of the oxidative stress-inducible *SWPA2* promoter caused AtYUCCA6-mediated auxin overproduction, which was linked with oxidative and drought stress resistance and inhibited storage root generation	Park et al. (2019)
Chickpea (*Cicer arietinum* L.)	Drought	Transcriptomic study was done in two drought-resistant genotypes of chickpea exposed to drought stress (PEG 6000). Amongst the 1624 genes differentially expressed under drought conditions, 97 exhibited common expression in both genotypes. The transcription factors AP2-EREBP, bHLH, bZIP, C3H, MYB, NAC, WRKY and MADS were observed to be involved	Kumar et al. (2019)
Korean fir (*Abies koreana*)	High temperature, high CO_2	Transcriptomic (RNA-seq) investigations in Korean fir (*A. koreana*) showed that high temperature changed the expression of a significant number of transcripts. Conversely, high CO_2 and high temperature stresses exerted simultaneously, resulted in few alterations in gene expression. Transcription factors found to be significant in abiotic stress response such as ERF, bHLH and NAC were identified in this study. The transcripts expressed during heat and CO_2 exposure were found to be associated with light responses, biotic and abiotic stress adaptations and development	Hwang et al. (2019)

(continued)

Table 7.1 (continued)

Plants	Types of stress	Transcriptomic studies	References
Tea (*Camellia sinensis*)	Heat	Transcriptomic analysis was performed for calcium mediated response to heat in *C. sinensis*. Overall 923 DEGs were identified. Amongst them 299 were upregulated and 624 were downregulated genes. Identified DEGs were associated with signal transduction, transcriptional control and post-translational alterations. A CsCML45 gene encoding a calmodulin-like protein was isolated. The expression of CsCML45 in *A. thaliana* elevated the resistance to heat stress in transgenic *A. thaliana* plants	Wang et al. (2019b)
Mustard (*Brassica juncea*)	Cadmium	Expression analysis revealed 11,294 DEGs to be significantly expressed in *B. juncea* under cadmium stress. Genes associated with chemical and oxidative stress, transport and secondary metabolic processes were upregulated, whereas those associated with multicellular organisms development and photosynthesis were downregulated in response to exposure to Cd. Cd-associated transporters such as metal transporter (Nramp1), metal tolerance proteins (MTPC2, MTP11), cadmium transporter ATPase and plant Cd resistance proteins (PCR2, PCR6) were upregulated. Conversely, Cadmium/zinc transporter ATPase (HMA2, HMA3 and HMA4), high-affinity calcium antiport protein (CAX1) and iron transporting protein (IRT1) were downregulated. It was observed that Cd treatment changed gene expression correlated with phytohormones, calcium control and MAP kinases	Thakur et al. (2019)
Arabidopsis thaliana	Zinc oxide (ZnO) nanoparticles	Transcriptomic and physiological investigations revealed that both zinc oxide nanoparticles (nZnO) and Zn^{2+} inhibited growth of the primary root (PR). nZnO adversely affected PR growth strongly in the elongation zone. The inhibitory effect was by damaging cell wall organization and architecture through physical interaction and transcriptional control, whereas Zn^{2+} has a stronger toxic effect on meristems of PR	Wan et al. (2019)

(continued)

Table 7.1 (continued)

Plants	Types of stress	Transcriptomic studies	References
Apple (*Malus × domestica* Borkh.)	Salt and osmotic	The transcription factor MdMYB46 is important in resistance of apples to salinity and osmotic stress. MdMYB46 positively affected secondary cell wall formation and lignin deposition by directly attaching to the promoter of genes corresponding to lignin biosynthesis. Another way through which MdMYB46 can enhance the salt and osmotic stress resistance of the apple plants is by direct activation of stress-responsive signals	Chen et al. (2019)
Maize (*Zea mays*)	Salt	Comparative salt transcriptome investigations of salt-sensitive and salt-tolerant maize inbred line (L87) exhibited that L87 has 1856 DEGs unique to it that were upregulated. These DEGs include Hsp70s and aquaporins. The DEGs associated with signal transduction of abscisic acid, ethylene, jasmonic acid and salicylic acid and WRKY transcription factors may be influencing the differential salt resistance response between two varieties DEGs related to reactive oxygen species scavenging may elevate salt resistance of L87 line	Wang et al. (2019a)
Eggplant (*Solanum melongena* L.)	Salt	Transcriptome analysis showed that response of eggplant to salinity stress is both genotype-specific and organ specific Genotype-specific DEGs in salt-tolerant variety were found to be important for its enhanced salt resistance. Transcription factors like members of C2C2-CO-like, WRKY, MYB and NAC group were observed in response to enhanced salt stress. AKT1, KAT1 and SOS1 upregulation was observed only in the leaves of salt-resistant variety. Expression of SmAKT1 in yeast and *Arabidopsis* akt1 mutants elevated their salt resistance	Li et al. (2019)

(continued)

Table 7.1 (continued)

Plants	Types of stress	Transcriptomic studies	References
Rice (*Oryza sativa*) *Arabidopsis*	Salt, drought, high temperature	OsRH58 expression was increased in rice during salinity, drought or elevated-temperature stress; however, it was downregulated during low-temperature, UV or ABA exposure. During high-salinity and drought stresses, the levels of many chloroplast proteins in the transgenic *Arabidopsis* plants expressing OsRH58 were increased OsRH58 transported by chloroplast having RNA chaperone function imparts stress resistance in plants by elevating translation of chloroplast mRNA	Nawaz and Kang (2019)
Rice (*Oryza sativa*)	Salt	Transcriptome study of rice roots and shoots exposed to salinity stress exhibited that salt stress had a major impact on carbohydrate and amino acid metabolism and stimulated secondary metabolite-associated genes. Phenolic and flavonoid compounds aggregated in roots exposed to salt stress	Chandran et al. (2019)
Sorghum [*Sorghum bicolor* (L.) Moench]	Drought	In transcriptome investigations, two genotypes of sorghum were exposed to pre-flowering and post-flowering drought stress from plant emergence to post-anthesis Genotypic differences in core photosynthesis and reactive oxygen species scavenging pathways were observed, which revealed possible mechanisms of drought resistance and delayed senescence, typically observed in the stay-green phenotype. A large-level decrease in the expression of genes important in arbuscular mycorrhizal (AM) symbiosis, with a related decrease in AM fungal mass in the plants' roots was observed	Varoquaux et al. (2019)

(continued)

Table 7.1 (continued)

Plants	Types of stress	Transcriptomic studies	References
Olea europaea, Vitis riparia Michx, Prunus mahaleb, Prunus persica, Phoenix dactylifera	Drought, salinity	In this investigation, identification and mapping of important genes expressed in root tissues during drought and salinity resistance amongst five fruit tree species, i.e., *O. europaea, V. riparia Michx, P. mahaleb, P. persica* and *P. dactylifera* was performed. Pathways coding for control of defence response, drug transmembrane transport and metal ion attachment are important molecular responses during exposure to drought and salt stress. Water shortage elevated many genes responsive to abscisic acid, gibberellin, brassinosteroids and the ethylene-stimulated signalling pathway. Salt stress has adverse effect on expression of genes encoding for important enzymes in signalling proteins in auxin-, gibberellin-(gibberellin 2 oxidase 8) and abscisic acid-related pathways (aldehyde oxidase 4, abscisic acid-responsive element-binding protein 3)	Benny et al. (2020)
Onion (*Allium cepa* L.)	Drought	Transcriptomic analysis of drought-tolerant (1656) and drought-sensitive (1627) onion genotypes was conducted for obtaining knowledge of differential response to drought stress at molecular level There was enhanced expression of 1189 genes and downregulation of 1180 genes in drought-tolerant (1656) genotype However, there was upregulation of 872 genes and downregulation of 1292 genes in drought-sensitive (1627) genotype. Genes encoding transcription factors, cytochrome P450, flavonoids, membrane transporters and linked to carbohydrate metabolic processes exhibited variable expression in the tolerant and susceptible genotypes	Ghodke et al. (2020)

(continued)

Table 7.1 (continued)

Plants	Types of stress	Transcriptomic studies	References
Rapeseed (*Brassica napus*)	Cold	Transcriptome and metabolome investigations of leaves of 5 spring *B. napus* L. ecotype lines and 5 winter *B. napus* ecotype lines treated at 4 and 28 °C were conducted. Overall 25,460 DEGs and 41 differentially expressed metabolites (DEMs) in spring oilseed ecotype and 28,512 DEGs and 47 DEMs in winter oilseed ecotype were identified by transcriptome and metabolome investigations after exposure to cold stress. This study indicated that lipid, ABA, secondary metabolism, signal transduction and transcription factors are possibly involved in the cold stress response of spring and winter ecotypes	Jian et al. (2020)
Peppers (*Capsicum annuum* L.)	Heat, cold, salinity and osmotic	RNA-seq study of peppers exposed to heat, cold, salinity and osmotic stress at variable temporal points was done. In this study, large amount of transcriptome data was verified by differentially expressed genes and gene ontology enrichment investigations, which may be helpful in development of stress tolerance in peppers	Kang et al. (2020)
Tomato (*Solanum lycopersicum cv. Micro-Tom*), rice (*Oryza sativa cv. Nipponbare*), Sorghum (*Sorghom bicolor*)	Salt and drought	The improvements exhibited in abiotic-stressed crops (tomato and sorghum exposed to salt stress and rice exposed to water-deficit stress), treated with carbon-based nanomaterials (CBNs) such as carbon nanotubes (CNTs) and graphene, can be due to induced full or partial restoration of gene expression. Majority of the genes with restored expression were associated with important stress responses and major transcriptional factors (aquaporins, dehydrins and heat shock proteins/co-chaperons, NAC, WRKY) and exhibited linkage with important stress-signalling pathways (ABA-InsP3 and MAPK signalling)	Cherati et al. (2021)

(continued)

Table 7.1 (continued)

Plants	Types of stress	Transcriptomic studies	References
Populus × canadensis 'Neva'	Cadmium (Cd) Stress	Research investigations were done by transcriptome sequencing on two Cd-treated cuttings of *P. × canadensis* 'Neva' to discover the important genes associated with Cd stress responses of *P. × canadensis* 'Neva'. Many DEGs identified were associated with catalytic activity, oxidoreductase, transferase activity and biosynthesis of secondary metabolites. This investigation may enhance the knowledge of molecular responsive mechanisms of *P. × canadensis* 'Neva' on Cd resistance	Li et al. (2021a, b)
Sorghum [*Sorghum bicolor* (L.) Moench]	Salinity stress	The de novo transcriptomic analysis of sorghum under various salt stress levels was done. Upregulation of transcripts encoding the NHX transporters, the aquaporin protein family, NAC, MYB and WRYK families, antioxidants and compatible osmolyte proteins and photosynthetic genes was observed. The tolerant line (SSG 59-3) was adopted to the enhanced salt stress levels by deploying metabolic processes, especially related to signal transduction, transport and influx of K^+ ions and osmotic homeostasis. This investigation will enhance the understanding of functions of NAC TFs in developing salt-tolerant sorghum lines at molecular level	Punia et al. (2021)

decreasing ethylene levels, which helped in better survival and growth of plants experiencing abiotic stress (Orozco-Mosqueda et al. 2020). Presence of secondary metabolites like carotenoids, flavonoids and phenylpropanoids in plants experiencing stress and inoculated with microorganisms like PGPR has been linked with decreasing ROS levels and thus helping in better survival of plants undergoing stress (Chukwuneme et al. 2021; Mohammadi et al. 2021). Additionally, PGPR can employ a cascade of signalling events to stimulate physiological processes, which cause priming and resistance to stress (Beckers and Conrath 2007; Mhlongo et al. 2018).

PGPR have been observed to stimulate plant immunity and defence system known as priming. Priming subsequently leads toward increased tolerance toward abiotic stresses. As compared to plants which have not undergone priming, primed plants exhibit enhanced and swift defence response in mitigating abiotic stress. Induced systemic resistance (ISR) and systemic acquired resistance (SAR) are the mechanisms, characteristic of this increased defence response. The stimulation of SAR pathway is dependent upon SA (Shah and Zeier 2013), while ethylene (ET) and jasmonic acid (JA) responses mediate ISR pathway for stress tolerance. ISR response is mediated by microorganisms such as PGPR, which reside in the root system of plant and stimulate defence response (Martínez-Medina et al. 2013). While SAR response involves localized infection and affects distal parts of plants (Ryals et al. 1996).

Priming mechanisms may be long or short term. Alterations at the transcript or protein level and in enzyme function act as short-term mechanisms (Hilker et al. 2016), while epigenetic changes like histone modification and methylation of DNA are long-term mechanisms (Tugizimana et al. 2018). During priming the long-lasting mechanisms are trained in such a way, causing generation of a molecular memory at epigenetic level, which is activated upon subsequent exposure to stress and resulting in swift and robust response to stress. Metabolic changes in plants taking place due to priming manifest in reconfiguring of primary metabolism and variable biosynthesis of secondary metabolites (Conrath et al. 2015; Tenenboim and Brotman 2016; Tugizimana et al. 2019). These metabolic changes form metabolic memory, which subsequently causes swift and good response upon exposure to stress (Schwachtje et al. 2019). Possible mechanisms behind priming are increased transcription of defence-related genes upon stress exposure through activation of transcription factors (Conrath et al. 2006), alterations at epigenetic level like changes in chromatin, histone and DNA (Madlung and Comai 2004) and aggregation of inactive proteins associated with signal amplification like MAPKs (Beckers et al. 2009).

7.6 Biostimulation of Stress Genes/Proteins by Plant Growth Promoting Rhizobacteria and Their Studies at Transcriptomic Level

Biostimulation occurs due to variable outside stimuli on the targeted cell's molecular machinery (Juárez-Maldonado et al. 2019). Biostimulants are biotic and abiotic agents that result in induction of biostimulation. UV radiation, nanomaterials and biological entities including pathogens, competitors and beneficial microorganisms come under the category of biostimulants (Juárez-Maldonado et al. 2019).

Two important biostimulants are beneficial bacteria and fungi (González-Morales et al. 2021). Plant biostimulants are substances or microorganisms resulting in improvement of nutrient utilization and quality parameters of plants and stress tolerance (du Jardin 2015). Biostimulants enhance absorption and utilization of nutrients, encourage soil microorganism's development and increase soil fertility, thus promoting overall plant development and plant's tolerance to stress (Calvo et al.

2014). Various abiotic stresses such as water shortage, low or high temperature and salinity can adversely affect yield of plants (Singhal et al. 2016). Biostimulants such as PGPR are being employed recently for enhancing plant's productivity and helping plants in mitigation of various stresses. The present part of chapter takes into account alterations at transcriptomic level of the effect of biostimulants on some important plant species exposed to different types of abiotic stresses.

PGPR are employed as biostimulants and cause priming of plants, which subsequently benefit plants during abiotic stresses. PGPR enter in symbiotic relationship with host plant to increase yield of plant. The interaction between PGPR and plant results in enhancement or initiation of various phenomenon such as increase in availability of mineral phosphates, nitrogen fixation and generation of plant growth regulators (Singh 2013; Verma et al. 2019). PGPR produce plant growth regulators, gibberellic acid (GA), ABA, ET, indole-3-acetic acid (IAA) and zeatin, secondary metabolites (lipopeptides, siderophores) and volatile organic compounds (2,3 butanediol, acetoin), which result in increase in yield of PGPR-infested plants. Sp245 lipopolysaccharide (LPS) in *Azospirillum brasilense* has been observed to enhance growth in *A. thaliana* (Méndez-Gómez et al. 2020). There are various types of plant biostimulants mainly according to their composition. These are humic substances (HS), hormone-containing products (HCP) and amino acid-containing products (AACP) besides PGPR (du Jardin 2015). Other types of biostimulants are seaweed extracts (SWE) (Battacharyya et al. 2015), protein hydrolysates (Colla et al. 2015), silicon (Savvas and Ntatsi 2015) and arbuscular mycorrhizal fungi (Rouphael et al. 2015). Knowledge of molecular mechanisms is very important for deeper understanding of plant-PGPR interactions, which are important for formulation of efficient biostimulants. Analysis at larger wholesome levels like transcriptomics, metabolomics, can reveal plant–PGPR interactions at molecular level and how regulatory pathways act during mitigation of stress in plants (Lephatsi et al. 2021). Some important transcriptomic research employing PGPR biostimulants in plants undergoing abiotic stress are discussed below:

PGPR-mediated resistance to salinity has been analyzed in rice (Chauhan et al. 2019), wheat (Safdarian et al. 2019) and *Arabidopsis* (Liu et al. 2017), employing transcriptomic approaches and these approaches gave knowledge about complex signalling pathways associated with salinity resistance in plants. In the study of Erice et al. (2017), salinity resistance mechanism was analyzed in *Arabidopsis* plants inoculated with PGPR *Bacillus megaterium*, employing genomic and proteomic approaches. Upregulation of JA metabolism (CYP94B3, lipoxygenase 4 and allene-oxide cyclase) was observed under salt stress. CYP94B3 gene involved in JA-Ile catabolism was found important in adaptation to salinity. Proteomic investigation revealed increased monodehydroascorbate reductase (MDHAR) content along with ATP synthase. *B. megaterium* helped *Arabidopsis* in mitigating salt stress.

In the study of Liu et al. (2017), transcriptome sequencing of *Arabidopsis* shoot tissues exposed to salinity stress and inoculated with or without plant growth promoting *Bacillus amyloliquefaciens* FZB42 was performed. Transcripts involved in processes like Na^+ translocation, auxin-associated, photosynthesis, SOS scavenging and osmoprotectant generation such as proline and trehalose were differentially expressed by FZB42 inoculation, which helped in mitigation of salt stress in

Arabidopsis. It was observed that FZB42 mediated plant salinity resistance involved activating plants ET/JA signalling but not ABA-dependent pathway. *Pseudomonas fluorescens* based biostimulant was employed to rice plants exposed to drought stress (Saakre et al. 2017). Six DEGs were found in the leaves. Amongst transcripts, AP2-EREBP, bZIP1, PKDP and COC1 belong to regulatory category. COC1, bZIP1 and AP2-EREBPs are considered as stress-responsive transcription factors mediating ABA signalling (Reeves et al. 2011; Seung et al. 2012). The PKDP gene encodes kinase function enzyme taking part in phosphorylation during signal transduction (Mohanta et al. 2015). The Hsp20 gene encodes a small heat shock protein harboring ATP-mediated chaperone function damaging denatured proteins (Nagaraju et al. 2020). The COX1 gene encodes cytochrome oxidase function enzyme mediating apoptosis [programmed cell death (PCD)] in plants. Stress has been observed to induce PCD and ROS play important role in starting PCD process (Foyer and Noctor 2005). The impact of *A. brasilense* and *Rhizobium tropici* treatment on maize plants experiencing NaCl-induced salt stress was observed by Fukami et al. (2018). These biostimulants altered antioxidant enzymes, proline and malondialdehyde (MDA) levels in leaves and roots. At the transcriptomics level, expressions of antioxidant response genes were increased. Expression of pathogenesis-related protein 1 gene (PR1) (Akbudak et al. 2020), proline-rich protein genes (prp2; prp4i) (Wilson and cooper 1994; Fowler et al. 1999) and hsp70 (Ul Haq et al. 2019) were decreased (Fukami et al. 2018). In the plants (*Vigna mungo* L. and *Pisum sativum* L.) treated with PGPR consortium (*Ochrobactrum pseudogrignonense*, *Pseudomonas* spp. and *Bacillus subtilis*), decreased expression of the 1-aminocyclopropane-1-carboxylate synthase (ACS) and 1-aminocyclopropane-1-carboxylate oxidase (ACO) genes was observed coupled with enhanced growth and stress tolerance under water shortage (Saikia et al. 2018). Conversely, in plants with stress exposure and no PGPR treatment, ACS and ACO genes were upregulated. ACC synthase and ACC oxidase enzymes are encoded by the ACS gene and the ACO gene, respectively. ACC synthase mediated processes lead to the generation of ACC, while ACC oxidase is involved in the generation of ethylene (Li et al. 2020). Burst in ethylene levels in plants exposed to stress adversely affects plants and initiates process of senescence (Czarny et al. 2006). PGPR found inside plants can change tryptophan (received from plants) to IAA, which is subsequently taken by plants. Process of decreased expression of ACS and ACO is initiated due to stimulation of the generation of ACC by IAA (from PGPR and plants) and feedback inhibition of IAA generation by elevated ACC (Glick 2012). ACC can be reduced by action of ACC deaminase generated by some PGPR. The decrease in ACC results in a corresponding decrease in quantities of ethylene, which otherwise in excess amounts can exert adverse effect on growth (Glick 2014). A study was done by Chauhan et al. (2019) to observe impact of biostimulant *B. amyloliquefaciens*-SN13 in rice plants exposed to salinity (Chauhan et al. 2019). Enhanced relative water, biomass, proline and sugar levels in plants were reported, while decreased electrolyte leakage, lipid peroxidation were observed. Modifications in expression of genes linked with stress response, growth regulators, photosynthesis, lipid metabolic processes and cell wall were stimulated. Biostimulant *Bacillus oryzicola* effect on *A. thaliana* plants exposed to salinity was evaluated by Baek et al. (2020). Treatment of *B. oryzicola* to *A. thaliana* exhibited enhanced lateral root numbers, enhanced fresh weight and chlorophyll content, while decreased storage of salt-induced MDA and *N*-acetyl L-cysteine (NAC) was reported. Elevated transcription of genes responsive to desiccation, dehydration, salinity, drought and ABA (RD29A, RD29B, RD20, RD22) (Kaur and Gupta 2005; Aubert et al. 2010; Wang et al. 2012; Bihmidine et al. 2013) and KIN1, ERD1 genes in shoots, roots was reported. KIN1 encodes protein stimulated by cold, water shortage, osmotic conditions and ABA. It may act like anti-freeze protein (Wang et al. 1995). ERD1 acts in chloroplasts where it damages wrongly folded or unassembled proteins (Peltier et al. 2001). Mellidou et al. (2021) investigated effects of seed inoculation on tomato seedlings under salt stress with a potential plant growth enhancing strain *Pseudomonas oryzihabitans* (AXSa06) (harboring ACC deaminase activity), employing transcriptomic and metabolomic approaches. Mitigation of salt stress upon inoculation with AXSa06 strain was associated with differences in levels of transcripts and metabolic compounds. Inoculation with AXSa06 strain induced primed state in tomato, subsequently causing enhanced response to salinity stress. This response is associated with decreasing toxic effects of Na^+ ions, efficient assimilation of carbon and nitrogen, activation of antioxidants and reducing intensity of stress signals. Leaf lipid peroxidation and ascorbate content prior to stress indicate primed state of plants. Important genes related to ABA or ET homeostasis, amino acids, serine, glutamate, threonine and aspartate, and ion antiporters were important compounds found associated with AXSa06-mediated salt tolerance in this study. Samaras et al. (2022) studied transcriptomic alterations in cucumber roots after inoculation with a PGPR *B. subtilis* MBI600 strain (used as biocontrol agent) to understand molecular basis of enhanced plant resistance by this PGPR species. The expression of more than 1000 genes was changed after 24 and 48 h of inoculation with Bs MBI600. Bs MBI600 caused stimulation of genes associated with SAR and ISR signalling. There was upregulation of genes involved in nutrient availability and plant growth regulator production. Activation of genes related to defence response such as endo-1,3(4)-beta-glucanase, peroxidase and PR-4 was more likely associated with biocontrol ability of Bs MBI600. In the study of Ji et al. (2023), changes at transcriptomics level in wheat, roots and leaves after inoculation with bacterial consortia (*B. subtilis* HG-15 and *Bacillus velezensis* JC-K3) were studied. Expression of numerous genes was altered significantly. The genes involved with ethylene-responsive transcription factor exhibited major upregulation pattern, while ethylene receptor 1 gene in wheat leaves exhibited significant downregulation pattern. The DEGs were associated with signal transduction pathways, in metabolism of amino acids and carbohydrates. Expression of cellular oxidant detoxification enrichment rate and linoleic acid metabolism exhibited significant upward pattern in roots. Expression of peroxisome size regulation and photosynthesis-antenna proteins was highest in leaves. *REF1* and *CYP98A* genes associated with flavonoid biosynthesis pathway exhibited upregulation pattern, while CCR, F5H, HCT and TOGT1-associated genes showed downregulation pattern. The phenylalanine ammonia lyase (PAL) gene was upregulated in wheat leaf cells, while CYP73A, 4CL and CCR exhibited downregulation. Compound microbial agents used in present investigation enhanced wheat growth under salinity and imparted tolerance to disease by activating immune pathway-related genes and controlling expression of metabolism-associated genes in wheat roots and leaves.

7.7 Gene Expression Databases

There are studies in which creation and maintenance of databases of expression of a specific group of genes were done. These databases may be helpful in other research works. Chandran et al. (2016) developed a rice kinase database (RKD 2.0), which can help in studying metadata obtained from National Center for Biotechnology Information Gene Expression Omnibus (NCBI GEO) and therefore can play important role in detailed transcriptomic analysis of kinase-encoding genes in rice tissues under diverse conditions. Sato et al. (2011) have created a transcriptional gene expression database (RiceXPro; http://ricexpro.dna.affrc.go.jp/) of different developmental stages and various spatio-temporal conditions in rice. This database can be helpful in analyzing organ/tissue and developmental stage-specific transcriptional profile of a specific gene. NetREx is a network-oriented rice expression analysis server harboring co-expression networks of rice (*O. sativa*) using mRNA sequencing data for investigating user-queried genes across various stress scenarios, hormonal treatments and tissues (https://bioinf.iiit.ac.in/netrex/index.html). Plant stress RNA-Seq nexus (PSRN) is a stress-oriented transcriptomic database of plants (Li et al. 2018) having expression profiles of coding-transcript/lncRNA and giving information regarding differential expression of transcripts between various stress-specific subsets (http://syslab5.nchu.edu.tw/PSRN). Similar databases have been created and maintained for other plants (Waese and Provart 2016; Hawkins et al. 2017; Lim et al. 2022; Zuluaga et al. 2023; http://intranet.icrisat.org/gt1/tog/homepage), which may be very relevant for further plant transcriptomics research works.

7.8 Future Prospects

In plants, induction of resistance to abiotic stress has been found to be more cumbersome task as compared to developing tolerance to biotic stress. Efforts are going on by concerned researchers to positively modify the genetic structure of plants for resistance to abiotic stress including systemizing genetic codes into improved combinations, more effective against abiotic stress. But for effective implementation of these approaches, more studies are needed at multidisciplinary level involving transcriptomics, proteomics and metabolomics (Javaid et al. 2022). It has been observed that combined and sequential stress scenarios may have some different mechanisms as compared to single stress scenarios, although there may be some common mechanism also between them. Therefore, stress signalling and sensing mechanisms need to be studied for combined and sequential stress scenarios also, which will be beneficial in developing crops resistant to actual multiple stress conditions encountered in field (Anwar et al. 2021). In the study of Biniaz et al. (2022), a meta-analysis was done to investigate differential and shared genes, pathways and transcriptomic mechanisms associated with *Arabidopsis* response to biotic and abiotic stresses. In the natural field conditions, plants are exposed to both abiotic and biotic stresses; therefore, the study will be helpful in identifying common and specific genes and pathways associated with abiotic and biotic stresses, which will be helpful in the development of crops resistant to these stresses and with enhanced productivity, Future studies are required for revealing the molecular mechanisms in detail and validating the function of DEGs, TFs and miRNAs that regulate plant response to abiotic and biotic stresses. Traditional crop-modification approaches are non-specific, time-taking and expensive. Modern approaches like genetic transformation can facilitate engineering of many traits in plants, thus rapidly producing genetically modified (GM) crops, but due to stigma attached with GM crops, they have not been accepted in many countries. A new genome-editing technology clustered regularly interspaced short palindromic repeat (CRISPR) has potential for development of crops with resistance to stress including abiotic stress. Multiplex genome-editing can be helpful in development of crops having tolerance against multiple stresses in a single event. But CRISPR technology has bottleneck of having the scarcity of functionally characterized genes involved in agronomic traits including stress tolerance, which needs to be addressed for development of stress-resistant crops through CRISPR technology (Ahmad et al. 2020; Rahman et al. 2022). Cherati et al. (2021) observed improvements in tomato, sorghum and rice exposed to abiotic stresses and treated with carbon-based nanomaterials (CBNs), which could be due to restoration of gene expression mainly in genes associated with stress responses and major transcriptional factors and exhibited linkage with important stress-signalling pathways. This study is one of the few studies exploring role of CBNs in mitigation of stress by treated plants but in future role of more such CBNs need to be tested in response to abiotic stress by plants. This study can be helpful in throwing light on mechanisms behind positive and negative effects of nano-sized materials in plant but further investigation is required for elucidating the mechanism of CBN-induced changes in stress-related gene expression. PGPR play important role in enhancing plant resistance to abiotic stress. Samaras et al. (2022) studied transcriptomic alterations in cucumber roots after inoculation with a PGPR *B. subtilis* MBI600 strain (used as biocontrol agent) to understand molecular basis of enhanced plant resistance by this PGPR species. It was observed that inoculation with Bs MBI600 strain caused stimulation of growth parameters and enhancement in preparedness of the plant for future stress encounters. Bs MBI600 can be very important in sustainable cultivation of cucumber, but plant–PGPR interactions need detailed investigations under field conditions. In the study of Ji et al. (2023), changes at transcriptomics level in wheat, roots and leaves after inoculation with compound microbial inoculants, consisting of *B. subtilis* HG-15 and *B. velezensis* JC-K3, were investigated. Compound microbial agents enhanced wheat growth and imparted tolerance to disease by activating immune pathway-related genes and controlling expression of metabolism-associated genes in wheat roots and leaves. This study enhanced knowledge of effects of microorganisms on improving salt resistance in wheat. But more studies are required at the transcriptomic, proteomic and genetic level to have better idea of role of microorganisms in imparting salt resistance to susceptible plants. Because diversity of PGPR is dependent on soil conditions; therefore, their physiological, biochemical and transcriptomics profile under different pH, temperature, drought and other abiotic stress conditions need to be studied to have an idea about their growth in these conditions. Studies should be done to elucidate the detailed mechanism of priming response elicited in plants by PGPR. Interaction of PGPR with other microorganisms in rhizosphere should also be studied. These studies will help in designing efficient PGPR strains and improving their biostimulation capacity for stress resistance induction in plants (Bhat et al. 2023).

7.9 Conclusion

Plants are exposed frequently in many global regions to abiotic stresses like drought, high temperature and salinity. These stresses have negative effects on the growth and subsequently yield of plants especially crop plants, which are very important in feeding the world's ever increasing population and thus solving the hunger problem. When exposed to varied kinds of stresses, plants respond and adapt to these scenarios by modifying the expression pattern of genes, proteins and metabolites. These changes have to be investigated thoroughly at transcriptomics level. Further validating the transcriptomic changes by other -omics technologies like metabolomics and proteomics could enhance the reliability of transcriptomic data. For identifying and characterizing these changes and adaptability capacity at deeper level, various tools and techniques such as microarrays, cDNA microarray analysis, RNA-seq analysis, genomic-scale ESTs, genomic sequencing, NGS and meta-analysis have been employed. Many researchers have studied transcriptomic alterations in plants in field where they are exposed to many types of stresses, but plant's reactions to abiotic stresses in field at various developmental stages have to be studied thoroughly to generate developmental stage-specific transcriptomic data. PGPR are an important group of microorganisms helpful in imparting stress resistance to plants exposed to abiotic stress. An interaction between PGPR and plants and knowledge of soil conditions needs to be understood in detail for designing better PGPR strains. Analyzing the data of stress-responsive genes, transcripts and associated machinery at a larger level such as system biology, and analyzing the networks and interactions of genes responding during stress will reveal the complex regulatory nature of stress responses in plants so that key stress regulators can be identified. The information obtained by such approaches could be very helpful in engineering crop plants whose yields will not be affected under challenging stress conditions especially abiotic stresses, generated nowadays by climate change and thus solving the ever increasing demand of food.

Acknowledgements Support and efforts done by friends and colleagues are gratefully acknowledged.

References

Adams MD, Kelley JM, Gocayne JD, Dubnick M, Polymeropoulos MH, Xiao H et al (1991) Complementary DNA sequencing: expressed sequence tags and human genome project. Science 252(5013):1651–1656

Aguayo P, Lagos C, Conejera D, Medina D, Fernandez M, Valenzuela S (2019) Transcriptome-wide identification of WRKY family genes and their expression under cold acclimation in *Eucalyptus globulus*. Trees 33:1313–1327

Ahmad P, Prasad MNV (2011) Environmental adaptations and stress tolerance of plants in the era of climate change. Springer Science & Business Media, Berlin

Ahmad N, Rahman MU, Mukhtar Z, Zafar Y, Zhang B (2020) A critical look on CRISPR-based genome editing in plants. J Cell Physiol 235:666–682

Akbudak MA, Yildiz S, Filiz E (2020) Pathogenesis related protein-1 (PR-1) genes in tomato (*Solanum lycopersicum* L.): bioinformatics analyses and expression profiles in response to drought stress. Genomics 112(6):4089–4099

Ambavaram MMR, Basu S, Krishnan A, Ramegowda V, Batlang U, Rahman L et al (2014) Coordinated regulation of photosynthesis in rice increases yield and tolerance to environmental stress. Nat Commun 5:5302

Andrews SJ, Rothnagel JA (2014) Emerging evidence for functional peptides encoded by short open reading frames. Nat Rev Genet 15(3):193–204

Anwar K, Joshi R, Dhankher OP, Singla-Pareek SL, Pareek A (2021) Elucidating the response of crop plants towards individual, combined and sequentially occurring abiotic stresses. Int J Mol Sci 22:6119

Ashraf MF, Hou D, Hussain Q, Imran M, Pei J, Ali M et al (2022) Entailing the next-generation sequencing and metabolome for sustainable agriculture by improving plant tolerance. Int J Mol Sci 23(2):651

Ashrafi-Dehkordi E, Alemzadeh A, Tanaka N, Razi H (2018) Meta-analysis of transcriptomic responses to biotic and abiotic stress in tomato. Peer J 6:e4631

Atkinson NJ, Lilley CJ, Urwin PE (2013) Identification of genes involved in the response of *Arabidopsis thaliana* to simultaneous biotic and abiotic stresses. Plant Physiol 162(4):2028–2041

Aubert Y, Vile D, Pervent M, Aldon D, Ranty B, Simonneau T et al (2010) RD20, a stress-inducible caleosin, participates in stomatal control, transpiration and drought tolerance in *Arabidopsis thaliana*. Plant Cell Physiol 51:1975–1987

Baek D, Rokibuzzaman M, Khan A, Kim MC, Park HJ, Yun DJ et al (2020) Plant-growth promoting *Bacillus oryzicola* YC7007 modulates stress response gene expression and provides protection from salt stress. Front Plant Sci 10:1646

Bashir K, Hanada K, Shimizu M, Seki M, Nakanishi H, Nishizawa NK (2014) Transcriptomic analysis of rice in response to iron deficiency and excess. Rice 7(1):18

Bashir K, Rasheed S, Matsui A, Lida K, Tanaka M, Seki M (2018) Monitoring transcriptomic changes in soil-grown roots and shoots of *Arabidopsis thaliana* subjected to a progressive drought stress. In: Ristova D, Barbez E (eds) Methods in molecular biology, vol 1761. Humana Press, New York, pp 223–230

Basim H, Basim E, Tombuloglu H, Unver T (2021) Comparative transcriptome analysis of resistant and cultivated tomato lines in response to *Clavibacter michiganensis* subsp. Genomics 113(4):2455–2467

Battacharyya D, Babgohari MZ, Rathor P, Prithiviraj B (2015) Seaweed extracts as biostimulants in horticulture. Sci Hortic 196:39–48

Beckers GJ, Conrath U (2007) Priming for stress resistance: from the lab to the field. Curr Opin Plant Biol 10:425–431

Beckers GJM, Jaskiewicz M, Liu Y, Underwood WR, He SY, Zhang S et al (2009) Mitogen-activated protein kinases 3 and 6 are required for full priming of stress responses in *Arabidopsis thaliana*. Plant Cell 21:944–953

Behjati S, Tarpey PS (2013) What is next generation sequencing? Arch Dis Child Educ Pract Ed 98:236–238

Beissinger M, Buchner J (1998) How chaperones fold proteins. Biol Chem 379(3):245–259

Benny J, Marchese A, Giovino A, Paolo Marra F, Perrone A, Caruso T et al (2020) Gaining insight into exclusive and common transcriptomic features linked to drought and salinity responses across fruit tree crops. Plants 9:1059

Bhat MA, Kumar V, Bhat MA, Wani IA, Dar FL, Farooq I et al (2020) Mechanistic insights of the interaction of plant growth-promoting rhizobacteria (PGPR) with plant roots toward enhancing plant productivity by alleviating salinity stress. Front Microbiol 11:1952

Bhat MA, Mishra AK, Jan S, Bhat MA, Kamal MA, Rahman S et al (2023) Plant growth promoting rhizobacteria in plant health: a perspective study of the underground interaction. Plants 12:629

Bihmidine S, Lin J, Stone JM, Awada T, Specht JE, Clemente TE (2013) Activity of the *Arabidopsis* RD29A and RD29B promoter elements in soybean under water stress. Planta 237:55–64

Biniaz Y, Tahmasebi A, Afsharifar A, Tahmasebi A, Poczai P (2022) Meta-analysis of common and differential transcriptomic responses to biotic and abiotic stresses in *Arabidopsis thaliana*. Plants 11:502

Brestic M, Zivcak M, Kunderlikova K, Allakhverdiev SI (2016) High temperature specifically affects the photoprotective responses of chlorophyll b-deficient wheat mutant lines. Photosynth Res 130:251–266

Cagirici HB, Alptekin B, Budak H (2017) RNA sequencing and co-expressed long non-coding RNA in modern and wild wheats. Sci Rep 7(1):10670

Calvo P, Nelson L, Kloepper JW (2014) Agricultural uses of plant biostimulants. Plant Soil 383:3–41

Chandran AKN, Yoo YH, Cao P, Sharma R, Sharma M, Dardick C et al (2016) Updated Rice Kinase Database RKD 2.0: enabling transcriptome and functional analysis of rice kinase genes. Rice (N Y) 9(1):40

Chandran AKN, Kim JW, Yoo YH, Park HL, Kim Y, Cho MH et al (2019) Transcriptome analysis of rice-seedling roots under soil-salt stress using RNA-Seq method. Plant Biotechnol Rep 13:567–578

Chang TW (1983) Binding of cells to matrixes of distinct antibodies coated on solid surface. J Immuno Methods 65(1–2):217–223

Chauhan PS, Lata C, Tiwari S, Chauhan AS, Mishra SK, Agrawal L et al (2019) Transcriptional alterations reveal *Bacillus amyloliquefaciens*-rice cooperation under salt stress. Sci Rep 9:1–13

Chekanova JA (2015) Long non-coding RNAs and their functions in plants. Curr Opin Plant Biol 27:207–216

Chen K, Song M, Guo Y, Liu L, Xue H, Dai H et al (2019) MdMYB46 could enhance salt and osmotic stress tolerance in apple by directly activating stress responsive signals. Plant Biotechnol J 17(12):2341–2355

Cherati SR, Shanmugam S, Pandey K, Khodakovskaya MV (2021) Whole transcriptome responses to environmental stresses in agricultural crops treated with carbon-based nanomaterials. ACS Appl Bio Mater 4(5):4292–4301

Chukwuneme CF, Ayangbenro AS, Babalola OO (2021) Impacts of land-use and management histories of maize fields on the structure, composition, and metabolic potentials of microbial communities. Curr Plant Biol 28:100228

Cohen SP, Leach JE (2019) Abiotic and biotic stresses induce a core transcriptome response in rice. Sci Rep 9:6273

Colla G, Nardi S, Cardarelli M, Ertani A, Lucini L, Canaguier R et al (2015) Protein hydrolysates as biostimulants in horticulture. Sci Hortic 196:28–38

Conrath U, Pieterse CMJ, Mauch-Mani B (2002) Priming in plant–pathogen interactions. Trends Plant Sci 7:210–216

Conrath U, Beckers GJM, Flors V, García-Agustín P, Jakab G, Mauch F et al (2006) Priming: getting ready for battle. Mol Plant-Microbe Interact 19:1062–1071

Conrath U, Beckers GJM, Langenbach CJG, Jaskiewicz MR (2015) Priming for enhanced defense. Annu Rev Phytopathol 53:97–119

Czarny JC, Grichko VP, Glick BR (2006) Genetic modulation of ethylene biosynthesis and signaling in plants. Biotechnol Adv 24:410–419

du Jardin P (2015) Plant biostimulants: definition, concept, main categories and regulation. Sci Hortic 196:3–14

Erice G, Ruiz-Lozano JM, Zamarreno AM, Garcia-Mina JM, Aroca R (2017) Transcriptomic analysis reveals the importance of JA-Ile turnover in the response of *Arabidopsis* plants to plant growth promoting rhizobacteria and salinity. Environ Exp Bot 143(4):10–19

Fatima T, Arora NK (2023) Structural, compositional characterization of novel exopolysaccharide from halotolerant strain of *Pseudomonas entomophila* PE3 and their role in growth promotion of sunflower under saline condition. Environ Exp Bot 213:105418

Filichkin SA, Hamilton M, Dharmawardhana PD, Singh SK, Sullivan C, Ben-Hur A et al (2018) Abiotic stresses modulate landscape of poplar transcriptome via alternative splicing, differential intron retention, and isoform ratio switching. Front Plant Sci 9:5

Fowler TJ, Bernhardt C, Tierney ML (1999) Characterization and expression of four proline-rich cell wall protein genes in *Arabidopsis* encoding two distinct subsets of multiple domain proteins. Plant Physiol 121:1081–1091

Foyer CH, Noctor G (2005) Redox homeostasis and antioxidant signaling: a metabolic interface between stress perception and physiological responses. Plant Cell 17:1866–1875

Fukami J, de la Osa C, Ollero FJ, Megías M, Hungria M (2018) Coinoculation of maize with *Azospirillum brasilense* and *Rhizobium tropici* as a strategy to mitigate salinity stress. Funct Plant Biol 45:328–339

Gálvez S, Mérida-García R, Camino C, Borrill P, Abrouk M, Ramírez-González RH et al (2019) Hotspots in the genomic architecture of field drought responses in wheat as breeding targets. Funct Integr Genomics 19(2):295–309

Gamalero E, Glick BR (2015) Bacterial modulation of plant ethylene levels. Plant Physiol 169:13–22

Gaur R, Azam S, Jeena G, Khan AW, Choudhary S, Jain M et al (2012) High-throughput SNP discovery and genotyping for constructing a saturated linkage map of chickpea (*Cicer arietinum* L.). DNA Res 19(5):357–373

Genitsaris S, Stefanidou N, Leontidou K, Matsi T, Karamanoli K, Mellidou I (2020) Bacterial communities in the rhizosphere and phyllosphere of halophytes and drought-tolerant plants in Mediterranean ecosystems. Microorganisms 8:1708

Ghodke P, Khandagale K, Thangasamy A, Kulkarni A, Narwade N, Shirsat D et al (2020) Comparative transcriptome analyses in contrasting onion (*Allium cepa* L.) genotypes for drought stress. PLoS One 15(8):e0237457

Glass GV (1976) Primary, secondary and meta-analysis of research. Educ Res 5:3–8

Glick BR (2012) Plant growth-promoting bacteria: mechanisms and applications. Scientifica 2012:1–15

Glick BR (2014) Bacteria with ACC deaminase can promote plant growth and help to feed the world. Microbiol Res 169:30–39

González-Morales S, Solís-Gaona S, Valdés-Caballero MV, Juárez-Maldonado A, Loredo-Treviño A, Benavides-Mendoza A (2021) Transcriptomics of biostimulation of plants under abiotic stress. Front Genet 12:58388

González-Schain N, Dreni L, Lawas LM, Galbiati M, Colombo L, Heuer S et al (2016) Genome-wide transcriptome analysis during anthesis reveals new insights into the molecular basis of heat stress responses in tolerant and sensitive rice varieties. Plant Cell Physiol 57(1):57–68

Habibpourmehraban F, Wu Y, Wu JX, Hamzelou S, Masoomi-Aladizgeh F, Kamath KS et al (2022) Multiple abiotic stresses applied simultaneously elicit distinct responses in two contrasting rice cultivars. Int J Mol Sci 23:1739

Hanada K, Higuchi-Takeuchi M, Okamoto M, Yoshizumi T, Shimizu M, Nakaminami K et al (2013) Small open reading frames associated with morphogenesis are hidden in plant genomes. Proc Natl Acad Sci U S A 110(6):2395–2400

Hasanuzzaman M, Parvin K, Bardhan K, Nahar K, Anee TI, Masud AAC et al (2021) Biostimulants for the regulation of reactive oxygen species metabolism in plants under abiotic stress. Cells 10:2537

Hawkins C, Caruana J, Li J, Zawora C, Darwish O, Wu J et al (2017) An eFP browser for visualizing strawberry fruit and flower transcriptomes. Hortic Res 4:17029

Hilker M, Schwachtje J, Baier M, Balazadeh S, Bäurle I, Geiselhardt S et al (2016) Priming and memory of stress responses in organisms lacking a nervous system. Biol Rev 91(4):1118–1133

Hiremath PJ, Farmer A, Cannon SB, Woodward J, Kudapa H, Tuteja R et al (2011) Large-scale transcriptome analysis in chickpea (*Cicer arietinum* L.), an orphan legume crop of the semi-arid tropics of Asia and Africa. Plant Biotechnol J 9(8):922–931

Holoch D, Moazed D (2015) RNA-mediated epigenetic regulation of gene expression. Nat Rev Genet 16(2):71–84

Hwang JE, Kim YJ, Jeong DY, Park HC (2019) Transcriptome analysis of Korean fir (*Abies koreana*) in response to elevated carbon dioxide and high temperature. Plant Biotechnol Rep 13:603–612

Javaid MH, Khan AR, Salam A, Neelam A, Azhar W, Ulhassan Z et al (2022) Exploring the adaptive responses of plants to abiotic stresses using transcriptome data. Agriculture 12:211

Ji C, Liang Z, Cao H, Chen Z, Kong X, Xin Z et al (2023) Transcriptome-based analysis of the effects of compound microbial agents on gene expression in wheat roots and leaves under salt stress. Front Plant Sci 14:1109077

Jian H, Xie L, Wang Y, Cao Y, Wan M, Lv D et al (2020) Characterization of cold stress responses in different rapeseed ecotypes based on metabolomics and transcriptomics analyses. Peer J 8:e8704

Juárez-Maldonado A, Ortega-Ortíz H, Morales-Díaz AB, González-Morales S, Morelos-Moreno Á, Cabrera-De la Fuente M et al (2019) Nanoparticles and nanomaterials as plant biostimulants. Int J Mol Sci 20:162

Kang Y, Khan S (2009) Climate change impacts on crop yield, crop water productivity and food security—a review. Prog Nat Sci 19(12):1665–1674

Kang WH, Sim YM, Koo N, Nam JY, Lee J, Kim N et al (2020) Transcriptome profiling of abiotic responses to heat, cold, salt, and osmotic stress of *Capsicum annuum* L. Sci Data 7:17

Kaur N, Gupta AK (2005) Signal transduction pathways under abiotic stresses in plants. Curr Sci 88:1771–1780

Kim JM, Sasaki T, Ueda M, Sako K, Seki M (2015) Chromatin changes in response to drought, salinity, heat, and cold stresses in plants. Front Plant Sci 6:114

Kim JM, To TK, Matsui A, Tanoi K, Kobayashi NI, Matsuda F et al (2017) Acetate-mediated novel survival strategy against drought in plants. Nat Plants 3:17097

Kiselev VY, Andrews TS, Hemberg M (2019) Challenges in unsupervised clustering of single-cell RNA-seq data. Nat Rev Genet 20(5):273–282

Krizek BA, Bequette CJ, Xu K, Blakley IC, Fu ZQ, Stratmann JW et al (2016) RNA-Seq links the transcription factors AINTEGUMENTA and AINTEGUMENTA-LIKE6 to cell wall remodeling and plant defense pathways. Plant Physiol 171(3):2069–2084

Kudapa H, Azam S, Sharpe AG, Taran B, Li R, Deonovic B et al (2014) Comprehensive transcriptome assembly of chickpea (*Cicer arietinum* L.) using sanger and next generation sequencing platforms: development and applications. PLoS One 9(1):e86039

Kumar M, Chauhan AS, Kumar M, Yusuf MA, Sanyal I, Chauhan PS (2019) Transcriptome sequencing of chickpea (*Cicer arietinum* L.) genotypes for identification of drought-responsive genes under drought stress condition. Plant Mol Biol Rep 37:186–203

Laloum T, Martín G, Duque P (2018) Alternative splicing control of abiotic stress responses. Trends Plant Sci 23(2):140–150

Larkindale J, Vierling E (2008) Core genome responses involved in acclimation to high temperature. Plant Physiol 146(2):748–761

Lephatsi MM, Meyer V, Piater LA, Dubery IA, Tugizimana F (2021) Plant responses to abiotic stresses and rhizobacterial biostimulants: metabolomics and epigenetics perspectives. Metabolites 11:457

Li YF, Wang Y, Tang Y, Kakani VG, Mahalingam R (2013) Transcriptome analysis of heat stress response in switchgrass (*Panicum virgatum* L.). BMC Plant Biol 13:153

Li JR, Liu CC, Sun CH, Chen YT (2018) Plant stress RNA-seq Nexus: a stress-specific transcriptome database in plant cells. BMC Genomics 19:966

Li J, Gao Z, Zhou L, Li L, Zhang J, Liu Y et al (2019) Comparative transcriptome analysis reveals K^+ transporter gene contributing to salt tolerance in eggplant. BMC Plant Biol 19(1):67–85

Li L, Shuai L, Sun J, Li C, Yi P, Zhou Z et al (2020) The role of 1-methylcyclopropene in the regulation of ethylene biosynthesis and ethylene receptor gene expression in *Mangifera indica* L. (mango fruit). Food Sci Nutr 8:1284–1294

Li PC, Yang XY, Wang HM, Ting PA, Yang JY, Wang YY et al (2021a) Metabolic responses to combined water deficit and salt stress in maize primary roots. J Integr Agric 20:109–119

Li X, Mao X, Xu Y, Li Y, Zhao N, Yao J et al (2021b) Comparative transcriptomic analysis reveals the coordinated mechanisms of *Populus* X *canadensis* 'Neva' leaves in response to cadmium stress. Ecotoxicol Environ Saf 216:112179

Libault M, Pingault L, Zogli P, Schiefelbein J (2017) Plant systems biology at the single-cell level. Trends Plant Sci 22(11):949–960

Lim PK, Zheng X, Goh JC, Mutwil M (2022) Exploiting plant transcriptomic databases: resources, tools and approaches. Plant Commun 3(4):100323

Liu S, Trapnell C (2016) Single-cell transcriptome sequencing: recent advances and remaining challenges. F1000Res 5:F1000 Faculty Rev-182

Liu W, Tai H, Li S, Gao W, Zhao M, Xie C et al (2014) *bHLH122* is important for drought and osmotic stress resistance in *Arabidopsis* and in the repression of ABA catabolism. New Phytol 201(4):1192–1204

Liu Z, Xin M, Qin J, Peng H, Ni Z, Yao Y, Sun Q (2015) Temporal transcriptome profiling reveals expression partitioning of homeologous genes contributing to heat and drought acclimation in wheat (*Triticum aestivum* L.). BMC Plant Biol 15:152

Liu S, Hao H, Lu X, Zhao X, Wang Y, Zhang Y et al (2017) Transcriptome profiling of genes involved in induced systemic salt tolerance conferred by *Bacillus amyloliquefaciens* FZB42 in *Arabidopsis thaliana*. Sci Rep 7(1):10795

Madlung A, Comai L (2004) The effect of stress on genome regulation and structure. Ann Bot 94:481–495

Martínez-Medina A, Fernández I, Sánchez-Guzmán MJ, Jung SC, Pascual JA, Pozo MJ (2013) Deciphering the hormonal signaling network behind the systemic resistance induced by *Trichoderma harzianum* in tomato. Front Plant Sci 4:1–12

Mellidou I, Ainalidou A, Papadopoulou A, Leontidou K, Genitsaris S, Karagiannis E et al (2021) Comparative transcriptomics and metabolomics reveal an intricate priming mechanism involved in PGPR-mediated salt tolerance in tomato. Front Plant Sci 12:713984

Méndez-Gómez M, Castro-Mercado E, Peña-Uribe CA, Reyes-de la Cruz H, López-Bucio J, García-Pineda E (2020) Target of rapamycin signaling plays a role in *Arabidopsis* growth promotion by *Azospirillum brasilense* Sp245. Plant Sci 293:110416

Mhlongo MI, Piater LA, Madala NE, Labuschagne N, Dubery IA (2018) The chemistry of plant–microbe interactions in the rhizosphere and the potential for metabolomics to reveal signaling related to defense priming and induced systemic resistance. Front Plant Sci 9:1–17

Misra S, Chauhan PS (2020) ACC deaminase-producing rhizosphere competent *Bacillus* spp mitigate salt stress and promote *Zea mays* growth by modulating ethylene metabolism. 3 Biotech 10(3):119

Mohammadi SS, Zibaeenezhad MJ, Sayadi M, Khorshidi S, Hadiyan E, Razeghian-Jahromi I (2021) The Impact of Smoking on Clinical Outcomes after Percutaneous Coronary Intervention in Women Compared to Men. J Interv Cardiol 2021:6619503. https://doi.org/10.1155/2021/6619503

Mohanta TK, Arora PK, Mohanta N, Parida P, Bae H (2015) Identification of new members of the MAPK gene family in plants shows diverse conserved domains and novel activation loop variants. BMC Genomics 16:58

Molina C, Rotter B, Horres R, Udupa SM, Besser B, Bellarmino L et al (2008) SuperSAGE: the drought stress responsive transcriptome of chickpea roots. BMC Genomics 9:553

Molina C, Zaman-Allah M, Khan F, Fatnassi N, Horres R, Rotter B et al (2011) The salt-responsive transcriptome of chickpea roots and nodules via deepSuperSAGE. BMC Plant Biol 11:31

Nagaraju M, Reddy PS, Kumar SA, Kumar A, Rajasheker G, Rao DM et al (2020) Genome-wide identification and transcriptional profiling of small heat shock protein gene family under diverse abiotic stress conditions in *Sorghum bicolor* (L.). Int J Biol Macromol 142:822–834

Nakaminami K, Okamoto M, Higuchi-Takeuchi M, Yoshizumi T, Yamaguchi Y, Fukao Y et al (2018) AtPep3 is a hormone-like peptide that plays a role in the salinity stress tolerance of plants. Proc Natl Acad Sci U S A 115(22):5810–5815

Nawaz G, Kang H (2019) Rice OsRH58, a chloroplast DEAD-box RNA helicase, improves salt or drought stress tolerance in *Arabidopsis* by affecting chloroplast translation. BMC Plant Biol 19(1):17–28

Ng IS, Tan SI, Kao PH, Chang YK, Chang JS (2017) Recent developments on genetic engineering of microalgae for biofuels and bio-based chemicals. Biotechnol J 12:1600644

Nguyen HM, Sako K, Matsui A, Suzuki Y, Mostofa MG, Ha CV et al (2017) Ethanol enhances high-salinity stress tolerance by detoxifying reactive oxygen species in *Arabidopsis thaliana* and rice. Front Plant Sci 8:1001

Orozco-Mosqueda MDC, Glick BR, Santoyo G (2020) ACC deaminase in plant growth-promoting bacteria (PGPB): an efficient mechanism to counter salt stress in crops. Microbiol Res 235:126439

Pajoro A, Severing E, Angenent GC, Immink RGH (2017) Histone H3 lysine 36 methylation affects temperature-induced alternative splicing and flowering in plants. Genome Biol 18(1):102

Palovaara J, Saiga S, Wendrich JR, van't Wout Hofland N, van Schayck JP, Hater F et al (2017) Transcriptome dynamics revealed by a gene expression atlas of the early *Arabidopsis* embryo. Nat Plants 3(11):894–904

Park SC, Kim HS, Lee HU, Kim YH, Kwak SS (2019) Overexpression of *Arabidopsis* YUCCA6 enhances environment stress tolerance and inhibits storage root formation in sweet potato. Plant Biotechnol Rep 13:345–352

Peltier JB, Ytterberg J, Liberles DA, Roepstorff P, Van Wijk KJ (2001) Identification of a 350-kDa ClpP protease complex with 10 different Clp isoforms in chloroplasts of *Arabidopsis thaliana*. J Biol Chem 276:16318–16327

Punia H, Tokas J, Malik A, Sangwan S, Rani A, Yashveer S et al (2021) Genome wide transcriptome profiling, characterization and functional identification of *NAC* transcription factors in Sorghum under salt stress. Antioxidants 10:1605

Qin D, Wu H, Peng H, Yao Y, Ni Z, Li Z et al (2008) Heat stress-responsive transcriptome analysis in heat susceptible and tolerant wheat (*Triticum aestivum* L.) by using wheat genome array. BMC Genomics 9:432

Rahman MU, Zulfiqar S, Raza MA, Ahmad N, Zhang B (2022) Engineering abiotic stress tolerance in crop plants through CRISPR genome editing. Cells 11:3590

Rasheed S, Bashir K, Matsui A, Tanaka M, Seki M (2016a) Transcriptomic analysis of soil-grown *Arabidopsis thaliana* roots and shoots in response to a drought stress. Front Plant Sci 7:180

Rasheed S, Bashir K, Nakaminami K, Hanada K, Matsui A, Seki M (2016b) Drought stress differentially regulates the expression of small open reading frames (sORFs) in *Arabidopsis* roots and shoots. Plant Signal Behav 11(8):e1215792

Rasheed S, Bashir K, Kim JM, Ando M, Tanaka M, Seki M (2018) The modulation of acetic acid pathway genes in *Arabidopsis* improves survival under drought stress. Sci Rep 8(1):7831

Rasmussen S, Barah P, Suarez-Rodriguez MC, Bressendorff S, Friis P, Costantino P et al (2013) Transcriptome responses to combinations of stresses in *Arabidopsis*. Plant Physiol 161(4):1783–1794

Reeves WM, Lynch TJ, Mobin R, Finkelstein RR (2011) Direct targets of the transcription factors ABA-insensitive (ABI)4 and ABI5 reveal synergistic action by ABI4 and several bZIP ABA response factors. Plant Mol Biol 75:347–363

Rouphael Y, Franken P, Schneider C, Schwarz D, Giovannetti M, Agnolucci M et al (2015) Arbuscular mycorrhizal fungi act as biostimulants in horticultural crops. Sci Hortic 196:91–108

Roychoudhury A, Chakraborty M (2013) Biochemical and molecular basis of varietal difference in plant salt tolerance. Annu Rev Res Biol 3(4):422–454

Roychoudhury A, Paul S, Basu S (2013) Cross-talk between abscisic acid-dependent and abscisic acid-independent pathways during abiotic stress. Plant Cell Rep 32(7):985–1006

Ryals JA, Neuenschwander UH, Willits MG, Molina A, Steiner HY, Hunt MD (1996) Systemic acquired resistance. Plant Cell 8:1809–1819

Saakre M, Baburao TM, Salim AP, Ffancies RM, Achuthan VP, Thomas G et al (2017) Identification and characterization of genes responsible for drought tolerance in rice mediated by *Pseudomonas fluorescens*. Rice Sci 24:291–298

Safdarian M, Askari H, Shariati JV, Nematzadeh G (2019) Transcriptional responses of wheat roots inoculated with *Arthrobacter nitroguajacolicus* to salt stress. Sci Rep 9:1–12

Saikia J, Sarma RK, Dhandia R, Yadav A, Bharali R, Gupta VK et al (2018) Alleviation of drought stress in pulse crops with ACC deaminase producing rhizobacteria isolated from acidic soil of Northeast India. Sci Rep 8:1–16

Samaras A, Kamou N, Tzelepis G, Karamanoli K, Menkissoglu-Spiroudi U, Karaoglanidis GS (2022) Root transcriptional and metabolic dynamics induced by the plant growth promoting rhizobacterium (PGPR) *Bacillus subtilis* Mbi600 on cucumber plants. Plants (Basel) 11(9):1218

Sarwar MB, Ahmad Z, Rashid B, Hassan S, Gregersen PL, Leyva MO et al (2019) *De novo* assembly of *Agave sisalana* transcriptome in response to drought stress provides insight into the tolerance mechanisms. Sci Rep 9(1):396

Sato Y, Antonio BA, Namiki N, Takehisa H, Minami H, Kamatsuki K et al (2011) RiceXPro: a platform for monitoring gene expression in *japonica* rice grown under natural field conditions. Nucleic Acids Res 39(Database issue):D1141–D1148

Savvas D, Ntatsi G (2015) Biostimulant activity of silicon in horticulture. Sci Hortic 196:66–81

Schwachtje J, Whitcomb SJ, Firmino AAP, Zuther E, Hincha DK, Kopka J (2019) Induced, imprinted, and primed responses to changing environments: does metabolism store and process information? Front Plant Sci 10:106

Seung D, Risopatron JPM, Jones BJ, Marc J (2012) Circadian clock dependent gating in ABA signalling networks. Protoplasma 249:445–457

Shaar-Moshe L, Hübner S, Peleg Z (2015) Identification of conserved drought adaptive genes using a cross-species meta-analysis approach. BMC Plant Biol 15:111

Shah J, Zeier J (2013) Long-distance communication and signal amplification in systemic acquired resistance. Front Plant Sci 4:1–16

Sharma R, De Vleesschauwer D, Sharma MK, Ronald PC (2013) Recent advances in dissecting stress-regulatory crosstalk in rice. Mol Plant 6:250–260

Shulse CN, Cole BJ, Ciobanu D, Lin J, Yoshinaga Y, Gouran M et al (2019) High-throughput single-cell transcriptome profiling of plant cell types. Cell Rep 27(7):2241–2247

Singh JS (2013) Potential microbes for sustainable agriculture. Reson J Sci Educ 18:275–281

Singhal P, Jan AT, Azam M, Haq QMR (2016) Plant abiotic stress: a prospective strategy of exploiting promoters as alternative to overcome the escalating burden. Front Life Sci 9:52–63

Sinha P, Singh VK, Bohra A, Kumar A, Reif JC, Varshney RK (2021) Genomics and breeding innovations for enhancing genetic gain for climate resilience and nutrition traits. Theor Appl Genet 134(7):1829–1843

Tahmasebi A, Ashrafi-Dehkordi E, Shahriari AG, Mazloomi SM, Ebrahimie E (2019) Integrative meta-analysis of transcriptomic responses to abiotic stress in cotton. Prog Biophys Mol Biol 146:112–122

Takahashi F, Shinozaki K (2019) Long-distance signaling in plant stress response. Curr Opin Plant Biol 47:106–111

Takahashi F, Suzuki T, Osakabe Y, Betsuyaku S, Kondo Y, Dohmae N et al (2018) A small peptide modulates stomatal control via abscisic acid in long-distance signalling. Nature 556(7700):235–238

Tardieu F, Simonneau T, Muller B (2018) The physiological basis of drought tolerance in crop plants: a scenario-dependent probabilistic approach. Annu Rev Plant Biol 69:733–759

Tavassoly I, Goldfarb J, Iyengar R (2018) Systems biology primer: the basic methods andapproaches. Essays in Biochemistry 62:487–500. https://doi.org/10.1042/EBC20180003

Tenenboim H, Brotman Y (2016) Omic relief for the biotically stressed: metabolomics of plant biotic interactions. Trends Plant Sci 21:781–791

Thakur S, Choudhary S, Bhardwaj P (2019) Comparative transcriptome profiling under cadmium stress reveals the uptake and tolerance mechanism in *Brassica juncea*. J Plant Growth Regul 38(3):1141–1152

Todaka D, Shinozaki K, Yamaguchi-Shinozaki K (2015) Recent advances in the dissection of drought-stress regulatory networks and strategies for development of drought-tolerant transgenic rice plants. Front Plant Sci 6:84

Tugizimana F, Mhlongo M, Piater L, Dubery I (2018) Metabolomics in plant priming research: the way forward? Int J Mol Sci 19:1759

Tugizimana F, Steenkamp PA, Piater LA, Labuschagne N, Dubery IA (2019) Unravelling the metabolic reconfiguration of the post-challenge primed state in *Sorghum bicolor* responding to *Colletotrichum sublineolum* infection. Metabolites 9:194

Tung CC, Kuo SC, Yang CL, Yu JH, Huang CE, Liou PC et al (2023) Single-cell transcriptomics unveils xylem cell development and evolution. Genome Biol 24:3

Ul Haq S, Khan A, Ali M, Khattak AM, Gai WX, Zhang HX et al (2019) Heat shock proteins: dynamic biomolecules to counter plant biotic and abiotic stresses. Int J Mol Sci 20:5321

Ullah N, Yüce M, Neslihan Öztürk Gökçe Z, Budak H (2017) Comparative metabolite profiling of drought stress in roots and leaves of seven Triticeae species. BMC Genomics 18(1):969

Varoquaux N, Cole B, Gao C, Pierroz G, Baker CR, Patel D et al (2019) Transcriptomic analysis of field-droughted sorghum from seedling to maturity reveals biotic and metabolic responses. Proc Natl Acad Sci U S A 116(52):27124–27132

Varshney RK, Shi C, Thudi M, Mariac C, Wallace J, Qi P et al (2017) Pearl millet genome sequence provides a resource to improve agronomic traits in arid environments. Nat Biotechnol 35(10):969–976

Verma M, Mishra J, Arora NK (2019) Plant growth-promoting rhizobacteria: diversity and applications. In: Sobti R, Arora N, Kothari R (eds) Environmental biotechnology: for sustainable future. Springer, Singapore, pp 129–173

Waese J, Provart NJ (2016) The bio-analytic resource: data visualization and analytic tools for multiple levels of plant biology. Curr Plant Biol 7–8:2–5

Wakasa Y, Oono Y, Yazawa T, Hayashi S, Ozawa K, Handa H et al (2014) RNA sequencing-mediated transcriptome analysis of rice plants in endoplasmic reticulum stress conditions. BMC Plant Biol 14:101

Wan J, Wang R, Wang R, Ju Q, Wang Y, Xu J (2019) Comparative physiological and transcriptomic analyses reveal the toxic effects of ZnO nanoparticles on plant growth. Environ Sci Technol 53(8):4235–4244

Wang H, Datla R, Georges F, Loewen M, Cutler AJ (1995) Promoters from kin1 and cor6.6, two homologous *Arabidopsis thaliana* genes: transcriptional regulation and gene expression induced by low temperature, ABA, osmoticum and dehydration. Plant Mol Biol 28:605–617

Wang W, Vinocur B, Altman A (2003) Plant responses to drought, salinity and extreme temperatures: towards genetic engineering for stress tolerance. Planta 218(1):1–14

Wang Z, Gerstein M, Snyder M (2009) RNA Seq: a revolutionary tool for transcriptomics. Nat Rev Genet 10(1):57–63

Wang H, Zhou L, Fu Y, Cheung MY, Wong FL, Phang TH et al (2012) Expression of an apoplast-localized BURP-domain protein from soybean (GmRD22) enhances tolerance towards abiotic stress. Plant Cell Environ 35:1932–1947

Wang M, Wang Y, Zhang Y, Li C, Gong S, Yan S et al (2019a) Comparative transcriptome analysis of salt-sensitive and salt-tolerant maize reveals potential mechanisms to enhance salt resistance. Genes Genomics 41(7):781–801

Wang M, Zhang X, Li Q, Chen X, Li X (2019b) Comparative transcriptome analysis to elucidate the enhanced thermotolerance of tea plants (*Camellia sinensis*) treated with exogenous calcium. Planta 249(3):775–786

Wang X, Li N, Li W, Gao X, Cha M, Qin L et al (2020) Advances in transcriptomics in the response to stress in plants. Global Med Genet 7:30–34

Webb AAR, Seki M, Satake A, Caldana C (2019) Continuous dynamic adjustment of the plant circadian oscillator. Nat Commun 10(1):550

Wilson RC, Cooper JB (1994) Characterization of PRP1 and PRP2 from *Medicago truncatula*. Plant Physiol 105:445–446

Wohlbach DJ, Quirino BF, Sussman MR (2008) Analysis of the *Arabidopsis* histidine kinase ATHK1 reveals a connection between vegetative osmotic stress sensing and seed maturation. Plant Cell 20(4):1101–1117

Xuan HD, Huang YZ, Zhou L, Deng SS, Wang CC, Xu JY et al (2022) Key soybean seedlings drought-responsive genes and pathways revealed by comparative transcriptome analysis of two cultivars. Int J Mol Sci 23:2893

Yang Z, Li JL, Liu LN, Xie Q, Sui N (2020) Photosynthetic regulation under salt stress and salt-tolerance mechanism of sweet sorghum. Front Plant Sci 10:1722

Yin H, Li M, Li D, Khan SA, Hepworth SR, Wang SM (2019) Transcriptome analysis reveals regulatory framework for salt and osmotic tolerance in a succulent xerophyte. BMC Plant Biol 19:88

Zhang ZF, Li YY, Xiao BZ (2016) Comparative transcriptome analysis highlights the crucial roles of photosynthetic system in drought stress adaptation in upland rice. Sci Rep 6:19349

Zhao S, Gao H, Jia X, Wang H, Ke M, Ma F (2020) The HD-Zip I transcription factor MdHB-7 regulates drought tolerance in transgenic apple (*Malus domestica*). Environ Exp Bot 180:104246

Zhu JK (2001) Cell signaling under salt, water and cold stresses. Curr Opin Plant Biol 4(5):401–406

Zhu JK (2016) Abiotic stress signaling and responses in plants. Cell 167(2):313–324

Zhu YN, Shi DQ, Ruan MB, Zhang LL, Meng ZH, Liu J et al (2013) Transcriptome analysis reveals crosstalk of responsive genes to multiple abiotic stresses in cotton (*Gossypium hirsutum* L.). PLoS One 8(11):e80218

Zhu X, Li X, Jiu S, Zhang K, Wang C, Fang J (2018) Analysis of the regulation networks in grapevine reveals response to waterlogging stress and candidate gene-marker selection for damage severity. R Soc Open Sci 5(6):172253

Zhu H, Zhou YY, Zhai H (2019) Transcriptome profiling reveals insights into the molecular mechanism of drought tolerance in sweet potato. J Integr Agric 18(1):9–23

Zuluaga DL, Blanco E, Mangini G, Sonnante G, Curci PL (2023) A survey of the transcriptomic resources in durum wheat: stress responses, data integration and exploitation. Plants 12:1267

Chapter 8
Exploring the Potential of Endophytes in Phytoremediation

Disha Changela, Ketaki Ramani, Kruti Dangar, and Bhavin Changela

Abstract Anthropogenic activities have led to an exponential rise in pollution. Pollutants like heavy metals and organic chemicals are foremost subjects of challenge. Phytoremediation has surfaced as a prospective tool to clean up polluted environment. It is an in situ process in which plants are employed for remediation of polluted soils. It is an appreciated technology for dealing with huge contaminated sites. Phytoremediation, however, is a time-consuming process and high levels of contaminants pose a toxicity threat to the plants. This condition is mitigated by investigating the partnership of endophytes and plants. Endophytes are the organisms that flourish within the plants in a mutualistic manner. Presence of certain endophytes has been shown to enhance stress tolerance and phytoremediation in plants. There is a boost in uptake of heavy metals and degradation of organic pollutants by plants when accompanied with endophytes. Even to tackle the hindrance of mixed contaminants, the endophytes have been reported to sequester heavy metals and degrade organic compounds. This chapter has highlighted the role of endophytes in phytoremediation of heavy metals, organic pollutants, and mixed wastes. The strategies employed by endophytes that assist the process of phytoremediation like phytotoxicity reduction, improved phytoremediation, bioaccumulation, etc. have been discussed. Also, comprehensive information on recent cases of endophyte aided phytoremediation of different pollutants has been tabularized.

Keywords Endophytes · Phytoremediation · Heavy metals · Organic pollutants

D. Changela · K. Dangar (✉)
Department of Biochemistry, Saurashtra University, Rajkot, Gujarat, India

K. Ramani
Department of Microbiology, School of Science, RK University, Rajkot, Gujarat, India

B. Changela
Department of Biomedical Engineering, A.V. Parekh Technical Institute, Rajkot, Gujarat, India

N. K. Arora, B. Bouizgarne (eds.), *Microbial Biotechnology for Sustainable Agriculture Volume 2*, Microorganisms for Sustainability 34,
https://doi.org/10.1007/978-981-97-2355-3_8

8.1 Introduction

Rapid industrialization, population explosion, and enormous technogenic activities have led to massive amounts of pollution (Appannagari 2017; Vardhan et al. 2019). Post World War II with advancement in science and technology, there has been an exponential rise in pollution. The chemicals that have alarmingly increased the rate of pollution include polycyclic aromatic hydrocarbons (PAHs), polychlorinated biphenyls (PCBs), petroleum hydrocarbons (PHs), volatile organic carbons (VOCs), heavy metals, etc. (Song et al. 2006). These chemicals persist in the environment for prolonged periods and severely affect the ecosystems.

In the current times, phytoremediation is harnessing considerable momentum. It is an eco-friendly and cost-effective technology that involves the use of plants to carry out remediation of polluted soils (Ali et al. 2013; Oladoye et al. 2022). In this technology, plants may carry out sequestration, extraction, transformation, and translocation of the pollutants and thus leading to remediation or restoration of the contaminated sites (Hooda 2007; Arora and Kumar 2019; Saxena et al. 2019). It is an in situ remediation technique involving least site disturbance and maintenance. For the treatment of large contaminated areas, phytoremediation has turned out to be valuable technology (Shen et al. 2022). Although phytoremediation has advantages, yet large scale application of this technology is facing some hurdles like high levels of contaminants turn out to be toxic for the organisms involved in remediation (Cristaldi et al. 2017).

Endophytes have been defined as organisms (usually bacteria or fungi) that ubiquitously occupy the internal tissues of plants with no symptoms of infection or harmful effects on the host (Ryan et al. 2008; Reinhold-Hurek and Hurek 2011; Khare et al. 2018). The existence of endophytes in plants is quite variable and transient. Because of the close links between the plant and the associated endophytes, they have evolved mutualistic interactions, which benefit both of them. Endophytes attain nutrients and a safe and protective niche to grow and flourish. The host plants, however, obtain a wide array of bioactive substances that promote the growth and development of plant (Strobel 2002; Brader et al. 2014). Endophytes have been isolated from several plant species and there are reports citing the potential of endophytes in promoting the growth and development of the host plant through mechanisms like biological control, nitrogen fixation, siderophore production, phosphate solubilization, production of plant growth regulators, etc. (Yaish et al. 2015; Santoyo et al. 2016). This plant growth promotes the potential of endophytes to increase the biomass of plants involved in phytoremediation (Kong and Glick 2017). They have also been proved useful in providing tolerance to high concentrations of heavy metal stress (Babu et al. 2013). As the interaction of endophytes with their host plant is profound, they can improve phytoremediation effectively. Thus, in the current scenario efforts are directed on the study of endophytes with the potential to degrade contaminants and/or promote plant growth. This chapter emphasizes the beneficial relationship between plants and endophytes as a strategy for phytoremediation.

8.2 Phytoremediation of Heavy Metals Through Endophytes

Overall plant growth is affected adversely due to heavy metal pollution (Athar and Ahmad 2002). Also, the persistent exposure to heavy metals induces changes in plant associated microflora. Various studies have cited the impact of heavy metals on the diversity and activity of endophytes. Generally, heavy metals are toxic to plants, but some plants are reported to be tolerant to heavy metals and are hyperaccumulators (Rascio and Navari-Izzo 2011). Due to the extended period of exposure of heavy metals to endophytes, they may be resistant to heavy metals. Due to its phytoremediation potential, the interactions between hyperaccumulator plants and endophytes have garnered a lot of attention. Hence a large number of studies emphasized on isolation of metal resistant endophytes from hyperaccumulator plants like *Alyssum bertolonii*, *Brassica napus*, *Nicotiana tabacum*, *Solanum nigrum*, etc. (Barzanti et al. 2007; Mastretta et al. 2009; Luo et al. 2011). Studies have shown the presence of a wide range of metal resistant bacterial endophytes, which include *Bacillus*, *Acinetobacter*, *Paenibacillus*, *Pseudomonas*, *Sphingomonas*, etc. (Sun et al. 2010; Shin et al. 2012; Chen et al. 2014a, b; Govarthanan et al. 2016) and endophytic fungi like *Phoma*, *Alternaria*, *Cladosporium*, *Verticillium*, *Peyronellaea*, etc. (Li et al. 2012a, b, 2016a, b; Khalid et al. 2021). Recent cases of endophyte aided phytoremediation have been summarized in Table 8.1. Khan et al. (2017) have highlighted the role of fungal endophytes of *S. nigrum*, which assisted in the phytoremediation of cadmium (Cd) contaminated soil. In another study by Wang et al. (2020), the impact of endophytes *Serratia* PRE01 and *Arthrobacter* PRE05 has been studied on vanadium contaminated soil. Both the endophytes significantly enhanced the phytoremediation post-inoculation in Indian mustard plants. Endophyte *Pseudomonas azotoformans* ASS1 was reported to enhance the copper (Cu), zinc (Zn), and nickel (Ni) uptake of *Trifolium arvense* under drought stress (Ma et al. 2017). Hence ASS1 protected the plants against abiotic stresses and accelerated phytoremediation against metal contamination. In a study by Liu et al. (2022), a rhizobox system was utilized to understand the impact of endophyte inoculation on bioavailability of phosphorus and Cd during the phytoremediation of contaminated soils (with Cd). Remarkable results were obtained showing enhanced phytoremediation efficiency of *Phytolacca acinosa* upon inoculation with endophyte PE31 *Bacillus cereus* by 52.70% and 46.73% in low and high Cd contaminated soils respectively.

8.3 Phytoremediation of Organic Pollutants Through Endophytes

Phytoremediation of soils that are contaminated with organic pollutants have drawn significant interest. Plants are photoautotrophic organisms and hence do not metabolize organic contaminants as energy or carbon sources (Burken 2003). This could

Table 8.1 Recent success stories of endophyte aided phytoremediation of different pollutants

Sr. no.	Plant	Endophytes	Strategies/mechanisms	Pollutants	Study area	References
1.	*Brachiaria mutica*, *Leptochloa fusca*	*Pseudomonas aeruginosa* BRRI54 *Acinetobacter* sp. strain BRSI56, *Klebsiella* sp. strain LCRI87	Crude oil degradation	Crude oil	Chakwal, Pakistan	Fatima et al. (2018)
2.	*Sedum alfredii*	*Burkholderia* sp. SaZR4, *Burkholderia* sp. SaMR10, *Sphingomonas* sp. SaMR12, *Variovorax* sp. SaNR1	Higher removal efficiencies of trace elements	Cd, lead (Pb)	Quzhou, China	Tang et al. (2020)
3.	*Ficus septica*	*Enterobacter cloacae*	Plant growth promotion, reduction in phytotoxicity	Chromium (Cr)	Malang City, East Java Province, Indonesia	Rohmah et al. (2020)
4.	*Cucumis sativus*	*Aureobasidium pullulans* BSS6	Improved metal tolerance, boost to antioxidant system	Cd, Pb	Nizwa, Oman	Ali et al. (2019)
5.	*Brachiaria mutica*	*Pantoea stewartii* ASI11, *Microbacterium arborescens* HU33, *Enterobacter* sp. HU38	Increased phytoaccumulation	Cd	Faisalabad, Pakistan	Ahsan et al. (2019)
6.	White sweet clover	*Flavobacterium* sp., *Pantoea* sp., *Pseudomonas* sp., *Stenotrophomonas* sp.	Increased plant tolerance, hydrocarbon degradation	Diesel	Canada	Mitter et al. (2019)
7.	White mustard	*Proteus vulgaris* H7, *Pseudomonas* sp. H15, *Pseudomonas helmanticensis* H16	Plant growth promotion, increased phytoextraction, phytoaccumulation	Cd, Zn	Katowice, Upper Silesia, Southern Poland	Płociniczak et al. (2019)
8.	*Pteris vittata*	*Serratia marcescens* PRE01	Heavy metal resistance, plant growth promotion	Vanadium, Cr, Cd	Yunxi County, Shiyan City, China	Wang et al. (2018)

9.	*Helianthus annuus*	*Pseudomonas lurida strain EOO26*	Increased Cu uptake, plant growth promotion	Cu	Karabash town, Chelyabinsk region, Russia	Kumar et al. (2021)
10.	Hybrid *Pennisetum*	*Bacillus megaterium* BM18-2	Increased tolerance, plant growth promotion	Cd	Nanjing, China	Wu et al. (2019a, b)
11.	*Helianthemum lippii* (Hl), *Zygophyllum album* (Za), *Bassia muricata* (Bm)	*Streptomyces* sp.	Petroleum degradation, plant growth promotion	Crude oil	Ouargla, Algeria	Baoune et al. (2018)
12.	*Juncus acutus*	*Sphingomonas* sp., *Bacillus* sp., *Ochrobactrum* sp.	Improved metal phytoextraction	Zn, Ni, Cd, bisphenol-A, ciprofloxacin, sulfamethoxazole	Greece	Syranidou et al. (2018)
13.	*Chloris virgata*	*Bacillus safensis* strain ZY16	Plant growth promotion, hydrocarbon degradation	Naphthalene, phenanthrene, and pyrene, *n*-undecane, *n*-hexadecane, *n*-octacosane, diesel oil	Shengli Oilfield, China	Wu et al. (2019a, b)
14.	*Sedum alfredii*	*Pseudomonas fluorescens Sasm05*	Plant growth promotion, increased chlorophyll content, elevated Zn uptake	Zn	China	Wang et al. (2019)
15.	*Phragmites australis*	*Pseudomonas aeruginosa*	Plant growth promotion, hydrocarbon degradation	Naphthalene, phenanthrene, pyrene, diesel oil	Shandong, China	Wu et al. (2018)
16.	*Clitoria ternatea*	*Bacillus cereus ERBP*	Increased expression of CYP83D1, dehydrogenases	Ethylbenzene	Thailand	Daudzai et al. (2018)
17.	*Dracaena sanderiana*	*Bacillus thuringiensis and Pantoea dispersa*	Maintained levels of indole-3-acetic acid, lower levels of reactive oxygen species	Bisphenol-A	Thailand	Suyamud et al. (2018)

(continued)

Table 8.1 (continued)

Sr. no.	Plant	Endophytes	Strategies/mechanisms	Pollutants	Study area	References
18.	*Robinia pseudoacacia*	*Mesorhizobium loti HZ76 and Agrobacterium radiobacter HZ6*	Relieved HM stress, plant growth promotion	Pb, Zn, Cd, Cu	China	Fan et al. (2018)
19.	*Trifolium repens* L.	*Pseudomonas putida* strain RE02	Improved soil fertility and enhanced heavy metal uptake	Cd, Cr, Pb	China	Liu et al. (2021)
20.	*Leptochloa fusca and Brachiaria mutica*	*Enterobacter* sp. *HU38, Microbacterium arborescens HU33, and Pantoea stewartii ASI11*	Increased uptake, translocation, accumulation of Cr	Cr	Pakistan	Ahsan et al. (2018)
21.	*Solanum nigrum*	*Kocuria* sp., *Enterobacter* sp., *Kosakonia* sp.	Improved plant growth, bioaccumulation, elevated glutathione levels	Arsenic	India	Mukherjee et al. (2018)
22.	*Brassica juncea*	*SerratiaPRE01*, *Arthrobacter PRE05*	Improving the rhizosphere and endosphere microecology	Vanadium	China	Wang et al. (2020)
23.	*Phytolacca acinosa*	PE31 *Bacillus cereus*	Enhanced bioavailability of P and Cd	Cd	China	Liu et al. (2022)
24.	*Cucurbita pepo*	*Sphingomonas taxi UH1, Methylobacterium radiotolerans UH1, Enterobacter aerogenes UH1*	Enhanced DDE accumulation	2,2-bis(*p*-chlorophenyl)-1,1-dichloro-ethylene (DDE)	Belgium	Eevers et al. (2018)
25.	*Pteris vittata*	*Serratia marcescens PRE01* and *Arthrobacter* sp. *PRE05*	Enhanced vanadium accumulation in roots	Vanadium	China	Wang et al. (2023)

be because plants were not under selective pressure during evolution to develop the potential of degrading recalcitrant molecules like microorganisms. Hence plants are able to metabolize/degrade only a limited range of chemicals. So, plants rely on their associated microorganisms for more proficient degradation. The growth of plants is impaired even if they can tolerate organic pollutants as these compounds are toxic and hydrophobic (Nwoko 2010).

As endophytes have biodegradative capacity, they can be utilized to improve the phytoremediation of organic pollutants (Van Aken et al. 2011; Arslan et al. 2017; Syranidou et al. 2018). So, the highlighted functions of endophytes are to provide fitness advantages, which include enhanced growth, stress tolerance, and efficient degradation of organic pollutants (Doty et al. 2017; Baoune et al. 2018; Singh et al. 2022). The competent degradative potential is attributed to the presence of catabolic genes in endophytes during soil phytoremediation (Yousaf et al. 2011; Fatima et al. 2018). Certain endophytes have been reported to produce biosurfactants, detoxifying enzymes, etc. for successful removal of contaminants from soil (Reichenauer and Germida 2008). After the uptake, organic contaminants can also be released into the atmosphere through stem or leaves by evapotranspiration (Weyens et al. 2013). The prominent endophytic families of bacteria found from polluted sites include Bacillaceae, Pseudomonadaceae, Burkholderiaceae, etc. (Weyens et al. 2009; Andreolli et al. 2013; Khan et al. 2014; Feng et al. 2018). Organic chemicals like trichloroethylene (TCE), naphthalene, catechol, phenol, etc. can be degraded by endophytes (Kang et al. 2012; Ho et al. 2012; Andreolli et al. 2013). Ho et al. (2013) have reported use of endophytic bacterium *Achromobacter xylosoxidans* for phytoremediation of phenolic pollutants like catechol and phenol using *Arabidopsis thaliana. Achromobacter xylosoxidans* helped plants to tolerate stress and improved phytoremediation. In a study by Syranidou et al. (2017), it was found that cultivable bacterial endophytes of *Juncus acutus* were found to be effective for remediating Bisphenol-A (BPA) contaminated groundwater. The study highlighted the potential of endophytes to use organic contaminants as carbon source.

In nutshell, the plant–endophyte partnership is very promising for enhancing remediation efficiencies and it strongly makes obvious the appropriateness of pollutant-degrading endophytes for the detoxification and degradation of organic contaminants.

8.4 Phytoremediation of Mixed Waste Pollutants Through Endophytes

The phytoremediation potential of organic pollutants is different from heavy metals. But in the majority of the contaminated sites, plants have to tackle with mixed contamination (Chirakkara et al. 2016). It has been found that the presence of toxic metals usually inhibits the potential to degrade organic contaminants (Sandrin and Maier 2003; Olaniran et al. 2013). Endophytes with the potential to degrade mixed contaminants have attributes of extracting heavy metals and degrading the organic pollutants (Gutiérrez-Ginés et al. 2014; Waigi et al. 2020).

Two strategies have been reported to combat the menace of mixed contamination. First, using consortia of organic pollutants degrading endophytes and those capable of detoxifying metals (Weyens et al. 2011); second using genetically engineered endophytes having the ability to degrade organics and sequester heavy metals (Weyens et al. 2010; Chirakkara et al. 2016).

8.5 Strategies Employed by Endophytes in Phytoremediation

8.5.1 Plant Growth Promotion

Phytoremediation is augmented by endophytes as they promote plant growth, reduce phytotoxicity, and improve metal tolerance (Babu et al. 2013; Durand et al. 2021). Endophytes enable plant growth in various ways (Fig. 8.1). These include production of phytohormones, nitrogen fixation, phosphate solubilization, and siderophore production (Dolatabad et al. 2017; Pham et al. 2017; Adhikari and Pandey 2019; Shahzad et al. 2019). Endophytic bacteria with strong associative nitrogen-fixing ability enable plants to survive easier in nitrogen-poor soil environment and play a major role in promoting plant health and growth, compared to rhizosphere microorganisms. Afzal et al. (2014) have reported endophytic *Burkholderia* sp. strain PsJN that resulted in improved plant (*Acacia ampliceps*, *Eucalyptus*

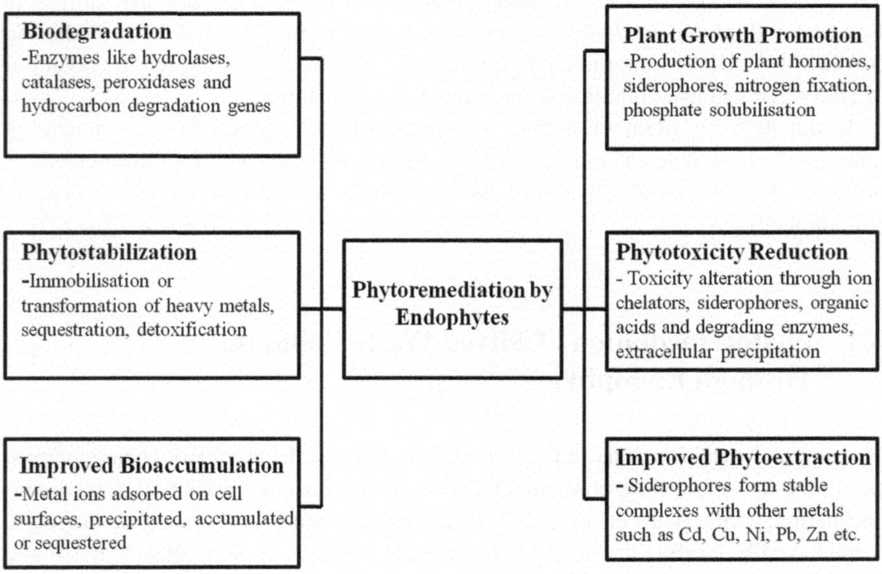

Fig. 8.1 Strategies employed by endophytes in phytoremediation

camaldulensis, and *Leucaena leucocephala*) growth and phytoremediation of soil irrigated with textile effluent. Post-inoculation plant biomass increased by up to 12% and contaminants removal was enhanced by 29%. In a study by Mesa et al. (2017), it was found that endophytic bacteria augmented the phytoremediation of industrial soils, which were contaminated by arsenic by *Betula celtiberica.* The results pointed out that the siderophore and indole-3-acetic acid generated by the endophytic consortium enhanced the arsenic deposition in the leaves and roots of *Betula celtiberica.* Plant growth promotion was also focused by Sun et al. (2019) where it was found that *Microbacterium* sp. BXGe71 from *Arabis alpina* appreciably improved plant growth under multi-heavy metal (HM) stress.

8.5.2 Phytotoxicity Reduction

Plants experience phytotoxic effects due to the pollutants like polycyclic aromatic hydrocarbons (PAHs), halogenated hydrocarbons, heavy metals and this leads to inhibition of phytoremediation (Harvey et al. 2002; Ghanem et al. 2010; Malar et al. 2014). Some endophytes have been reported to alter the toxicity of the contaminants through ion chelators, siderophores, organic acids, and degrading enzymes (Rajkumar et al. 2010; Chen et al. 2014a, b). Some studies have also demonstrated extracellular precipitation as a means to reduce phytotoxicity (Babu et al. 2015).

8.5.3 Improved Phytoextraction

Endophytes have been implicated to enhance phytoextraction through modification of the solubility and availability of heavy metals through releasing organic acids, chelators, and bringing about redox changes (Ma et al. 2015; Zamani et al. 2015). Siderophores act as solubilizing agents for iron and enhance its efficient uptake under conditions of iron limitation. Siderophores can also form stable complexes with other metals such as Cd, Cu, Ni, Pb, Zn, etc. (Visioli et al. 2015; Shi et al. 2017). Siderophore binding to a metal enhances the concentration of soluble metal (Rajkumar et al. 2010). There is ample literature citing the importance of siderophore-producing endophytes from various plants and their impact on plant growth in low nutrition environments (Yuan et al. 2014).

8.5.4 Metal Accumulation in Plants (Bioaccumulation)

It has been demonstrated that endophytes can also contribute in reducing metal phytotoxicity through uptake and accumulation in plants (Mello et al. 2020; Wang et al. 2023). Metal ions have been speculated to be adsorbed on the cell surfaces by interacting with cell surface functional groups like hydroxyl, carbonyl, carboxyl, sulfhydryl, thioether, etc. The metals can also be transported inside the cells by living cells (active transport) (Soleimani et al. 2010). Inside the cell, the metals can be bound, precipitated, accumulated, or sequestered within specific intracellular organelles (Ma et al. 2015). Thus, endophytes possessing outstanding bioaccumulation potential can assist in detoxification of heavy metals and advance the efficiency of phytoremediation (Wu et al. 2019a, b; Shahabivand et al. 2020). Recently, Ullah et al. (2022) explored the role of endophytic bacteria *Serratia* sp. AI001 and *Klebsiella* sp. strain AI002 in phytoremediation of Cd and heavy metal ATPase gene expression. Post-inoculation with both the strains, there was significant increase in expression of heavy metal ATPase genes *HMA2*, *HMA3,* and *HMA4* leading to Cd translocation within the plants.

8.5.5 Phytostabilization

To thrive under metal stressed conditions, endophytes have advanced to myriad mechanisms. Heavy metals are either immobilized or transformed to become inactive, promoting plants to tolerate the uptake of heavy metals (Pinheiro et al. 2013; Li et al. 2016a, b). One of the strategies is to actively export metal out of the cell or physically sequester the metal by binding to extracellular polymers. At times the detoxification of metal is also carried out to render it less effective (Rajkumar et al. 2013; Burges et al. 2018; Shahabivand et al. 2019).

8.5.6 Biodegradation

Varieties of enzymes are produced by endophytes that enhance the process of phytoremediation. These include hydrolases, glutathione-*S*-transferases, cytochrome P450 monooxygenases, catalases, peroxidases, and polyphenol peroxidases (Fu et al. 2018). Organic pollutants undergo oxygenation, dehydrogenation, and covalent modification in the presence of these enzymes that promote increase in polarity and hence water solubility. Organic pollutants become more degradable and hence plants and endophytes are able to assimilate more (Li et al. 2012a, b). Kukla et al. (2014) reported endophytic bacteria isolated from *Lolium perenne* L. The strains exhibited plant growth promoting potential and possessed hydrocarbon degradation genes like *alkH*, *alkB*, *C23O*, *P450,* and *pah.* In another study by Pawlik et al. (2017), 40% of the isolates were found to have the *P450* and *acdS* genes encoding for cytochrome P450-type alkane hydroxylase (CYP153) and 1-aminocyclopropane-1-carboxylic acid deaminase respectively.

8.6 Future Prospects

In a nutshell, production of degradative enzymes, organic acids, chelators, and other bioactive compounds by endophytes promotes phytoremediation. This stands in as a promise to wrestle the problem of pollution. The isolation and characterization of novel endophytes with above mentioned potential are worth and important. We foresee a greater emphasis on discovery of molecular mechanisms that endorse better phytoremediation by endophytes. However, at present the diversity of plants that assist phytoremediation and their corresponding endophytic niche needs to be explored. This will open a new dimension on understanding the production of a myriad of biomolecules and their mechanisms in phytoremediation leading to better environmental sustainability. The future of endophyte-assisted phytoremediation lies in the identification and development of specific endophyte–plant combinations for different types of pollutants and contaminated environments. By understanding the interactions between endophytes and plants, scientists can select or engineer endophytes that are specialized in targeting particular contaminants. This approach can maximize the efficiency and effectiveness of phytoremediation strategies and make them more tailored to specific pollution scenarios. Despite the promising future prospects, there are still some challenges that need to be addressed. These include optimizing the selection and application of endophytes, understanding the complex interactions between endophytes and plants, ensuring the safety of genetically modified endophytes, and scaling up endophyte-assisted phytoremediation to field-scale applications. However, with ongoing research and technological advancements, endophyte-based phytoremediation approaches hold great promise for addressing environmental pollution in a sustainable and cost-effective manner.

8.7 Conclusion

Endophyte-assisted phytoremediation has received substantial awareness due to their plant growth promoting activities and tolerance against a variety of contaminants. Thus, a diverse endophytic population is extremely promising in boosting the pollutant removal efficiency of plant species. Endophytes are thus hasting up the process of phytoremediation by alleviating the rate of mobilization, absorption, and accumulation of pollutants. Besides that, they also promote plant growth and nutrient uptake. As per the above exploration, we can deduce that endophyte-assisted phytoremediation is a process that we can look forth upon in the future. Utilization of this partnership can play a momentous role in low-cost sustainable approach of phytoremediation.

References

Adhikari P, Pandey A (2019) Phosphate solubilization potential of endophytic fungi isolated from *Taxus wallichiana* Zucc. roots. Rhizosphere 9:2–9

Afzal M, Shabir G, Tahseen R, Islam EU, Iqbal S, Khan QM et al (2014) Endophytic *Burkholderia* sp. strain PsJN improves plant growth and phytoremediation of soil irrigated with textile effluent. Clean 42(9):1304–1310

Ahsan MT, Saeed A, Mustafa T, Afzal M (2018) Augmentation with potential endophytes enhances phytostabilization of Cr in contaminated soil. Environ Sci Pollut Res 25(7):7021–7032

Ahsan MT, Tahseen R, Ashraf A, Mahmood A, Arslan M, Afzal M (2019) Effective plant-endophyte interplay can improve the cadmium hyperaccumulation in *Brachiaria mutica*. World J Microbiol Biotechnol 35(12):188

Ali H, Khan E, Sajad MA (2013) Phytoremediation of heavy metals—concepts and applications. Chemosphere 91(7):869–881

Ali A, Bilal S, Khan AL, Mabood F, Al-Harrasi A, Lee IJ (2019) Endophytic *Aureobasidium pullulans* BSS6 assisted developments in phytoremediation potentials of *Cucumis sativus* under Cd and Pb stress. J Plant Interact 14(1):303–313

Andreolli M, Lampis S, Poli M, Gullner G, Biró B, Vallini G (2013) Endophytic *Burkholderia fungorum* DBT1 can improve phytoremediation efficiency of polycyclic aromatic hydrocarbons. Chemosphere 92(6):688–694

Appannagari RR (2017) Environmental pollution causes and consequences: a study. North Asian Int Res J Soc Sci Humanit 3:1–12

Arora NK, Kumar N (2019) In: Arora NK (ed) Phyto and rhizo remediation, Microorganisms for sustainability, vol 9. Springer, Singapore

Arslan M, Imran A, Khan QM, Afzal M (2017) Plant–bacteria partnerships for the remediation of persistent organic pollutants. Environ Sci Pollut Res 24(5):4322–4336

Athar R, Ahmad M (2002) Heavy metal toxicity: effect on plant growth and metal uptake by wheat, and on free living Azotobacter. Water Air Soil Pollut 138(1–4):165–180

Babu AG, Kim JD, Oh BT (2013) Enhancement of heavy metal phytoremediation by *Alnus firma* with endophytic *Bacillus thuringiensis* GDB-1. J Hazard Mater 250:477–483

Babu AG, Shea PJ, Sudhakar D, Jung IB, Oh BT (2015) Potential use of *Pseudomonas koreensis* AGB-1 in association with Miscanthus sinensis to remediate heavy metal(loid)-contaminated mining site soil. J Environ Manag 151:160–166

Baoune H, El Hadj-Khelil AO, Pucci G, Sineli P, Loucif L, Polti MA (2018) Petroleum degradation by endophytic Streptomyces spp. isolated from plants grown in contaminated soil of southern Algeria. Ecotoxicol Environ Saf 147:602–609

Barzanti R, Ozino F, Bazzicalupo M, Gabbrielli R, Galardi F, Gonnelli C et al (2007) Isolation and characterization of endophytic bacteria from the nickel hyperaccumulator plant *Alyssum bertolonii*. Microb Ecol 53(2):306–316

Brader G, Compant S, Mitter B, Trognitz F, Sessitsch A (2014) Metabolic potential of endophytic bacteria. Curr Opin Biotechnol 27:30–37

Burges A, Alkorta I, Epelde L, Garbisu C (2018) From phytoremediation of soil contaminants to phytomanagement of ecosystem services in metal contaminated sites. Int J Phytoremediation 20(4):384–397

Burken JG (2003) Uptake and metabolism of organic compounds: green-liver model. In: Phytoremediation: transformation and control of contaminants, vol 59. Wiley, New York, pp 59–84

Chen L, Luo S, Li X, Wan Y, Chen J, Liu C (2014a) Interaction of Cd-hyperaccumulator *Solanum nigrum* L. and functional endophyte *Pseudomonas* sp. Lk9 on soil heavy metals uptake. Soil Biol Biochem 68:300–308

Chen B, Shen J, Zhang X, Pan F, Yang X, Feng Y (2014b) The endophytic bacterium, *Sphingomonas* SaMR12, improves the potential for zinc phytoremediation by its host, *Sedum alfredii*. PLoS One 9(9):e106826

Chirakkara RA, Cameselle C, Reddy KR (2016) Assessing the applicability of phytoremediation of soils with mixed organic and heavy metal contaminants. Rev Environ Sci Biotechnol 15(2):299–326

Cristaldi A, Conti GO, Jho EH, Zuccarello P, Grasso A, Copat C et al (2017) Phytoremediation of contaminated soils by heavy metals and PAHs: a brief review. Environ Technol Innov 8:309–326

Daudzai Z, Treesubsuntorn C, Thiravetyan P (2018) Inoculated *Clitoria ternatea* with *Bacillus cereus* ERBP for enhancing gaseous ethylbenzene phytoremediation: plant metabolites and expression of ethylbenzene degradation genes. Ecotoxicol Environ 164:50–60

Dolatabad HK, Javan-Nikkhah M, Shier WT (2017) Evaluation of antifungal, phosphate solubilisation, and siderophore and chitinase release activities of endophytic fungi from *Pistacia vera*. Mycol Prog 16(8):777–790

Doty SL, Freeman JL, Cohu CM, Burken JG, Firrincieli A, Simon A et al (2017) Enhanced degradation of TCE on a superfund site using endophyte-assisted poplar tree phytoremediation. Environ Sci Technol 51(17):10050–10058

Durand A, Leglize P, Benizri E (2021) Are endophytes essential partners for plants and what are the prospects for metal phytoremediation? Plant Soil 460:1–30

Eevers N, Hawthorne JR, White JC, Vangronsveld J, Weyens N (2018) Endophyte-enhanced phytoremediation of DDE-contaminated using Cucurbita pepo: a field trial. Int J Phytoremediation 20(4):301–310

Fan M, Liu Z, Nan L, Wang E, Chen W, Lin Y et al (2018) Isolation, characterization, and selection of heavy metal-resistant and plant growth-promoting endophytic bacteria from root nodules of *Robinia pseudoacacia* in a Pb/Zn mining area. Microbiol Res 217:51–59

Fatima K, Imran A, Amin I, Khan QM, Afzal M (2018) Successful phytoremediation of crude-oil contaminated soil at an oil exploration and production company by plants-bacterial synergism. Int J Phytoremediation 20(7):675–681

Feng NX, Yu J, Mo CH, Zhao HM, Li YW, Wu BX (2018) Biodegradation of di-n-butyl phthalate (DBP) by a novel endophytic *Bacillus megaterium* strain YJB3. Sci Total Environ 616:117–127

Fu W, Xu M, Sun K, Hu L, Cao W, Dai C (2018) Biodegradation of phenanthrene by endophytic fungus *Phomopsis liquidambari* in vitro and in vivo. Chemosphere 203:160–169

Ghanem A, D'Orazio V, Senesi N (2010) Phytotoxicity assay of selected plants to pyrene contaminated soil. In: Proceedings of the 19th world congress of soil science, soil solutions for a changing world, Brisbane

Govarthanan M, Mythili R, Selvankumar T, Kamala-Kannan S, Rajasekar A, Chang YC (2016) Bioremediation of heavy metals using an endophytic bacterium *Paenibacillus* sp. RM isolated from the roots of *Tridax procumbens*. 3 Biotech 6(2):242

Gutiérrez-Ginés MJ, Hernández AJ, Pérez-Leblic MI, Pastor J, Vangronsveld J (2014) Phytoremediation of soils co-contaminated by organic compounds and heavy metals: bioassays with *Lupinus luteus* L. and associated endophytic bacteria. J Environ Manag 143:197–207

Harvey PJ, Campanella BF, Castro PM, Harms H, Lichtfouse E, Schäffner AR et al (2002) Phytoremediation of polyaromatic hydrocarbons, anilines and phenols. Environ Sci Pollut Res 9(1):29–47

Ho YN, Mathew DC, Hsiao SC, Shih CH, Chien MF, Chiang HM (2012) Selection and application of endophytic bacterium *Achromobacter xylosoxidans* strain F3B for improving phytoremediation of phenolic pollutants. J Hazard Mater 219:43–49

Ho YN, Hsieh JL, Huang CC (2013) Construction of a plant–microbe phytoremediation system: combination of vetiver grass with a functional endophytic bacterium, *Achromobacter xylosoxidans* F3B, for aromatic pollutants removal. Bioresour Technol 145:43–47

Hooda V (2007) Phytoremediation of toxic metals from soil and waste water. J Environ Biol 28(2):367

Kang JW, Khan Z, Doty SL (2012) Biodegradation of trichloroethylene by an endophyte of hybrid poplar. Appl Environ Microbiol 78(9):3504–3507

Khalid M, Saeed UR, Hassani D, Hayat K, Pei ZH, Nan HU (2021) Advances in fungal-assisted phytoremediation of heavy metals: a review. Pedosphere 31(3):475–495

Khan Z, Roman D, Kintz T, delas Alas M, Yap R, Doty S (2014) Degradation, phytoprotection and phytoremediation of phenanthrene by endophyte *Pseudomonas putida*, PD1. Environ Sci Technol 48(20):12221–12228

Khan AR, Waqas M, Ullah I, Khan AL, Khan MA, Lee IJ (2017) Culturable endophytic fungal diversity in the cadmium hyperaccumulator *Solanum nigrum* L. and their role in enhancing phytoremediation. Environ Exp Bot 135:126–135

Khare E, Mishra J, Arora NK (2018) Multifaceted interactions between endophytes and plant: developments and prospects. Front Microbiol 9:2732. https://doi.org/10.3389/fmicb.2018.02732

Kong Z, Glick BR (2017) The role of plant growth-promoting bacteria in metal phytoremediation. In: Advances in microbial physiology, vol 71. Academic, New York, pp 97–132

Kukla M, Płociniczak T, Piotrowska-Seget Z (2014) Diversity of endophytic bacteria in *Lolium perenne* and their potential to degrade petroleum hydrocarbons and promote plant growth. Chemosphere 117:40–46

Kumar A, Voropaeva O, Maleva M, Panikovskaya K, Borisova G, Rajkumar M (2021) Bioaugmentation with copper tolerant endophyte *Pseudomonas lurida* strain EOO26 for improved plant growth and copper phytoremediation by *Helianthus annuus*. Chemosphere 266:128983

Li HY, Li DW, He CM, Zhou ZP, Mei T, Xu HM (2012a) Diversity and heavy metal tolerance of endophytic fungi from six dominant plant species in a Pb–Zn mine wasteland in China. Fungal Ecol 5(3):309–315

Li HY, Wei DQ, Shen M, Zhou ZP (2012b) Endophytes and their role in phytoremediation. Fungal Divers 54(1):11–18

Li X, Li W, Chu L, White JF Jr, Xiong Z, Li H (2016a) Diversity and heavy metal tolerance of endophytic fungi from *Dysphania ambrosioides*, a hyperaccumulator from Pb–Zn contaminated soils. J Plant Interact 11(1):186–192

Li Y, Wang Q, Wang L, He LY, Sheng XF (2016b) Increased growth and root Cu accumulation of *Sorghum sudanense* by endophytic *Enterobacter* sp. K3-2: implications for *Sorghum sudanense* biomass production and phytostabilization. Ecotoxicol Environ Saf 124:163–168

Liu C, Lin H, Li B, Dong Y, Menzembere ER (2021) Endophyte *Pseudomonas putida* enhanced *Trifolium repens* L. growth and heavy metal uptake: a promising in-situ non-soil cover phytoremediation method of nonferrous metallic tailing. Chemosphere 272:129816

Liu C, Lin H, Dong Y, Li B (2022) Increase of P and Cd bioavailability in the rhizosphere by endophytes promoted phytoremediation efficiency of *Phytolacca acinosa*. J Hazard Mater 431:128546

Luo SL, Chen L, Chen JL, Xiao X, Xu TY, Wan Y et al (2011) Analysis and characterization of cultivable heavy metal-resistant bacterial endophytes isolated from Cd-hyperaccumulator *Solanum nigrum* L. and their potential use for phytoremediation. Chemosphere 85(7):1130–1138

Ma Y, Oliveira RS, Nai F, Rajkumar M, Luo Y, Rocha I et al (2015) The hyperaccumulator *Sedum plumbizincicola* harbors metal-resistant endophytic bacteria that improve its phytoextraction capacity in multi-metal contaminated soil. J Environ Manag 156:62–69

Ma Y, Rajkumar M, Moreno A, Zhang C, Freitas H (2017) Serpentine endophytic bacterium *Pseudomonas azotoformans* ASS1 accelerates phytoremediation of soil metals under drought stress. Chemosphere 185:75–85

Malar S, Manikandan R, Favas PJ, Sahi SV, Venkatachalam P (2014) Effect of lead on phytotoxicity, growth, biochemical alterations and its role on genomic template stability in *Sesbania grandiflora*: a potential plant for phytoremediation. Ecotoxicol Environ Saf 108:249–257

Mastretta C, Taghavi S, Van Der Lelie D, Mengoni A, Galardi F, Gonnelli C et al (2009) Endophytic bacteria from seeds of *Nicotiana tabacum* can reduce cadmium phytotoxicity. Int J Phytoremediation 11(3):251–267

Mello IS, Targanski S, Pietro-Souza W, Stachack FF, Terezo AJ, Soares MA (2020) Endophytic bacteria stimulate mercury phytoremediation by modulating its bioaccumulation and volatilization. Ecotoxicol Environ Saf 202:110818
Mesa V, Navazas A, González-Gil R, González A, Weyens N, Lauga B et al (2017) Use of endophytic and rhizosphere bacteria to improve phytoremediation of arsenic-contaminated industrial soils by autochthonous *Betula celtiberica*. Appl Environ Microbiol 83(8):e03411-16
Mitter EK, Kataoka R, de Freitas JR, Germida JJ (2019) Potential use of endophytic root bacteria and host plants to degrade hydrocarbons. Int J Phytoremediation 21(9):928–938
Mukherjee G, Saha C, Naskar N, Mukherjee A, Mukherjee A, Lahiri S et al (2018) An endophytic bacterial consortium modulates multiple strategies to improve arsenic phytoremediation efficacy in *Solanum nigrum*. Sci Rep 8(1):1–6
Nwoko CO (2010) Trends in phytoremediation of toxic elemental and organic pollutants. Afr J Biotechnol 9(37):6010–6016
Oladoye PO, Olowe OM, Asemoloye MD (2022) Phytoremediation technology and food security impacts of heavy metal contaminated soils: a review of literature. Chemosphere 288:132555
Olaniran AO, Balgobind A, Pillay B (2013) Bioavailability of heavy metals in soil: impact on microbial biodegradation of organic compounds and possible improvement strategies. Int J Mol Sci 5:10197–10228
Pawlik M, Cania B, Thijs S, Vangronsveld J, Piotrowska-Seget Z (2017) Hydrocarbon degradation potential and plant growth-promoting activity of culturable endophytic bacteria of *Lotus corniculatus* and *Oenothera biennis* from a long-term polluted site. Environ Sci Pollut Res 24(24):19640–19652
Pham VT, Rediers H, Ghequire MG, Nguyen HH, De Mot R, Vanderleyden J (2017) The plant growth-promoting effect of the nitrogen-fixing endophyte *Pseudomonas stutzeri* A15. Arch Microbiol 199(3):513–517
Pinheiro JC, Marques CR, Pinto G, Bouguerra S, Mendo S, Gomes NC et al (2013) The performance of *Fraxinus angustifolia* as a helper for metal phytoremediation programs and its relation to the endophytic bacterial communities. Geoderma 202:171–182
Płociniczak T, Chodór M, Pacwa-Płociniczak M, Piotrowska-Seget Z (2019) Metal-tolerant endophytic bacteria associated with Silene vulgaris support the Cd and Zn phytoextraction in non-host plants. Chemosphere 219:250–260
Rajkumar M, Ae N, Prasad MN, Freitas H (2010) Potential of siderophore-producing bacteria for improving heavy metal phytoextraction. Trends Biotechnol 28(3):142–149
Rajkumar M, Ma Y, Freitas H (2013) Improvement of Ni phytostabilization by inoculation of Ni resistant *Bacillus megaterium* SR28C. J Environ Manage 128:973–980
Rascio N, Navari-Izzo F (2011) Heavy metal hyperaccumulating plants: how and why do they do it? And what makes them so interesting? Plant Sci 180(2):169–181
Reichenauer TG, Germida JJ (2008) Phytoremediation of organic contaminants in soil and groundwater. ChemSusChem 1(8–9):708–717
Reinhold-Hurek B, Hurek T (2011) Living inside plants: bacterial endophytes. Curr Opin Plant Biol 14(4):435–443
Rohmah NS, Jatmiko YD, Siswanto D, Mustafa I (2020) The potency of endophytic bacteria isolated from Ficus septica as phytoremediation promoting agent of Cr (VI) contaminated soil. Biodiversitas 21(5):1920–1927
Ryan RP, Germaine K, Franks A, Ryan DJ, Dowling DN (2008) Bacterial endophytes: recent developments and applications. FEMS Microbiol Lett 278(1):1–9
Sandrin TR, Maier RM (2003) Impact of metals on the biodegradation of organic pollutants. Environ Health Perspect 111(8):1093–1101
Santoyo G, Moreno-Hagelsieb G, del Carmen Orozco-Mosqueda M, Glick BR (2016) Plant growth-promoting bacterial endophytes. Microbiol Res 183:92–99
Saxena G, Purchase D, Mulla SI, Saratale GD, Bharagava RN (2019) Phytoremediation of heavy metal-contaminated sites: eco-environmental concerns, field studies, sustainability issues, and

future prospects. In: de Voogt P (ed) Reviews of environmental contamination and toxicology, vol 249. Springer, Cham, pp 71–131

Shahabivand S, Tayebnia V, Aliloo AA (2019) Impact of endophyte fungus *Serendipita indica* on fungus-assisted phyto-stabilization and performance of *Carthamus tinctorius* in a lead polluted soil. J Plant Res 31(4):812–825

Shahabivand S, Parvaneh A, Aliloo AA (2020) Different response of *Alyssum montanum* and *Helianthus annuus* to cadmium bioaccumulation mediated by the endophyte fungus *Serendipita indica*. Acta Ecol Sin 40(4):315–322

Shahzad R, Khan AL, Waqas M, Ullah I, Bilal S, Kim YH et al (2019) Metabolic and proteomic alteration in phytohormone-producing endophytic *Bacillus amyloliquefaciens* RWL-1 during methanol utilization. Metabolomics 15(2):16

Shen X, Dai M, Yang J, Sun L, Tan X, Peng C et al (2022) A critical review on the phytoremediation of heavy metals from environment: performance and challenges. Chemosphere 291:132979

Shi Y, Xie H, Cao L, Zhang R, Xu Z, Wang Z et al (2017) Effects of Cd- and Pb-resistant endophytic fungi on growth and phytoextraction of *Brassica napus* in metal-contaminated soils. Environ Sci Pollut Res 24(1):417–426

Shin MN, Shim J, You Y, Myung H, Bang KS, Cho M et al (2012) Characterization of lead resistant endophytic Bacillus sp. MN3-4 and its potential for promoting lead accumulation in metal hyperaccumulator *Alnus firma*. J Hazard Mater 199:314–320

Singh T, Awasthi G, Tiwari Y (2022) Recruiting endophytic bacteria of wetland plants to phytoremediate organic pollutants. Int J Environ Sci Technol 19(9):9177–9188

Soleimani M, Hajabbasi MA, Afyuni M, Mirlohi A, Borggaard OK, Holm PE (2010) Effect of endophytic fungi on cadmium tolerance and bioaccumulation by *Festuca arundinacea* and *Festuca pratensis*. Int J Phytoremediation 12(6):535–549

Song YF, Wilke BM, Song XY, Gong P, Zhou QX, Yang GF (2006) Polycyclic aromatic hydrocarbons (PAHs), polychlorinated biphenyls (PCBs) and heavy metals (HMs) as well as their genotoxicity in soil after long-term wastewater irrigation. Chemosphere 65(10):1859–1868

Strobel GA (2002) Rainforest endophytes and bioactive products. Crit Rev Biotechnol 22(4):315–333

Sun LN, Zhang YF, He LY, Chen ZJ, Wang QY, Qian M et al (2010) Genetic diversity and characterization of heavy metal-resistant-endophytic bacteria from two copper-tolerant plant species on copper mine wasteland. Bioresour Technol 101(2):501–509

Sun W, Xiong Z, Chu L, Li W, Soares MA, White JF Jr et al (2019) Bacterial communities of three plant species from Pb-Zn contaminated sites and plant-growth promotional benefits of endophytic Microbacterium sp. (strain BXGe71). J Hazard Mater 370:225–231

Suyamud B, Thiravetyan P, Panyapinyopol B, Inthorn D (2018) *Dracaena sanderiana* endophytic bacteria interactions: effect of endophyte inoculation on bisphenol A removal. Ecotoxicol Environ Saf 157:318–326

Syranidou E, Christofilopoulos S, Politi M, Weyens N, Venieri D, Vangronsveld J et al (2017) Bisphenol-A removal by the halophyte *Juncus acutus* in a phytoremediation pilot: characterization and potential role of the endophytic community. J Hazard Mater 323:350–358

Syranidou E, Thijs S, Avramidou M, Weyens N, Venieri D, Pintelon I et al (2018) Responses of the endophytic bacterial communities of *Juncus acutus* to pollution with metals, emerging organic pollutants and to bioaugmentation with indigenous strains. Front Plant Sci 9:1526

Tang L, Hamid Y, Zehra A, Sahito ZA, He Z, Beri WT et al (2020) Fava bean intercropping with *Sedum alfredii* inoculated with endophytes enhances phytoremediation of cadmium and lead co-contaminated field. Environ Pollut 29:114861

Ullah I, Mateen A, Ahmad MA, Munir I, Iqbal A, Alghamdi KM et al (2022) Heavy metal ATPase genes (HMAs) expression induced by endophytic bacteria, "AI001, and AI002" mediate cadmium translocation and phytoremediation. Environ Pollut 293:118508

Van Aken B, Tehrani R, Schnoor JL (2011) Endophyte-assisted phytoremediation of explosives in poplar trees by *Methylobacterium populi* BJ001 T. In: Pirttilä A, Frank A (eds) Endophytes of forest trees. Springer, Dordrecht, pp 217–234

Vardhan KH, Kumar PS, Panda RC (2019) A review on heavy metal pollution, toxicity and remedial measures: current trends and future perspectives. J Mol Liq 290:111197

Visioli G, Vamerali T, Mattarozzi M, Dramis L, Sanangelantoni AM (2015) Combined endophytic inoculants enhance nickel phytoextraction from serpentine soil in the hyperaccumulator *Noccaea caerulescens*. Front Plant Sci 6:638

Waigi MG, Wang J, Yang B, Gudda FO, Ling W, Liu J et al (2020) Endophytic bacteria in planta organopollutant detoxification in crops. Rev Environ Contam Toxicol 252:1–50

Wang L, Lin H, Dong Y, He Y, Liu C (2018) Isolation of vanadium-resistance endophytic bacterium PRE01 from *Pteris vittata* in stone coal smelting district and characterization for potential use in phytoremediation. J Hazard Mater 341:1–9

Wang Q, Ye J, Wu Y, Luo S, Chen B, Ma L et al (2019) Promotion of the root development and Zn uptake of *Sedum alfredii* was achieved by an endophytic bacterium Sasm05. Ecotoxicol Environ Saf 172:97–104

Wang L, Lin H, Dong Y, Li B, He Y (2020) Effects of endophytes inoculation on rhizosphere and endosphere microecology of Indian mustard (*Brassica juncea*) grown in vanadium-contaminated soil and its enhancement on phytoremediation. Chemosphere 240:124891

Wang L, Liao X, Dong Y, Lin H (2023) Vanadium-resistant endophytes modulate multiple strategies to facilitate vanadium detoxification and phytoremediation in *Pteris vittata*. J Hazard Mater 443:130388

Weyens N, Taghavi S, Barac T, van der Lelie D, Boulet J, Artois T et al (2009) Bacteria associated with oak and ash on a TCE-contaminated site: characterization of isolates with potential to avoid evapotranspiration of TCE. Environ Sci Pollut Res 16(7):830–843

Weyens N, Croes S, Dupae J, Newman L, van der Lelie D, Carleer R et al (2010) Endophytic bacteria improve phytoremediation of Ni and TCE co-contamination. Environ Pollut 158(7):2422–2427

Weyens N, Truyens S, Saenen E, Boulet J, Dupae J, Taghavi S et al (2011) Endophytes and their potential to deal with co-contamination of organic contaminants (toluene) and toxic metals (nickel) during phytoremediation. Int J Phytoremediation 13(3):244–255

Weyens N, van der Lelie D, Taghavi S, Vangronsveld J (2013) The poplar endophyte *Pseudomonas putida* W619 as a key to a successful phytoremediation of volatile organic contaminants. In: Molecular microbial ecology of the rhizosphere, vol 1. Wiley, New York, pp 429–423

Wu T, Xu J, Xie W, Yao Z, Yang H, Sun C et al (2018) *Pseudomonas aeruginosa* L10: a hydrocarbon-degrading, biosurfactant-producing, and plant-growth-promoting endophytic bacterium isolated from a reed (*Phragmites australis*). Front Microbiol 9:1087

Wu J, Kamal N, Hao H, Qian C, Liu Z, Shao Y et al (2019a) Endophytic *Bacillus megaterium* BM18-2 mutated for cadmium accumulation and improving plant growth in hybrid Pennisetum. Biotechnol Rep 24:e00374

Wu T, Xu J, Liu J, Guo WH, Li XB, Xia JB et al (2019b) Characterization and initial application of endophytic Bacillus safensis strain ZY16 for improving phytoremediation of oil-contaminated saline soils. Front Microbiol 10:991

Yaish MW, Antony I, Glick BR (2015) Isolation and characterization of endophytic plant growth-promoting bacteria from date palm tree (*Phoenix dactylifera* L.) and their potential role in salinity tolerance. Antonie Van Leeuwenhoek 107(6):1519–1532

Yousaf S, Afzal M, Reichenauer TG, Brady CL, Sessitsch A (2011) Hydrocarbon degradation, plant colonization and gene expression of alkane degradation genes by endophytic *Enterobacter ludwigii* strains. Environ Pollut 159(10):2675–2683

Yuan M, He H, Xiao L, Zhong T, Liu H, Li S et al (2014) Enhancement of Cd phytoextraction by two Amaranthus species with endophytic Rahnella sp. JN27. Chemosphere 103:99–104

Zamani N, Sabzalian MR, Khoshgoftarmanesh A, Afyuni M (2015) Neotyphodium endophyte changes phytoextraction of zinc in *Festuca arundinacea* and *Lolium perenne*. Int J Phytoremediation 17(5):456–463

Chapter 9
Tomato Bacterial Canker Disease Management: Molecular Interactions and Role of Biocontrol Agents (BCAs) Against *Clavibacter michiganensis*

Brahim Bouizgarne and Mohamed Bakki

Abstract Tomato bacterial canker disease, caused by *Clavibacter michiganensis* subsp. *michiganensis* (*Cmm*) represents a burden on tomato cultivation around the world. No efficient chemical treatment and no resistant varieties are available. In recent years, the study of soil bacteria, particularly plant growth promoting rhizobacteria (PGPR), antagonistic toward phytopathogens has made great progress. The most studied bacteria belong to *Pseudomonas* and *Bacillus*, which are used as biocontrol agents (BCAs) against several plant diseases. Researches on screening for antagonistic bacteria went from isolating in vitro effective antagonists to application under greenhouses and field conditions. However, few are effective under greenhouse against *Cmm*. Thus, only little headway has so far been made to develop formulated BCAs. Most studies have focused on direct interactions displayed by BCAs through inhibition of *Cmm* growth and development, while relatively little attention has been paid to indirect interactions including those inducing reinforcement of tomato resistance to subsequent attack by the pathogen. It is also not clear which mechanisms take the lead in combatting this pathogen in soil. Therefore, in this chapter, we focused on the following questions: how *C. michiganensis* infect tomato plants and what are, molecular traits underlying interactions between tomato and *C. michiganensis*? What are the main bacterial canker disease management strategies? How to detect early the presence of *Cmm* in tomato seeds, plants, and soil? What are the most effective antagonistic microorganisms, especially PGPR to control tomato bacterial canker? What are the mechanisms displayed by antagonistic soil bacteria and how induction of tomato resistance could be a promising tool? Finally, our experience of work project "BIO-BIZ-01/2017," aiming at selecting PGPR with multiple traits as effective antagonists toward *C. michiganensis* and inducers of tomato resistance is highlighted.

B. Bouizgarne (✉)
Department of Biology, Ibn Zohr University, Agadir, Morocco

M. Bakki
Faculty of Sciences, Department of Biology, University Ibn Zohr (UIZ), Agadir, Morocco

N. K. Arora, B. Bouizgarne (eds.), *Microbial Biotechnology for Sustainable Agriculture Volume 2*, Microorganisms for Sustainability 34,
https://doi.org/10.1007/978-981-97-2355-3_9

Keywords Tomato · PGPR · Bacterial canker · Antagonism · *Clavibacter michiganensis*

9.1 Introduction

Currently, the world's population is constantly increasing and is expected to reach ten billion by 2050 (FAO 2017). Unfortunately, world's population is outpacing food production rate. Therefore, it is estimated that food production will be required to increase by 60% by 2050 to feed the expected population on Earth (FAO 2019). Besides, increasing pressures on natural resources in addition to climate change will make global food security precarious (Bouizgarne 2022; Arora and Mishra 2022). Another thread to food security is represented by the outbreaks of many plant pathogens, which are considered a real challenge to agriculture and food security. Indeed, plant diseases affect food security by affecting access, availability, utilization, and stability of food (Savary et al. 2019). Modern life has made plant disease transmission faster with increased global food trade networks. Spread of plant disease agents could affect human health, survival, and stability in a situation like what world of the nineteenth century lived during Irish potato famine caused by *Phytophthora infestans* (1845–1852). In addition, pathogens' deleterious effects are often influenced by environmental conditions. According to Food and Agriculture Organization of the United Nations (FAO, https://www.fao.org/fao-stories/article/en/c/1507753/ accessed April 12, 2023), the spread of plant diseases and pests is on the rise due to changing climate. Indeed, climate change is exacerbating the situation as it affects crops either directly through plant physiology or indirectly through plant–pathogen interactions. Climate change impacts concern pathogen severity, development, and survival, emergence of new pathogen lineages, in addition to plant-susceptibility to pathogens (Hunjan and Lore 2020; Ristaino et al. 2021). Most crops are affected by one or more phytopathogens resulting in substantial losses. Losses due to plant pests and diseases could result in crop yield losses by averages of 30% in rice, 22.5% in maize, 21.5% in wheat, 21.4% in soybean, 17.2% in potato, and up to 30% in tomato (Savary et al. 2019; Liu and Wang 2020). Globally, it is estimated that yield losses represent around 20% of crop production. Thus, combatting phytopathogens is becoming a cornerstone of sustainable agriculture strategies.

Tomato (*Solanum lycopersicum* L.), belonging to family Solanaceae, is a widely cultivated horticultural crop. It is the world's second most important and most cultivated vegetable crop after potato (*Solanum tuberosum*). Tomato fruit is important for human nutrition as it contains vital compounds such as lycopene and β-carotene. Tomato is also a rich source of vitamins (vitamin B and C), minerals (iron and phosphorus), dietary fibers, and amino acids. Globally, the culture of tomato covers over 5 million hectares with a production of 181 million tons per year according to FAO (2021). In 2021, China was the largest producer of tomatoes 67.53 Million tones (Mt) followed by India (21.18 Mt), Turkey (13.09 Mt), and the USA (10.4Mt) (FAO 2021). In the Mediterranean region, the major producers are Turkey, Italy, Egypt,

Spain, Algeria, Tunisia, Morocco, and Syria (Fig. 9.1). In Morocco, tomato production was 1,388,542 tons in 2022 with a harvested area of 14,956 ha, (FAO Data, available online: https://www.fao.org/faostat/en/#data/QCL (accessed on 20 May 2024)). The cultivation areas are located mainly near the city of Agadir in the Souss-Massa region, South of Morocco.

Unfortunately, the production of tomatoes is hampered by numerous abiotic and biotic constrains either under greenhouse or at postharvest levels, causing considerable tomato losses. Tomato diseases are caused by various pests and phytopathogens including insects, fungi, bacteria, and viruses. Most common fungal diseases are caused by *Alternaria*, *Botrytis*, *Cladosporium*, *Colletotrichum*, *Fusarium*, *Oidium*, *Phytophthora, Pythium Rhizoctonia*, *Septoria*, *Sclerotinia*, *Sclerotium*, *Stemphylium,* and *Verticillium*. Most common bacterial diseases are caused by *Clavibacter*, *Erwinia*, *Pseudomonas*, *Ralstonia*, and *Xanthomonas*, while viruses causing disease in tomato are pepino mosaic virus (PepMV), tobacco mosaic virus (TMV), tomato mosaic virus (ToMV), tomato brown rugose fruit virus (ToBRFV), tomato chlorosis virus (ToCV), and tomato yellow leaf curl virus (TYLCV). Bacterial canker disease of tomato also called vascular wilt of tomato is a wilt disease caused by the bacterium *Clavibacter michiganensis* subsp. *michiganensis* (*Cmm*), which is mainly a soil-borne and seed-borne pathogen. It is considered the most widespread and devastating diseases of tomato crops in different countries including Morocco. This disease is considered a quarantine organism in Europe, Canada, the USA, Caribbean, and other countries in Africa and Asia (Eichenlaub and Gartemann 2011; EPPO 2016; Thapa et al. 2017). Bacterial canker disease is also considered a seed-borne disease and infected seeds could constitute one of the most risky transmission vectors (de León et al. 2011). Exported contaminated seeds are considered a real threat as the yield losses could reach 100% if the disease is not controlled.

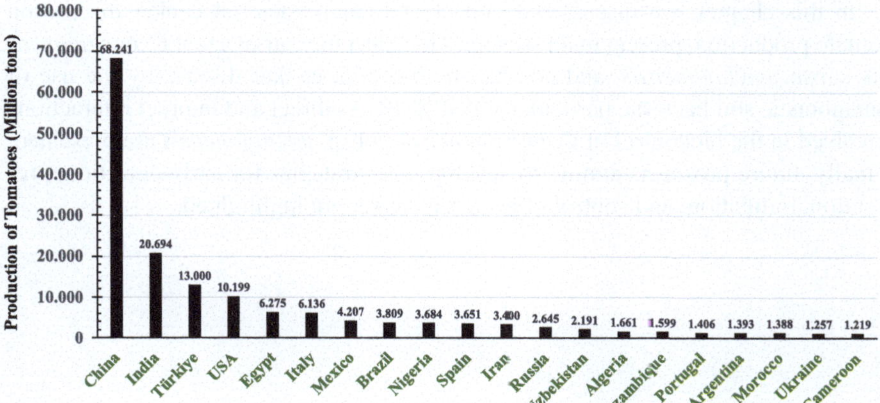

Fig. 9.1 Top twenty countries for production of tomatoes in 2022 (FAO data, available online: http://www.fao.org/faostat/en/#rankings/countries_by_commodity, accessed on 20 May 2024)

The variety of disease agents that attack important crops led to a worldwide use of large amounts of different chemical pesticides, which cause many ecological problems. Although numerous antibacterial chemical pesticides are available, no one is effective against tomato bacterial canker disease. In addition, no resistant tomato cultivars are commercially available although attempts at breeding resistant cultivars (Abebe et al. 2022) or to develop GMO tolerant tomato plants were performed resulting in alleviation or absence of disease symptoms and reduced *Clavibacter michiganensis* subsp. *michiganensis* population (Balaji and Smart 2012; Wittmann et al. 2016). Treatment of seeds by chemicals such as acidified nitrite, copper hydroxide has proved its effectiveness (Kasselaki et al. 2011). As the disease causes the disintegration of vascular tissues, chemical pesticides are also ineffective to control the disease in soil. Hence, only pathogen free seeds and prophylaxis are relatively effective available strategies (de León et al. 2011). The importance of using soil microbia and particularly plant growth promoting rhizobacteria (PGPR) with antagonistic effects toward phytopathogens as an effective alternative and an ecofriendly strategy is well documented (Bouizgarne 2013, 2022; Bouizgarne and Ouhdouch 2017). PGPR generally are a group of soil and rhizosphere free-living bacteria, which are able to colonize plant roots competitively and exert a beneficial effect on plant growth (Bouizgarne 2013). PGPR beneficial effects toward the plants are either direct mechanisms such as phosphate solubilization and phytohormone synthesis or indirect mechanisms that promote plant growth by reducing pathogen populations and disease severity (Bouizgarne 2013; Pieterse et al. 2014; Backer et al. 2018; Orozco-Mosqueda et al. 2023; Salwan et al. 2023). Soil bacteria and PGPR are used as biocontrol agents (BCAs) to control tomato bacterial canker in growth chambers and under greenhouses (Paulin et al. 2017; Takishita et al. 2018; Laird et al. 2020; Bouizgarne 2022; Bakki et al. 2024). BCAs display two types of interactions toward the pathogen: direct interactions target the pathogen itself, while indirect interactions are mainly aimed to reinforce plant resistance to the pathogen.

In this chapter, we discuss the impact of tomato bacterial canker disease on tomato production, present main strategies to detect the causal agent *C. michiganensis* subsp. *michiganensis*, and combat the bacterial canker disease by the use of antagonistic soil bacteria, particularly PGPR. BCAs direct and indirect interactions involved in the biocontrol of *C. michiganensis* subsp. *michiganensis* are presented. Finally, future prospects aiming to develop new strategies for early detection, prevention, mitigation, and control of bacterial canker are highlighted.

9.2 Diversity and Taxonomy Among Subspecies of *C. michiganensis*

Currently, more than 200 phytopathogenic bacterial species have been described, of which the majority belong to the Gram-negative phylum Proteobacteria (Eichenlaub and Gartemann 2011; Van der Wolf and De Boer 2016). Among Gram-positive bacteria, most pathogens belong to the phylum Actinobacteria and firmicutes. *Clavibacter* is a non-motile, non-spore forming, Gram-positive actinobacterium of the coryneform group (Phylum Actinobacteria, family Microbacteriaceae). The family Microbacteriaceae comprises three other bacterial phytopathogens: *Curtobacterium*, *Leifsonia*, and *Rathayibacter* (Tampakaki et al. 2009). *Clavibacter* was originally named *Corynebacterium* on the basis of its morphological patterns resulting from its particular mode of division leading to pairs of cells with a diphtheroid arrangement (V shape). The genus was later reclassified as *Clavibacter* on the basis of chemotaxonomic and phylogenetic analysis of 16S rRNA gene sequences. Genus *Clavibacter* consists of only one species *C. michiganensis*, which encompass biotrophic phytopathogen of economically important plants. Strains of the species *C. michiganensis* are pleomorphic rods, mesophile, oxidase-negative obligate aerobes and nitrate and nitrite reductase negative (Davis et al. 1984). They possess cell walls with DL diaminobutyrate type (β2γ type peptidoglycan), containing also rhamnose, galactose, mannose, and unsaturated menaquinones (MK-9) (Collins and Jones 1980; Park et al. 1993), Their fatty acids are straight saturated chains, anteiso-branched (ai-17:0 and ai-15:0) (Collins and Jones 1980) and saturated iso-methyl branched acids. Their phospholipids are mainly phosphatidylglycerol and diphosphatidylglycerol. Their Chargaff coefficient (G + C content of the DNA) is about 69–78 mol%. The species includes seven subspecies, which are highly host-specific phytopathogens and thus were assigned according to host specificity (Eichenlaub et al. 2006), i.e., each causing disease in a particular crop. *C. michiganensis* subsp. *capsici* (bacterial canker disease in pepper; *Capsicum annum*), *C. michiganensis* subsp. *insidiosus* (bacterial wilt disease of lucerne; *Medicago sativa*), *C. michiganensis* subsp. *michiganensis* (bacterial canker disease of tomato; *Solanum lycopersicum*), *C. michiganensis* subsp. *nebraskensis* (bacterial wilt disease of corn; *Zea mays*), *C. michiganensis* subsp. *phaseoli* (leaf yellowing of bean; *Phaseolus vulgaris*), *C. michiganensis* subsp. *sepedonicus* (ring rot disease of potato; *Solanum tuberosum*), and *C. michiganensis* subsp. *tessellarius* (leaf spot or mosaic disease of wheat; *Triticum aestivum*) (Eichenlaub and Gartemann 2011; González and Trapiello 2014; Oh et al. 2016). Other new *C. michiganensis* subspecies were discovered but show no symptoms on plants: *C. michiganensis* subsp. *californiensis* and *C. michiganensis* subsp. *chilensis* (Yasuhara-Bell and Alvarez 2015). Recently, subspecies assignment based on host specificity was questioned by modern high throughput taxonomic molecular methods. Next generation sequencing by whole-genome analysis (using average nucleotide identity (ANI) value) has revolutionized the identification of bacteria in general and was applied to establish the relationships between *C. michiganensis* subspecies. Waleron et al. (2011)

analyzed polymorphism of three housekeeping genes (recA, rpoB, and rpoD), by PCR-RFLP RFLP fingerprints, showed usefulness for differentiating between subspecies of *C. michiganensis*. Recent work by Tambong (2017), using digital DNA–DNA hybridization (dDDH) and average nucleotide identity (ANI) found that the subspecies should be reclassified as separate species. More evidences are currently available demonstrating the monophyletic line of the subspecies. Analysis of genomes from 11 *C. michiganensis* subsp. *michiganensis* strains including pathogenic and endophytic strains (nonpathogenic to tomato, but pathogenic to other plants) revealed a diversity among the strains, demonstrated the monophyletic nature of this pathogen (Thapa et al. 2017). In a study aiming to compare 25 Chilean *C. michiganensis* subsp. *michiganensis* strains with a worldwide collection of strains, Valenzuela et al. (2018), by using housekeeping genes MLSA and MLST (multilocus sequence analysis and typing) in addition to multilocus variable number of tandem repeats (VNTR) analysis (MLVA) found that the Chilean strains showed low diversity, suggesting that they tend to be clonal. Moreover, most strains are close to strains from China and Taiwan while the remaining strains shared single sequence type with strains from Algeria, Belgium, France, the Netherlands, Brazil, Uruguay, and the USA, confirming the genetic similarity of worldwide *C. michiganensis* subsp. *michiganensis* collection. Hence, strong evidences separated *C. michiganensis* subsp. *michiganensis* subspecies from the other *Clavibacter* subspecies (Nouioui et al. 2018), suggesting that it should be reclassified as new species from the subspecies (Méndez et al. 2020). Moreover, based on the genomic patterns and host specificity, Méndez et al. (2020) suggested that *C. michiganensis* subsp. *michiganensis* and other subspecies to be elevated to species. Also, works by Li et al. (2018) and Arizala et al. (2022), using whole-genome and multilocus sequence analyses (MLSA), suggested to elevate some of the subspecies to species-level. Hence, *C. californiensis, C. capsici, C. chilensis, C. insidiosus, C. nebraskensis, C. tessellarius,* and *C. sepedonicus* were proposed (Li et al. 2018; Arizala et al. 2022). Thanks to these new techniques, a new species *C. zhangzhiyongii* causing leaf brown spot and decline of barley was recently identified by Tian et al. (2021). Finally, based on ANI and dDDH, Nouioui et al. (2018) emended nomenclature of *C. michiganensis* subsp. *michiganensis* subspecies and suggested the new name of the species as *C. michiganensis*.

Considering all these findings, we strongly recommend using the species name *C. michiganensis* instead of the subspecies name *C. michiganensis* subsp. *michiganensis* to researchers dealing with this pathogen. Also, in the rest of this chapter, we will use species name *C. michiganensis* to designate the causal agent of tomato bacterial canker. Although not universally adopted at the moment, an overview of recent publications reveals that this denomination is likely to be more and more adopted (Barda et al. 2015; Peritore-Galve et al. 2020; Thapa et al. 2020: Solano-Alvarez et al. 2021; Tsitsekian et al. 2021; Abebe et al. 2022; Oh et al. 2022; Xiao et al. 2022).

9.3 Molecular Traits Underlying Interactions Between Tomato and *C. michiganensis*

9.3.1 Molecular Traits Underlying Infection by C. michiganensis

Primary *C. michiganensis* inoculum consists essentially of contaminated seeds, seedlings, and transplants or debris (Franken et al. 1993). The pathogen was reported as able to survive several months in tomato stems (Vega and Romero 2016). It was also found that the pathogen's survival on tomato stems depends on the geographic areas. *Cmm* was recovered from tomato stems after a period ranging from 2-6 months in Morocco to about 1-2 years in Ohio and Iowa (USA) (Fatmi and Schaad 2002). Also, it was reported that the bacteria can survive in the seed coat for several years (Kado 2010). The pathogen could infect tomato plants at root level via root wounds (Chalupowicz et al. 2012; Xu et al. 2012; Lelis et al. 2014). It can also penetrate leaves through stomata and hydathodes. Labeling a virulent *C. michiganensis* with green fluorescent protein (GFP) demonstrated that penetration occurs mainly through trichomes (Chalupowicz et al. 2017). The role of extracellular polymeric substance (EPS) in root colonization, including polysaccharides and lipids as protective compounds was suggested by Chalupowicz et al. (2012). The role of cultivar type, seedling age, pathogen inoculum, cultural conditions, and various environmental factors was also reported on the infection and symptoms by *C. michiganensis* (Chang et al. 1992; Franken et al. 1993; de León et al. 2011; Xu et al. 2012; Sharabani et al. 2014).

After infection, the pathogen progresses systemically throughout the stem and leaves via the vascular system (xylem) where it proliferates extensively. First, the symptoms appear in one side of the leaves and plant and lately expand to the whole plant. Internally, infected tomato plants show vascular discoloration, stem canker, and necrosis due to high concentrations of *C. michiganensis* inside the vascular tissues and consecutive tissue–pathogen interactions (Thapa et al. 2017). Externally, tomato infected leaves show localized blister-like spots or stems with marginal necrosis symptoms (Chalupowicz et al. 2017). However, the most remarkable localized symptoms of the disease are necrotic spots on fruits referred to as bird's eye symptoms corresponding to dark spots surrounded by white halos. Wilting of leaves is followed by stem necrosis symptoms. Ultimately, the whole plant shows canker and wilting symptoms. It has been suggested that colonization by *C. michiganensis* is facilitated by the suppression of plant defenses (Chalupowicz et al. 2010). Consequently, ineffective defense mechanisms could not stop colonization and this leads to plant death. Figure 9.2 shows main symptoms due to *C. michiganensis* infection in tomato.

The pathogenomics of *C. michiganensis* was extensively studied. The chromosome of the pathogen holds multiple genes including the 129-kb *chp/tomA* chromosomal pathogenicity island (PAI), with a low GC content near the chromosomal origin of replication shown to be necessary for pathogenicity (Gartemann et al.

Fig. 9.2 Bacterial canker of tomato. Up and middle: wilt symptoms on seedlings showing damaged fruit, below: production under greenhouse showing removed plants from rows in order to stop pathogen progression

2008). PAI region contains genes encoding for tomatinase (gene *tomA*), seven Chp family proteases (genes *chpA, chpB, chpC, chpD, chpE, chpF*, and *chpG*), six chymotrypsin-like serine proteases (genes *ppaA, ppaB1, ppaB2 ppaC, ppaD*, and *ppaE*), two pectate lyases (*genes pelA1 and pelA2*), and one subtilase protease (*sbtA*). In addition, four other chymotrypsin-like serine proteases (*ppaF, ppaG, ppaH,* and *ppaI*) and two subtilase proteases (*sbtB* and *sbtC*) genes are not on the

PAI region. In addition to PAI pectinases (*pelA1, pelA2*), *C. michiganensis* chromosome contains genes for secretion of other carbohydrate-active enzymes (CAZymes), not located in PAI region and involved in degradation of plant cell wall constituents (pectin cellulose, xylan, and glucanes): pectinases (pgaA), cellulases (*celB*), β-1,4 xylanases (*xysA, xysB*), and endoglucanases (*endX/Y*). *C. michiganensis* possess two plasmids: pCM1 (31–59 kbp) and pCM2 (64–109 kbp) (Nandi et al. 2018). Plasmid pCM1 holds *celA* gene encoding a cellulase with endo-β-1,4-glucanase, and genes encoding for the chymotrypsin-like serine proteases (*ppaJ*) and for α-expansin (*Cm*EXLX1), while pCM2 holds *pat-1* gene and two other genes (*phpA, phpB*), all of them encoding serine proteases.

It is well established that *C. michiganensis* pathogenicity determinants are located on the chromosome. They are also located on the circular plasmids, which carry pathogenicity genes which lack in endophytic, non-virulent forms of *C. michiganensis* (Thapa et al. 2017). It was reported that plasmid pCM2 *pat-1* region encoding serine proteases and pCM1 *celA* encoding endo-β-1,4-glucanase were involved in *C. michiganensis* virulence (Meletzus et al. 1993; Jahr et al. 2000; Chalupowicz et al. 2017; Thapa et al. 2017). Valenzuela et al. (2021) found that an asymptomatic strain of *C. michiganensis,* lacking cellulase activity also lacked the *celA* gene, confirming its essential role in pathogenicity. Hence, pCM1 and pCM2 plasmids are critical for systemic infection leading to canker and wilting. The role of virulence plasmids was demonstrated by DNA sequence analysis of *pat-1*, which evidenced that this region is common in virulent *C. michiganensis* strains (Dreier et al. 1997). The importance of the virulence plasmids was demonstrated in isolates with deletion of *pat-1* region, which resulted in a reduced virulence, whereas *pat-1* cloning into plasmid-free endophytic isolates rendered them virulent, resulting in tomato wilting (Meletzus et al. 1993). The role of virulence plasmids was also demonstrated in strains for which plasmids are lacking, turning the biotrophic *C. michiganensis* into a simple endophyte (Eichenlaub and Gartemann 2011). However, although the role of pCM1 and pCM2 plasmids in systemic infection and canker was demonstrated, they are not required for the locally blister-like spots on leaves (Chalupowicz et al. 2017).

Other pathogenetic genes are chromosomal, located in the chromosomal PAI (Gartemann et al. 2008; Załuga et al. 2014; Chalupowicz et al. 2017). Załuga et al. (2014) compared whole-genome sequence of the nonpathogenic *C. michiganensis* strain with pathogenic strains and concluded that nonpathogenic strains were lacking most of the PAI genes. Moreover, it was demonstrated that mutants for *chp/tomA* genomic island, included in PAI region were ineffective pathogens (Gartemann et al. 2008; Chalupowicz et al. 2017). This finding was corroborated by Thapa et al. (2017) who found that tomato endophytic *C. michiganensis* strains, nonpathogenic to tomato, are lacking subregions of *chp* and *tomA*, the two are universally found in *C. michiganensis*. *Chp/tomA* includes genes necessary for colonization and virulence (Stork et al. 2008; Thapa et al. 2017). Among the seven putative serine protease (*chpA–chpG*) genes, which are homologous to the plasmid virulence factor pat-1, *chpC* was reported to be the chromosomal gene required for host colonization and affects the interaction of the pathogen with the host plant (Stork et al. 2008).

Indeed, it was demonstrated that in planta, the titer of *chpC* lacking mutant was drastically reduced, causing only weak disease symptoms, while complementation of the mutant by this gene restored full virulence (Stork et al. 2008). In addition, it was reported that expression of the host-defense genes (chitinase and pathogenesis-related protein) was triggered in the absence of the PAI region at early stages of infection (Stork et al. 2008; Chalupowicz et al. 2010), suggesting defense mechanisms in tomato plants as effective in suppressing the pathogen. Particularly, the *chpC* gene of PAI *chp* subregion is involved in the suppression of those defense mechanisms (Chalupowicz et al. 2017). *chp* contains genes involved in virulence molecular mechanisms in addition to putative protease-encoding genes (Nandi et al. 2018). Among 11 genes encoding chymotrypsin-related serine proteases (the Ppa family), the 6 genes (*ppaA, ppaB1, ppaB2, ppaC, ppaD and ppaE*) present in the PAI *chp* region are suggested as virulent factors having a role in host colonization (Chalupowicz et al. 2017). Attempts to explain the potential role of *ppaA* or *ppaC* in the severity of symptoms were performed by Chalupowicz et al. (2017) by using mutants. In addition to Ppa family serine proteases, perception by tomato plant is linked to another serine proteases, named subtilase proteases family (Savidor et al. 2012; Chalupowicz et al. 2017), including a *sbtA* gene, located on the PAI *chp* subregion (Chalupowicz et al. 2017). Among other chromosomal genes encoding cell wall degrading enzymes, the chromosomal *celA*-like gene *pelA1* (pectate lyase) was reported to be putatively responsible for tissue maceration (Chalupowicz et al. 2017). In addition to *chp*, *tomA* also located in the PAI region encodes a tomatinase, which is involved in α-tomatine degradation, resulting in hampering one of the basal tomato defense mechanisms. However, its role as a virulent factor was questioned as mutants for *tomA* maintain their virulence (Kaup et al. 2005). Savidor et al. (2014) identified two transcription virulence factors (*vatr1* and *vatr2*) involved in ethylene synthesis and wilting. A delay in the wilting was observed following infection of ethylene-insensitive never ripe (*Nr*) and in low-ethylene tomato plants mutants compared with wild-type plants (Balaji et al. 2008) suggesting the role of *C. michiganensis*-triggered synthesis of this phytohormone as crucial for infection and symptom development (Savidor et al. 2012), probably through its effect in vascular tissue senescence. Colonization by *C. michiganensis* involves close interactions between the pathogen and the host plant. In order to better understand the genetic virulence in *C. michiganensis,* a worldwide collection of pathogenic and nonpathogenic strains was investigated by using a polyphasic approach based on multilocus sequence analysis and typing (MLSA and MLST) of six housekeeping genes (*atpD*, *dnaK*, *gyrB*, *ppK*, *recA,* and *rpoB*), 16S rRNA gene sequencing, the pathogenicity determinants *celA*, *chpC*, *pat*-1, and *ppaA* in addition to phenotype; Jacques et al. (2012) demonstrated that pathogenic *C. michiganensis* are phylogenetically distinct from nonpathogenic *C. michiganensis* neighbor strains. In addition, it was demonstrated that successful colonization depends on an interplay of chromosomal and plasmid genes, as gene expression of plasmid *celA* and *pat-1* was reduced in the absence of the chromosomal PAI region, and conversely for chromosomal *chpC*, *ppaA*, *celB,* and *pelA1,* their expression was reduced in the absence of the plasmids (Chalupowicz et al. 2010). Figure 9.3 presents a schematic diagram

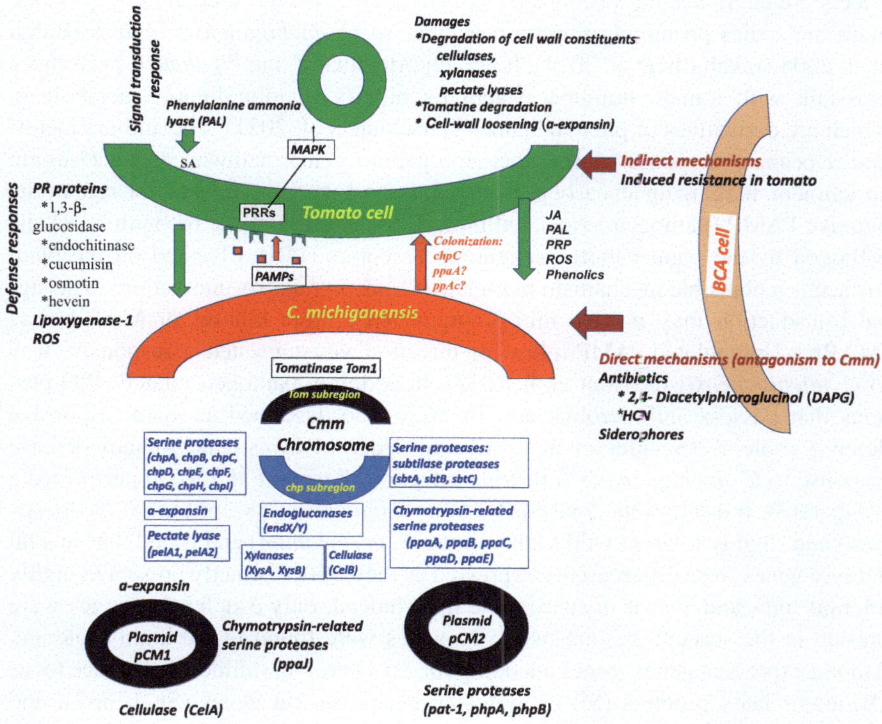

Fig. 9.3 Schematic diagram showing genes involved in the colonization and virulence by *C. michiganensis*, and damages to tomato cells inflicted by the pathogen, main defense mechanisms displayed by tomato plant cells following infection and those induced by BCAs. PAI: the chromosomal PAI region (*chp*/*tom*) contains two subregions: *chp* and *tom* subregions. Chrom: genes located on other chromosomal regions different than PAI. CAT: catalase, chs: chalcone synthase, HCN: hydrogen cyanide, JA: jasmonic acid, MAPK: mitogen-activated protein kinases (MAP kinase), Phe: phenylalanine, PAL: phenylalanine ammonia-lyase, PAMP: pathogen associated molecular patterns, PRP: pathogenesis-related proteins, PRRs: plant pattern recognition receptors, ROS: reactive oxygen species, SA: salicylic acid, SOD: superoxide dismutase

showing roles of *C. michiganensis* genes involved in the colonization and virulence, and ingresses caused to tomato cells.

9.3.2 *Molecular Traits Underlying Response of Tomato to* C. michiganensis *Ingress*

In response to various abiotic and biotic stresses, plants respond by activating reactive oxygen species (ROS) leading to various harmful effects on cell viability and function. In response to pathogens, plants react by a transduced signal leading to the biosynthesis of several defense compounds, including those with antimicrobial effects. Signal molecules jasmonate (JA), ethylene (ET), and salicylic acid (SA) are main molecules prompting tomato resistance to *C. michiganensis* ingress (Balaji et al. 2008, Takahashi et al. 2014). It was reported that *C. michiganensis* performs a crosstalk with tomato immune system by modifying monolignols metabolism, which are derivatives of phenylalanine (Tsitsekian et al. 2021). The authors identified a primary response of the phenylpropanoid genes pathway and a G-lignin enrichment, making up an early protective barrier. Like major plants, the role of the putative PAMP (pathogen associated molecular patterns) in the recognition of the pathogen by the plant pattern recognition receptors (PRRs) located on the plant surface is a plausible mechanism in tomato–*C. michiganensis* interactions. The signal transduction may involve mitogen-activated protein kinases or MAP kinase (MAPK). Upregulated PAMP following infection was suggested as responsive way to *C. michiganensis* (Savidor et al. 2012). In addition, pathogen-related (PR) proteins that possess antimicrobial activity were also described as main responsive defense proteins (Savidor et al. 2012). In order to investigate tomato defense response to *C. michiganensis* following infection, Basim et al. (2021) performed a comparative transcriptome analysis in *C. michiganensis*-susceptible (*S. lycopersicum*) and -highly tolerant wild tomato lines (*S. peruvianum*) and found that several defense genes were differentially expressed as they were distinctly present in highly tolerant lines and absent in susceptible line. Indeed, only 3 defensive genes were present in the susceptible line, while 26 genes were found in the highly tolerant. Among expressed genes, genes encoding for MLO protein (mildew resistance locus O), major latex proteins (MLP), defensin genes, snakin genes (Sn-1/Sn-2), and pathogenesis-related genes (PR genes) were identified. The authors suggest relationship between tomato tolerance and resistance genes expression and their regulation during plant–*C. michiganensis* interactions. Yokotani et al. (2021) analyzed transcriptome sequences of infected tomato cotyledons and revealed that tomato responds to the pathogen in a common way as all plants do against pathogens. Tomato plant responds by stimulated SA accumulation in tomato cotyledons. It was hence demonstrated that tomato upgrades defense genes involved in the phenylalanine ammonia-lyase (PAL) pathway (genes *PAL*, *4Cl*, and *AIM1*) in response to *C. michiganensis,* leading to SA accumulation (Yokotani et al. 2021). The role of SA in the response of plants to pathogens is well documented as SAR (systemic acquired resistance). Thereafter, the role of SA in *C. michiganensis*–tomato interactions was confirmed by exogenous application of SA resulting in expression of SA-dependent genes and particularly the expression of WRKYs gene family, including SlWRKY8, which could promote the expression of pathogenesis-related protein genes (Yokotani et al. 2021). Treatment also resulted in acceptable *C. michiganensis* growth inhibition. Taken together, results demonstrated relationship between bacterial growth, disease symptoms, and the role of SA in tomato response through expression of defense genes (Yokotani et al. 2021). Although several molecules could be expressed during plant–pathogen interactions including the production and scavenging of free oxygen radicals (Savidor et al. 2012), the most important genes expressed upon tomato systemic infection by *C. michiganensis* are those encoding for pathogenesis-related proteins: 1,3-β-glucosidase (encoded by *GluA*), endochitinase, cucumisin, osmotin, (thaumatin like protein), and hevein, which are defense induced SA-dependent PR proteins, and lipoxygenase-1 (LOX1, encoded by *LoxA*), which is responsible for the biosynthesis of oxylipins with antimicrobial activity (Savidor et al. 2012; Basim et al. 2021). Indeed, while SA-dependent pathway relatively succeeded to inhibit *C. michiganensis* growth in tomato cotyledons, ET mediated response, on the contrary, contributes to the development of wilt symptoms during tomato plants infection without affecting the pathogen growth (Balaji et al. 2008; Yokotani et al. 2021). It seems that in the case of tomato, ET interactions are not in favor of defense as they are in other plants. Besides, it was also reported that lack of resistance or susceptibility is linked to gene suppression. It was observed a downregulation of the class III peroxidases (a class of the defense related genes), suggesting the suppression of the tomato defense by *C. michiganensis* (Savidor et al. 2012). Figure 9.3 also presents ingresses caused by *C. michiganensis* and the main defense mechanisms displayed by tomato plant cells following infection.

9.4 Management Strategies of Tomato Bacterial Canker Disease

9.4.1 Selection of Resistant Cultivars

Although efforts were made to selecting and breeding genetically resistant tomato lines to *C. michiganensis,* obtained results are insufficient. Currently, there are no commercially available resistant cultivars due to the genetic diversity of *C. michiganensis.* Another considerable factor is its own genomic heterogeneity. Tolerant cultivars are tomato lines which do not belong to the species *Solanum lycopersicon*, but to wild species such as *S. pimpinellifolium, S. peruvianum, S. arcanum, S. habrochaites* (Sen et al. 2015; Peritore-Galve et al. 2020; Basim et al. 2021). Some *S. lycopersicon* derived lines are also tolerant but not commercially successful (Sen et al. 2015), and the above-mentioned wild lines hold genetic loci conferring only tolerance to the pathogen. Although reduced intravascular spread results in less severe symptoms in wild tomato species, their fruits are still susceptible and show severity unrelated to vascular system infection. Lelis et al. (2014) developed a pathogenicity assay based on in vitro axenically grown tomato plants infected with GFP-tagged strain of *C. michiganensis* for fast screening of plant material for resistance and the *C. michiganensis*–tomato interactions. Compared with potted plants, this pathogenicity assays have the advantages to be safe, less time consuming, with rapid results at low cost and could be performed in small controlled units. It was reported that phenotypic traits including stem morphology and the vascular systems structures and functions are different between wild and commercial tomato cultivars and may have a role in tomato resistance to *C. michiganensis* (Coaker and Francis 2004). Molecular markers were investigated and marker-assisted breeding against tomato bacterial canker was developed. It was hypothesized that a chromosomal gene, QTL Rcm 2.0 conferred resistance to *C. michiganensis* by altering vascular morphology and development (Coaker and Francis 2004). Recently, QTL-Seq analysis for identification resistance *loci* to bacterial canker in disease tolerant tomato was performed by Abebe et al. (2022) and Celik (2023), suggesting the importance of detected genes as valuable targets for mapping molecular markers in order to select tomatoes for *C. michiganensis* resistance. Works by Wittmann et al. (2016) aimed to develop resistant *S. lycopersicum* cultivars by transferring into tomato, the bacteriophage CMP1 endolysin gene (lys), encoding a peptidase, which hydrolyzes *C. michiganensis* cell wall peptidoglycan. The transformation led to absence of disease symptoms in transgenic tomato plants after infection. Balaji and Smart et al. (2012) showed that overexpression of two tomato proteins snakin-2 (SN2) and extensin-like (ELP) protein genes restricts *C. michiganensis* invasiveness and confers resistance in transgenic *S. lycopersicum.* Furthermore, in transgenic plants, *C. michiganensis* populations were significantly reduced (suggesting potential in vivo antibacterial activity of SN2 and ELP as defense proteins).

9.4.2 Early Detection of **C. michiganensis**

Bacterial canker disease dissemination occurs generally through the seeds. Early detection of *C. michiganensis* in seeds is often used to prevent the disease. For significant results, various methods were used to detect *C. michiganensis* in infected materials. Thus, sensitive, reliable, non-destructive fast, and inexpensive detection methods are to be encouraged. For significant detection, Hadas et al. (2005) stated that the extraction method is important for sensitivity and demonstrated that the extraction of *C. michiganensis* in seeds by grinding was more effective than soaking. Current knowledge about detection methods was developed by the use of methods including (a) classical dilution-plating methods by using selective and semi-selective media, (b) serological methods (enzyme-linked immunosorbent assay, ELISA; immunofluorescence colony staining, IFC; immunofluorescence, IF; and immunomagnetic bead separation, IMS), bioassays (on seed and plant material), and (c) molecular methods. Thus methods like Biolog (Kaneshiro et al. 2006), immunofluorescence microscopy (Franken et al. 1993), ELISA (Kaneshiro et al. 2006) were developed. Among serological methods, direct immunofluorescence colony (IFC) staining, using fluorophore-labeled antibodies is highly sensitive as it can detect *C. michiganensis* concentrations as low as 2×10^2 CFU/g of seeds (Veena and van Vuurde 2002). Immunofluorescent assay (IF) and ELISA allowed to detect 10^3 and up to 10^4 CFU mL^{-1}, respectively (de León et al. 2008). Immunomagnetic bead separation: IMS (or IMS-plating), which uses immunomagnetic beads coated with specific antibodies, allowing to target *C. michiganensis* cells and removing of non-target cells, is the most sensitive serological method. This technique could detect concentrations lower than 10 CFU/mL, even in contaminated seed extracts by high populations of non-target bacteria (de León et al. 2008). Molecular methods are based on the detection of specific sequences based on PCR (de León et al. 2009; Waleron et al. 2011) and real-time PCR (Luo et al. 2008). qPCR (real-time PCR) allows to detect *C. michiganensis* in pure cultures with a detection threshold of 10^3 cfu mL^{-1} (Zhao et al. 2007; Luo et al. 2008).

However, each of these methods could present some disadvantages such as false positive or negative responses. Direct isolation on semi-selective media is time consuming. In addition, in case of stress conditions, *C. michiganensis* enters a *viable but nonculturable* (VBNC) state, making conventional cultural techniques as limited, as *C. michiganensis* population can be underestimated by these techniques. Serological methods are fast, inexpensive and commercial kits are available. However, except IMS, they suffer from limited sensitivity and false positive/negative results. ELISA and IF gave satisfactory detection when monoclonal antibodies are used. However, it suffers from cross reactions with other bacteria and cannot distinguish between dead and viable cell (Franken et al. 1993). Indeed, simultaneous detection of dead and viable cells is the main drawback of other serological such as IF, ELISA, and molecular methods such as PCR (Veena and van Vuurde 2002). Immunofluorescence colony staining (IFC) has the advantage of detecting viable bacteria and is highly sensitive compared to IF (Sen et al. 2015). However, IFC has the drawback of being an expensive technique. Most PCR techniques use insufficiently specific primers. In addition, they are only qualitative. Real-time quantitative PCR (qPCR) could be used to detect and to evaluate pathogen concentration. However, both PCR and qPCR techniques could not distinguish between dead and viable cells. Regarding this drawback, Bio-PCR, in which viable cells are allowed to grow before PCR is advantageous and is the standardized method by the European and Mediterranean Plant Protection Organization (EPPO) (Hadas et al. 2005; de León et al. 2011). TaqMan RT-PCR also could differentiate between dead and viable cells (Zhao et al. 2007; Sen et al. 2012). Recently, Han et al. (2018) developed an improved qPCR technique (PMA-qPCR) based on the use of propidium monoazide (PMA) as pretreatment prior to DNA extraction, followed by qPCR. PMA could bind to the DNA of dead cells and block PCR amplification, allowing quantifying selectively the living cells (and even the VBNC). More recently, a multiplex PCR (mPCR) assay for detection of the pathogen based on two chromosomal genes (*rhuM* and *tomA*) with high sensitivity, specificity, and reliability was developed by Thapa et al. (2020). However although it allowed separation between *C. michiganensis* and other *Clavibacter* species, detection threshold did not exceed those already provided by qPCR (10^3 cfu mL^{-1}) (Luo et al. 2008). Currently, more innovative techniques are proposed such as the loop-mediated isothermal amplification (Yasuhara-Bell et al. 2013) and spectral indices (Cordon et al. 2022). A technique based on the reduction in the water due to *C. michiganensis* adverse effects on leaves xylem conductivity was developed by Cordon et al. (2022). This technique allowed early detection of *C. michiganensis* by spectral indices based on stomata closure in asymptomatic tomato plants, even before appearance of wilting symptoms.

9.5 Biocontrol Agents (BCAs) Antagonistic to *C. michiganensis*

Although numerous biocontrol agents were described, no commercialized microorganism based product is available. Biological control by soil bacteria of the so-called PGPR group can help to limit the progress of canker disease by either direct mechanisms including production of biocidal substances or indirect mechanisms through inducing plant systemic resistance. Numerous biocontrol agents with antagonistic activity toward *C. michiganensis* were selected from soil or tomato rhizosphere. Antagonistic fluorescent *Pseudomonas* (Bouizgarne et al. 2023), *Serratia* (Bouizgarne et al. 2023), *Bacillus* (Abo-Elyousr et al. 2019) have been reported. Direct mechanisms displayed by these bacteria are either competition or antagonism. Competition could be due to siderophores (Takishita et al. 2018; Abo-Elyousr et al. 2019), while antagonism is mainly due to production of antibiotics including diffusible compounds such as 2,4-diacetylphloroglucinol (DAPG), phenazines (PCA), and volatile compounds such as hydrogen cyanide (HCN) (Lanteigne et al. 2012; Deng et al. 2015; Paulin et al. 2017; Abo-Elyousr et al. 2019; Takishita et al. 2018; Mishra et al. 2022). In order to induce a decrease in the incidence of tomato bacterial canker, liquid cultures of the antagonistic *Bacillus* spp. strains were used for inoculation of the tomato seeds, resulting in 100% pathogen free seed (Kasselaki et al. 2011). Antibiotics such as DAPG and the volatile HCN are well known as inhibitory agents, effective against phytopathogens (Mishra et al. 2022). Lanteigne et al. (2012) described a DAPG and HCN producing *Pseudomonas brassicacearum* LBUM300 with inhibitory effects in vitro against *C. michiganensis* was effective at reducing pathogen population (assessed by TaqMan) in the plant rhizosphere and significantly reducing bacterial canker development. Moreover, it appeared that *C. michiganensis* contributes to significantly increase rhizosphere populations of the BCA as demonstrated by a greater colonization through increased biofilm formation accompanied by an overexpression of *phlD* gene expression compared to *hcnC* expression, suggesting the leading role of DAPG in the biocontrol effect (Paulin et al. 2017). Other molecules which proved their efficacy against *C. michiganensis* are lipopeptides, including surfactins: surfactin C14 (surfactin B) and surfactin C15 (surfactin C), produced by *Bacillus velezensis* 9D-6, which alters *Cmm* membrane by interference with protein biosynthesis and secretion, solubilizing the phospholipid bilayer, leading to pores and ion leakage (Grady et al. 2019). Therefore, recent works by Laird et al. (2020) used surfactin producing *B. velezensis* 1B-23 and *Bacillus* sp. 1D-12 as BCAs for controlling bacterial canker disease. Works aiming to confirm the role of these antagonistic mechanisms usually use mutants lacking genes for the appropriate mechanism (Lanteigne et al. 2012; Paulin et al. 2017). Paulin et al. (2017) found that non-producing *Pseudomonas brassicacearum* isogenic mutants for DAPG (LBUM300*ΔphlD*) were deficient in forming considerable biofilm and also inefficient to control the pathogen. However, most works focused on in vitro inhibition and genome characterization of the BCAs and few works succeeded at greenhouse or field trials. Indeed, only a small proportion of such biological agents succeed in field trials, and, to the best of our knowledge, none of these agents is formulated and commercialized.

9.6 Biocontrol by Induction of Systemic Resistance to *C. michiganensis*

9.6.1 Induced Resistance by External Application of Synthetic Compounds

It is obvious that any external treatment that could alleviate oxidative stress is in favor of plant growth promotion and plant overall health. This goal could be attained by enhancing defense mechanisms in plants by either synthetic products or microbial (Bouizgarne 2013; Pieterse et al. 2014; Tripathi et al. 2022; Salwan et al. 2023). In order to trigger such enhanced mechanisms against phytopathogens, synthetic products proved their effectiveness (Baysal et al. 2003; Choi et al. 2013; Tripathi et al. 2022; Bektas 2022) and particularly against bacterial canker (Baysal et al. 2003; Bektas 2021; Tripathi et al. 2022; Yokotani et al. 2021). To achieve the goal of protecting tomato plants against bacterial canker, synthetic elicitors such as SA (Yokotani et al. 2021; Tripathi et al. 2022), acibenzolar-*S*-methyl (Baysal et al. 2003; Soylu et al. 2003), 2,6-dichloro-isonicotinic acid (INA) (Bektas 2021), and 2,4-dichloro-6-{(*E*)-[(3-methoxyphenyl)imino]methyl} phenol (DPMP) (Bektas 2021) were investigated. For instance, Tripathi et al. (2022) used 1,2,3-benzothiadiazole (BTH), a bicyclic aromatic chemical and showed that defense mechanisms including enzymatic activity, hydrogen peroxide, and lignin production were significantly higher in BTH-treated plants in comparison with the control. The treatment resulted in low disease incidence of 27.88%.

9.6.2 Induced Resistance by Application of BCAs

BCAs are also promising tools for achieving this goal by inducing systemic resistance in plants. In general, their effects result in induced systemic resistance (ISR) modulated by either JA or ET pathways (Bouizgarne 2013). In order to enhance tomato immunity against *C. michiganensis*, the BCA strain, *Pseudomonas entomophila* 23S with PGPR traits (phosphate solubilizer, indole acetic acid (IAA), HCN and siderophore producer), succeeded to trigger systemic defense mechanisms in tomato plants and effectiveness to control bacterial canker disease (Takishita et al. 2018; Takishita et al. 2021). In their work, Takishita et al. (2018) used *P. entomophila* 23S, a rhizobacterium with PGPR and biocontrol abilities, and studied three marker genes involved in systemic resistance of tomato plants against *C. michiganensis*: (a) the pathogenesis-related protein *PR1a* (a marker gene of salicylic acid, SA), (b) the proteinase inhibitor *PI2* (induced by JA), and (c) the 1-aminocyclopropane-1-carboxylic acid oxidase *ACO* gene (enzyme related to ET synthesis). They found that treated plants with *Pseudomonas* sp. 23S resulted in more growth and nutrient uptake, due to the PGPR effect of the BCA bacterium. When tomato plants were post challenged with *C. michiganensis*, 5 days after treatment with the BCA, they showed a significantly delayed progression of the bacterial canker, concomitant to a lower *C. michiganensis* population in tomato stems. In addition, in protected plants, real-time quantitative PCR (qRT-PCR) revealed an upgrading of *PR1a* and *ACO* expression, suggesting that the BCA acts through the well-known SA-dependent pathway, resembling the classical mechanism leading to pathogenesis-related protein synthesis, normally found when plants are challenged by their specific pathogens (Yokotani et al. 2021). The role of SA-pathway, JA-pathway, PAL, pathogenesis-related proteins, and phenolic compounds as key molecules in the BCAs–*C. michiganensis* interactions was demonstrated by using bacteria belonging to the genera *Pseudomonas* and *Bacillus* (Aksoy et al., 2017; Solano-Alvarez et al. 2021). Actually, the role of these molecules in transduction mechanisms is not exclusively detected in tomato–*C. michiganensis* interactions but universally activated when microbial biocontrol agents were applied to elicit resistance against diseases caused by phytopathogens (Alzandi and Naguib 2019; Konappa et al. 2020; Elsharkawy et al. 2022).

Those interactions were reported, for instance, by Barda et al. (2015), who used NahG transgenic tomato line (salicylic hydroxylase-expressing plants with impaired SA accumulation) in order to demonstrate induction of resistance by epiphytic fungus *Pseudozyma aphidis*. In this work, it was shown that the fungus triggered upregulation of PR1a and other pathogenesis-related marker genes in a SA independent manner. These mechanisms were also found by Solano-Alvarez et al. (2021), who used *Bacillus cereus* strain "Amazcala" (B.c-A) and showed an elicitation of tomato defense against *C. michiganensis* through increase in PAL and chalcone synthase (chs) gene expression, both responsible of the biosynthesis of the flavonoids phenolic compounds, a well-known compound involved in plant defense as biochemical barriers. In addition, antioxidant enzymes such as SOD and CAT activities were significantly increased suggesting that the effect of *B. cereus* strain "Amazcala" could trigger multiple antioxidative and defense reactions. Aksoy et al. (2017) reported changes in tomato foliar phenolic compounds, both qualitatively and quantitatively in bacterized tomato seedlings with *Pseudomonas putida* (CKPp9) and challenged by *C. michiganensis*. The authors showed that protection against bacterial canker by CKPp9 induced systemic resistance was correlated to catechin contents. Figure 9.3 shows mechanisms displayed by BCAs acting as bioprotectors against *C. michiganensis*.

However, most works focused on *Pseudomonas* and *Bacillus* and the bacterial reservoir with potential antagonistic effects and protection against tomato bacterial canker is only poorly harnessed. Indeed, still yet many taxa are worth to be tested. In Morocco, the APPHOS biotechnology project "BIO-BIZ-01/2017," recently sponsored by Office Chérifien des Phosphates (OCP) Foundation, Mohammed VI Polytechnic University, National Center of Scientific and Technical Research

(CNRST), and Ministry of Higher Education, Scientific Research and Professional Training of Morocco, allowed to select bacteria belonging not only from the above-mentioned taxa but also from *Serratia* and *Streptomyces* (Bouizgarne 2022; Bouizgarne et al. 2023; Bakki et al. 2024). Figure 9.4 shows our research workchart, describing steps from sampling soil for isolation of *C. michiganensis* and antagonistic BCAs, their identification, and application in tomato cultivated in pots under greenhouse. Moroccan phosphate mines and tomato rhizosphere soils, from cultures in areas with no symptoms were sampled for PGPR bacteria antagonistic toward *C. michiganensis*. Virulent *C. michiganensis* strains were isolated from tomato plants showing bacterial canker symptoms and tested for its pathogenicity. PGPR bacteria showing high phosphate solubilization potential (mainly by production of organic acids) and other activities (potash and zinc solubilization, IAA, gibberellic acid (GA), siderophores, and ACC deaminase synthesis), in addition to in vitro antagonism toward a virulent strain of *C. michiganensis,* were selected and identified by phenotypic and molecular methods. Selected bacteria were further used as biofertilizers in order to enhance tomato growth and nutrient uptake, mainly phosphorus and also as biopesticides for the biocontrol of bacterial canker disease under greenhouse.

In these works, two bacteria isolated from Moroccan phosphate mines, *Serratia* strain Bg22c and *Pseudomonas* strains (Bg32c and 15d), were also able to solubilize insoluble potassium and zinc forms (P, K, and Zn solubilizers) and produce IAA. In vitro, *Serratia* strain Bg22c and *Pseudomonas* strain Bg15d were able to inhibit *C. michiganensis. Pseudomonas* strain Bg32c, either alone or in combination with *Serratia* strain Bg22c showed remarkable ability to promote tomato growth and yield. Under greenhouse conditions growth of whole plant, height, root length, shoot and root weight, number of leaves and fruits, as well as fruit fresh weight, induced stomatal conductance, total soluble phenolic compounds, sugars, protein, phosphorus were improved. Results concerning total soluble phenolic compounds suggest their effectiveness in controlling the tomato bacterial canker by inducing defense responses (Bouizgarne et al. 2023). Experiments aiming to check their potential to induce defense mechanisms were performed. Therefore, we confirmed that two isolates showing high rhizosphere competence and endophytic abilities, applied at the root system, were able to trigger induced systemic resistance in *C. michiganensis* inoculated tomato seedlings. With those strains, high protection rate was attained, correlated with an increase in total soluble phenolic compounds, PAL, and H_2O_2 (Bouizgarne 2022; Bakki et al. 2024). Figure 9.5 shows inhibition of *C. michiganensis* growth by three bacterial isolates with PGPR potential, belonging to *Pseudomonas* (Bg32c and Bg15d) and *Serratia* (Bg22c) (Bouizgarne et al. 2023).

Table 9.1 summarizes the main bacteria reported to be efficiently used as BCAs to control tomato bacterial canker during the last 12 years (2012–2023), their eventual plant growth promoting effects, their mechanisms of action against *C. michiganensis* and main findings.

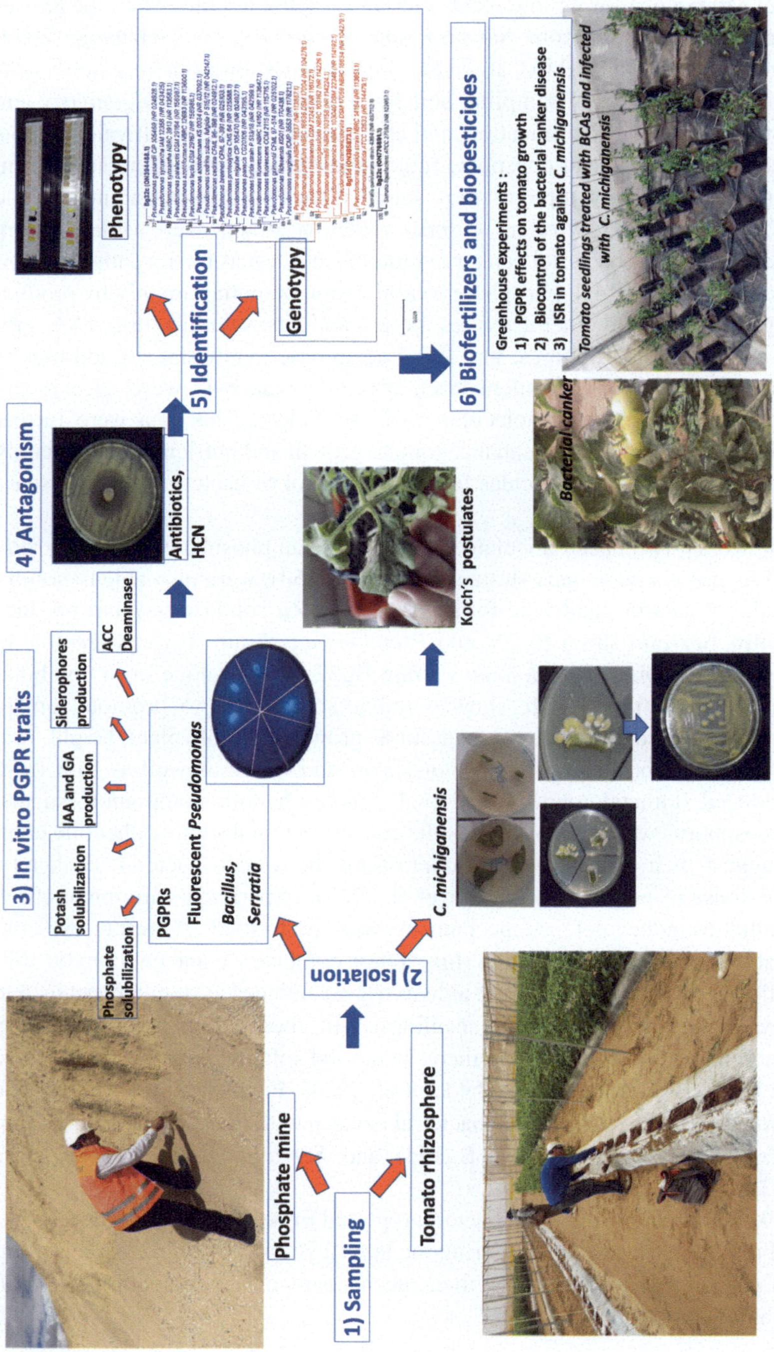

Fig. 9.4 Research work chart describing steps from sampling soil for isolation of *C. michiganensis* and antagonistic BCAs with PGPR effects, their identification, and application as biofertilizers and biopesticides for tomato cultivated in pots under greenhouse

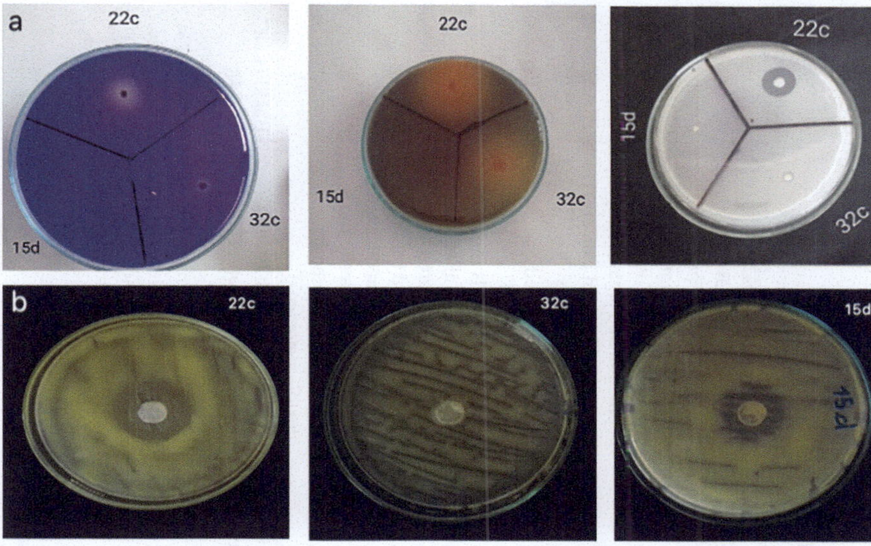

Fig. 9.5 Inhibition of *C. michiganensis* growth by three bacterial isolates with PGPR potential, belonging to *Pseudomonas* (Bg32c and Bg15d) and *Serratia* (Bg22c): (**a**) solubilization of insoluble forms of phosphorus, potassium, and zinc by the bacterial isolates, (**b**) antagonism against *C. michiganensis*. (Adapted from Bouizgarne et al. 2023) (Copyright © 2023 Bouizgarne et al. 2023)

9.7 Conclusion

C. michiganensis is still the threat for tomato crops, and tomato bacterial canker disease continues to cause significant losses around the world. As no resistant *S. lycopersicum* cultivars are available, and no effective conventional control methods are effective, breakthrough technology for the early detection of pathogen and the use of efficient ecofriendly methods such as BCAs are needed. Reliable bioassays aiming to early detect the pathogen and to provide pathogen free seeds and seedling are encouraged. These methods should be reliable, sensitive, non-destructive, and as fast as possible. They also should be cost-effective. Serological methods (such as ELISA) are reliable and very sensitive. New DNA based methods are evolving for accurate detection of the pathogen. Development of early detection methods, such as the spectral indices based on stomata closure upon water stress of the leaves, imposed by *C. michiganensis* is required. Also, methods based specific sensors are promising. All those innovative methods are encouraged in order to effectively prevent secondary progress of the pathogen at the right time, thus reducing subsequent economic losses. Management of the disease could benefit from the great progress in knowledge about plant–soil microbial interactions. Some of these microorganisms are multifaceted and could display PGP effects, enhancing the growth parameters of the plant while showing antagonistic effects against the pathogen but also could enhance the defense status of the plant. Enhancement of

Table 9.1 Selected relevant works with biocontrol agents (BCAs) against bacterial canker disease during the last twelve years (2012–2023)

Biocontrol agents (BCAs)	Experimental conditions	PGPR effects	Biocontrol effects		References
			Mechanism of inhibition/ protection	Main results	
Pseudomonas brassicacearum LBUM300	Growth chamber experiments under controlled conditions (photoperiod: 12 light/12 h dark, temperature: 25 °C, humidity: 80%)	ND	Antagonism by diacetylphloroglucinol (DAPG) and HCN	In planta, *Pseudomonas brassicacearum* LBUM300 (DAPG and HCN producer) reduced *C. michiganensis* rhizosphere *population* population and significantly decreased bacterial canker severity symptoms	Lanteigne et al. (2012), Paulin et al. (2017)
Bacillus subtilis, B. amyloliquefaciens, Pseudomonas fluorescens, P. aeruginosa	Greenhouse experiments under controlled conditions (photoperiod: 11 h light/13 h dark, temperature: 30 ± 5 °C, relative humidity: 68–80%)	In vitro production of siderophores, hydrogen cyanide, and IAA Treatments with *B. amyloliquefaciens* significantly increased seed germination, plant shoot and root length, and vigor index	Antagonism by siderophores and HCN	Treatments retarded multiplication of *C. michiganensis* cells in xylem tissue by 2 weeks after inoculation compared to the control. Talc-based formulation of the BCAs reduced bacterial canker disease severity (up to 74.4% by formulated *B. amyloliquefaciens*)	Abo-Elyousr et al. (2019)
Bacillus sp. *Pseudomonas* sp. *Serratia* sp.	Greenhouse experiments under natural conditions (photoperiod: average 16 light/8 h dark, temperature: average 32 °C)	Growth promotion of tomato: enhancement in whole plant height, root length, shoot and root weight, number of leaves and fruits, fruit fresh weight. Induced stomatal conductance. Sugars, protein, phosphorus	Antagonism and induced systemic resistance	Antagonism by antibiotic production and biofilm inhibition Induced systemic resistance, potentially through total soluble phenolic compounds, phenylalanine ammonia-lyase and H_2O_2	Bakki et al. (2024), Bouizgarne et al. (2023), Bouizgarne (2022)

Pseudomonas entomophila 23S	Growth chamber under controlled conditions (photoperiod : 14 h light/10 h dark, temperature: 25/23 °C)		Antagonism by siderophores and HCN/ISR in tomato plants	*P. entomophila* 23S trigger salicylic acid (SA)-dependent pathway by *PR1a* gene expression. Treatment with 23S resulted in a significant reduction in the number of wilted leaves and a delay of the disease progression. In addition, a lower *C. michiganensis* population in tomato stems was observed	Takishita et al. (2018)
Bacillus velezensis 1B-23 *Bacillus* sp. 1D-12	Hydroponic tank and pots in greenhouse (18 h light/6 h dark, temperature: 26 °C/18 °C, relative humidity: 65%)	ND	Antagonism towards a *Cmm* strain *Cmm*98–1 by cyclic lipopeptides (surfactin A, surfactin B, and surfactin C)	Inoculation of *B. velezensis* 1B-23 or *Bacillus* sp. 1D-12 to infected tomato plants reduced significantly disease incidence	Laird et al. (2020)
Bacillus cereus strain "Amazcala" (B.c-A)	Greenhouse under controlled conditions (photoperiod: 12 h light/12 h dark, temperature: 29 ± 3 °C, relative humidity: 70%)	PGPB effects: phosphate solubilization and production of gibberellic acid (GA3). Treatment by B.c-A induced increases in height, stem width, dry weight, and total chlorophyll content	Antagonism and induced systemic resistance	Treatment by *Bacillus cereus* B.c-A significantly decreased the severity of bacterial canker disease on tomato in preventive disease assays Elicitation of tomato defense: expression PAL and chalcone synthase (chs) and biosynthesis of flavonoids	Solano-Alvarez et al. (2021)

(continued)

Table 9.1 (continued)

Biocontrol agents (BCAs)	Experimental conditions	PGPR effects	Biocontrol effects		References
			Mechanism of inhibition/ protection	Main results	
Alcaligenes faecalis *Acinetobacter* sp.	Greenhouse (photoperiod: 12-h light/12 h dark, temperature: 25–30 °C, relative humidity: 70–90%)	*A. faecalis* and *Acinetobacter* sp. significantly promoted the growth of tomato plants	Antagonism	*A. faecalis* and *Acinetobacter* sp. delayed canker development and reduced the disease incidence by up to 50.0%	Oloyede et al. (2021)
Bacillus amyloliquefaciens strain S1	Net house (temperature: 25–35 °C, relative humidity: 70–100%)	P-solubilization, IAA production, and ability to fix atmospheric nitrogen	– Antagonism by siderophore production – Possible involvement of ISR by cell wall degrading enzymes production (lipase and protease)	*B. amyloliquefaciens* S1 application resulted in reduction of bacterial canker disease incidence by 70%	Gautam et al. (2019, 2020)
Pseudomonas putida strain CKPp9	Pots in growth chamber (temperature: 24–26 °C with photoperiod 16 h light/8 h dark, relative humidity: 70%)	Plant growth promoting effects		Bacterized tomato seedlings with *P. putida* (CKPp9) showed protection against bacterial canker Induction of systemic resistance correlated to disease resistance, through changes in tomato foliar phenolic compounds, both qualitatively and quantitatively In bacterized plants by CKPp9 and infected by *C. michiganensis*, high catechin contents were observed in comparison with controls	Aksoy et al. (2017)

ND not determined

resistance could also be achieved by microorganisms derived products such as harpin (HrpN) used for combatting fire blight disease caused by *Erwinia amylovora* at least at foliar and fruit levels for bacterial canker disease. BCAs showing multiple effects are the best candidates for biocontrol programs. In this chapter, we mentioned works by our team regarding isolation of antagonistic PGPR from either tomato rhizosphere or phosphate mines. Our program for combatting *C. michiganensis* by selected phosphate solubilizing bacteria belonging to *Bacillus*, *Pseudomonas,* and *Serratia,* showing multiple effects, i.e.. competition, antibiosis, and induced resistance is in progress.

Acknowledgement The authors would like to acknowledge the support through the R&D Initiative—Call for projects around phosphates APPHOS—sponsored by OCP (OCP Foundation, R&D OCP, Mohammed VI Polytechnic University, National Center of Scientific and Technical Research (CNRST), Ministry of Higher Education, Scientific Research and Professional Training of Morocco (MESRSFC)) under the project entitled "Bioformulation of a Consortium of Phosphate Solubilizing Microorganisms: Beneficial Effects on Plant Growth and Protection," project ID "BIO-BIZ-01/2017."

References

Abebe AM, Oh C-S, Kim HT, Choi G, Seo E, Yeam I, Lee JM (2022) QTL-Seq analysis for identification of resistance loci to bacterial canker in tomato. Front Plant Sci 12:809959. https://doi.org/10.3389/fpls.2021.809959

Abo-Elyousr KAM, Khalil Bagy HMM, Hashem M, Alamri SAM, Mostafa YS (2019) Biological control of the tomato wilt caused by *Clavibacter michiganensis* subsp. *michiganensis* using formulated plant growth-promoting bacteria. Egypt J Biol Pest Control 29:54. https://doi.org/10.1186/s41938-019-0152-6

Aksoy HM, Kaya Y, Ozturk M, Secgin Z, Onder H, Okumus A (2017) *Pseudomonas putida*-induced response in phenolic profile of tomato seedlings (*Solanum lycopersicum* L.) infected by *Clavibacter michiganensis* subsp. *michiganensis*. Biol Control 105:6–12

Alzandi AA, Naguib DM (2019) *Pseudomonas fluorescens* metabolites as biopriming agent for systemic resistance induction in tomato against *Fusarium* wilt. Rhizosphere 11:100168. https://doi.org/10.1016/j.rhisph.2019.100168

Arizala D, Dobhal S, Alvarez AM, Arif M (2022) Elevation of *Clavibacter michiganensis* subsp. *californiensis* to species level as *Clavibacter californiensis* sp. nov., merging and reclassification of *Clavibacter michiganensis* subsp. *chilensis* and *Clavibacter michiganensis* subsp. *phaseoli* as *Clavibacter phaseoli* sp. nov. based on complete genome in silico analyses. Int J Syst Evol Microbiol 72. https://doi.org/10.1099/ijsem.0.005427

Arora NK, Mishra I (2022) Current scenario and future directions for sustainable development goal 2: a roadmap to zero hunger. Environ Sustain 5:129–133. https://doi.org/10.1007/s42398-022-00235-8

Backer R, Rokem JS, Ilangumaran G, Lamont J, Praslickova D, Ricci E, Subramanian S, Smith DL (2018) Plant growth-promoting rhizobacteria: context, mechanisms of action, and roadmap to commercialization of biostimulants for sustainable agriculture. Front Plant Sci 9:1473. https://doi.org/10.3389/fpls.2018.01473

Bakki M, Banane B, Marhane O, Esmaeel Q, Hatimi A, Barka EA, Azim K, Bouizgarne B (2024) Phosphate solubilizing Pseudomonas and Bacillus combined with rock phosphates promoting tomato growth and reducing bacterial canker disease. Front. Microbiol. 15:1289466. https://doi.org/10.3389/fmicb.2024.1289466

Balaji V, Smart CD (2012) Over-expression of snakin-2 and extensin-like protein genes restricts pathogen invasiveness and enhances tolerance to *Clavibacter michiganensis* subsp. *michiganensis* in transgenic tomato (*Solanum lycopersicum*). Transgenic Res 21:23–37. https://doi.org/10.1007/s11248-011-9506-x

Balaji V, Mayrose M, Sherf O, Jacob-Hirsch J, Eichenlaub R, Iraki N, Manulis-Sasson S, Rechavi G, Barash I, Sessa G (2008) Tomato transcriptional changes in response to *C. michiganensis* subsp. *michiganensis* reveal a role for ethylene in disease development. Plant Physiol 146:1797–1809

Barda O, Shalev O, Alster S, Buxdorf K, Gafni A, Levy M (2015) *Pseudozyma aphidis* induces salicylic-acid-independent resistance to *Clavibacter michiganensis* in tomato plants. Plant Dis 99:621–626. https://doi.org/10.1094/PDIS-04-14-0377-RE

Basim H, Basim E, Tombuloglu H, Unver T (2021) Comparative transcriptome analysis of resistant and cultivated tomato lines in response to *Clavibacter michiganensis* subsp. *michiganensis*. Genomics 113:2455–2467. https://doi.org/10.1016/j.ygeno.2021.05.033

Baysal Ö, Soylu EM, Soylu S (2003) Induction of defence-related enzymes and resistance by the plant activator acibenzolar-S-methyl in tomato seedlings against bacterial canker caused by *Clavibacter michiganensis* ssp. *michiganensis*. Plant Pathol 52:747–753. https://doi.org/10.1111/j.1365-3059.2003.00936.x

Bektas Y (2021) The synthetic elicitors 2,6-dichloro-isonicotinic acid (INA) and 2,4-dichloro-6-{(E)-[(3-methoxyphenyl)imino]methyl}phenol (DPMP) enhances tomato resistance against bacterial canker disease with different molecular mechanisms. Physiol Mol Plant Pathol 116:101740. https://doi.org/10.1016/j.pmpp.2021.101740

Bektas Y (2022) FytoSol, a Promising Plant Defense Elicitor, controls early blight (*Alternaria solani*) disease in the tomato by inducing host resistance-associated gene expression. Horticulturae 8:484. https://doi.org/10.3390/horticulturae8060484

Bouizgarne B (2013) Bacteria for plant growth promotion and disease management. In: Maheshwari DK (ed) Bacteria in agrobiology: disease management. Springer, Berlin, pp 15–47. https://doi.org/10.1007/978-3-642-33639-3_2

Bouizgarne B (2022) Phosphate and potash solubilizing bacteria, from phosphate mines: beneficial effects towards Moroccan economically important crops. In: International conference on "Environmental sustainability & biotechnology: opportunities & challenges", Goa, India, 16-19 November, 2022

Bouizgarne B, Ouhdouch Y (2017) Management of tomato foot and root rot (TFRR) by biocontrol agents with emphasis on factors affecting its effectiveness. In: Kumar V et al (eds) Probiotics and plant health. Singapore Pte Ltd, Springer, pp 1–19. https://doi.org/10.1007/978-981-10-3473-2_1

Bouizgarne B, Bakki M, Boutasknit A, Banane B, El Ouarrat H, Ait El Maalem S, Amenzou A, Ghousmi A, Meddich A (2023) Phosphate and potash solubilizing bacteria from Moroccan phosphate mine showing antagonism of bacterial canker agent and inducing effective tomato growth promotion. Front Plant Sci 14:970382. https://doi.org/10.3389/fpls.2023.970382

Celik I (2023) *In silico* integrated analysis of genomic, transcriptomic, and proteomic data reveals QTL-specific genes for bacterial canker resistance in tomato (*Solanum lycopersicum* L.). Curr Issues Mol Biol 45:1387–1395. https://doi.org/10.3390/cimb45020090

Chalupowicz L, Cohen-Kandli M, Dror O, Eichenlaub R, Gartemann K-H, Sessa G, Barash I, Manulis-Sasson S (2010) Sequential expression of bacterial virulence and plant defense genes during infection of tomato with *Clavibacter michiganensis* subsp. *michiganensis*. Phytopathology 100:252–261

Chalupowicz L, Zellermann E-M, Fluegel M, Dror O, Eichenlaub R, Gartemann K-H, Savidor A, Sessa G, Iraki N, Barash I, Manulis-Sasson S (2012) Colonization and movement of GFP-labeled *Clavibacter michiganensis* subsp. *michiganensis* during tomato infection. Phytopathology 102:23–31

Chalupowicz L, Barash I, Reuven M, Dror O, Sharabani G, Gartemann KH, Eichenlaub R, Sessa G, Manulis-Sasson S (2017) Differential contribution of *Clavibacter michiganensis* ssp. *michiganensis* virulence factors to systemic and local infection in tomato. Mol Plant Pathol 18:336–346. https://doi.org/10.1111/mpp.12400

Chang RJ, Ries SM, Pataky JK (1992) Effects of temperature, plant-age, inoculum concentration, and cultivar on the incubation period and severity of bacterial canker of tomato. Plant Dis 76:1150–1155. https://doi.org/10.1094/PD-76-1150

Choi MS, Kim W, Lee C, Oh CS (2013) Harpins, multifunctional proteins secreted by gram-negative plant-pathogenic bacteria. Mol Plant-Microbe Interact 26:1115–1122. https://doi.org/10.1094/MPMI-02-13-0050-CR

Coaker G, Francis D (2004) Mapping, genetic effects, and epistatic interaction of two bacterial canker resistance QTLs from *Lycopersicon hirsutum*. Theor Appl Genet 108:1047–1055

Collins MD, Jones D (1980) Lipids in the classification and identification of coryneform bacteria containing peptidoglycans based on 2,4-diaminobutyric acid (DAB). J Appl Bacteriol 48:459–470

Cordon G, Andrade C, Barbara L, Romero AM (2022) Early detection of tomato bacterial canker by reflectance indices. Inf Process Agric 9:184–194. https://doi.org/10.1016/j.inpa.2021.06.004

Davis M, Gillaspie A, Vidaver A, Harris R (1984) *Clavibacter*: a new genus containing some phytopathogenic coryneform bacteria, including *Clavibacter xyli* subsp. *xyli* sp. nov., and *Clavibacter xyli* subsp. *cynodontis* subsp. nov., pathogens that cause ratoon stunting disease of sugarcane and Bermudagrass. Int J Syst Bacteriol 34:107–117

de León L, Rodriguez A, Lopez MM, Siverio F (2008) Evaluation of the efficacy of immunomagnetic separation for the detection of *C. michiganensis* subsp. *michiganensis* in tomato seeds. J Appl Microbiol 104:776–786

de Leon L, Rodriguez A, Llop P, Lopez MM, Siverio F (2009) Comparative study of genetic diversity of *Clavibacter michiganensis* subsp. *michiganensis* isolates from the Canary Islands by RAPD-PCR, BOX-PCR and AFLP. Plant Pathol 58:862–871

de León L, Siverio F, López MM, Rodríguez A (2011) *Clavibacter michiganensis* subsp. *michiganensis*, a seedborne tomato pathogen: healthy seeds are still the goal. Plant Dis 95:1328–1338

Deng P, Wang X, Baird SM, Lu SE (2015) Complete genome of *Pseudomonas chlororaphis* strain UFB2, a soil bacterium with antibacterial activity against bacterial canker pathogen of tomato. Stand Genom Sci 10:117. https://doi.org/10.1186/s40793-015-0106-x

Dreier J, Meletzus D, Eichenlaub R (1997) Characterization of the plasmid encoded virulence region pat-1 of the phytopathogenic *Clavibacter michiganensis subsp. michiganensis*. Mol Plant-Microbe Interact 10:195–206

Eichenlaub R, Gartemann K-H (2011) The *Clavibacter michiganensis* subspecies: molecular investigation of gram-positive bacterial plant pathogens. Annu Rev Phytopathol 49:445–464

Eichenlaub R, Gartemann KH, Burger A (2006) *Clavibacter michiganensis*, a group of gram-positive phytopathogenic bacteria. In: Gnanamanickam SS (ed) Plant-associated bacteria. Springer, Dordrecht, pp 385–422

Elsharkawy MM, El-Okkiah S, Elsadany AY, Bedier MY, Omara RI, Behiry SI, Hassan S, Abdelkhalek A (2022) Systemic resistance induction of tomato plants against tomato mosaic virus by microalgae. Egypt J Biol Pest Control 32:37. https://doi.org/10.1186/s41938-022-00538-2

EPPO: European and Mediterranean Plant Protection Organization (2016) PM 7/42 (3) *Clavibacter michiganensis* subsp. *michiganensis*. EPPO Bull 46:202–225

FAO: Food and Agriculture Organization of the United Nations (2017) The future of food and agriculture-trends and challenges. FAO, Rome. ISBN 978-92-5-109551-5

FAO: Food and Agriculture Organization of the United Nations (2019) The state of food and agriculture: moving forward on food loss and waste reduction. Food and Agricultural Organization, Rome

FAO: Food and Agriculture Organization of the United Nations (2021). http://www.fao.org/faostat/en/#data/QC. Accessed 14 Apr 2023

Fatmi M, Schaad NW (2002) Survival of *Clavibacter michiganensis* ssp. *michiganensis* in infected tomato stems under natural field conditions in California, Ohio and Morocco. Plant Pathol 51:149–154. https://doi.org/10.1046/j.1365-3059.2002.00675.x

Franken AAJM, Kamminga GC, Snyders W, Van Der Zouwen PS, Birnbaum YE (1993) Detection of *Clavibacter michiganensis* ssp *michiganensis* in tomato seeds by immunofluorescence microscopy and dilution plating. Neth J Plant Pathol 99:125–137

Gartemann K-H, Abt B, Bekel T, Burger A, Engemann J, Flügel M, Gaigalat L, Goesmann A, Gäafen I, Kalinowski J, Kaup O, Kirchner O, Krause L, Linke B, McHardy A, Meyer F, Pohle S, Rückert C, Schneiker S, Zellermann EM, Pühler A, Eichenlaub R, Kaiser O, Bartels D (2008) The genome sequence of the tomato-pathogenic actinomycete *Clavibacter michiganensis* subsp. *michiganensis* NCPPB382 reveals a large Island involved in pathogenicity. J Bacteriol 190:2138–2149

Gautam S, Chauhan A, Sharma R, Sehgal R, Shirkot CK (2019) Potential of *Bacillus amyloliquefaciens* for biocontrol of bacterial canker of tomato incited by *Clavibacter michiganensis* ssp. *michiganensis*. Microb Pathog 130:196–203. https://doi.org/10.1016/j.micpath.2019.03.006

Gautam S, Sharma R, Chauhan A, Shirkot CK, Kaushal R (2020) Biocontrol activities of rhizobacteria associated with apple, apricot and kiwi rhizosphere against bacterial canker caused by *Clavibacter michiganensis*. Indian Phytopathol 73:45–56. https://doi.org/10.1007/s42360-019-00172-3

González AJ, Trapiello E (2014) *Clavibacter michiganensis* subsp. *phaseoli* subsp. nov., pathogenic in bean. Int J Sys Evol Microbiol 64:1752–1755. https://doi.org/10.1099/ijs.0.058099-0

Grady EN, MacDonald J, Ho MT, Weselowski B, McDowell T, Solomon O, Renaud J, Yuan ZC (2019) Characterization and complete genome analysis of the surfactin-producing, plant-protecting bacterium *Bacillus velezensis* 9D-6. BMC Microbiol 19:1–14

Hadas R, Kritzman G, Klietman F, Gefen T, Manulis S (2005) Comparison of extraction procedures and determination of the detection threshold for *Clavibacter michiganensis* ssp. *michiganensis* in tomato seeds. Plant Pathol 54:643–649

Han S, Jiang N, Lv Q, Kan Y, Hao J, Li J, Luo L (2018) Detection of *Clavibacter michiganensis* subsp. *michiganensis* in viable but nonculturable state from tomato seed using improved qPCR. PLoS One 13:e0196525. https://doi.org/10.1371/journal.pone.0196525

Hunjan MS, Lore JS (2020) Climate change: impact on plant pathogens, diseases, and their management. In: Jabran K, Florentine S, Chauhan B (eds) Crop protection under changing climate. Springer, Cham. https://doi.org/10.1007/978-3-030-46111-9_4

Jacques MA, Durand K, Orgeur G, Balidas S, Fricot C, Bonneau S, Quillévéré A, Audusseau C, Olivier V, Grimault V, Mathis R (2012) Phylogenetic analysis and polyphasic characterization of *Clavibacter michiganensis* strains isolated from tomato seeds reveal that nonpathogenic strains are distinct from *C. michiganensis* subsp. *michiganensis*. Appl Environ Microbiol 78:8388–8402. https://doi.org/10.1128/AEM.02158-12

Jahr H, Dreier J, Meletzus D, Bahro R, Eichenlaub R (2000) The endo-beta-1,4-glucanase CelA of *Clavibacter michiganensis* subsp. *michiganensis* is a pathogenicity determinant required for induction of bacterial wilt of tomato. Mol Plant-Microbe Interact 13:703–714. https://doi.org/10.1094/MPMI.2000.13.7.703

Kado CI (2010) Plant bacteriology. The American Phytopathological Society, St. Paul. 336 p

Kaneshiro WS, Mizumoto CY, Alvarez AM (2006) Differentiation of Clavibacter *michiganensis* subsp *michiganensis* from seed-borne saprophytes using ELISA, Biolog and 16S rDNA sequencing. Eur J Plant Pathol 116:45–56

Kasselaki AM, Goumas D, Tamm L, Fuchs J, Cooper J, Leifert C (2011) Effect of alternative strategies for the disinfection of tomato seed infected with bacterial canker (*Clavibacter michiganensis* subsp. *michiganensis*). NJAS-Wagening J Life Sci 58:145–147

Kaup O, Grafen I, Zellermann EM, Eichenlaub R, Gartemann KH (2005) Identification of a tomatinase in the tomato-pathogenic actinomycete *C. michiganensis* subsp. *michiganensis* NCPPB382. Mol Plant-Microbe Interact 18:1090–1098

Konappa N, Krishnamurthy S, Arakere UC, Chowdappa S, Ramachandrappa NS (2020) Efficacy of indigenous plant growth-promoting rhizobacteria and *Trichoderma* strains in eliciting resistance against bacterial wilt in a tomato. Egypt J Biol Pest Control 30:106. https://doi.org/10.1186/s41938-020-00303-3

Laird M, Piccoli D, Weselowski B, McDowell T, Renaud J, MacDonald J, Yuan ZC (2020) Surfactin-producing *Bacillus velezensis* 1B-23 and *Bacillus* sp. 1D-12 protect tomato against bacterial canker caused by *Clavibacter michiganensis* subsp. *michiganensis*. J Plant Pathol 102:451–458

Lanteigne C, Gadkar VJ, Wallon T, Novinscak A, Filion M (2012) Production of DAPG and HCN by *Pseudomonas* sp. LBUM300 contributes to the biological control of bacterial canker of tomato. Phytopathology 102:967–973. https://doi.org/10.1094/PHYTO-11-11-0312

Lelis FMV, Czajkowski R, de Souza RM, Ribeiro Daniel H, van der Wolf JM (2014) Studios on the colonization of axenically grown tomato plants by a GFP-tagged strain of *Clavibacter michiganensis* subsp. *michiganensis*. Eur J Plant Pathol 139:53–66. https://doi.org/10.1007/s10658-013-0362-9

Li X, Tambong J, Yuan KX, Chen W, Xu H, Lévesque CA, De Boer SH (2018) Re-classification of *Clavibacter michiganensis* subspecies on the basis of whole-genome and multi-locus sequence analyses. Int J Syst Evol Microbiol 68:234–240. https://doi.org/10.1099/ijsem.0.002492

Liu J, Wang X (2020) Tomato diseases and pests detection based on improved Yolo V3 Convolutional Neural Network. Front Plant Sci 11:898. https://doi.org/10.3389/fpls.2020.00898

Luo LX, Walters C, Bolkan H, Liu XL, Li JQ (2008) Quantification of viable cells of *Clavibacter michiganensis* subsp. *michiganensis* using a DNA binding dye and a real-time PCR assay. Plant Pathol 57:332–337. https://doi.org/10.1111/j.1365-3059.2007.01736.x

Meletzus D, Bermpohl A, Dreier J, Eichenlaub R (1993) Evidence for plasmid-encoded virulence factors in the phytopathogenic bacterium *Clavibacter michiganensis* subsp *michiganensis* NCPPB382. J Bacteriol 175:2131–2136. https://doi.org/10.1128/jb.175.7.2131-2136

Méndez V, Valenzuela M, Salvà-Serra F, Jaén-Luchoro D, Besoain X, Moore ERB, Seeger M (2020) Comparative genomics of pathogenic *Clavibacter michiganensis* subsp. *michiganensis* strains from Chile reveals potential virulence features for tomato plants. Microorganisms 29: 8:1679. https://doi.org/10.3390/microorganisms8111679

Mishra J, Mishra I, Arora NK (2022) 2,4-Diacetylphloroglucinol producing *Pseudomonas fluorescens* JM-1 for management of ear rot disease caused by *Fusarium moniliforme* in *Zea mays* L. 3 Biotech 12:138. https://doi.org/10.1007/s13205-022-03201-7

Nandi M, Macdonald J, Liu P, Weselowski B, Yan ZC (2018) *Clavibacter michiganensis* ssp. *michiganensis*: bacterial canker of tomato, molecular interactions and disease management. Mol Plant Pathol 19:2036–2050. https://doi.org/10.1111/mpp.12678

Nouioui I, Carro L, García-López M, Meier-Kolthoff JP, Woyke T, Kyrpides NC, Pukall R, Klenk H-P, Goodfellow M, Göker M (2018) Genome-based taxonomic classification of the phylum actinobacteria. Front Microbiol 9:2007. https://doi.org/10.3389/fmicb.2018.02007

Oh EJ, Bae C, Lee HB, Hwang IS, Lee HI, Yea MC, Yim KO, Lee S, Heu S, Cha JS, Oh CS (2016) *Clavibacter michiganensis* subsp. *capsici* subsp. nov., causing bacterial canker disease in pepper. Int J Sys Evol Microbiol 66:4065–4070. https://doi.org/10.1099/ijsem.0.001311

Oh E-J, Hwang IS, Park IW, Oh C-S (2022) Comparative genome analyses of *Clavibacter michiganensis* type strain LMG7333T reveal distinct gene contents in plasmids from other *Clavibacter* species. Front Microbiol 12:793345. https://doi.org/10.3389/fmicb.2021.793345

Oloyede AR, Ogbuagor CJ, Afolabi CG, Akintokun AK (2021) Biological control of bacterial canker of tomato (*Lycopersicon esculentum* Mill.) by use of non-native strains of plant growth-promoting rhizobacteria. Arch Phytopathol Plant Protect 54:1182–1203. https://doi.org/10.1080/03235408.2021.1895477

Orozco-Mosqueda MC, Santoyo G, Glick BR (2023) Recent advances in the bacterial phytohormone modulation of plant growth. Plants 12:606. https://doi.org/10.3390/plants12030606

Park Y-H, Suzuki K-I, Yim D-G, Lee K-C, Kim E, Yoon J-S, Kim S-J, Kho Y-H, Goodfellow M, Komogata K (1993) Suprageneric classification of peptidoglycan group B actinomycetes by nucleotide sequencing of 5S ribosomal RNA. Antonie Van Leeuwenhoek 64:307–313

Paulin MM, Novinscak A, Lanteigne C, Gadkar VJ, Filion M (2017) Interaction between 2,4-diacetylphloroglucinol- and hydrogen cyanide-producing *Pseudomonas brassicacearum* LBUM300 and *Clavibacter michiganensis* subsp. *michiganensis* in the tomato rhizosphere. Appl Environ Microbiol 83:e00073-17. https://doi.org/10.1128/AEM.00073-17

Peritore-Galve FC, Miller C, Smart CD (2020) Characterizing colonization patterns of *Clavibacter michiganensis* during infection of tolerant wild solanum species. Phytopathology 110:574–581. https://doi.org/10.1094/PHYTO-09-19-0329-R

Pieterse CM, Zamioudis C, Berendsen RL, Weller DM, Van Wees SC, Bakker PA (2014) Induced systemic resistance by beneficial microbes. Annu Rev Phytopathol 52:347–375. https://doi.org/10.1146/annurev-phyto-082712-102340

Ristaino JB, Anderson PK, Bebber DP, Brauman KA, Cunniffe NJ, Fedoroff NV, Finegold C, Garrett KA, Gilligan CA, Jones CM, Martin MD, MacDonald GK, Neenan P, Records A, Schmale DG, Tateosian L, Wei Q (2021) The persistent threat of emerging plant disease pandemics to global food security. Proc Natl Acad Sci U S A 118:e2022239118. https://doi.org/10.1073/pnas.2022239118

Salwan R, Sharma M, Sharma A, Sharma V (2023) Insights into plant beneficial microorganism-triggered induced systemic resistance. Plant Stress 7:100140. https://doi.org/10.1016/j.stress.100140

Savary S, Willocquet L, Pethybridge SJ, Esker P, McRoberts N, Nelson A (2019) The global burden of pathogens and pests on major food crops. Nat Ecol Evol 3:430–439. https://doi.org/10.1038/s41559-018-0793-y

Savidor A, Teper D, Gartemann KH, Eichenlaub R, Chalupowicz L, Manulis-Sasson S, Barash I, Tews H, Mayer K, Giannone RJ, Hettich R, Sessa G (2012) The *Clavibacter michiganensis* subsp. *michiganensis*-tomato interactome reveals the perception of pathogen by the host and suggests mechanisms of infection. J Proteome Res 1:736–750. https://doi.org/10.1021/pr200646a

Savidor A, Chalupowicz L, Teper D, Gartemann K-H, Eichenlaub R, Manulis-Sasson S, Barash I, Sessa G (2014) *C. michiganensis* subsp. *michiganensis* Vatr1 and Vatr2 transcriptional regulators are required for virulence in tomato. Mol Plant-Microbe Interact 10:1035–1047

Sen Y, Zhu F, Vandenbroucke H, van der Wolf J, Visser R, van Heusden A (2012) Screening for new sources of resistance to *C. michiganensis* subsp. *michiganensis* (Cmm) in tomato. Euphytica 190:309–317. https://doi.org/10.1007/s10681-012-0802-1

Sen Y, van der Wolf J, Visser RGF, van Heusden S (2015) Bacterial canker of tomato: current knowledge of detection, management, resistance, and interactions. Plant Dis 99:4–13. https://doi.org/10.1094/PDIS-05-14-0499-FE

Sharabani G, Manulis-Sasson S, Chalupowicz L, Borenstein M, Shulhani R, Lofthouse M, Sofer M, Frenkel O, Dror O, Shtienberg D (2014) Temperature at the early stages of *Clavibacter michiganensis* subsp. *michiganensis* infection affects bacterial canker development and virulence gene expression. Plant Pathol 63:1119–1129

Solano-Alvarez N, Valencia-Hernández JA, Rico-García E, Torres-Pacheco I, Ocampo-Velázquez RV, Escamilla-Silva EM, Romero-García AL, Alpuche-Solís ÁG, Guevara-González RG (2021) A novel isolate of *Bacillus cereus* promotes growth in tomato and inhibits *Clavibacter michiganensis* infection under greenhouse conditions. Plants 10:506. https://doi.org/10.3390/plants10030506

Soylu EM, Soylu S, Baysal Ö (2003) Induction of disease resistance and antioxidant enzymes by acibenzolar-S-methyl against bacterial wilt disease (*Clavibacter michiganensis* ssp. *michiganensis*) in tomato. J Plant Pathol 85:175–181

Stork I, Gartemann KH, Burger A, Eichenlaub R (2008) A family of serine proteases of *C. michiganensis* subsp. *michiganensis*: *chpC* plays a role in colonization of the host plant tomato. Mol Plant Pathol 9:599–608

Takahashi H, Nakaho K, Ishihara T, Ando S, Wada T, Kanayama Y, Asano S, Yoshida S, Tsushima S, Hyakumachi M (2014) Transcriptional profile of tomato roots exhibiting *Bacillus thuringiensis*-induced resistance to *Ralstonia solanacearum*. Plant Cell Rep 33:99–110

Takishita Y, Charron JB, Smith DL (2018) Biocontrol rhizobacterium *Pseudomonas* sp. 23S induces systemic resistance in tomato (*Solanum lycopersicum* L.) against bacterial canker *Clavibacter michiganensis* subsp. *michiganensis*. Front Microbiol 9:2119. https://doi.org/10.3389/fmicb.2018.02119

Takishita Y, Souleimanov A, Bourguet C, Ohlund LB, Arnold AA, Sleno L, Smith DL (2021) *Pseudomonas entomophila* 23S produces a novel antagonistic compound against *Clavibacter michiganensis* subsp. *michiganensis*, a pathogen of tomato bacterial Canker. Appl Microbiol 1:60–73. https://doi.org/10.3390/applmicrobiol1010006

Tambong JT (2017) Comparative genomics of *Clavibacter michiganensis* subspecies, pathogens of important agricultural crops. PLoS One 12:e0172295. https://doi.org/10.1371/journal.pone.0172295

Tampakaki AP, Hatziloukas E, Panopoulos NJ (2009) Plant pathogens, bacterial. In: Schaechter M (ed) Encyclopedia of microbiology, 3rd edn. Academic Press, New York, pp 655–677. https://doi.org/10.1016/B978-012373944-5.00346-1. ISBN 9780123739445

Thapa SP, Pattathil S, Hahn MG, Jacques MA, Gilbertson RL, Coaker G (2017) Genomic analysis of *Clavibacter michiganensis* reveals insight into virulence strategies and genetic diversity of a gram-positive bacterial pathogen. Mol Plant-Microbe Interact 30:786–802. https://doi.org/10.1094/MPMI-06-17-0146-R

Thapa SP, O'Leary M, Jacques MA, Gilbertson RL, Coaker G (2020) Comparative genomics to develop a specific multiplex PCR assay for detection of *Clavibacter michiganensis*. Phytopathology 110:556–566. https://doi.org/10.1094/PHYTO-10-19-0405-R

Tian Q, Chuan J, Sun X, Zhou A, Wang L, Zou J, Zhao W, Li X (2021) Description of *Clavibacter zhangzhiyongii* sp. nov., a phytopathogenic actinobacterium isolated from barley seeds, causing leaf brown spot and decline. Int J Syst Evol Microbiol 71:004786. https://doi.org/10.1099/ijsem.0.004786

Tripathi R, Vishunavat K, Tewari R, Kumar S, Minkina T, De Corato U, Keswani C (2022) Defense inducers mediated mitigation of bacterial canker in tomato through alteration in oxidative stress markers. Microorganisms 10:2160. https://doi.org/10.3390/microorganisms10112160

Tsitsekian D, Daras G, Karamanou K, Templalexis D, Koudounas K, Malliarakis D, Koufakis T, Chatzopoulos D, Goumas D, Ntoukakis V, Hatzopoulos P, Rigas S (2021) *Clavibacter michiganensis* downregulates photosynthesis and modifies monolignols metabolism revealing a crosstalk with tomato immune responses. Int J Mol Sci 22:8442. https://doi.org/10.3390/ijms22168442

Valenzuela M, Besoain X, Durand K, Cesbron S, Fuentes S, Claverías F, Jacques MA, Seeger M (2018) *Clavibacter michiganensis* subsp. *michiganensis* strains from Central Chile exhibit low genetic diversity and sequence types match strains in other parts of the world. Plant Pathol 67:1944–1954. https://doi.org/10.1111/ppa.12911

Valenzuela M, González M, Velásquez A, Dorta F, Montenegro I, Besoain X, Salvà-Serra F, Jaén-Luchoro D, Moore ERB, Seeger M (2021) Analyses of virulence genes of *Clavibacter michiganensis* subsp. *michiganensis* strains reveal heterogeneity and deletions that correlate with pathogenicity. Microorganisms 9:1530. https://doi.org/10.3390/microorganisms9071530

Van der Wolf J, De Boer SH (2016) Phytopathogenic bacteria. In: Principles of plant-microbe interactions. Springer, Cham, pp 65–77

Veena MS, van Vuurde JWL (2002) Indirect immunofluorescence colony staining method for detecting bacterial pathogens of tomato. J Microbiol Methods 49:11–17. https://doi.org/10.1016/s0167-7012(01)00346-3

Vega D, Romero AM (2016) Survival of *Clavibacter michiganensis* subsp. *michiganensisin* tomato debris under greenhouse conditions. Plant Pathol 65:545–550

Waleron M, Waleron K, Kamasa J, Przewodowski W, Lojkowska E (2011) Polymorphism analysis of housekeeping genes for identification and differentiation of *Clavibacter michiganensis* subspecies. Eur J Plant Pathol 131:341–354. https://doi.org/10.1007/s10658-011-9812-4

Wittmann J, Brancato C, Berendzen KW, Dreiseikelmann B (2016) Development of a tomato plant resistant to *Clavibacter michiganensis* using the endolysin gene of bacteriophage CMP1 as a transgene. Plant Pathol 65:496–502

Xiao M, Ren X, Cui J, Li R, Liu Z, Zhu L, Kong Q, Fu X, Mou H (2022) A novel glucofucobiose with potential prebiotic activity prepared from the exopolysaccharides of *Clavibacter michiganensis* M1. Food Chem 377:132001. https://doi.org/10.1016/j.foodchem.2021.132001

Xu X, Rajashekara G, Paul PA, Miller SA (2012) Colonization of tomato seedlings by bioluminescent *Clavibacter michiganensis* subsp. *michiganensis* under different humidity regimes. Phytopathology 102:177–184. https://doi.org/10.1094/PHYTO-03-11-0090

Yasuhara-Bell J, Alvarez AM (2015) Seed-associated subspecies of the genus *Clavibacter* are clearly distinguishable from *Clavibacter michiganensis* subsp. *michiganensis*. Int J Syst Evol Microbiol 65:811–826. https://doi.org/10.1099/ijs.0.000022

Yasuhara-Bell J, Kubota R, Jenkins DM, Alvarez AM (2013) Loop-mediated amplification of the *Clavibacter michiganensis* subsp. *michiganensis* micA gene is highly specific. Phytopathology 103:1220–1226. https://doi.org/10.1094/PHYTO-03-13-0078-R

Yokotani N, Hasegawa Y, Sato M, Hirakawa H, Kouzai Y, Nishizawa Y, Yamamoto E, Naito Y, Isobe S (2021) Transcriptome analysis of *Clavibacter michiganensis* subsp. *michiganensis*-infected tomatoes: a role of salicylic acid in the host response. BMC Plant Biol 21:476. https://doi.org/10.1186/s12870-021-03251-8

Załuga J, Stragier P, Baeyen S, Haegeman A, Van Vaerenbergh J, Maes M, De Vos P (2014) Comparative genome analysis of pathogenic and non-pathogenic *Clavibacter* strains reveals adaptations to their lifestyle. BMC Genomics 15:1–14

Zhao W-J, Chen H-Y, Zhu SF, Xia M-X, Tan T-W (2007) One step detection of *Clavibacter michiganensis* subsp. *michiganensis* in symptomless tomato seeds using a TaqMan probe. J Plant Pathol 89:349–351

Chapter 10
Recovery of Tomato Plant Residues Through Composting: Potential and Challenges to Promote Circular Economy

Ilyass Tabrika, Brahim Bouizgarne, El Hassan Mayad, Mina Zaafrani, and Khalid Azim

Abstract The increase in tomato production especially in modern agriculture generally results in a significant amount of plant residues representing one of the greatest challenges for sustainable agriculture such as economic losses of valuable biomass and pollution. Valorization of tomato waste as a potential source of organic matter and plant nutrients is promising to improve soil fertility. Hence, an effective composting process, which enables the production of good quality compost at an economical cost, can meet the criteria for sustainable tomato waste management. This chapter reports tomato waste as composting feedstock, methods for composting process recommended for tomato waste along with the most relevant challenges in composting process and compost quality.

Keywords Tomato waste · Valorization · Compost · Economy

I. Tabrika (✉)
Research Unit "Integrated Plant Production" Regional Centre of Agricultural Research, Agadir, Morocco

Laboratory of Mechanic, Process Energy and Environment, National School of Applied Sciences, Ibn Zohr University, Agadir, Morocco

B. Bouizgarne
Department of Biology, Ibn Zohr University, Agadir, Morocco

E. H. Mayad
Laboratory of Biotechnology and Valorisation of Natural Resources, Faculty of Sciences of Agadir, Ibn Zohr University, Agadir, Morocco

M. Zaafrani
Laboratory of Mechanic, Process Energy and Environment, National School of Applied Sciences, Ibn Zohr University, Agadir, Morocco

K. Azim
Research Unit "Integrated Plant Production" Regional Centre of Agricultural Research, Agadir, Morocco

N. K. Arora, B. Bouizgarne (eds.), *Microbial Biotechnology for Sustainable Agriculture Volume 2*, Microorganisms for Sustainability 34,
https://doi.org/10.1007/978-981-97-2355-3_10

10.1 Introduction

Tomato (*Solanum lycopersicum* L., family Solanaceae) has become, in the past 50 years, one of the most important grown horticultural crop worldwide. According to United Nations Food and Agriculture Organization (FAO) data of 2019, more than 182 million ton of tomato was produced globally reaching an increase by more than 40 million ton in the decade between 2008 and 2018. Asia is the most important tomato producer, followed by America and Africa, representing 62%, 13%, and 12% of total production respectively. Although this plant is grown in many countries over the world.

Morocco is one of the leading countries for tomato production because of favorable climatic conditions characterized by a temperature average between 19 °C in winter and 27 °C in summer (Malki et al. 2017). According to data of ORMVA, a public establishment placed under the supervision of the Ministry of Agriculture of Morocco, 5141 ha of greenhouse in Souss Massa region is made of tomato crops, produce 830,000 ton of tomato resulting in a yield of 161 ton/ha. This success in tomato yield is due to greenhouse conditions. This protected crop system is characterized by its high efficiency in the use of resources and high productivity per land unit year-round (Payen et al. 2015; Rufí-Salís et al. 2020). However, tomato crops generate an important amount of plant residues during the season due to trimming of the stems and by pulling the entire plant at the end of season after the loss of the economic value of crops. Generation of leaves and stems per cycle has been estimated to be 40–60 ton/ha fresh plant residues characterized by high content of carbon and plant nutrients. By taking into consideration the production areas of tomato greenhouse in the world, an important amount of tomato plant residues is wasted each year implicating a high carbon and nutrient loss in the environment. Within the framework of the circular economy, agriculture is strongly involved in the sustainable management by minimizing the amount of external inputs for agricultural production and reducing negative impacts to the environment by carbon sequestration and preventing loss of nutrient (Azim et al. 2017; Toop et al. 2017). However, sustainability is included as tomato plant residues are reused or recycled. In the scientific literature, many researches investigate tomato plant residues by developing ecofriendly solutions that could help to fix the carbon captured by tomato plants and reduce resources depletion (Azim et al. 2018a; Gavilanes-Terán et al. 2016; Moreno et al. 2021). On global perspective, valorization by composting is one of the best options, thanks to its low cost and technology. Composting is a biological transformation of heterogeneous organic material into biologically stable humus like substances by action of successive microbial populations under controlled conditions of aeration, heating, moisture, and nutritional balance in managed composting pile (Lerch et al. 2019). After composting process, compost as biological stable organic amendment is ready to use for soil improvement and carbon sequestration (Guerrini et al. 2017; Bernal et al. 2009; Morales et al. 2016). When compost is applied to soil, it initiates a positive cascade of changes in soil physical, chemical, and biological properties and nutrients become gradually available for plants (Sánchez et al. 2017).

Composting efficiency depends on feedstock quality and nutrient balance for microbial activity especially C/N ratio (Bernal et al. 1996). In this context, several studies have shown that some chemical characteristics of tomato plant residues are not adequate for composting as unique feedstock and could limit the efficiency of composting process. These include recalcitrant nature of carbon, potentially leading to low availability of carbon; nitrogen is available in tomato plant residues, resulting in low C/N ratio than the optimum range for composting (20–30) and high moisture content, which largely influences porosity and consequentially affects oxygen availability, and thus has a central influence on the biodegradation of organic matter (Azim et al. 2018a, b; Gavilanes-Terán et al. 2016; Moreno et al. 2021). The objective of this chapter is to generate a comprehensive diagnosis on tomato plant residues as composting feedstock and how chemical variability and microbial content of feedstock affect composting efficiency.

10.2 The History and Current State of the Composting in Morocco

Composting research has a long history and can be traced back since 1920. The first recorded document dealing with composting was found to be in the late 1920s in the Scopus database with the Wiltshire work entitled "Cooperative Experiments for the Composting of Phosphate Rock and Sulfur," published in October 1920 in fourth issue of Soil Science Journal. At the end of the twentieth century, a consistent number of publications have increased and reached an average of 40–100 publications per year between 1990 and 2000. This increase may be due to the rise of the sustainable development movement and efforts to minimize pollution. Since then, the number of articles has regularly increased over the years and reached between 100 and 276 publications in the period 2000–2010. But since 2010, the number of publications related to composting saw a strong increase and has recorded 568 publications worldwide in 2021 only. In Morocco, composting research began in the 2003 and the first publication in Scopus database is "Characterization of humic acids extracted from sewage sludge during composting and of their Sephadex® gel fractions" published by Amir et al. (2003) followed by "Characterization and identification of polyphenols during waste treatment of olive oil mills by composting" published by Ait Baddi et al. (2003b). The research has primarily focused on composting of sewage sludge (Amir et al. 2003), olive mill waste (Ait Baddi et al. 2003a; Barje et al. 2008), date-palm wastes (Chakroune et al. 2005; Hakkou et al. 2011; El Fels et al. 2014), and municipal wastes (Lhadi et al. 2006; Barje et al. 2012; Makanet al. 2013). Furthermore, composting processes and monitoring were assessed including microbial dynamics characterization (El Fels et al. 2014; Biyada et al. 2020), humification (Droussi et al. 2009; Barje et al. 2012; El Ouaqoudi et al. 2014; Aguelmous et al. 2020), and compost phytotoxicity (Kabil et al. 2016 Azim et al. 2018a). Composting as a component recycling of nutrients and carbon sequestration in agriculture is reported; studies have focused on horticultural waste as composting feedstock and evaluation of their composting efficiency and compost quality (Azim et al. 2017, 2018a; Tabrika et al. 2019, 2021).

10.3 Tomato Plant Residues as a Valuable Source of Organic Matter and Plant Nutrient

Researchers reported that every year, millions of ton of tomatoes are cultivated worldwide (Table 10.1), leading to a large quantities of tomato waste (Azim et al. 2017; Encinar et al. 2008; Gavilanes-Terán et al. 2016; Tabrika et al. 2019). The definition of tomato waste has rarely been debated. However, for this review, it is defined as any plant residues generated from pruning and maintenance operation during the growing cycle such as leaves axillary buds and plant residues coming from the end of growing cycle after pulling the main stem, branches, and leaves. Generation of tomato plant residues is estimated to be 49 ton/ha of fresh weight (Gavilanes-Terán et al. 2016) and assuming a dry waste of 9 ton/ha (Llorach-Massana et al. 2017). Tomato waste is characterized by high moisture and ash content (Table 10.2).

Tomato plant residues (TPR) had moisture content as high as 75% at the harvesting. Therefore, water could be identified as the main component of this tomato waste. Apart from water, organic matter of tomato plant showed more than 78% of dry weight, of which a significant number of lignocellulosic fibers with cellulose, hemicelluloses, and lignin represent 33%, 11% and 8%, respectively. However, the recalcitrant nature of lignocellulose hinders the utilization of cellulose and hemicellulose, decreasing the bioconversion efficiency of lignocellulose (Wu et al. 2022). The organic matter content of tomato plant residues is 78% of total dry tomato plant weight (Tabrika et al. 2019). Consequently, tomato waste generated per hectare would contain approximately 7 ton of organic matter (Nisticò et al. 2017).

Table 10.1 Physicochemical composition of tomato plant residues

	C/N ratio	TOC	TN	K	P	
Plant residues	(%)					References
Tomato waste	13.2	31	2.35	2.94	0.62	Gavilanes-Terán et al. (2016)
	8.89	33	3.71	ND	0.897	Sülük et al. (2017)
	10.0	31	3.1	5.76	1.05	Alkoaik and Ghaly (2006)

TOC total organic carbon, *TN* total nitrogen; *K* potassium, *P* phosphorus

Table 10.2 Tomato waste generation in some countries in comparison with Morocco

Country	Tomato waste (ton)	References
Florida	396,000	Amose and Jeyakumar (2018)
Spain	470,000	Moreno et al. (2021)
Turkey	90,000	Ulusoy et al. (2009)
Morocco	290,000	Azim et al. (2017)
Europe	3,000,000	Løvdal et al. (2019)

10.4 Composting as Sustainable Recycling Strategy for Tomato Plant Residues

TPR representing a source of organic matter, macro and micronutrient (Tabrika et al. 2019) and their application as composting feedstock may represent a sustainable recycling strategy that affects plant nutrients and global carbon-nitrogen cycling in soil–plant systems. Composting is a biological transformation of heterogeneous organic material into biologically stable humus like substances by action of successive microbial populations under controlled conditions of aeration, heating, moisture, and nutritional balance in a managed composting pile (Lerch et al. 2019) (Fig. 10.1). Controlled conditions and feedstock formulation make difference between composting and natural decomposition of organic residues. After the composting process, compost as biological stable organic amendment is ready to be used as fertilizer and soil conditioner (Bernal et al. 2009; Morales et al. 2016). In composting process, microbial population is generally indigenous in feedstock, yet certain microbial consortium could be introduced into feedstock or enriched by the use of recycled compost at the starting of the process (Choińska-Pulit et al. 2019). Composting feedstock serves as biofilm support and sources of nutrients, where carbon and nitrogen are the most crucial microbial nutrients. Composting process involves essentially three different phases: (a) a mesophilic or moderate-temperature phase, followed by a (b) thermophilic or high temperature phase, and finally a (c) psychrophilic or maturation phase (Mehta et al. 2014). Few days after formation of the pile, thermophilic microorganisms become more active and thermophilic phase occurs when temperatures reach the values of 45–70 °C (Fig. 10.2). During this phase, microbial communities contribute to exothermal degradation of biodegradable organic compounds and consequently generating a large thermal energy (Choińska-Pulit et al. 2019; Smith et al. 2017).

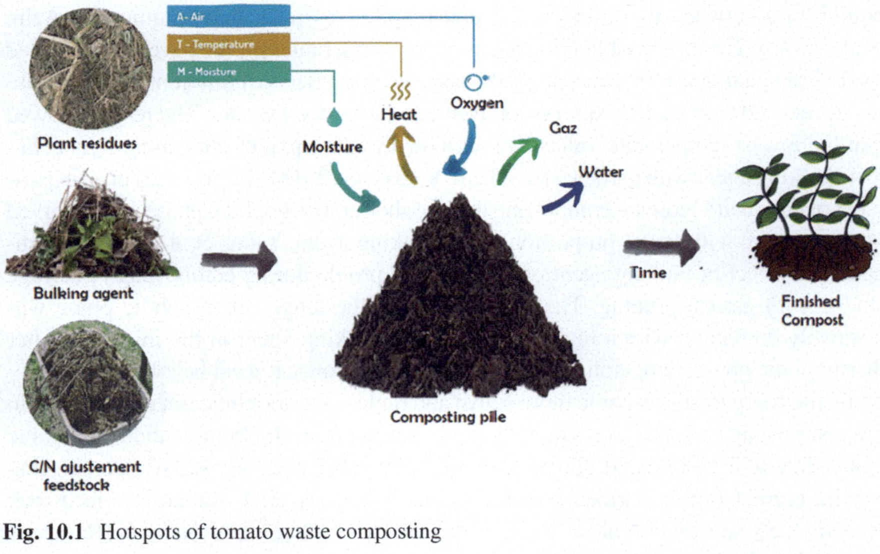

Fig. 10.1 Hotspots of tomato waste composting

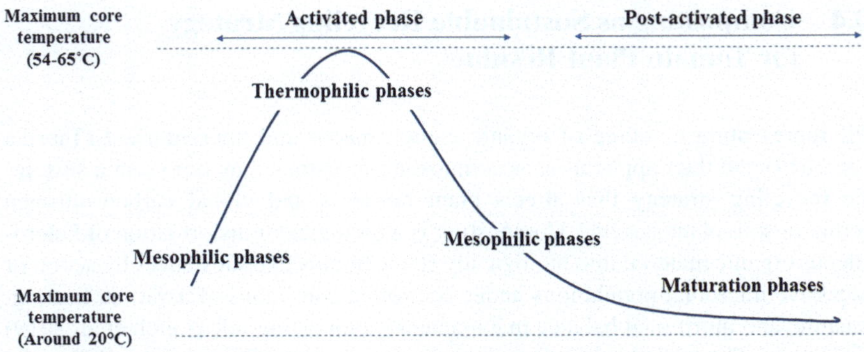

Fig. 10.2 Different stages during composting as function of microbial activity and temperature

During thermophilic phase, simple carbohydrates, lipid molecules, proteins, and some organic compounds are catabolized by microbial population to support active biomass growth. Consequently, carbon is exchanged during composting through global carbon cycle (Tuomela et al. 2000). According to Luo et al. (2019), more than 79% of protein and carbohydrate were transformed by heteroorganotrophic bacteria in the thermophilic phase. Wang et al. (2018) studied the succession of organics metabolic functions of bacterial community during composting process. The research found that fungal communities rapidly degraded carbohydrate polymers at the thermophilic phase and stronger ability to degrade amines, amino acids, and carbohydrates in the maturing phase.

Regarding temperature range of composting phases, there is no standardized values, but it is generally assumed that the top value of mesophilic phase is 40 °C and for the thermophilic phase from 45 to 70 °C (Choińska-Pulit et al. 2019; Sundberg and Jönsson 2008). Time and temperature of thermophilic phase depend on experimental variants of composting, especially feedstock, aeration rate, and presence of bulking agent. The bulking agent is important to ensure oxygen availability in the mixture by maintaining equilibrium between its diffusion and consumption by microbial community (Azim et al. 2018b). The effects of bulking agent on the temperature profile have been reported by different authors. Morales et al. (2016) assessed the effects of different types of bulking agents and their particle size on composting of agro-food wastes. The results showed that the lowest temperature values are attributed to the mixture containing lignocellulosic bulking agent with particle size <3 cm. Vico et al. (2018) also reported in an experiment of date-palm biomass composting that the shorter thermophilic phase was observed in the mixture with lower proportion of the bulking agent. Yañez et al. (2009) investigated the effect of bulking agent on temperature profile during composting of sewage sludge with garden pruning. They suggested that the longer thermophilic phase was positively correlated with a higher proportion of bulking agent in the mixtures. After thermophilic phase, temperatures start to decrease and once it went below 40 °C, mesophilic microorganisms became more active and replace thermophilic microorganisms in the composting process. Consequently, they contribute to the humification of organic matter that will be released almost as humic-like substances. Nowadays, composting can be carried out in various procedures depending on land availability, feedstock resources, economic conditions, and time required to reach compost maturity (Rasapoor et al. 2016) (Table 10.3).

Table 10.3 Main characteristics of plant residues composting and final product

	Composting mixture					Composting process characteristics				Final compost			References
Composting process	Composting feedstock	Mixture mass	Starting C/N	Moisture (%)	OM (%)	Aeration	T° max (°C)	T >50 °C	Composting time (days)	C/N	Germination index (%)	OM (%)	Gavilanes-Terán et al. (2016)
Windrows	Broccoli waste + laying hen manure + sawdust	1000 kg	29.8	>40	69.5	Mechanical turning	>55	49–56	69	18.1	86.6	24.1	Gavilanes-Terán et al. (2016)
Windrows	Tomato waste + waste + laying hen manure + sawdust	1000 kg	28.1	>40	64.5	Mechanical turning	>56	28–35	69	18.2	70.3	31.9	Azim et al. (2018a, b)
In vessel	Tomatoes waste + melons waste + sheep manure + olive mill pomace	200–250 L	25–40	50–65	nd	Passive aeration + manual turning	nd	nd	169	12.41–19.1	70–95	nd	Tabrika et al. (2019)
In vessel	Tomato waste + sheep manure	200–250 L	16	50–60	72–77%	Passive aeration + manual turning	>57	7–22	60	16–17	67–83	37.27–50.01	Wei et al. (2014)
In vessel	Tomato stalk + chicken manure + biochar + peat bog	nd	25–30	60	nd	Forced air ventilator	56	Up to 7 days		>13	>80	nd	Şevik et al. (2016)
In vessel	Olive pomace, sewage sludge, and tomato stalks + dairy manure	100 L	20–41.28	48.5–74.7	83.5–94.0	Forced air ventilator	60.3	3.24 for T° >50 °C	32	17–34.56	nd	80.1–94.1	Matei et al. (2016)
In vessel	Vine waste + hen manure	1200 kg	14–20	>50		Mechanical turning		Up to 7 days	60	12–22	>100	nd	Matei et al. (2016)

OM organic matter, *T* temperature

10.5 Composting Methods

Two types of composting systems exist, windrow and in-vessel systems. In-vessel composting requires less space and little turning. Thus, windrow composting is the more appropriate for large volumes of waste.

10.5.1 Windrow Systems

Windrow composting system is an open-air pile in which composting is carried out into long rows usually in trapezoidal form with maximum height and width of 1–1.6 and 2–4 m, respectively. The row size is large enough to maintain temperatures and small enough to allow oxygen flow into the windrow's core (Arriaga et al. 2017; Zhang and Sun 2016). Windrow composting can be classified into two categories (Fig. 10.3): Dynamic windrows (with turning), in which aeration can be achieved by mechanical turning of windrow piles, and static windrow (without turning), in which air is supplied to the composting feedstock through perforated piping that delivers oxygen to windrow. Before setting up a windrow composting, it is important to consider how the space will be utilized, soils permeability accessibility for

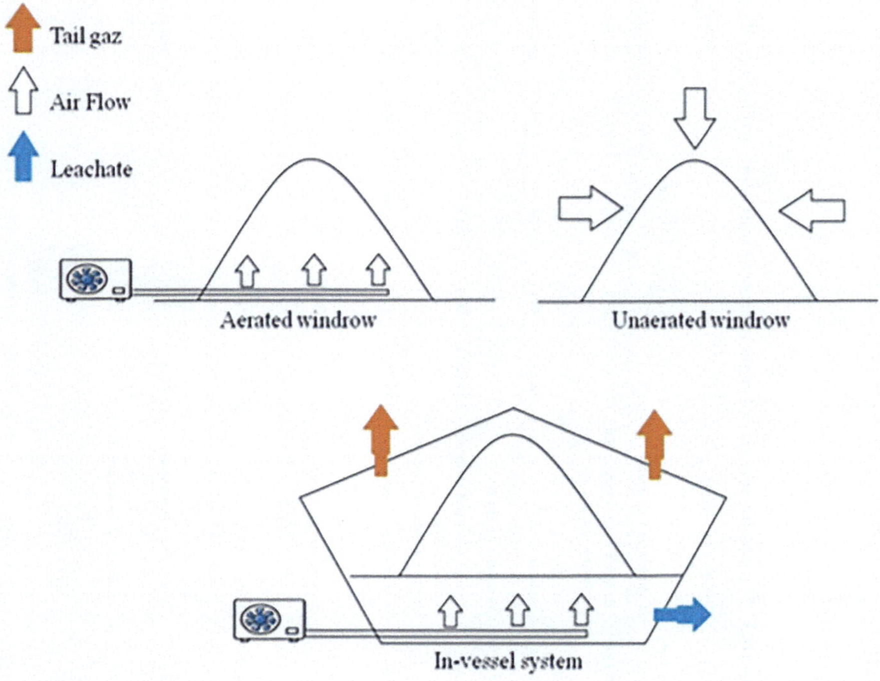

Fig. 10.3 Typical composting systems

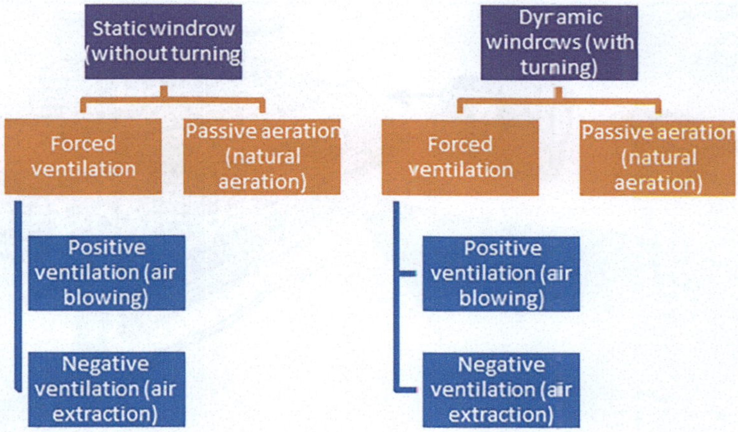

Fig. 10.4 Different windrow composting systems

Table 10.4 Approximate number of microorganisms during different phases of composting (Miller 1994)

Organism	Number
Bacteria in mesophilic stage	10^9–10^{13} g^{-1} substrate
Bacteria in thermophilic stage	10^8–10^{12} g^{-1} substrate
Actinomycetes, thermophilic stage	10^7–10^9 g^{-1} substrate
Actinomycetes, mesophilic stage	10^8–10^{12} g^{-1} substrate
Fungi, average value	10^5–10^8 g^{-1} substrate

turning operations, and water resources availability (Arriaga et al. 2017). Windrow composting is suited for large volumes of waste such as agro-industrial wastes (Abu Qdais and Al-Widyan 2016), horticultural wastes (Gavilanes-Terán et al. 2016), municipal solid wastes generated by entire communities and food-processing plants (De Silva and Yatawara 2017) (Fig. 10.4). According to Donaldson et al. (2013), windrow system is appropriate to generate thermophilic conditions without insulation facilities for pathogen destruction. Cifuentes et al. (2013) showed in their study of composting waste sugar cane and press mud mixtures that the totality of *Escherichia coli* and *Salmonella* was reduced in the first week and ascarids were found non-viable in maturity phase. In terms of compost quality, compost obtained by windrows systems showed low phytotoxicity and high level of humic substances (Cifuentes et al. 2013) (Table 10.4).

10.5.2 In-Vessel Composting

In-vessel composting system is relatively more complicated than windrow as it requires a bioreactor with permanent control and high technicity and proficiency (Fig. 10.5). The raw composting material is introduced into the reactor through an inlet at one end, while the compost produced is recovered from the other end by a

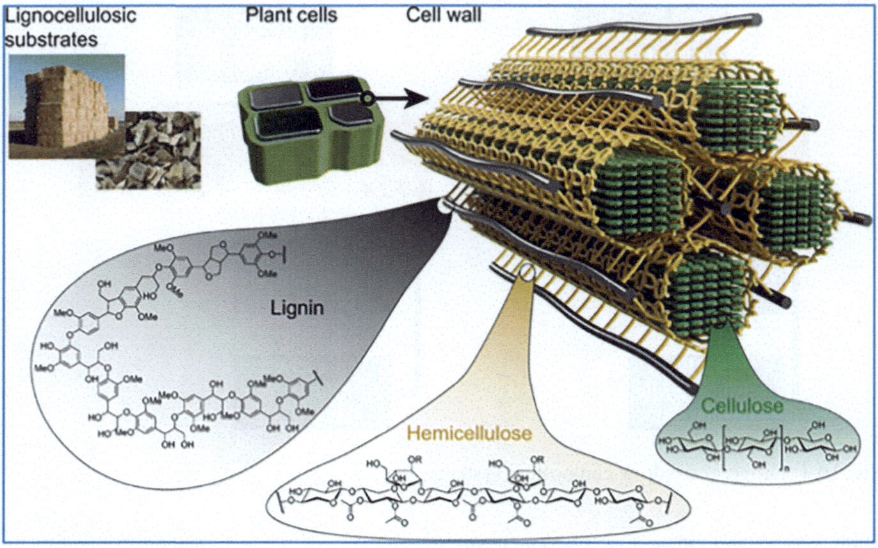

Fig. 10.5 Components and structure of lignocellulosic biomass. Lignocellulosic plant cell walls are mainly composed of cellulose, hemicellulose, and lignins (Brethauer et al. 2020)

movable bottom of the hydraulic cylinder. Although more expensive, the in-vessel composting process is considerably shorter than other composting systems for which the retention time takes approximately 14–28 days (Amlinger et al. 2008) and while compost stability may be reached after 30–45 days according to Waqas et al. (2018).

In addition, the in-vessel composter is not demanding in terms of work area and is more easily controlled in terms of environmental conditions such as temperature, air flow, oxygen concentration and a has a lower environmental impact resulting from a lower gas emission, which makes it suitable for raw materials that cannot be treated in the open area. Because of these advantages, there is a global interest to use in-vessel composting in last few years. According to Areikin et al. (2012), the percentage of waste treated by in-vessel composting in UK increased 17% between 2008 and 2010.

10.6 Decomposition of Organic Matter During Composting

Because of their high content of recalcitrant carbon such as cellulose, hemicellulose, and lignin, horticultural plant residues play a crucial role in the global carbon nutrition cycle (Wu et al. 2022). Plant residues when applied without composting to the soil may lead to temporary lock up of nutrients as a result of impaired C:N ratio and may not be beneficial to crop. Hence, composting of organic wastes is necessary to reduce the lignin and cellulose contents, thereby the nutrient availability is improved.

The composition of lignocellulosic biomass varies with the biomass source (i.e., hardwoods, softwoods, agricultural residues, and energy crops) and is affected by origin, age, climatic conditions, harvesting, and storage processes. The lignocellulosic biomass is not easily degradable even by the microorganisms. This is mainly because the cellulose and hemicellulose in lignocellulose are protected by lignin, which forms a complex structure. Lignin is a three-dimensional complex aromatic heteropolymer, while cellulose and hemicellulose are composed of different types of polysaccharides (Aswathy Jayakumar et al. 2023). Hence lignin, which is itself difficult to get hydrolyzed by enzymes because of its difficult structure, also hampers the decomposition of cellulose and hemicellulose. Therefore, pretreatment of lignocellulosic biomass is an essential way to destroy the structure of refractory lignocellulose and then depolymerize lignin (Bhatia et al. 2020).

After formation of composting pile, microbes start to use soluble organic carbon as energy sources. Further, microbes hydrolyze lignocellulose by releasing lignocellulase enzyme. The result of lignocellulose degradation are amino acids, polyphenols, carboxylic acids, and polysaccharides. These are important precursors of humic substances. The decomposition of lignocellulosic wastes occurs in the thermophilic phase and continues until all the available material is exhausted. The end products of lignocellulose degradation polymerize to form humic substances during the maturation phase when the temperature comes down (Guo et al. 2019).

The mineralization of organic matter during composting is an important parameter for measuring composting efficiency and to design optimization strategies for composting process management. Mineralization of organic matter has been calculated from the initial and final ash content using Eq. (10.1) of Viel et al. (1987) or using the profiles of OM mass of volatile solids consumption against time as shown in Eq. (10.2) of Huiliñir and Villegas (2014)

$$\mathrm{OM}_{\mathrm{consumed},t}\left(\%\right)=100-\frac{100-\left(\mathrm{Ash}_i\left(100-\mathrm{Ash}_t\right)\right)}{\mathrm{Ash}_t\left(100-\mathrm{Ash}_i\right)} \tag{10.1}$$

$$\mathrm{OM}_{\mathrm{consumed},t}\left(\%\right)=\frac{\mathrm{OM}_0-\mathrm{OM}_t}{\mathrm{OM}_0} \tag{10.2}$$

where OM is the mass of volatile solids (kg), Ash is total solid content, and t and i represent the initial at any time of composting process. Different kinetic models were reported in scientific literature to fit loss of OM as a function of time but the first-order kinetic model is generally used and expressed by Eq. (10.3): (Haug 1996; Huiliñir and Villegas 2014; Kulcu and Yaldiz 2004):

$$\frac{d\left(\mathrm{OM}\right)}{dt}=-kt.\mathrm{OM} \tag{10.3}$$

where OM is the mass of biodegradable volatile solids (kg) as a function of time and k_t is the decomposition rate constant of reaction in d^{-1}. In order to calculate kt, many

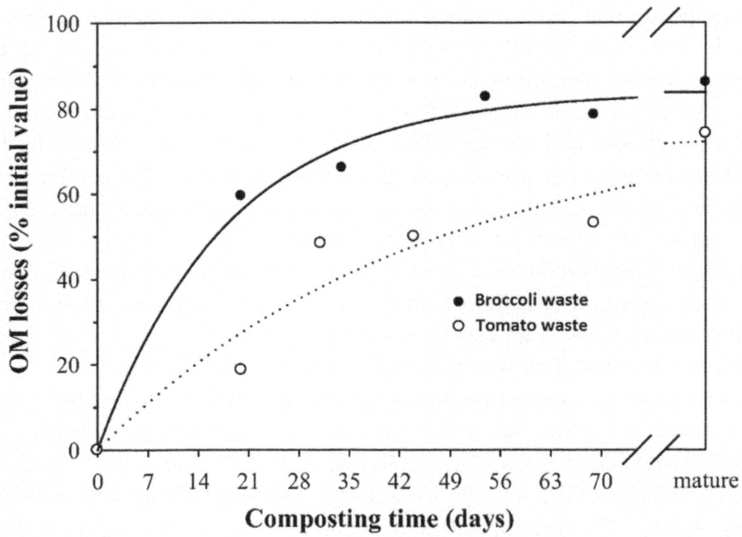

Fig. 10.6 OM loss of tomato and broccoli waste during composting (Gavilanes-Terán et al. 2016)

equations were applied using many experimental data such as moisture content, temperature, electrical conductivity, O_2 and CO_2 concentration in composting reactor.

The data corresponding to OM loss (Eq. 10.4) was fitted to a first-order kinetic model by the Marquardt–Levenberg algorithm, where A_0 is the maximum loss (%) of OM

$$(\text{OM})\text{Loss} = A_0\left(1 - \mathrm{e}^{-kt}\right) \tag{10.4}$$

Generally, OM mineralization is substantial during thermophilic phase that corresponds to the maximum rate of biological reactions since the lowest value of OM mineralization was recorded after dropping of temperature or during the maturation phase (Paredes et al. 2002; Bustamante et al. 2008; Gavilanes-Terán et al. 2016). Gavilanes-Terán et al. (2016) studied OM mineralization during composting of tomatoes and broccoli waste in windrow system. This research found that lignocellulosic content of tomato wastes substantially reduced mineralization of OM compared to broccoli waste (Fig. 10.6).

10.7 Microorganisms as Composting Drivers

Previous studies indicated that the abundance and diversity of bacterial and fungal communities depend on the initial composting materials and the composting phases (Hernández-Lara et al. 2022). Temperature is the primary determinant that shapes the microbial community structure and metabolic activity levels during the oxidative stage of the composting process. Microbial population increase in the mesophilic phase decreases rapidly in the thermophilic phase and stabilizes in the maturation phase (Joo et al. 2007; Pantet al. 2012). Wefers et al. (2017) reported that fungi are the most effective microorganisms for degradation of carbohydrate polymers in lignocellulosic feedstock by action of various ligninolytic extracellular oxidative enzymes. Fungal species like *Aspergillus niger*, *Trichoderma viride,* and several white-rot fungi, such as *Phanerochaete chrysosporium, Coriolus versicolor, Trametes hirsuta, Pleurotus ostreatus, Ceriporiopsis subvermispora, and Paecilomyces* spp. have attracted scientists attention due to their accelerated degradation of recalcitrant lignocellulosic feedstock (Gong et al. 2017; Heidarzadeh et al. 2019; Voběrková et al. 2017; Yang et al. 2019).

Lignocellulose, as the principal form of carbon in horticultural waste, influences the composting process, contributes to poor biodegradable capability, and reduces nutrient content in compost products (Xu et al. 2019). Also, low population of indigenous community in horticultural biomass affects decomposition efficiency and may require a long time to obtain mature compost. These problems are attempted to be solved by improving composting efficiency through addition of livestock manure and optimization of ratio between composting feedstock to providing a favorable environment for microbial community (Tabrika et al. 2020).

Bioaugmentation through introducing microbial strains or consortia has been shown to be effective for degrading complex carbohydrates and promoting humification during composting. This approach aims to optimize the composting process and the production of compost rich in humic substances. Bioaugmentation is a conventional method for improvement of the degradative capacity of composting process by introduction of specific competent strains or consortia of microorganisms to regulate and modify the composting process (Wei et al. 2014). Bioaugmentation was reported to be able to improve degradation of lignocellulose (Awasthi et al. 2020), accelerate the heating phase (Wang et al. 2019b), phosphorus solubilization (Billah and Bano 2015), reduction of nitrogen loss (Koyama et al. 2022), and detoxification of organic compounds (Bouhia et al. 2021).

Many researches revealed the influence of the microbial consortia inoculation on the decomposition of lignocellulosic materials during composting of plant residues. According to research (Du et al. 2021), white-rot fungi such as *C. versicolor* and *Phanerochaete chrysosporium* are the most efficient ligninolytic organisms described to date.

Gong et al. (2017) also investigated the effect of inoculation with fungi *Trametes versicolor* or *P. chrysosporium* in different stages of green waste composting. The authors found that the anthropogenic microbial inoculation highly increased temperatures in the pile's core meeting the sanitation criteria and increased the quality of the final composts. Kausar et al. (2011) examined the possibility of using lignocellulolytic actinobacteria isolated from rice straw as a potential agent for rapid composting. Results indicated that *Micromonospora carbonacea* was able to degrade cellulose and hemicelluloses after 3 weeks of in vitro incubation of compost of rice straw by producing ligninolytic enzymes at the beginning of the process. Also, inoculation by bacterial consortia accelerates biodegradation of organic matter such as amylolytic and cellulolytic bacteria during composting. Tran et al. (2015) assessed the effects of homofermentative lactic acid bacteria *Pediococcus acidilactici* on composting processes. Results showed that the inoculation enhanced the proliferation of fungi by their ability to produce available carbon. In addition, inoculation by *Bacillus* and *Geobacillus* species showed contrasting effects on degradation of lignocellulosic compound during composting process (Abdel-Rahman et al. 2016; Tran et al. 2015). There are many different factors that affect inoculation efficiency such as type of applied microorganisms, inoculation moment, properties of raw materials, environmental conditions as well as the influence of the competition between inoculated and indigenous microbes (Huang et al. 2015; Wang and Zeng 2018; Xi et al. 2015). Therefore, to understand the efficiency of composting, Xi et al. (2015) investigated the impact of multi-stage inoculation by bacteria on deodorizing, cellulose and lignin decomposition and microorganisms (bacterial and fungal communities) during composting. The research found that time of inoculation affects the temperature profile and temperature was extended by the multi-stage inoculation treatment. Similarly, Huang et al. (2015) investigated the effect of inoculation by white-rot fungus, *P. chrysosporium* at different phases on OM humification during composting of agricultural waste. They found that when the white-rot fungus inoculation was carried after thermophilic stage, an increase in aromatization and polymerization of humic acid was observed. Conversely, there was no change on the humification degree of humic acid when the inoculation was performed before thermophilic stage. Regarding the effect of feedstock properties on inoculation efficiency, Yadav et al. (1982) investigated the effect of fungi inoculation on decomposition of wheat straw and vegetable waste composting. The research found that the inoculation with fungi enhanced the decomposition of wheat straw and vegetable waste, while there was no effect on the decomposition of the mixed materials (Table 10.5).

Inoculation viability and competition between inoculated and indigenous microbes is important to promote bioaugmentation efficiency. Therefore, the selection of inoculants is important in promoting the degradation of lignocellulose and thus overall composting efficiency. Also, microbial interaction between inoculum and indigenous microbe plays a crucial role that affects bioaugmentation efficiency. Interactions between inoculum and indigenous microbe among composting must be synergistic rather than antagonistic. Inoculum efficiency is related to the habitat in which they are isolated and phase of inoculation.

Table 10.5 Impact of microbial inoculum on composting process and compost quality

Composting feedstock	Bioaugmentation inoculum	Effect on composting	References
Citrus peel	Compost bacteria	The addition of microbial inoculum to the CP could increase the temperature of the pile, accelerate the degradation of pectin and cellulose	Wang et al. (2019a)
Municipal solid waste	*P. chrysosporium T. versicolor Fomes fomentarius*	*T. versicolor* and *Fomes fomentarius* led to a higher degrading ratio and a better degree of maturity	Voběrková et al. (2017)
Olive mill solid sludge	Thermophilic inoculum specific obtained naturally from a windrow of composting	Microbial inoculum could promote detoxification of organic compounds such as polyphenols and total lipids	Bouhia et al. (2021)
Agricultural wastes	*P. chrysosporium*	Inoculation accelerates the humification and improves the compost maturity	Taccari et al. (2009)
		Efficiency of inoculation affected by phases of composting	
Maize stover	*Luteimonas* sp. *Aspergillus flavus*	Inoculation promotes nutrient transformation and shortens compost maturity time	
Organic sludge	*Neurospora* sp.	Organic matter decomposition was significantly higher in the sludge pretreated with *Neurospora* sp. Inoculation improves ammonia conversion rate	Koyama et al. (2022)
Chicken manure	*Bacillus subtilis, Bacillus licheniformis, Trichoderma koningii, Saccharomyces cerevisiae, and P. chrysosporium*	Inoculation accelerates organic aromatization and polymerization for humus formation Inoculation reduces organic fermentation and nitrate respiration, resulting in a decrease in GHGs emission	Gao et al. (2021)

10.8 Challenges in Composting of Tomato Waste

Tomato plant residues have features in common such as high recalcitrant carbon with low density and thus lead to low bioavailable carbon for microbial activity for biorefinery (Doublet et al. 2011). Also, typically inadequate C/N ratios, acidic pH, high salinity, less solid content (10–30%), high moisture content are also common (Faust et al. 2018; Fernández-Gómez et al. 2011).

10.8.1 Moisture Content

Moisture and air-filled porosity are two important initial conditions. They are involved in both physical and biological aspects of the composting process. Moisture affects many aspects of the composting process and its management requires a balance between tree factors: oxygen supply, microbial growth rate, and biodegradation kinetics (Cáceres et al. 2018). The moisture content of the compost and raw material is determined gravimetrically by drying samples at 105 °C in hot air oven.

The moisture content below 25% can have adverse effects on the microbial activity and higher than 70% may create challenges for aeration and can lead to methanogenesis (Kim et al. 2018). According to Ekinci et al. (2019), moisture content up to 67.67% adversely affects oxygen movement within compost in the composting pile. The recommendable moisture content during composting of organic wastes is usually 65–75% (Guo et al. 2012), while some other materials can be effectively composted outside this range (Han et al. 2014). The moisture content of plant residues as collected typically ranges between 70% and 90% wet basis (Chen et al. 2019b). Therefore, the moisture content must be carefully managed to allow oxygenation of the composting pile. In the literature, different strategies were investigated to maintain composting efficiency with higher range of moisture content (70–90%). In fact, Jain and Kalamdhad (2018) added sawdust as a bulking agent to adjust the moisture content and rotation was used for uniformity and homogeneity during the composting. The authors found maximum recorded loss in moisture content was 15%. Recently considerations have been raised for the addition of biochar as a bulking agent Jain et al. (2019). The authors examined the effects of biochar on moisture content dynamics during composting of lignocellulose waste. The study showed that the addition of biochar can result in a 17% reduction in moisture content by its adsorption.

10.8.2 Porosity

Porosity can also be influenced by moisture content management. In terms of porosity, two parameters should be distinguished: total porosity and air-filled porosity (AFP). Total porosity of composting feedstock is defined as the ratio of pores volume to the total volume of sample. Some of those pores will be filled with water and the rest with air. AFP, also referred as free air space (FAS), is defined as the ratio of air volume to total volume of the sample (Han et al. 2014; Ruggieri et al. 2009). The optimal starting levels of AFP should be in the range of 45–60% to satisfy an optimum oxygen concentration, to remove carbon dioxide and limit excessive heat accumulation (Haug 1996). The minimum recommended value for the initial PFA in a composting process is 30%, while values higher than 60–70% seem to cause over-aeration and therefore restrictive to reaching thermophilic temperatures in the compost pile (Ruggieri et al. 2009). Although bulking agents prevent excessive compaction of the composting substrate, physicochemical properties of raw materials are influenced by their characteristics. Research showed that large amount of organic matter in bulking agent substrates may significantly influence C/N ratio of composting matrix, microbial composition, and metabolism characteristics (Dias et al. 2010; Wang and Zeng 2018). Therefore, it is essential to eliminate the adverse effect of organic bulking agents for best composting process monitoring. Wu et al. (2017) studied the effects of rice straw and maize bulking agents on composting process in terms of organic matter decomposition, mobility of heavy metal, and urease activity. The research found that using maize straw has greatly enhanced organic matter but has little effects on the mobility of heavy metal and that urease activity was not affected during composting. In recent years, there is an increasing interest on using biochar as bulking agent because of its ability to change the physicochemical properties during composting process, microbial activities, mineralization of organic matter, and humification process (Dias et al. 2010). The authors evaluated the use of biochar of *Eucalyptus* biomass as bulking agent. They found that biochar underwent high organic matter degradation and the final compost was characterized by a high polymerization degree of the humic-like substances, with a relative high proportion of humic acids compared to fulvic acids (Kumar Awasthi et al. 2017). Researchers have reported a significant increase in alkali extractable carbon, humic-like substances, and fulvic acids from the composting piles amended with biochar, and consequently an increase in the percentage of humic substances in final compost (Jindo et al. 2016; Zhang et al. 2014). Biochar is obtained by pyrolysis of biomass, and an important advantage of biochar is that it is also a bulking agent in addition to its role of carbon source for the microbial community (Adhikari et al. 2010). Its high cation exchange capacity plays role in retention of nutrients and enhances environmental conditions for microbial growth (Sanchez-Monedero et al. 2018). In general, organic bulking agents have been widely used in intensive composting but their availability is starting to decrease. In this context, the challenge is to find a suitable alternative (Table 10.6), thereby the organic degradation and the compost maturity could be improved while the production cost would be reduced.

Table 10.6 Physicochemical composting of some bulking agents

Bulking agent	C/N	Moisture (%)	TOC	OM	TN	References
Sawdust	285.7	36.4	51.4	98.8	0.18	Gavilanes-Terán et al. (2016)
	86.88	10.2	45.18	94.15	0.52	Awasthi et al. (2015)
	89.8	ND	52.9	95.22	0.59	Zhou et al. (2018)
	384	12.64	46.09	ND	0.12	Jurado et al. (2014)
Strawberry substrate	35.5	ND	ND	88.2	1.3	Viaene et al. (2017)
Tomato substrate	27.4	ND	ND	78	1.59	Viaene et al. (2017)
Maize straw	97.3	30	46.7	99.2	ND	Wu et al. (2017)
Rice straw	58.7	8.8	42.3	97.1	ND	Wu et al. (2017)
	50.7	8.92	45.1	86.1	0.89	Chen et al. (2019a)
Biochar	82.6	7.1	49.8	94.6	0.61	Dias et al. (2010)
Eucalyptus biochar						
Olive husks	152	ND	45.6	65	0.3	Abid and Sayadi (2006)
Wood shavings	>240	8.0	48.0	ND	<0.02	Alkoaik and Ghaly (2006)

OM organic matter, *TOC* total organic matter, *TN* total nitrogen

10.8.3 C/N Ratio

The C/N ratio is among the parameters used for feedstock characterization and the most important factors influencing compost quality because carbon and nitrogen are the two most important components of the composting material. In composting, microbial population uses carbon as the basic source of energy, while nitrogen is essential for their protein synthesis. The C/N ratio of starting mixture can be calculated according to Md Khudzari et al. (2016) by substituting physicochemical characteristics of each raw material according to Eq. 10.5:

$$\mathrm{C:N\,Ratio} = \frac{\sum_{i=1}^{n} Q_i \times \left(C_i \times \left(100 - M_i\right)\right)}{\sum_{i=1}^{n} Q_i \times \left(N_i \times \left(100 - M_i\right)\right)} \quad (10.5)$$

where Q_i is the mass of material *i*, C_i is the total organic carbon (% in dry basis) of material *i*, N_i is the nitrogen (% in dry basis) of material *i*, and M_i is the moisture content (% in wet basis) of material *i*. For compost sample, C/N ratio can be calculated after determination of organic matter content and total nitrogen of the sample. Then C/N ratio of the sample can be calculated using Eq. 10.6: (Jolanun and Towprayoon 2010)

$$\mathrm{C:N\,Ratio} = \frac{0.58 \times \mathrm{MO}}{\mathrm{TN}} \quad (10.6)$$

In general, initial C/N ratio of 25–30 is considered ideal for composting (Kumar et al. 2010). However, recently some researchers have successfully carried out composting at different initial C/N ratios (Table 10.6). Ekinci et al. (2019) assessed the effects of initial C/N ratio on organic matter degradation during composting of rose oil processing solid wastes, which included carbon loss (C-loss) and nitrogen loss (N-loss). The results showed that the initial C/N ratio of the mixture could be greater than 30 in order to reduce nitrogen loss, while the initial C/N ratio up to 21 was suggested to reduce carbon loss. In the same context, Azim et al. (2018a, b) have successfully co-composted tomato waste at C/N ratio values between 25 and 40 and found that all initial C/N ratios led to high carbon loss (up to 40%), whereas compost with the initial C/N ratio of 35 showed the highest humification rate with the lowest phytotoxic effects. Tabrika et al. (2019) investigated the composting of tomato waste at low C/N ratio (around 17) using sheep manure as co-composting feedstock and found that sheep manure improved composting of tomatoes waste at low C/N ratio.

10.8.4 Persistence of Plant Pathogens and Contaminant

It has been reported that chemical fertilizers and pesticides contain trace amount of heavy metals as impurities and their application either by addition to the soil or as a foliar spray may be a potential risk to plants by their bioaccumulation (Wuana and Okieimen 2011). Gavilanes-Terán et al. (2016) identified the heavy metals content in horticultural waste (tomato and broccoli waste) and they found that tomato waste contains around 251, 74, 56, 62, 0.23 0.11, 0.52, 0.12 ppm of Cu, Zn, Cr, Ni, Cd, Pb, As, and Se, respectively. The content of Cu is higher than the content limits suggested by the European Standards and that may be due to Cu fungicides used for crops protection and a low content in heavy metals. A summary of standards concerning heavy metal content in composting feedstock and compost is shown in Table 10.7.

Composting is considered a useful tool for eliminating phytopathogenic bacteria and fungi in horticultural residues. Phytopathogens are eliminated by this process suggesting that it could be an effective alternative for the management of plant wastes after harvesting (Suárez-Estrella et al. 2012). However, not all composts are phytopathogen free and numerous studies reported the persistence of pathogenic fungi, bacteria, viruses, and nematodes in plant residues (Bollen et al. 1989). Particularly, fungi reportedly cause most soil-borne diseases and survival of these plant pathogens constitutes a real challenge for plant waste management as these residues could constitute an inoculum source of phytopathogens.

Compost microbial communities' inhabitant could originate from the residues used during composting. In compost process, phytopathogen elimination is mainly due to the high temperatures. Noble and Roberts (2004) have reviewed survival and suppressiveness of soil-borne plant pathogens and nematodes during composting.

Table 10.7 Heavy metal content in compost and composting feedstock (Gavilanes-Terán et al. 2016)

Regulation	Cadmium (Cd)	Chromium (Cr)	Lead (Pb)	Copper (Cu)	Zinc (Zn)	Nickel (Ni)	Cobalt (Co)	Selenium (Se)	Arsenic (As)
EU guideline	1	100	100	100	300	50	–	1.5	–
European compost Eco-Label	1	100	100	100	300	50	–	–	–
European Compost Network	1.3	60	130	110	400	40	–	–	–
NF U44-295	3	120	180	300	600	60	–	12	18
US guidelines	39		300	1500	2800	420		100	41
Canadian guideline	3–20	210	150–500	400	700–1850	62–180	34–150	2–14	13–75

They found that temperature and time exposure combination is a crucial parameter. However, temperature of 64–70 °C during 21 days was sufficient to reduce the number of nematodes and microbial pathogens (fungal, oomycete, and bacterial-plant pathogens). Temperature for a long time is not always maintained as it requires technical conditions especially mixture engineering and good monitoring of composting process. Other mechanisms such as interactions between microorganisms, production of secondary vegetable metabolites like phenolic compounds or triggering systemic resistance in plants may also contribute, to a lesser extent, to pathogen elimination. In addition, numerous studies showed that persistence and survival of phytopathogenic bacteria on horticultural debris depend also on bacterial-plant species. While *Erwinia carotovora* subsp. *carotovora*, *Xanthomonas campestris* pv. *vesicatoria*, and *Pseudomonas syringae* pv. *syringae* disappeared in less than 60 h of composting, a range of horticultural wastes remained (Elorrieta et al. 2003). While *Phytophthora infestans* survived in tomato waste for less than 36 days under greenhouse and field conditions (Lima et al. 2009), *Clavibacter michiganensis* subsp. *michiganensis* (Cmm) was reported as able to survive several months in tomato stems (Fatmi et al. 2002). Steinmöller et al. (2013) checked the risk of dissemination of *C. michiganensis* sp. *sepedonicus* through potato residues of different industries. The authors studied sanitized residues before use in agriculture, e.g., by composting or pasteurization, and concluded that these sanitation treatments are not sufficient to inactivate the pathogen. The residues were still of high risk as the pathogen could be isolated after composting for 6 days at maximum temperatures at 70 °C, 13 days at 55 °C, and 90 min pasteurization at 70 °C. It seems also that environmental and soil conditions (physical, chemical, and biological properties) and geography could influence this ability to persist in plant debris (Lima et al. 2009; Andrivon 1995). *P. infestans* is more likely to survive on plant debris kept in dry than in wet soil conditions. It was also found that Cmm survival on tomato stems depends on the geographic areas. The survival time of the pathogen was ranging from 2 months in Morocco to 2 years in Iowa (USA) (Fatmi et al. 2002). Temperature could influence the survival of several pathogens. It was reported that *P. infestans* is favored by cool temperature and constant dry conditions (Harrison 1995). While it is recommended to maintain the residual piles at above 55 °C for at least 5 days, to eliminate pathogens. It was reported that phytopathogens like *Fusarium oxysporum* f. sp. *melonis* were eradicated at 65 °C (Suárez-Estrella et al. 2012). *Plasmodiophora brassicae*, *F. oxysporum* f. sp. *lycopersici*, and *Macrophomina phaseolina* are tolerant to temperatures above that value as they survived a peak compost temperature of 74 °C for 21 days, while tobacco mosaic virus (TMV) survives at above 68 °C for more than 20 days (Noble and Roberts 2004). All these facts suggest to recommend that composts should be properly evaluated for microbial content in order to avoid phytopathogenic introduction into soils and to reduce plants disease outbreaks.

10.9 Conclusions and Future Prospects

Development of technology and innovation in horticultural industry leads to the generation of large quantities of plant residues that can be serious environmental issues such as pollution in landfills, if no control measures are adopted. Due to its low cost and simple technical requirement, composting is considered as a suitable option for plant residues management and valorization, and compost obtained can be used as soil improver or fertilizers. The role that composting can play in sustainability of horticultural industry had led to a global interest on crucial aspects related to plant residues as composting feedstock and to the quality of the produced compost. There is lack of statistical data regarding generation of plant residues in horticulture about generation rate and characterization and where they are available. This challenge may often represent a source of uncertainty for stakeholders who may want to invest in composting plants. The quality and the availability of feedstock are a key challenge to the composting industry. Plant residues are heterogeneous with a high organic to ash ratio, low porosity, and high moisture content. Plant residues may also contain high nitrogen, which largely depends on type of crop. This can result in loss of nitrogen in the form of ammonia causing air pollution and decline in the agronomic value of the final product (Wang and Zeng 2018). Composting of plant residues alone can lead to various challenges such as nitrogen loss and lack of aeration, due to a lower C/N ratio and higher moisture content. For this reason, it is not desirable to compost plant residues alone. Thus, using other organic wastes and bulking agent in the mixture is necessary for suitable composting. The effects of such peculiarities on the composting of feedstock and also potential contamination by plant pathogens have been addressed by some researchers. There is also risk that improper monitoring of composting processes can lead to insufficient temperature–time combinations for plant pathogen eradication. Potential presence of several soilborne plant pathogens and plant parasitic nematodes in plant debris is one of the most important challenges that may affect compost quality and can have adverse consequences on composting hygienization and compost quality. Compost producers should be cautious about this problem. However, suppressive compost is an environmentally friendly agricultural technique characterized by its cheap imputes as constituted by agricultural wastes and used as biofertilizer and/or biopesticide for major crops. Yet, still rapid methods need to be developed that can predict its effectiveness against targeted plant pathogens.

References

Abdel-Rahman MA, Nour El-Din M, Refaat BM et al (2016) Biotechnological application of thermotolerant cellulose-decomposing bacteria in composting of rice straw. Ann Agric Sci 61:135–143. https://doi.org/10.1016/j.aoas.2015.11 006

Abid N, Sayadi S (2006) Detrimental effects of olive mill wastewater on the composting process of agricultural wastes. Waste Manag 26:1099–1107. https://doi.org/10.1016/j.wasman.2005.06.015

Abu Qdais H, Al-Widyan M (2016) Evaluating composting and co-composting kinetics of various agro-industrial wastes. Int J Recycl Org Waste Agric 5:273–280. https://doi.org/10.1007/s40093-016-0137-3

Adhikari BK, Trémier A, Martinez J, Barrington S (2010) Home and community composting for on-site treatment of urban organic waste: perspective for Europe and Canada. Waste Manag Res 28:1039–1053. https://doi.org/10.1177/0734242X10373801

Aguelmous A, Zegzouti Y, Khadra A, El Fels L, Souabi S, Hafidi M (2020) Landfilling and composting efficiency to reduce genotoxic effect of petroleum sludge. Environ Technol Innov 20. https://doi.org/10.1016/j.eti.2020.101047

Ait Baddi G, Hafidi M, Gilard V, Revel J-C (2003a) Characterization of humic acids produced during composting of olive mill wastes: elemental and spectroscopic analyses (FTIR and 13C-NMR). 23(7):661–666. https://doi.org/10.1051/agro:2003042

Ait Baddi G, Hafidi M, Merlina G, Revel JC (2003b) Characterization and identification of polyphenols during waste treatment of olive oil mills by composting. 47(5–6):161–172. https://www.scopus.com/inward/record.uri?eid=2-s2.0-0842279860&partnerID=40&md5=347cf766caffcec9b451d80760b1838e

Ait Baddi G, Hafidi M, Ouatmane A, Revel J-C (2004) Characterization of humic acids extracted during composting of wastes from olive oil mills on Sephadex Gel. 48(1–2):13–24. https://www.scopus.com/inward/record.uri?eid=2-s2.0-3142607373&partnerID=40&md5=b8e11a0ca9e37e1d2359a360ca12cf5e

Ait Baddi G, Antonio Alburquerque J, Gonzálvez J et al (2004) Chemical and spectroscopic analyses of organic matter transformations during composting of olive mill wastes. Int Biodeterior Biodegradation 54:39–44. https://doi.org/10.1016/j.ibiod.2003.12.004

Ali M, Kazmi AA, Ahmed N (2014) Study on effects of temperature, moisture and pH in degradation and degradation kinetics of aldrin, endosulfan, lindane pesticides during full-scale continuous rotary drum composting. Chemosphere 102:68–75. https://doi.org/10.1016/j.chemosphere.2013.12.022

Alkoaik F, Ghaly AE (2006) Influence of dairy manure addition on the biological and thermal kinetics of composting of greenhouse tomato plant residues. Waste Manag 26:902–913. https://doi.org/10.1016/j.wasman.2005.11.023

Amir S, Hafidi M, Bailly J-R, Revel J-C (2003) Characterization of humic acids extracted from sewage sludge during composting and of their Sephadex® gel fractions. 23(3):269–275. https://doi.org/10.1051/agro:2002090

Amlinger F, Peyr S, Cuhls C (2008) Green house gas emissions from composting and mechanical biological treatment. Waste Manag Res 26(1):47–60. https://doi.org/10.1177/0734242X07088432

Amose T, Jeyakumar K (2018) Will growth of technology lift up the economic status of farmers? Int J Environ Agric Biotechnol 3(3):264346

Andrivon D (1995) Biology, ecology, and epidemiology of the potato late blight pathogen Phytophthora infestans in soil. Phytopathology 85(10):1053–1056

Areikin E, Horne J, Scholes P, Mines U, Briggs L, Brown B, Dyson B, Resource W (2012) A survey of the UK organics recycling industry in 2010. Waste and Resources Action Programme (WRAP), Banbury

Arriaga H, Viguria M, López DM, Merino P (2017) Ammonia and greenhouse gases losses from mechanically turned cattle manure windrows: a regional composting network. J Environ Manag 203:557–563. https://doi.org/10.1016/j.jenvman.2017.06.006

Awasthi MK, Pandey AK, Bundela PS, Khan J (2015) Co-composting of organic fraction of municipal solid waste mixed with different bulking waste: characterization of physicochemical parameters and microbial enzymatic dynamic. Bioresour Technol 182:200–207. https://doi.org/10.1016/j.biortech.2015.01.104

Awasthi MK, Duan Y, Awasthi SK, Liu T, Zhang Z (2020) Effect of biochar and bacterial inoculum additions on cow dung composting. Bioresour Technol 297:122407. https://doi.org/10.1016/J.BIORTECH.2019.122407

Azim K, Komenane S, Soudi B (2017) Agro-environmental assessment of composting plants in Southwestern of Morocco (Souss-Massa Region). Int J Recycl Org Waste Agric 6:107–115. https://doi.org/10.1007/s40093-017-0157-7

Azim K, Faissal Y, Soudi B et al (2018a) Elucidation of functional chemical groups responsible of compost phytotoxicity using solid-state 13C NMR spectroscopy under different initial C/N ratios. Environ Sci Pollut Res 25:3408–3422. https://doi.org/10.1007/s11356-017-0704-9

Azim K, Soudi B, Boukhari S et al (2018b) Composting parameters and compost quality: a literature review. Org Agric 8:141–158. https://doi.org/10.1007/s13165-017-0180-z

Barje F, Amir S, Winterton P, Pinelli E, Merlina G, Cegarra J, Revel J-C, Hafidi M (2008) Phospholipid fatty acid analysis to monitor the co-composting process of olive oil mill wastes and organic household refuse. 154(1–3):682–687. https://doi.org/10.1016/j.jhazmat.2007.10.089

Barje F, El Fels L, El Hajjouji H, Amir S, Winterton P, Hafidi M (2012) Molecular behaviour of humic acid-like substances during co-composting of olive mill waste and the organic part of municipal solid waste. Int Biodeterior Biodegrad 74:17–23. https://doi.org/10.1016/j.ibiod.2012.07.004

Bernal MP, Navarro AF, Roig A, Cegarra J, García D (1996) Carbon and nitrogen transformation during composting of sweet sorghum bagasse. Biol Fertil Soils 22(1):141–148. https://doi.org/10.1007/BF00384446

Bernal MP, Alburquerque JA, Moral R (2009) Composting of animal manures and chemical criteria for compost maturity assessment. A review. Bioresour Technol 100:5444–5453. https://doi.org/10.1016/j.biortech.2008.11.027

Bhatia SK, Jagtap SS, Bedekar AA, Bhatia RK, Patel AK, Pant D, Rajesh Banu J, Rao CV, Kim YG, Yang YH (2020) Recent developments in pretreatment technologies on lignocellulosic biomass: effect of key parameters, technological improvements, and challenges. Bioresour Technol 300:122724. https://doi.org/10.1016/J.BIORTECH.2019.122724

Billah M, Bano A (2015) Role of plant growth promoting rhizobacteria in modulating the efficiency of poultry litter composting with rock phosphate and its effect on growth and yield of wheat. Waste Manag Res 33(1):63–72. https://doi.org/10.1177/0734242X14559593

Biyada S, Merzouki M, Demčenko T, Vasiliauskiene D, Urbonavičius J, Marčiulaitiene E, Vasarevičius S, Benlemlih M (2020) Evolution of microbial composition and enzymatic activities during the composting of textile waste. Appl Sci (Switzerland) 10(11). https://doi.org/10.3390/app10113758

Bollen GJ, Volker D, Wijnen AP (1989) Inactivation of soil-borne plant pathogens during small-scale composting of crop residues. Neth J Plant Pathol 95(1):19–30. https://doi.org/10.1007/BF01974281

Bouhia Y, Lyamlouli K, Fels LE, Youssef Z, Ouhdouch Y, Hafidi M (2021) Effect of microbial inoculation on lipid and phenols removal during the co-composting of olive mill solid sludge with green waste in bioreactor. Waste Biomass Valoriz 12(3):1417–1429. https://doi.org/10.1007/s12649-020-01077-3

Brethauer S, Shahab RL, Studer MH (2020) Impacts of biofilms on the conversion of cellulose. Appl Microbiol Biotechnol 104(12):5201–5212. https://doi.org/10.1007/s00253-020-10595-y

Bustamante MA, Paredes C, Marhuenda-Egea FC et al (2008) Co-composting of distillery wastes with animal manures: carbon and nitrogen transformations in the evaluation of compost stability. Chemosphere 72:551–557. https://doi.org/10.1016/j.chemosphere.2008.03.030

Cáceres R, Malińska K, Marfà O (2018) Nitrification within composting: a review. Waste Manag 72:119–137. https://doi.org/10.1016/j.wasman.2017.10.049

Chakroune K, Bouakka M, Hakkou A (2005) Incidence of aeration on the treatment by composting of date palm by-products contaminated with *Fusarium oxysporum* f.sp. *albedinis*. 51(1):69–77. https://doi.org/10.1139/w04-109

Chen X, Liu R, Hao J, Li D, Wei Z, Teng R, Sun B (2019a) Protein and carbohydrate drive microbial responses in diverse ways during different animal manures composting. Bioresour Technol 271:482–486. https://doi.org/10.1016/j.biortech.2018.09.096

Chen X, Zhao Y, Zeng C et al (2019b) Assessment contributions of physicochemical properties and bacterial community to mitigate the bioavailability of heavy metals during composting based on structural equation models. Bioresour Technol 289:121657. https://doi.org/10.1016/j.biortech.2019.121657

Choińska-Pulit A, Łaba W, Rodziewicz A (2019) Enhancement of pig bristles waste bioconversion by inoculum of keratinolytic bacteria during composting. Waste Manag 84:269–276. https://doi.org/10.1016/j.wasman.2018.11.052

Cifuentes R, de León R, Porres C, Rolz C (2013) Windrow composting of waste sugar cane and press mud mixtures. Sugar Tech 15:406–411. https://doi.org/10.1007/s12355-013-0217-x

De Silva S, Yatawara M (2017) Assessment of aeration procedures on windrow composting process efficiency: a case on municipal solid waste in Sri Lanka. Environ Nanotechnol Monit Manag 8:169–174. https://doi.org/10.1016/j.enmm.2017.07.008

Dias BO, Silva CA, Higashikawa FS et al (2010) Use of biochar as bulking agent for the composting of poultry manure: effect on organic matter degradation and humification. Bioresour Technol 101:1239–1246. https://doi.org/10.1016/j.biortech.2009.09.024

Donaldson BM, Smith GP, Kweon Y-J, Sriranganathan N (2013) An analysis of leachate constituents and pathogen destruction in deer mortality static windrow composting. Water Air Soil Pollut 224:1431. https://doi.org/10.1007/s11270-012-1431-y

Dorais M, Dubé Y (2011) Managing greenhouse organic wastes: a holistic approach. Acta Hortic 893:183–197. https://doi.org/10.17660/ActaHortic.2011.893.12

Doublet J, Francou C, Poitrenaud M, Houot S (2011) Influence of bulking agents on organic matter evolution during sewage sludge composting; consequences on compost organic matter stability and N availability. Bioresour Technol 102:1298–1307. https://doi.org/10.1016/j.biortech.2010.08.065

Droussi Z, D'Orazio V, Hafidi M, Ouatmane A (2009) Elemental and spectroscopic characterization of humic-acid-like compounds during composting of olive mill by-products. 163(2–3):1289–1297. https://doi.org/10.1016/j.jhazmat.2008.07.136

Du X, Li B, Chen K, Zhao C, Xu L, Yang Z, Sun Q, Chandio FA, Wu G (2021) Rice straw addition and biological inoculation promote the maturation of aerobic compost of rice straw biogas residue. Biomass Convers Biorefinery 11(5):1885–1896. https://doi.org/10.1007/s13399-019-00587-y

Ekinci K, Tosun İ, Bıtrak B et al (2019) Effects of initial C/N ratio on organic matter degradation of composting of rose oil processing solid wastes. Int J Environ Sci Technol 16:5131–5140. https://doi.org/10.1007/s13762-018-1947-9

El Fels L, Lemee L, Ambles A, Hafidi M (2014) Identification and biotransformation of lignin compounds during co-composting of sewage sludge-palm tree waste using pyrolysis-GC/MS. Int Biodeterior Biodegrad 92:26–35. https://doi.org/10.1016/j.ibiod.2014.04.001

El Ouaqoudi FZ, El Fels L, Winterton P, Lemée L, Amblès A, Hafidi M (2014) Study of humic acids during composting of ligno-cellulose waste by infra-red spectroscopic and Thermogravimetric/Thermal differential analysis. Compost Sci Util 22(3):188–198. https://doi.org/10.1080/1065657X.2014.910148

Elorrieta MA, Suárez-Estrella F, López MJ, Vargas-García MC, Moreno J (2003) Survival of phytopathogenic bacteria during waste composting. Agric Ecosyst Environ 96(1):141–146. https://doi.org/10.1016/S0167-8809(02)00170-6

Encinar JM, González JF, Martínez G (2008) Energetic use of the tomato plant waste. Fuel Process Technol 89(11):1193–1200. https://doi.org/10.1016/J.FUPROC.2008.05.011

Fatmi M, Schaad NW (2002) Survival of Clavibacter michiganensis ssp. michiganensis in infected tomato stems under natural field conditions in California, Ohio and Morocco. Plant Pathol 51(2):149–154. https://doi.org/10.1046/j.1365-3059.2002.00675.x

Faust S, Kaiser K, Wiedner K et al (2018) Comparison of different methods for determining lignin concentration and quality in herbaceous and woody plant residues. Plant Soil 433:7–18. https://doi.org/10.1007/s11104-018-3817-0

Fernández-Gómez MJ, Romero E, Nogales R (2011) Impact of imidacloprid residues on the development of Eisenia fetida during vermicomposting of greenhouse plant waste. J Hazard Mater 192:1886–1889. https://doi.org/10.1016/j.jhazmat.2011.06.077

Fernández-Gómez MJ, Díaz-Raviña M, Romero E, Nogales R (2013) Recycling of environmentally problematic plant wastes generated from greenhouse tomato crops through vermicomposting. Int J Environ Sci Technol 10(4):697–708. https://doi.org/10.1007/s13762-013-0239-7

Gao X, Xu Z, Li Y, Zhang L, Li G, Nghiem LD, Luo W (2021) Bacterial dynamics for gaseous emission and humification in bio-augmented composting of kitchen waste. Sci Total Environ 801. https://doi.org/10.1016/j.scitotenv.2021.149640

Gavilanes-Terán I, Jara-Samaniego J, Idrovo-Novillo J et al (2016) Windrow composting as horticultural waste management strategy—a case study in Ecuador. Waste Manag 48:127–134. https://doi.org/10.1016/j.wasman.2015.11.026

Gong X, Li S, Sun X et al (2017) Maturation of green waste compost as affected by inoculation with the white-rot fungi *Trametes versicolor* and *Phanerochaete chrysosporium*. Environ Technol 38:872–879. https://doi.org/10.1080/09593330.2016.1214622

Guerrini IA, Croce CGG, de Carvalho Bueno O, Jacon CPRP, Nogueira TAR, Fernandes DM, Ganga A, Capra GF (2017) Composted sewage sludge and steel mill slag as potential amendments for urban soils involved in afforestation programs. Urban For Urban Green 22:93–104. https://doi.org/10.1016/j.ufug.2017.01.015

Guo X, Gu J, Gao H et al (2012) Effects of Cu on metabolisms and enzyme activities of microbial communities in the process of composting. Bioresour Technol 108:140–148. https://doi.org/10.1016/j.biortech.2011.12.087

Guo X, Liu H, Wu S (2019) Humic substances developed during organic waste composting: formation mechanisms, structural properties, and agronomic functions. Sci Total Environ 662:501–510. https://doi.org/10.1016/j.scitotenv.2019.01.137

Hakkou A, Chakroune K, Bouakka M, Souna F, Cotxarrera L, Trillas MI (2011) Effect of nitrogen sources on the composting of date palm (*Phoenix dactylifera*) by-products infected by *Fusarium oxysporum* f.sp. *albedinis*. Adv Environ Biol 5(7):1638–1646

Han W, Clarke W, Pratt S (2014) Composting of waste algae: a review. Waste Manag 34:1148–1155. https://doi.org/10.1016/j.wasman.2014.01.019

Harrison JG (1995) Factors involved in the development of potato late blight disease (*Phytophthora infestans*). In: Haverkort AJ, MacKerron DKL (eds) Potato ecology and modelling of crops under conditions limiting growth: proceedings of the second international potato modeling conference, held in Wageningen 17–19 May, 1994. Springer Netherlands, pp 215–236. https://doi.org/10.1007/978-94-011-0051-9_14

Haug RT (1996) Composting plant design and process management. In: de Bertoldi M, Sequi P, Lemmes B, Papi T (eds) The science of composting. Springer, Dordrecht, pp 60–70

Heidarzadeh MH, Amani H, Javadian B (2019) Improving municipal solid waste compost process by cycle time reduction through inoculation of *Aspergillus niger*. J Environ Health Sci Eng 17:295–303. https://doi.org/10.1007/s40201-019-00348-z

Hernández-Lara A, Ros M, Cuartero J, Bustamante MÁ, Moral R, Andreu-Rodríguez FJ, Fernández JA, Egea-Gilabert C, Pascual JA (2022) Bacterial and fungal community dynamics during different stages of agro-industrial waste composting and its relationship with compost suppressiveness. Sci Total Environ 805:150330. https://doi.org/10.1016/J.SCITOTENV.2021.150330

Huang H, Zeng G, Luo L et al (2015) Effect of inoculation during different phases of agricultural waste composting on spectroscopic characteristics of humic acid. J Cent South Univ 22:4177–4183. https://doi.org/10.1007/s11771-015-2965-0

Huiliñir C, Villegas M (2014) Biodrying of pulp and paper secondary sludge: kinetics of volatile solids biodegradation. Bioresour Technol 157:206–213. https://doi.org/10.1016/j.biortech.2014.01.109

Jain MS, Kalamdhad AS (2018) Efficacy of batch mode rotary drum composter for management of aquatic weed (*Hydrilla verticillata* (L.f.) Royle). J Environ Manag 221:20–27. https://doi.org/10.1016/j.jenvman.2018.05.055

Jain MS, Paul S, Kalamdhad AS (2019) Utilization of biochar as an amendment during lignocellulose waste composting: impact on composting physics and realization (probability) amongst physical properties. Process Saf Environ Prot 121:229–238. https://doi.org/10.1016/j.psep.2018.10.031

Jayakumar A, Radoor S, Siengchin S, Shin GH, Kim JT (2023) Recent progress of bioplastics in their properties, standards, certifications and regulations: a review. Sci Total Environ 878:163156. https://doi.org/10.1016/j.scitotenv.2023.163156

Jindo K, Sonoki T, Matsumoto K et al (2016) Influence of biochar addition on the humic substances of composting manures. Waste Manag 49:545–552. https://doi.org/10.1016/j.wasman.2016.01.007

Jolanun B, Towprayoon S (2010) Novel bulking agent from clay residue for food waste composting. Bioresour Technol 101:4484–4490. https://doi.org/10.1016/j.biortech.2010.01.116

Joo HS, Shoda M, Phae CG (2007) Degradation of diesel oil in soil using a food waste composting process. Biodegradation 18(5):597–605. https://doi.org/10.1007/S10532-006-9092-4

Jurado MM, Suárez-Estrella F, Vargas-García MC et al (2014) Evolution of enzymatic activities and carbon fractions throughout composting of plant waste. J Environ Manag 133:355–364. https://doi.org/10.1016/j.jenvman.2013.12.020

Kabil EM, Semlali LA, Aajjane A, Assobhei O (2016) Phytotoxicity of composts obtained by accelerated composting on crops cultivated in Doukkala region, Morocco | Phytotoxicité de composts obtenus par compostage accéléré sur des plantes cultivées dans la région des Doukkala, Maroc. J Mater Environ Sci 7(12):4828–4838

Kausar H, Sariah M, Mohd Saud H et al (2011) Isolation and screening of potential actinobacteria for rapid composting of rice straw. Biodegradation 22:367–375. https://doi.org/10.1007/s10532-010-9407-3

Khapte PS, Kumar P, Burman U, Kumar P (2019) Deficit irrigation in tomato: agronomical and physio-biochemical implications. Sci Hortic 248:256–264. https://doi.org/10.1016/j.scienta.2019.01.006

Kim E, Lee S, Jo H et al (2018) Solid-state anaerobic digestion of dairy manure from a sawdust-bedded pack barn: moisture responses. Energies 11:484. https://doi.org/10.3390/en11030484

Koyama M, Kakiuchi A, Syukri F, Toda T, Tran QNM, Nakasaki K (2022) Inoculation of *Neurospora* sp. for improving ammonia production during thermophilic composting of organic sludge. Sci Total Environ 802. https://doi.org/10.1016/j.scitotenv.2021.149961

Kulcu R, Yaldiz O (2004) Determination of aeration rate and kinetics of composting some agricultural wastes. Bioresour Technol 93:49–57. https://doi.org/10.1016/j.biortech.2003.10.007

Kumar M, Ou YL, Lin JG (2010) Co-composting of green waste and food waste at low C/N ratio. Waste Manag 30(4):602–609

Kumar Awasthi M, Wang M, Pandey A et al (2017) Heterogeneity of zeolite combined with biochar properties as a function of sewage sludge composting and production of nutrient-rich compost. Waste Manag 68:760–773. https://doi.org/10.1016/j.wasman.2017.06.008

Lerch TZ, Dignac MF, Thevenot M, Mchergui C, Houot S (2019) Chemical changes during composting of plant residues reduce their mineralisation in soil and cancel the priming effect. Soil Biol Biochem 136:107525. https://doi.org/10.1016/J.SOILBIO.2019.107525

Lhadi EK, Tazi H, Aylaj M, Genevini PL, Adani F (2006) Organic matter evolution during co-composting of the organic fraction of municipal waste and poultry manure. 97(16):2117–2123. https://doi.org/10.1016/j.biortech.2005.09.012

Lima MA, Maffia LA, Barreto RW, Mizubuti ESG (2009) *Phytophthora infestans* in a subtropical region: survival on tomato debris, temporal dynamics of airborne sporangia and alternative hosts. Plant Pathol 58(1):87–99. https://doi.org/10.1111/j.1365-3059.2008.01951.x

Llorach-Massana P, Lopez-Capel E, Peña J, Rieradevall J, Montero JI, Puy N (2017) Technical feasibility and carbon footprint of biochar co-production with tomato plant residue. Waste Manag 67:121–130. https://doi.org/10.1016/J.WASMAN.2017.05.021

Løvdal T, Van Droogenbroeck B, Eroglu EC, Kaniszewski S, Agati G, Verheul M, Skipnes D (2019) Valorization of tomato surplus and waste fractions: a case study using Norway, Belgium, Poland, and Turkey as Examples. Foods 8(7). https://doi.org/10.3390/foods8070229

Luo X, Liu J, Zheng P, Li M, Zhou Y, Huang L, Chen L, Shuai L (2019) Promoting enzymatic hydrolysis of lignocellulosic biomass by inexpensive soy protein. Biotechnol Biofuels 12(1):51. https://doi.org/10.1186/s13068-019-1387-x

Malki M, Bouchaou L, Hirich A, Ait Brahim Y, Choukr-Allah R (2017) Impact of agricultural practices on groundwater quality in intensive irrigated area of Chtouka-Massa, Morocco. Sci Total Environ 574:760–770. https://doi.org/10.1016/j.scitotenv.2016.09.145

Maniadakis K, Lasaridi K, Manios Y, Kyriacou M, Manios T (2004) Integrated waste management through producers and consumers education: composting of vegetable crop residues for reuse in cultivation. J Environ Sci Health B 39(1):169–183. https://doi.org/10.1081/PFC-120027447

Matei PM, Sánchez-Báscones M, Bravo-Sánchez CT et al (2016) Hygienization and control of *Diplodia seriata* fungus in vine pruning waste composting and its seasonal variability in open and closed systems. Waste Manag 58:126–134. https://doi.org/10.1016/j.wasman.2016.08.002

Md Khudzari J, Tartakovsky B, Raghavan GSV (2016) Effect of C/N ratio and salinity on power generation in compost microbial fuel cells. Waste Manag 48:135–142. https://doi.org/10.1016/j.wasman.2015.11.022

Mehta CM, Palni U, Franke-Whittle IH, Sharma AK (2014) Compost: its role, mechanism and impact on reducing soil-borne plant diseases. Waste Manag 34:607–622. https://doi.org/10.1016/j.wasman.2013.11.012

Miller FC (1994) Conventional composting system. Agaricus compost miracle. Australian Mushroom Grower Association, Windsor, NSW, pp 1–18

Morales AB, Bustamante MA, Marhuenda-Egea FC et al (2016) Agri-food sludge management using different co-composting strategies: study of the added value of the composts obtained. J Clean Prod 121:186–197. https://doi.org/10.1016/j.jclepro.2016.02.012

Moreno AD, Duque A, González A, Ballesteros I, Negro MJ (2021) Valorization of greenhouse horticulture waste from a biorefinery perspective. Foods 10(4). https://doi.org/10.3390/foods10040814

Nisticò R, Evon P, Labonne L, Vaca-Medina G, Montoneri E, Vaca-Garcia C, Negre M (2017) Post-harvest tomato plants and urban food wastes for manufacturing plastic films. J Clean Prod 167:68–74. https://doi.org/10.1016/J.JCLEPRO.2017.08.160

Noble R, Roberts SJ (2004) Eradication of plant pathogens and nematodes during composting: a review. Plant Pathol 53:548–568. https://doi.org/10.1111/j.0032-0862.2004.01059.x

Onwosi CO, Igbokwe VC, Odimba JN, Eke IE, Nwankwoala MO, Iroh IN, Ezeogu LI (2017) Composting technology in waste stabilization: on the methods, challenges and future prospects. J Environ Manag 190:140–157. https://doi.org/10.1016/j.jenvman.2016.12.051

Pane C, Celano G, Piccolo A, Villecco D, Spaccini R, Palese AM, Zaccardelli M (2015) Effects of on-farm composted tomato residues on soil biological activity and yields in a tomato cropping system. Chem Biol Technol Agric 2(1):4. https://doi.org/10.1186/s40538-014-0026-9

Paredes C, Bernal MP, Cegarra J, Roig A (2002) Bio-degradation of olive mill wastewater sludge by its co-composting with agricultural wastes. Bioresour Technol 85:1–8. https://doi.org/10.1016/S0960-8524(02)00078-0

Payen S, Basset-Mens C, Perret S (2015) LCA of local and imported tomato: an energy and water trade-off. J Clean Prod 87:139–148. https://doi.org/10.1016/j.jclepro.2014.10.007

Rasapoor M, Adl M, Pourazizi B (2016) Comparative evaluation of aeration methods for municipal solid waste composting from the perspective of resource management: a practical case study in Tehran, Iran. J Environ Manage 184:528–534. https://doi.org/10.1016/j.jenvman.2016.10.029

Rufí-Salís M, Petit-Boix A, Villalba G, Ercilla-Montserrat M, Sanjuan-Delmás D, Parada F, Arcas V, Muñoz-Liesa J, Gabarrell X (2020) Identifying eco-efficient year-round crop combinations for rooftop greenhouse agriculture. Int J Life Cycle Assess 25(3):564–576. https://doi.org/10.1007/s11367-019-01724-5

Ruggieri L, Gea T, Artola A, Sánchez A (2009) Air filled porosity measurements by air pycnometry in the composting process: a review and a correlation analysis. Bioresour Technol 100:2655–2666. https://doi.org/10.1016/j.biortech.2008.12.049

Sánchez ÓJ, Ospina DA, Montoya S (2017) Compost supplementation with nutrients and microorganisms in composting process. Waste Manag 69:136–153. https://doi.org/10.1016/j.wasman.2017.08.012

Sanchez-Monedero MA, Cayuela ML, Roig A et al (2018) Role of biochar as an additive in organic waste composting. Bioresour Technol 247:1155–1164. https://doi.org/10.1016/j.biortech.2017.09.193

Şevik F, Tosun İ, Ekinci K (2016) Composting of olive processing wastes and tomato stalks together with sewage sludge or dairy manure. Int J Environ Sci Technol 13:1207–1218. https://doi.org/10.1007/s13762-016-0946-y

Şevik F, Tosun İ, Ekinci K (2018) The effect of FAS and C/N ratios on co-composting of sewage sludge, dairy manure and tomato stalks. Waste Manag 80:450–456. https://doi.org/10.1016/J.WASMAN.2018.07.051

Smith MM, Aber JD, Rynk R (2017) Heat recovery from composting: a comprehensive review of system design, recovery rate, and utilization. Compost Sci Util 25:S11–S22. https://doi.org/10.1080/1065657X.2016.1233082

Steinmöller S, Müller P, Bandte M, Büttner C (2013) Risk of dissemination of *Clavibacter michiganensis* ssp. *sepedonicus* with potato waste. Eur J Plant Pathol 137(3):573–584. https://doi.org/10.1007/s10658-013-0271-y

Suárez-Estrella F, Bustamante MA, Moral R, Vargas-García MC, López MJ, Moreno J (2012) In vitro control of fusarium wilt using agroindustrlal subproduct-based composts. J Plant Pathol 94(1):59–70. http://www.jstor.org/stable/45156009

Sülük K, Tosun İ, Ekinci K (2017) Co-composting of two-phase olive-mill pomace and poultry manure with tomato harvest stalks. Environ Technol 38:923–932. https://doi.org/10.1080/09593330.2016.1217279

Sundberg C, Jönsson H (2008) Higher pH and faster decomposition in biowaste composting by increased aeration. Waste Manag 28:518–526. https://doi.org/10.1016/j.wasman.2007.01.011

Tabrika I, Azim K, Mayad EH, Zaafrani M (2019) Composting of tomato plant residues: improvement of composting process and compost quality by integration of sheep manure. Org Agric 10:229. https://doi.org/10.1007/s13165-019-00268-0

Tabrika I, Mayad EH, Zaafrani M et al (2020) Optimization of solid phosphate sludge composting by integration of horticultural waste. Org Agric 11:319. https://doi.org/10.1007/s13165-020-00302-6

Tabrika I, Mayad EH, Furze JN, Zaafrani M, Azim K (2021) Optimization of tomato waste composting with integration of organic feedstock. Environ Sci Pollut Res 28(45):64140–64149. https://doi.org/10.1007/s11356-020-12303-9

Taccari M, Stringini M, Comitini F, Ciani M (2009) Effect of Phanerochaete chrysosporium inoculation during maturation of co-composted agricultural wastes mixed with olive mill wastewater. Waste Manag 29(5):1615–1621. https://doi.org/10.1016/j.wasman.2008.12.014

Thomas B, Raj MC, Athira KB, Rubiyah MH, Joy J, Moores A, Drisko GL, Sanchez C (2018) Nanocellulose, a versatile green platform: from biosources to materials and their applications. Chem Rev 118(24):11575–11625. https://doi.org/10.1021/acs.chemrev.7b00627

Toop TA, Ward S, Oldfield T, Hull M, Kirby ME, Theodorou MK (2017) AgroCycle – developing a circular economy in agriculture. Energy Procedia 123:76–80. https://doi.org/10.1016/j.egypro.2017.07.269

Tran QNM, Mimoto H, Nakasaki K (2015) Inoculation of lactic acid bacterium accelerates organic matter degradation during composting. Int Biodeterior Biodegradation 104:377–383. https://doi.org/10.1016/j.ibiod.2015.07.007

Tuomela M, Vikman M, Hatakka A, Itävaara M (2000) Biodegradation of lignin in a compost environment: a review. Bioresour Technol 72:169–183. https://doi.org/10.1016/S0960-8524(99)00104-2

Ulusoy Y, Ulukardeşler AH, Alibaş K (2009) Analysis of biogas production in Turkey utilising three different materials and two scenarios. Afr J Agric Res 4(10):996–1003

Viaene J, Agneessens L, Capito C et al (2017) Co-ensiling, co-composting and anaerobic co-digestion of vegetable crop residues: product stability and effect on soil carbon and nitrogen dynamics. Sci Hortic (Amsterdam) 220:214–225. https://doi.org/10.1016/j.scienta.2017.03.015

Vico A, Pérez-Murcia MD, Bustamante MA et al (2018) Valorization of date palm (*Phoenix dactylifera* L.) pruning biomass by co-composting with urban and agri-food sludge. J Environ Manag 226:408–415. https://doi.org/10.1016/j.jenvman.2018.08.035

Viel M, Sayag D, Peyre A, André L (1987) Optimization of in-vessel co-composting through heat recovery. Biol Wastes 20:167–185. https://doi.org/10.1016/0269-7483(87)90152-2

Voběrková S, Vaverková MD, Burešová A et al (2017) Effect of inoculation with white-rot fungi and fungal consortium on the composting efficiency of municipal solid waste. Waste Manag 61:157–164. https://doi.org/10.1016/j.wasman.2016.12.039

Wainwright H, Jordan C, Day H (2014) Environmental impact of production horticulture BT - horticulture: plants for people and places, volume 1: production horticulture. (G. R. Dixon & D. E. Aldous, Eds). Springer Netherlands, pp 503–522. https://doi.org/10.1007/978-94-017-8578-5_15

Wang S, Zeng Y (2018) Ammonia emission mitigation in food waste composting: a review. Bioresour Technol 248:13–19. https://doi.org/10.1016/j.biortech.2017.07.050

Wang S-P, Zhong X-Z, Wang T-T, Sun Z-Y, Tang Y-Q, Kida K (2017) Aerobic composting of distilled grain waste eluted from a Chinese spirit-making process: the effects of initial pH adjustment. Bioresour Technol 245:778–785. https://doi.org/10.1016/j.biortech.2017.09.051

Wang K, Mao H, Li X (2018) Functional characteristics and influence factors of microbial community in sewage sludge composting with inorganic bulking agent. Bioresour Technol 249:527–535. https://doi.org/10.1016/j.biortech.2017.10.034

Wang J, Liu Z, Xia J, Chen Y (2019a) Effect of microbial inoculation on physicochemical properties and bacterial community structure of citrus peel composting. Bioresour Technol 291. https://doi.org/10.1016/j.biortech.2019.121843

Wang J, Liu Z, Xia J, Chen Y (2019b) Effect of microbial inoculation on physicochemical properties and bacterial community structure of citrus peel composting. Bioresour Technol 291:121843. https://doi.org/10.1016/J.BIORTECH.2019.121843

Waqas M, Almeelbi T, Nizami A-S (2018) Resource recovery of food waste through continuous thermophilic in-vessel composting. Environ Sci Pollut Res 25:5212–5222. https://doi.org/10.1007/s11356-017-9358-x

Wefers D, Dong J, Abdel-Hamid AM et al (2017) Enzymatic mechanism for Arabinan degradation and transport in the thermophilic bacterium *Caldanaerobius polysaccharolyticus*. Appl Environ Microbiol 83:e00794-17. https://doi.org/10.1128/AEM.00794-17

Wei L, Shutao W, Jin Z, Tong X (2014) Biochar influences the microbial community structure during tomato stalk composting with chicken manure. Bioresour Technol 154:148–154. https://doi.org/10.1016/j.biortech.2013.12.022

Wu J, Zhao Y, Zhao W et al (2017) Effect of precursors combined with bacteria communities on the formation of humic substances during different materials composting. Bioresour Technol 226:191–199. https://doi.org/10.1016/j.biortech.2016.12.031

Wu D, Wei Z, Mohamed TA, Zheng G, Qu F, Wang F, Zhao Y, Song C (2022) Lignocellulose biomass bioconversion during composting: mechanism of action of lignocellulase, pretreatment methods and future perspectives. Chemosphere 286. https://doi.org/10.1016/j.chemosphere.2021.131635

Wuana RA, Okieimen FE (2011) Heavy metals in contaminated soils: a review of sources, chemistry, risks and best available strategies for remediation. ISRN Ecol 2011:402647. https://doi.org/10.5402/2011/402647

Xi B, He X, Dang Q et al (2015) Effect of multi-stage inoculation on the bacterial and fungal community structure during organic municipal solid wastes composting. Bioresour Technol 196:399–405. https://doi.org/10.1016/j.biortech.2015.07.069

Xu J, Jiang Z, Li M, Li Q (2019) A compost-derived thermophilic microbial consortium enhances the humification process and alters the microbial diversity during composting. J Environ Manag 243:240–249. https://doi.org/10.1016/j.jenvman.2019.05.008

Yadav KS, Mishra MM, Kapoor KK (1982) The effect of fungal inoculation on composting. Agric Wastes 4:329–333. https://doi.org/10.1016/0141-4607(82)90029-4

Yañez R, Alonso JL, Díaz MJ (2009) Influence of bulking agent on sewage sludge composting process. Bioresour Technol 100:5827–5833. https://doi.org/10.1016/j.biortech.2009.05.073

Yang L, Yuan H, Yang Y et al (2019) Enhanced lignin degradation in tobacco stalk composting with inoculation of white-rot fungi *Trametes hirsuta* and *Pleurotus ostreatus*. Waste Biomass Valorization 11:3525. https://doi.org/10.1007/s12649-019-00692-z

Yu K, Li S, Sun X, Cai L, Zhang P, Kang Y, Yu Z, Tong J, Wang L (2019) Application of seasonal freeze-thaw to pretreat raw material for accelerating green waste composting. J Environ Manag 239:96–102. https://doi.org/10.1016/j.jenvman.2019.02.128

Zhang L, Sun X (2016) Improving green waste composting by addition of sugarcane bagasse and exhausted grape marc. Bioresour Technol 218:335–343. https://doi.org/10.1016/j.biortech.2016.06.097

Zhang J, Lü F, Shao L, He P (2014) The use of biochar-amended composting to improve the humification and degradation of sewage sludge. Bioresour Technol 168:252–258. https://doi.org/10.1016/j.biortech.2014.02.080

Zhou H, Meng H, Zhao L et al (2018) Effect of biochar and humic acid on the copper, lead, and cadmium passivation during composting. Bioresour Technol 258:279–286. https://doi.org/10.1016/j.biortech.2018.02.086

Chapter 11
The Evolution of Date Palm Tolerance to Salinity Stress via Arbuscular Mycorrhizal Fungi and Compost Application

Mohamed Ait-El-Mokhtar, Raja Ben-Laouane, Abderrahim Boutasknit, Mohamed Anli, Abdessamad Fakhech, Youssef Ait-Rahou, Toshiaki Mitsui, Said Wahbi, Marouane Baslam, and Abdelilah Meddich

M. Ait-El-Mokhtar (✉)
Laboratory of Biochemistry, Environment & Agri-food URAC 36, Department of Biology, Faculty of Science and Techniques – Mohammedia, Hassan II University of Casablanca, Mohammedia, Morocco

Laboratory of Agro-Food, Biotechnologies and Valorization of Plant Bioresources (AGROBIOVAL), Department of Biology, Faculty of Science – Semlalia, Cadi Ayyad University, Marrakesh, Morocco

Centre d'Agrobiotechnologie et Bioingénierie, Unité de Recherche Labellisée CNRST (Centre AgroBiotech-URL-CNRST-05), Cadi Ayyad University, Marrakesh, Morocco

R. Ben-Laouane
Laboratory of Agro-Food, Biotechnologies and Valorization of Plant Bioresources (AGROBIOVAL), Department of Biology, Faculty of Science – Semlalia, Cadi Ayyad University, Marrakesh, Morocco

Centre d'Agrobiotechnologie et Bioingénierie, Unité de Recherche Labellisée CNRST (Centre AgroBiotech-URL-CNRST-05), Cadi Ayyad University, Marrakesh, Morocco

Laboratory of Environment and Health, Department of Biology, Faculty of Science and Techniques, Boutalamine, Errachidia, Morocco

A. Boutasknit
Laboratory of Agro-Food, Biotechnologies and Valorization of Plant Bioresources (AGROBIOVAL), Department of Biology, Faculty of Science – Semlalia, Cadi Ayyad University, Marrakesh, Morocco

N. K. Arora, B. Bouizgarne (eds.), *Microbial Biotechnology for Sustainable Agriculture Volume 2*, Microorganisms for Sustainability 34,
https://doi.org/10.1007/978-981-97-2355-3_11

Centre d'Agrobiotechnologie et Bioingénierie, Unité de Recherche Labellisée CNRST (Centre AgroBiotech-URL-CNRST-05), Cadi Ayyad University, Marrakesh, Morocco

Department of Biology, Multidisciplinary Faculty of Nador, Mohamed First University, Nador, Morocco

M. Anli
Laboratory of Agro-Food, Biotechnologies and Valorization of Plant Bioresources (AGROBIOVAL), Department of Biology, Faculty of Science – Semlalia, Cadi Ayyad University, Marrakesh, Morocco

Centre d'Agrobiotechnologie et Bioingénierie, Unité de Recherche Labellisée CNRST (Centre AgroBiotech-URL-CNRST-05), Cadi Ayyad University, Marrakesh, Morocco

Department of Life, Earth and Environmental Sciences, University of Comoros, Patsy University Center, Moroni, Comoros

A. Fakhech
Laboratory of Agro-Food, Biotechnologies and Valorization of Plant Bioresources (AGROBIOVAL), Department of Biology, Faculty of Science – Semlalia, Cadi Ayyad University, Marrakesh, Morocco

Y. Ait-Rahou
Laboratory of Agro-Food, Biotechnologies and Valorization of Plant Bioresources (AGROBIOVAL), Department of Biology, Faculty of Science – Semlalia, Cadi Ayyad University, Marrakesh, Morocco

Centre d'Agrobiotechnologie et Bioingénierie, Unité de Recherche Labellisée CNRST (Centre AgroBiotech-URL-CNRST-05), Cadi Ayyad University, Marrakesh, Morocco

Higher Institute of Nursing Professions and Health Techniques, Ministry of Health, Ouarzazate, Morocco

T. Mitsui
Laboratory of Biochemistry, Faculty of Agriculture, Niigata University, Niigata, Japan

Department of Life and Food Sciences, Graduate School of Science and Technology, Niigata University, Niigata, Japan

S. Wahbi · A. Meddich
Laboratory of Agro-Food, Biotechnologies and Valorization of Plant Bioresources (AGROBIOVAL), Department of Biology, Faculty of Science – Semlalia, Cadi Ayyad University, Marrakesh, Morocco

Centre d'Agrobiotechnologie et Bioingénierie, Unité de Recherche Labellisée CNRST (Centre AgroBiotech-URL-CNRST-05), Cadi Ayyad University, Marrakesh, Morocco

M. Baslam
Laboratory of Agro-Food, Biotechnologies and Valorization of Plant Bioresources (AGROBIOVAL), Department of Biology, Faculty of Science – Semlalia, Cadi Ayyad University, Marrakesh, Morocco

Centre d'Agrobiotechnologie et Bioingénierie, Unité de Recherche Labellisée CNRST (Centre AgroBiotech-URL-CNRST-05), Cadi Ayyad University, Marrakesh, Morocco

Department of Life and Food Sciences, Graduate School of Science and Technology, Niigata University, Niigata, Japan

Abstract Our study investigated the overtime mitigation of NaCl stress negative impact on growth performances of *Phoenix dactylifera* through the application of compost and/or arbuscular mycorrhizal fungi (AMF). The experiment was designed in eight treatments that include a control treatment (no AMF and compost application) and AMF and compost treatments supplemented alone or in dual application under 0 mM NaCl (absence of salt stress) and 240 mM NaCl (presence of salt stress) after two periods of cultivation (10 and 14 months). Our results indicate that date palm grown under 240 mM NaCl recorded enhanced sodium, chlorine, soluble sugars, proline, H_2O_2, and malondialdehyde (MDA) content; all together, they showed reduced growth fitness, leaf water potential, mycorrhizal colonization, phosphorus (P), nitrogen (N), potassium (K), calcium (Ca), and photosynthetic pigment contents. AMF and compost supplementation, applied alone or in combination, lessened the negative effect of salinity via various strategies including the improvement of N, P, K, and Ca uptake, water status, photosynthetic pigment concentrations, stomatal conductance, photosystem II efficiency, soluble sugars, proline content, antioxidant defense, and the decline of H_2O_2 and MDA production. The results also reveal that palms tolerance to salinity was gradually enhanced over time in supplemented plants and particularly when both biofertilizers were applied. Between the two periods of cultivation (10 and 14 months), growth traits and nutrients content were improved, and stress markers and toxic ions concentrations were reduced, while osmolyte content and antioxidant enzyme activity were regulated. In conclusion, this study pointed out the potential effect of the dual supplementation of compost and native AMF in alleviating the adverse impact of salinity stress on date palm with improved tolerance mechanisms over time.

Keywords Mycorrhizal symbiosis · Compost · Date palm · Tolerance · Salt · Oxidative stress · Nutrient content

11.1 Introduction

Climate change is disrupting the ecosystems functioning and inducing substantial soil salinization (Ait-El-Mokhtar et al. 2019a, Ait-El-Mokhtar et al. 2021). Salt-affected soils represent around 8% of the earth's surface with an evident expanding tendency (Shankar and Evelin 2019). Salinity figures among the most serious ecological problems alter arable lands and reduce agricultural yields by more than 20% (Muchate and Nikalje 2016). Soil salinity leads to osmotic, ionic, and oxidative stresses that induce negative impact on crops (Mishra et al. 2022). As a consequence, salinity has deleterious effects on physiological and metabolic functioning including osmotic adjustment, mineral homeostasis, photosynthesis, respiration, and biomolecules synthesis (Ben-Laouane et al. 2020; Ait-El-Mokhtar et al. 2022a, b; Toubali et al. 2022). The induced stresses stimulate reactive oxygen species

(ROS) accumulation that induces cellular toxicity and alters various biomolecules, such proteins, lipids, and DNA, and can lead to cell apoptosis (Evelin et al. 2019). To alleviate salinity-induced damages on different cell components, plants use many adaptive strategies. Osmotic adjustment figures among the most efficient defense strategies that prevent plants from osmotic and ionic stress impact via the exclusion of toxic ions from cell or its compartmentalization in the vacuole (Parida et al. 2018). Plants can also tolerate salinity via their antioxidant defense including peroxidase (POD), superoxide dismutase (SOD), catalase (CAT), polyphenoloxidase (PPO), and ascorbate peroxidase (APX) that scavenge ROS (Foyer 2018; Ait-El-Mokhtar et al. 2019b). In addition, salinity tolerance may also take place through enhancing the levels of several organic molecules like glycine betaine, proline, and soluble sugars (Ait-El-Mokhtar et al. 2020a, b; Evelin et al. 2019).

Arbuscular mycorrhizal fungi (AMF) are beneficial microorganisms able to promote plant adaptation to saline soils. They are symbiotic soil fungi in association with plant roots (Smith and Read 2008). The established mutualistic relation may mitigate the deleterious impact of salinity on crops, even if AMF are altered by soil salinity (Ait-El-Mokhtar et al. 2020a, b; Meddich et al. 2015a, 2017). In saline environments, AMF improve soil rhizospheric conditions and plant water uptake and mineral nutrition (Anli et al. 2022; Ben-Laouane et al. 2021; Boutasknit et al. 2021a). In addition, AMF colonization is well known to lessen toxic ions absorption (Ait-El-Mokhtar et al. 2020a; Toubali et al. 2020) and to activate the photosynthesis and the antioxidant system (Ait Rahou et al. 2020; Anli et al. 2020a, b, c; Benaffari et al. 2022; Boutasknit et al. 2023; Lahbouki et al. 2021).

In arid and semi-arid areas, the supplementation of soils with compost is a crucial ecological alternative for soil rehabilitation (Anli et al. 2020a, c; Boutasknit et al. 2020a; Meddich et al. 2019). Many investigations have shown the use of organic amendment in salt-affected soils restoration (Guangming et al. 2017; Mbarki et al. 2020). Compost may mitigate salinity impact in plants through soil fertility promotion (Ben-Laouane et al. 2020), promoting nutrient availability and plant growth (Trivedi et al. 2017; Ait-El-Mokhtar et al. 2022a, b), stimulating photosynthetic apparatus and the activity of antioxidant molecules (Mbarki et al. 2018; Ben-Laouane et al. 2020).

Phoenix dactylifera L. figures among the most common plants in the arid and semi-arid area of the Middle East and North Africa with nutritional, socio-economic, and environmental values (Chao and Krueger 2007). So far, various scientific reports have explored date palm response in saline soils (Ait-El-Mokhtar et al. 2019b, 2020a, 2022a, b; Al Kharusi et al. 2019; Meddich et al. 2018; Toubali et al. 2020) and few researches have examined the effects of compost application and/or AMF inoculation in improving date palm tolerance to salinity (Ait-El-Mokhtar et al. 2020a; Toubali et al. 2020).

To date, little is known about the effect of the dual supplementation of compost and oases native AMF in date palm resistance to salinity and the evolution of the involved processes over time. Thereby, our research was performed to explore the role of these two biofertilizers in separate or combined application on date palm seedlings growth, photosynthesis, nutrient content, antioxidant system and water status in the presence of salinity, and the evolution of resistance mechanisms of this plant between 10 and 14 months of cultivation.

11.2 Materials and Methods

11.2.1 Plant Material and Biofertilizers Treatments

Boufeggous cultivar, a salinity sensitive date palm variety of economic interest, was used in this experiment. Seeds were sterilized with a 10% bleach solution for 10 min followed by rinsing with distilled water. The seeds were placed in plastic basins containing previously sterilized sand (at 180 °C for 3 h) and germination was carried out for 3 weeks at 38 °C. Date palm transplants were then transferred to black plastic bags filled with 2.3 kg of the aforementioned sterile sandy soil. The characteristics of the used substrate are as follows: phosphorus (P) (0.001%), potassium (K) (0.001%), magnesium (Mg) (0.006%), iron (Fe) (0.012%), calcium (Ca) (0.01%), sodium (Na) (0.002%), aluminum (Al) (0.01%), silicon (Si) (0.002%), pH (9.31), and electrical conductivity (EC) (0.29 dS cm^{-1}).

The mycorrhizal inoculum used is the native Aoufous mycorrhizal consortium (NAMC), collected from *P. dactylifera* rhizospheric soil in the Aoufous oasis (Southeast of Morocco). The consortium contains three genera: *Glomus* sp. (60%), *Sclerocystis* sp. (36%), and *Acaulospora* sp. (4%) (Meddich et al. 2015b).

The multiplication of the NAMC was performed for 3 months under controlled greenhouse conditions (16 h day/8 h night photoperiod, 19.5–32 °C temperature, 60–75% relative humidity, and 410–490 μm^{-2} s^{-1} illumination) using maize as a host plant. Rhizospheric soil of the propagating crop was applied as the mycorrhizal inoculum containing mycorrhizal root fragments, spores, and hyphae. Infectivity of the inoculum was assessed by the most probable number test to adjust the rates to be added. The inoculum had 1056 infectious propagules/100 g of multiplication soil (Ait-El-Mokhtar et al. 2020b). AMF inoculum was applied to the corresponding bags by adding 10 g of multiplicated inoculum in each bag (about 11 spores g^{-1} inoculum trapping soil), when transplanting the seedlings. The same quantity and quality of autoclaved mycorrhizal inoculum was added to the uninoculated pots.

The used organic amendment was produced from composted quack grass waste (Meddich et al. 2016). Compost physico-chemical parameters were as follows: pH 6.87, organic carbon 306.5 g kg^{-1}, available P 0.25 mg g^{-1}, total N 21.9 g kg^{-1}, NO^{-}_{3} 0.07 10^{-3} mg g^{-1}, NH^{+}_{4} 0.03 10^{-3} mg g^{-1}, NH^{+}_{4}/NO^{-}_{3} ratio 0.44, C/N ratio 14.0, organic matter 527.2 g kg^{-1}, ashes 490 g kg^{-1}, bacterial population 2.12 10^{5} CFU g^{-1}, and fungal population 9.75 10^{4} CFU g^{-1}. The organic amendment was added at 5% (w/w) rate to the subsequent bags when seedlings were transplanted.

11.2.2 Application of Saline Stress

After seed germination, Boufeggous seedlings were subjected to normal water irrigation (75% of field capacity) for 5 months, then two salinity levels (0 and 240 mM NaCl) were applied. The NaCl level was progressively enhanced to avoid acute osmotic stress. Pot irrigation with saline solution and distilled water was based on EC measurements to keep the desired level of salinity. Excess solution was collected using small plastic basins placed under each pot. This amount of solution was re-added to the corresponding pots. All cultivated transplants received no fertilizer throughout the experiment lifespan.

11.2.3 Experimental Design and Growth Conditions

The experimentation was made in eight treatments with two compost levels (−compost and + compost), two AMF inoculation levels (−AMF and +AMF), and two salt levels (0 and 240 mM NaCl). For each treatment, 10 biological replicates were used (80 pots in total), which were randomly arranged. The experiment lasted for 14 months under greenhouse conditions, which are the same as mentioned above.

11.2.4 Growth Assessment

After 10 and 14 months of cultivation, plants were harvested and root parts were pulled out from the shoot parts and biometric measurements such as plant elongation, shoot and root dry weight, leaf area, and number of leaves were determined.

11.2.5 Mycorrhization Parameters

Thin 1-cm root segments of date palm were observed using Zeiss Axioskop 40 microscope. Root segments with mycorrhizal structures were counted up and used to calculate AMF root colonization. The intensity and frequency of mycorrhization were calculated as explained by McGonigle et al. (1990).

11.2.6 Mineral Elements Content

The determination of Na^+, K^+, and Ca^{2+} content was performed as described by Wolf (1982). The ashes were obtained after incinerating 0.2 g of grinded dry matter (DM) in an oven at 500 °C for 6 h and were recovered in HCl. Na^+, K^+, and Ca^{2+} concentrations were determined by a flame spectrophotometer (JENWAY, PFP7). Phosphorus

(P) levels were determined using sodium molybdate method (Olsen and Dean 1965). Solution samples (0.25 mL) were used for the measurement of Na^+, K^+, and Ca^{2+} content and were added to 1 mL of distilled water, 1.25 mL of sodium molybdate (2.5%), and hydrazine sulfate (0.15%) and the mixture solution was boiled for 10 min. Each sample was left to cool prior to reading the absorbance at 820 nm. Chlorine (Cl^-) concentration was measured by digesting 100 mg DM samples in a mixture of 0.1 N nitric acid (HNO_3) and 10% (w/v) acetic acid (3:1) for 12 h (Boursier et al. 1987). The Cl^- ions were then titrated with a 0.1 N silver nitrate ($AgNO_3$) solution in the presence of potassium dichromate as an indicator. The end of the reaction was marked by a change in color from yellow to brick red. Nitrogen concentration measurement was performed following the Kjeldahl method. Shoot and root dry samples undergo mineralization by sulfuric acid in the presence of the Kjeldahl catalyst, then a distillation was carried out to displace the ammonia formed by NaOH (40%). Then boric acid was used to fix the ammonia and was titrated with sulfuric acid (N/50). The values were provided in mg of nitrogen per g of DM.

11.2.7 Physiological and Water Status Parameters

Stomatal conductance (g_s) measurements were taken via a portable porometer (Leaf Porometer LP1989). The leaf is attached to a relative humidity sensor. The water vapor emitted by tranrspiration from the leaf causes an increase in relative humidity at the sensor. Measurements of g_s are made on a sunny day. Measurements of leaf water potential (LWP) were performed following the Scholander et al. (1965) method. The leaf, freshly harvested, is inserted into a plug to seal it and immediately placed in the chamber. The cut end protrudes a few millimeters out of the plug so that the sap can be observed to be flush with the section of the vessels. The pressure applied is gently increased in the chamber until the sap appears. The pressure that causes the first drop to come out, expressed in MPa, is the opposite of the water potential of the leaf. For more precision, the observation is made with a magnifying glass. Chlorophyll fluorescence determination was carried out on a fully developed young leaf using a portable fluorometer (Opti-sciences OS1 30p). Before measurements were taken, the leaf was acclimatized to 30 min darkness using leaf clips. Three chlorophyll fluorescence parameters were recorded Initial (F_0), maximum (F_m), variable ($F_v = F_m - F_0$) fluorescence, and F_v/F_m ratio (Baker 2008). For relative water content (RWC) measurement, leaf materials were taken from five replicates per treatment to measure fresh weight (FW), dry weight (DW), and turgid weight (TW). These parameters were used to determine RWC values as described by Barrs and Weatherly (1962). The extraction of chlorophyll a, chlorophyll b, and carotenoids was carried out following the method of Arnon (1949). Leaf samples were grinded in 80% acetone, then extracts were centrifuged for 10 min at 10,000×g. Supernatants absorbance was measured at 480, 645, and 663 nm on the spectrophotometer. A blank containing 80% acetone was used as a control.

11.2.8 Hydrogen Peroxide and Malondialdehyde Content

The method of Velikova et al. (2000) was used to assess hydrogen peroxide concentrations. 0.25 g sample of fresh shoot and root material was grinded in 5 mL of 10% (w/v) trichloroacetic acid (TCA) and then centrifuged at 15,000×g for 15 min at 4 °C. 0.5 mL of the obtained supernatant was added to 1 mL potassium iodide and 0.5 mL potassium phosphate buffer (10 mM, pH 7). After 1 h incubation in darkness, the absorbance was recorded at 390 nm. 10% TCA solution was used as the blank.

The malondialdehyde (MDA) concentration was assessed as described by Dhindsa et al. (1981). 0.25 g of fresh leaf and root material was grinded in 10 mL TCA (0.1%) and centrifuged for 10 min at 18,000×g. The solution mixture contains 2 mL supernatant aliquot and 2 mL of TCA (20%) containing thiobarbituric acid (TBA) (0.5%). The solution was boiled for 30 min at 100 °C, rapidly cooled, and then centrifuged for 10 min at 10,000×g for clarification. Obtained supernatant absorbance was measured at 532 nm (A532) and the non-specific turbidity was adjusted by subtracting A600 from A532.

11.2.9 Free Proline and Total Soluble Sugar Content

The determination of free proline content was performed following Bates et al. (1973). Aliquots of finely ground powder of the plant material (0.1 g) were homogenized in 4 mL of 3% sulfosalicylic acid solution. The extracts were centrifuged at 15,000×g for 15 min. The obtained supernatants (2 mL) were mixed with 2 mL of 2.5% ninhydrin solution and 2 mL of glacial acetic acid and boiled for 1 h. The reaction solution was extracted with 3 mL toluene and the absorbance was recorded at 520 nm using L-proline as standard and toluene as a blank.

Total soluble sugar (TSS) content determination was performed as described by Dubois et al. (1956). 0.1 g fresh leaf and root samples were grinded in 4 mL of ethanol (80%) and then the mixture was centrifuged for 10 min at 5000 rpm. The obtained supernatants were collected, and the pellets were resuspended in 2 mL of ethanol for a second centrifugation. The mixture of the two supernatants was used to determine the amount of soluble sugars. Supernatant sample (1 mL) was mixed with 1 mL of phenol solution (5%) and 5 mL of concentrated sulfuric acid. After a 5 min incubation, the absorbance was recorded at 485 nm in a spectrophotometer. Glucose solution was used as standard.

11.2.10 Antioxidant Enzyme Activity

The extraction of the antioxidant enzymes was carried out as described by Ait-El-Mokhtar et al. (2019b). Aliquots of leaves and root fresh materials (0.1 g) were ground in a 4 mL solution mixture containing 0.1 M potassium phosphate buffer (pH 7.0), 0.1 mmol ethylenediaminetetraacetic acid (EDTA), and 0.1 g polyvinylpolypyrrolidone (PVPP). The mixture was centrifuged for 10 min at 18,000×g at 4 °C and the supernatants were maintained at −20 °C for antioxidant enzymes analysis. Superoxide dismutase (SOD, EC 1.15.1.1) activity was measured according to Beyer and Fridovich (1987) method. The assessment of this activity is based on the measurement of the ability of the enzyme to inhibit the photochemical reduction of the nitroblue tetrazolium (NBT). The amount of enzyme required to inhibit 50% NBT reduction at 25 °C is considered as one SOD unit. The results were given in units of min^{-1} mg^{-1} of protein. The activity of catalase (CAT, EC 1.11.1.6) was measured as a decrease in the absorbance at 240 nm for 3 min after H_2O_2 decomposition following Aebi (1984) method. The reaction solution (2 mL volume) contained potassium phosphate buffer (0.1 M, pH 7.0), 0.1 M EDTA, 20 mM H_2O_2, and 100 μL enzyme extract. The activity of ascorbate peroxidase (APX, EC 1.11.1.11) was determined as a decrease in the absorbance at 290 nm for 1 min according to the method of Amako et al. (1994). The reaction mixture contained 100 μL enzyme extract, 50 mM potassium phosphate buffer (pH 7.0), 0 1 mM ascorbate, and 0.5 mM H_2O_2. Peroxidase activity (POD, EC 1.11.1.1.7) was determined using the guaiacol method (Polle et al. 1994). The enzyme activity was assessed in a reaction mixture containing 0.15 mL enzyme extract, 100 mM potassium phosphate buffer (pH 7.0), and 20 mM guaiacol. The reaction was started by the addition of 100 μL of H_2O_2 (10 mM) and then the variation in the absorbance was measured by spectrophotometry at 470 nm for 3 min.

11.2.11 Statistical Analysis

The results are presented as mean values using five biological replicates ± SE (standard error) per treatment. To perform statistical analysis, SPSS 23.0 software for Windows was used. A three-way analysis of variance (ANOVA) was performed to assess the significance of the differences between date palm responses to different treatments and the control. Duncan's test (DMRT) calculated at $P < 0.05$ was used for comparisons between means.

11.3 Results

11.3.1 Growth and AMF Colonization Traits

According to Table 11.1, salinity application significantly reduced plant growth after 10 and 14 months of cultivation. The application of the different biofertilizers, applied in combination or alone, significantly enhanced plants' growth traits under both levels of salt. Under 240 mM NaCl, the combined use of compost and AMF

Table 11.1 Effects of AMF and compost applied separately or in combination on date palm growth parameters after 10 and 14 months of cultivation in the absence (0 mM NaCl) and presence (240 mM NaCl) of salinity stress

		10 months of cultivation					14 months of cultivation				
NaCl level	Treatments	Plant height (cm)	Leaves number	Leaf area (cm²)	Shoot dry weight (g)	Root dry weight (g)	Plant height (cm)	Leaves number	Leaf area (cm²)	Shoot dry weight (g)	Root dry weight (g)
0 mM	Control	30.06 ± 0.39^{gh}	3.2 ± 0.45^{h}	18.95 ± 0.68^{i}	1.55 ± 0.06^{gh}	1.11 ± 0.10^{h}	34.52 ± 0.88^{ded}	4.2 ± 0.45^{fg}	24.19 ± 0.85^{g}	2.82 ± 0.11^{f}	1.92 ± 0.18^{ef}
	Compost	33.74 ± 0.43^{ef}	4.8 ± 0.45^{ef}	31.67 ± 0.27^{d}	3.59 ± 0.37^{e}	2.01 ± 0.10^{e}	38.76 ± 1.26^{c}	6.2 ± 0.45^{c}	35.71 ± 1.05^{c}	6.65 ± 0.26^{b}	3.28 ± 0.36^{b}
	AMF	35.66 ± 0.54^{d}	5.4 ± 0.55^{de}	32.40 ± 0.65^{d}	4.99 ± 0.89^{d}	3.36 ± 0.53^{b}	42.54 ± 1.18^{a}	7.2 ± 0.45^{b}	38.00 ± 0.31^{b}	6.32 ± 0.27^{bc}	4.13 ± 0.30^{a}
	Compost+AMF	33.12 ± 0.54^{f}	5.8 ± 0.45^{cd}	34.85 ± 1.45^{c}	5.06 ± 0.28^{d}	2.45 ± 0.52^{cd}	40.70 ± 1.06^{b}	8.2 ± 0.45^{a}	40.29 ± 0.65^{a}	8.83 ± 0.44^{a}	4.06 ± 0.44^{a}
240 mM	Control	27.32 ± 0.40^{i}	3.2 ± 0.45^{h}	11.52 ± 1.20^{l}	1.09 ± 0.14^{h}	0.66 ± 0.09^{i}	29.62 ± 1.21^{h}	3.8 ± 0.45^{j}	14.46 ± 0.39^{k}	1.85 ± 0.12^{g}	1.11 ± 0.06^{h}
	Compost	30.10 ± 0.64^{h}	4.2 ± 0.45^{fg}	17.01 ± 0.55^{j}	2.57 ± 0.50^{f}	1.29 ± 0.29^{gh}	33.24 ± 1.33^{f}	5.8 ± 0.45^{cd}	21.21 ± 0.51^{h}	4.51 ± 0.58^{d}	2.00 ± 0.31^{e}
	AMF	31.54 ± 0.34^{h}	4.2 ± 0.45^{fg}	20.62 ± 0.55^{h}	3.51 ± 0.47^{e}	1.68 ± 0.22^{efg}	34.76 ± 1.66^{de}	5.8 ± 0.45^{cd}	27.04 ± 0.50^{f}	4.63 ± 0.55^{d}	2.66 ± 0.33^{c}
	Compost+AMF	29.92 ± 0.80^{h}	$4.8 \pm 0.45e^{f}$	23.31 ± 0.43^{g}	4.50 ± 0.27^{d}	1.53 ± 0.28^{fg}	33.96 ± 1.96^{ef}	6.2 ± 0.45^{c}	27.97 ± 0.33^{e}	6.04 ± 0.74^{c}	2.09 ± 0.11^{de}

The values of each column denoted by a different letter show significant differences as determined by Duncan's test ($P < 0.05$) following the completion of the three-way ANOVA

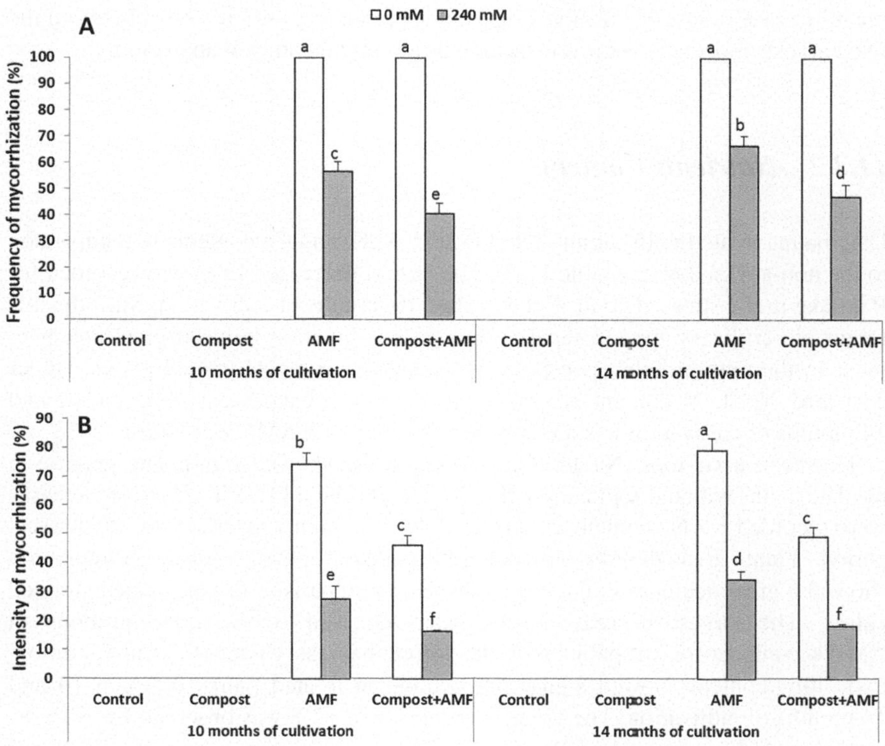

Fig. 11.1 Effects of AMF and compost applied separately or in combination on date palm frequency (**a**) and intensity (**b**) of mycorrhization after 10 and 14 months of cultivation in the absence (0 mM NaCl) and presence (240 mM NaCl) of salinity stress. The bars denoted by a different letter show significant differences as determined by Duncan's test ($P < 0.05$) following the completion of the three-way ANOVA

exhibited the highest rise in leaves number, leaf area, and shoot dry matter (50%, 102%, and 313% respectively, after 10 months of cultivation and 63%, 93%, and 226% respectively, after 14 months of cultivation). However, the mycorrhized date palms subjected to 240 mM NaCl showed the greatest improvement of plant height and root dry biomass compared to the untreated palms grown under the same conditions, with 15% and 156% improvement respectively, after 10 months of cultivation and 17% and 139% improvement respectively, after 14 months of cultivation. Under salinity, the largest increase in these traits, when comparing the two harvests, was observed in compost-treated plants where a 76% improvement was recorded in comparison to the obtained values after 10 months of cultivation.

The roots of uninoculated date palms showed no AMF colonization (Fig. 11.1). Mycorrhization frequency and intensity were negatively affected under salinity. The greatest decline in the percentages of colonization was observed during the first harvest in roots of salt-affected mycorrhizal-inoculated plants (59% and 64% respectively, for AMF colonization frequency and intensity). Under saline condition, a great increase in mycorrhization frequency and intensity was recorded between the two harvests especially for mycorrhizal palms with no compost application.

11.3.2 Nutrients Content

Date palms treated with salinity had lower P, N, K, and Ca contents in comparison to the non-stressed ones (Table 11.2). The largest decrease (43%) was recorded for P uptake in the stressed control plants after 10 months of cultivation. Still, the different biofertilizers applied separately or in combination induced great enhancement in the nutrient's levels especially in AMF + compost-treated palms. Under 240 mM NaCl, N content recorded the highest enhancement between 10 and 14 months of cultivation when plants were treated with AMF + compost.

However, toxic ions (Na and Cl) concentrations showed significant increase in the date palms with the application of saline stress (Table 11.2). The largest increases were recorded for Na content in untreated stressed palms regardless of the harvest period. Plants treated with biofertilizers applied separately or in combination showed a great decrease in these ions levels in comparison to non-treated stressed palms. The largest decrease (30%) was observed in Na concentrations in AMF + compost-treated palms with the second harvest. Under stressful condition, toxic ions content showed significant decline in treated palms between 10 and 14 months of cultivation. The greatest reduction of 17% was observed for Na content when AMF were applied alone.

As indicated in Table 11.3, salinity application also induced a clearly significant decline in Ca/Na and K/Na ratios in stressed palms in comparison to the untreated ones. This decline was maximal in stressed control plants, which recorded 90% and 89% decline respectively, in shoot Ca/Na and K/Na ratios after 10 months of cultivation and 78% and 70% decline respectively, in root Ca/Na and K/Na ratios after 14 months of cultivation. Under 240 mM NaCl, biofertilizers, applied alone or combined, significantly improved both parameters, particularly in AMF+compost-treated plants after 14 months of cultivation. Under these conditions, the two ratios were mainly improved between 10 and 14 months of cultivation with a maximum increase (41%) in K/Na ratio in the shoot of palms treated with AMF alone.

11.3.3 Water Status and Physiological Parameters

Salt-affected date palms recorded a clear decline in physiological and water status traits compared to palms treated with 0 mM NaCl (Table 11.4 and Fig. 11.2). The largest reduction was observed for the content of chlorophyll a in the untreated stressed plants after 14 months of cultivation with a decrease of 58%. Under salinity, the supplementation with the different biofertilizers, alone or combined, significantly improved these traits in both harvests in comparison to the control. The great

Table 11.2 Effects of AMF and compost applied separately or in combination on date palm phosphorus (P), nitrogen (N), potassium (K), calcium (Ca), sodium (Na), and chlorine (Cl) uptake after 10 and 14 months of cultivation in the absence (0 mM NaCl) and presence (240 mM NaCl) of salinity stress

		10 months of cultivation						14 months of cultivation					
NaCl level	Treatments	P	N	K	Ca	Na	Cl	P	N	K	Ca	Na	Cl
0 mM	Control	2.97 ± 0.41^{jk}	2.61 ± 0.24^{h}	15.08 ± 1.32^{g}	16.07 ± 0.98^{f}	29.26 ± 1.30^{j}	42.93 ± 2.39^{i}	3.61 ± 0.22^{j}	2.64 ± 0.10^{h}	13.41 ± 1.35^{hi}	13.92 ± 0.76^{g}	28.60 ± 1.58^{j}	44.72 ± 3.83^{hi}
	Compost	8.16 ± 0.77^{f}	5.49 ± 0.06^{a}	29.55 ± 2.32^{a}	20.12 ± 0.29^{b}	36.21 ± 1.42^{h}	49.72 ± 1.48^{g}	6.87 ± 0.30^{g}	4.51 ± 0.22^{b}	25.55 ± 1.13^{b}	17.99 ± 0.69^{d}	34.88 ± 1.78^{hi}	52.89 ± 3.73^{f}
	AMF	10.86 ± 0.99^{d}	4.35 ± 0.08^{c}	21.64 ± 0.66^{cd}	20.91 ± 0.28^{b}	38.91 ± 3.96^{g}	49.17 ± 3.16^{g}	9.94 ± 2.24^{e}	4.02 ± 0.28^{d}	21.64 ± 1.48^{cd}	20.78 ± 1.79^{b}	32.57 ± 2.39^{i}	47.28 ± 3.48^{gh}
	Compost+AMF	19.49 ± 1.01^{a}	4.62 ± 0.08^{b}	20.40 ± 1.14^{d}	23.71 ± 0.61^{a}	40.24 ± 3.58^{g}	54.22 ± 2.39^{f}	16.11 ± 0.63^{b}	4.64 ± 0.36^{b}	22.73 ± 1.37^{c}	20.71 ± 0.61^{b}	34.91 ± 3.08^{hi}	53.22 ± 3.00^{f}
240 mM	Control	1.69 ± 0.18^{l}	1.94 ± 0.10^{i}	9.47 ± 0.81^{j}	12.13 ± 0.64^{h}	91.03 ± 2.06^{a}	89.75 ± 3.18^{a}	2.60 ± 0.36^{k}	1.82 ± 0.06^{i}	7.80 ± 1.02^{k}	10.13 ± 0.60^{i}	92.36 ± 4.05^{a}	87.20 ± 5.58^{a}
	Compost	5.44 ± 1.06^{h}	3.30 ± 0.06^{f}	16.87 ± 1.58^{f}	16.97 ± 0.99^{e}	79.43 ± 1.12^{b}	80.57 ± 1.92^{b}	4.70 ± 0.15^{i}	2.76 ± 0.13^{h}	17.20 ± 1.15^{f}	13.31 ± 0.83^{g}	72.76 ± 3.81^{c}	79.08 ± 3.62^{bc}
	AMF	6.61 ± 0.06^{g}	3.33 ± 0.08^{f}	13.20 ± 0.75^{i}	17.96 ± 1.17^{d}	77.27 ± 1.86^{b}	76.92 ± 0.21^{c}	5.63 ± 0.10^{h}	3.38 ± 0.21^{f}	14.53 ± 0.93^{gh}	17.63 ± 0.38^{de}	63.93 ± 1.34^{e}	72.15 ± 2.72^{d}
	Compost+AMF	16.29 ± 0.06^{b}	3.00 ± 0.08^{g}	16.35 ± 0.72^{f}	19.16 ± 0.50^{c}	67.43 ± 2.43^{d}	69.65 ± 3.37^{d}	13.61 ± 0.62^{c}	3.56 ± 0.24^{e}	19.01 ± 2.21^{e}	17.49 ± 1.28^{de}	58.43 ± 2.72^{f}	64.35 ± 1.49^{e}

The values of each column denoted by a different letter show significant differences as determined by Duncan's test ($P < 0.05$) following the completion of the three-way ANOVA

Table 11.3 Effects of AMF and compost applied separately or in combination on date palm shoot and root K/Na and Ca/Na ratios after 10 and 14 months of cultivation in the absence (0 mM NaCl) and presence (240 mM NaCl) of salinity stress

		10 months of cultivation				14 months of cultivation			
		Shoot		Root		Shoot		Root	
NaCl level	Treatments	K/Na	Ca/Na	K/Na	Ca/Na	K/Na	Ca/Na	K/Na	Ca/Na
0 mM	Control	1.61 ± 0.03[b]	1.27 ± 0.04[a]	0.23 ± 0.04[cd]	0.36 ± 0.01[ef]	1.19 ± 0.04[f]	0.88 ± 0.10[d]	0.24 ± 0.01[cd]	0.36 ± 0.02[f]
	Compost	1.98 ± 0.08[a]	1.03 ± 0.05[c]	0.47 ± 0.05[a]	0.42 ± 0.02[d]	1.58 ± 0.12[b]	0.86 ± 0.07[d]	0.44 ± 0.01[a]	0.40 ± 0.01[de]
	AMF	1.54 ± 0.05[bc]	1.00 ± 0.04[c]	0.26 ± 0.03[c]	0.40 ± 0.02[de]	1.29 ± 0.05[e]	0.86 ± 0.06[d]	0.39 ± 0.01[b]	0.54 ± 0.03[a]
	Compost+AMF	1.48 ± 0.04[cd]	1.13 ± 0.05[b]	0.27 ± 0.03[c]	0.46 ± 0.04[c]	1.41 ± 0.03[d]	0.86 ± 0.06[d]	0.38 ± 0.02[b]	0.50 ± 0.03[b]
240 mM	Control	0.16 ± 0.02[j]	0.14 ± 0.01[h]	0.06 ± 0.01[g]	0.13 ± 0.01[jk]	0.13 ± 0.01[j]	0.11 ± 0.01[h]	0.05 ± 0.01[g]	0.11 ± 0.01[k]
	Compost	0.31 ± 0.03[hi]	0.26 ± 0.01[fg]	0.16 ± 0.01[e]	0.19 ± 0.01[hi]	0.38 ± 0.02[h]	0.23 ± 0.01[g]	0.15 ± 0.01[ef]	0.15 ± 0.01[ij]
	AMF	0.26 ± 0.01[i]	0.26 ± 0.02[fg]	0.11 ± 0.01[f]	0.22 ± 0.01[h]	0.32 ± 0.01[hi]	0.30 ± 0.01[efg]	0.16 ± 0.01[e]	0.26 ± 0.01[g]
	Compost+AMF	0.38 ± 0.02[h]	0.32 ± 0.01[ef]	0.16 ± 0.01[e]	0.26 ± 0.02[g]	0.52 ± 0.04[g]	0.35 ± 0.01[e]	0.21 ± 0.02[d]	0.27 ± 0.01[g]

The values of each column denoted by a different letter show significant differences as determined by Duncan's test ($P < 0.05$) following the completion of the three-way ANOVA

Table 11.4 Effects of AMF and compost applied separately or in combination on date palm physiological and water status parameters after 10 and 14 months of cultivation in the absence (0 mM NaCl) and presence (240 mM NaCl) of salinity stress

		10 months of cultivation				14 months of cultivation			
NaCl level	Treatments	g_s (mmol m^{-2} s^{-1})	LWP (Bar)	RWC (%)	F_v/F_m	g_s (mmol m^{-2} s^{-1})	LWP (Bar)	RWC (%)	F_v/F_m
0 mM	Control	17.63 ± 1.66^{g}	-23.70 ± 0.98^{g}	73.72 ± 1.35^{b}	0.47 ± 0.03^{f}	24.60 ± 1.22^{ef}	-24.64 ± 0.54^{g}	71.02 ± 1.25^{b}	0.55 ± 0.02^{e}
	Compost	26.50 ± 1.71^{de}	-16.28 ± 3.42^{cd}	80.31 ± 3.22^{a}	0.77 ± 0.02^{ab}	44.50 ± 2.11^{c}	-17.27 ± 2.05^{de}	75.15 ± 0.39^{b}	0.77 ± 0.01^{ab}
	AMF	45.77 ± 4.31^{c}	-15.96 ± 1.44^{cd}	82.81 ± 1.78^{a}	0.78 ± 0.03^{ab}	61.63 ± 1.15^{a}	-14.85 ± 1.75^{cd}	83.51 ± 4.12^{a}	0.79 ± 0.01^{a}
	Compost+AMF	53.43 ± 2.80^{b}	-11.68 ± 1.59^{ab}	80.84 ± 2.12^{a}	0.77 ± 0.01^{ab}	50.60 ± 0.69^{b}	-10.46 ± 1.30^{a}	73.89 ± 1.12^{b}	0.77 ± 0.01^{ab}
240 mM	Control	10.87 ± 1.72^{i}	-31.47 ± 2.81^{h}	53.03 ± 0.76^{e}	0.39 ± 0.01^{h}	14.27 ± 0.85^{h}	-29.84 ± 2.51^{h}	52.71 ± 1.36^{e}	0.41 ± 0.03^{g}
	Compost	18.10 ± 1.93^{gi}	-24.50 ± 0.89^{g}	61.74 ± 3.72^{cd}	0.72 ± 0.01^{d}	22.2 ± 1.44^{f}	-20.61 ± 0.50^{f}	58.55 ± 2.33^{d}	0.74 ± 0.01^{cd}
	AMF	23.00 ± 1.20^{f}	-21.97 ± 1.06^{fg}	65.92 ± 1.14^{c}	0.73 ± 0.01^{d}	26.53 ± 0.60^{de}	-19.61 ± 0.75^{ef}	63.31 ± 1.01^{c}	0.78 ± 0.01^{ab}
	Compost+AMF	26.67 ± 1.72^{de}	-15.30 ± 0.66^{cd}	61.83 ± 2.69^{cd}	0.73 ± 0.01^{cd}	29.57 ± 1.29^{d}	-14.07 ± 0.77^{bc}	57.83 ± 4.25^{d}	0.76 ± 0.01^{bc}

F_v/F_m: photosystem II efficiency, g_s: stomatal conductance, *LWP:* leaf water potential, *RWC:* Leaf relative water content. The values of each column denoted by a different letter show significant differences as determined by Duncan's test ($P < 0.05$) following the completion of the three-way ANOVA

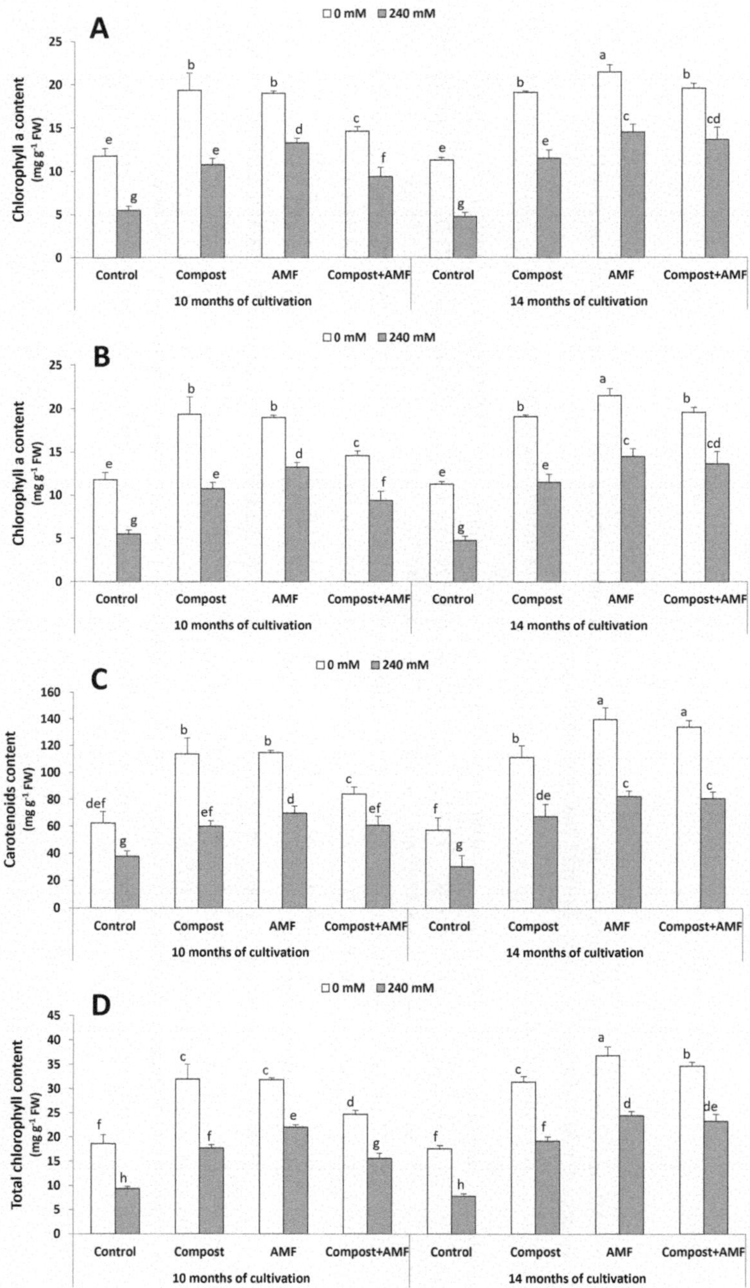

Fig. 11.2 Effects of AMF and compost applied separately or in combination on date palm photosynthetic pigment content (chlorophyll a (**a**), chlorophyll b (**b**), carotenoids (**c**), and total chlorophyll (**d**)) after 10 and 14 months of cultivation in the absence (0 mM NaCl) and presence (240 mM NaCl) of salinity stress. The bars denoted by a different letter show significant differences as determined by Duncan's test ($P < 0.05$) following the completion of the three-way ANOVA

increase (221%) was noticed for chlorophyll b content with the application of AMF alone after 14 months of cultivation. When comparing the recorded values of these parameters in salt-affected palms between 10 and 14 months of cultivation, the content of chlorophyll b showed the highest increase (56%) in stressed AMF + compost-treated plants.

11.3.4 Stress Markers Content

Figure 11.3 indicates an obvious enhancement in H_2O_2 and MDA contents in date palm shoot and root when applying salt stress compared to the untreated palms. A maximum increase (39%) was observed for roots H_2O_2 of the stressed control date palms after 10 months of cultivation. However, biofertilizers application significantly lessened H_2O_2 and MDA levels in date palms under 240 mM NaCl condition. Under the same condition, the lowest values of H_2O_2 and MDA were recorded in AMF + compost-treated plants, except for shoot H_2O_2, which showed the lowest one in AMF treated plants. Under the same conditions and for most of the treated plants, MDA and H_2O_2 contents were significantly reduced between the two harvests.

11.3.5 Organic Osmolytes Content

Stressed plants recorded significant increase in proline and soluble sugar concentrations in shoot and root than the control palms. Compost and/or AMF application further increased these two organic osmolytes concentrations under 240 mM NaCl, especially in AMF + compost-treated plants (Table 11.5). The highest increase (90%) was recorded after 14 months of cultivation for root soluble sugar content when AMF was applied alone. The concentrations of these two parameters showed significant differences in most treatments under salt stress between 10 and 14 months of cultivation. Root TSS concentration recorded the highest variation since this trait was increased by 28% in compost-treated plants.

11.3.6 Activity of Antioxidant Enzymes

Figures 11.4 and 11.5 indicate that the exposure of date palms to salinity significantly increased APX, SOD, CAT, and POD activities in roots as well as in the shoots. Under saline condition, these activities were further stimulated by the application of different biofertilizers where CAT activity recorded the highest improvement (100%) in the shoot of AMF + compost-treated plants after 10 months of cultivation. Under salt stress, these enzymes activity was mainly reduced between

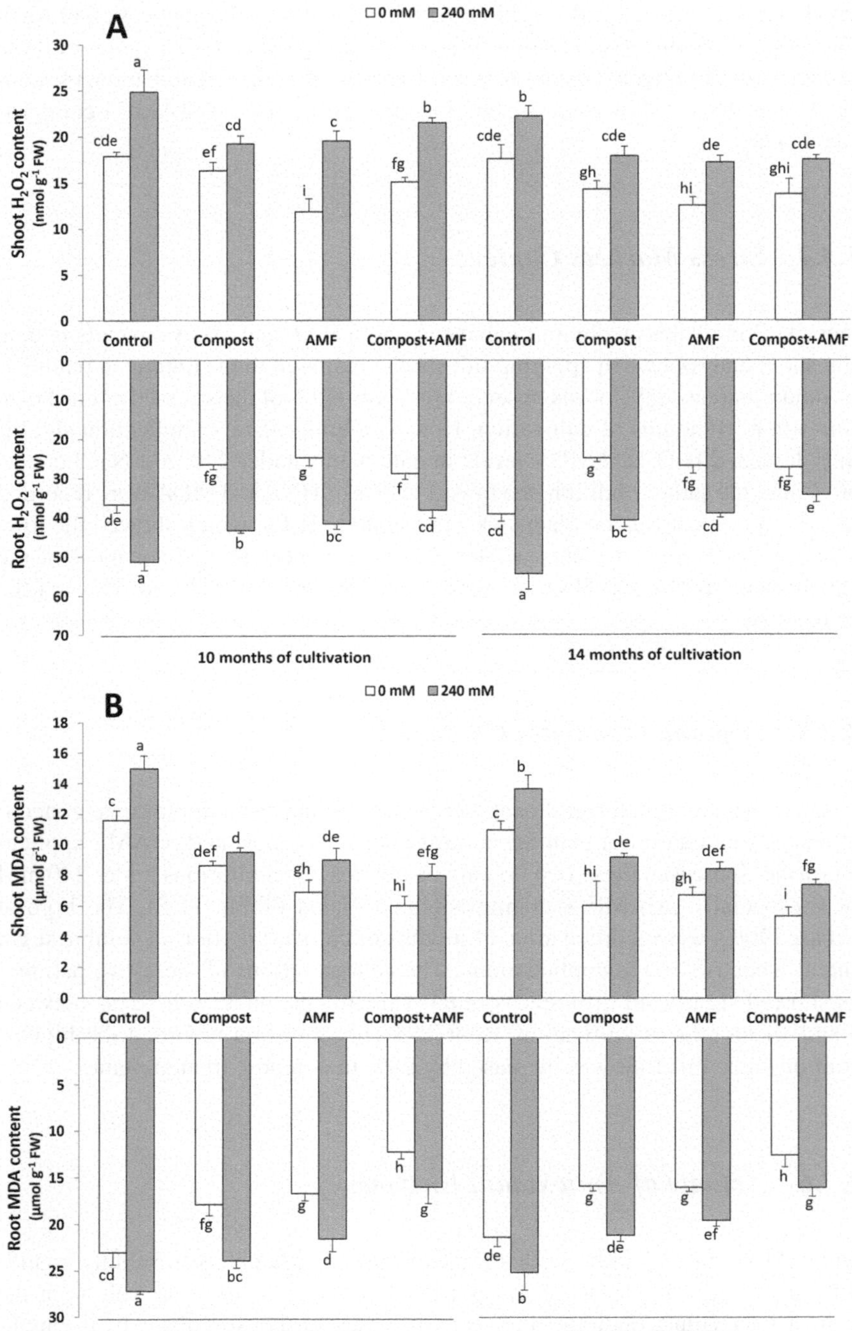

Fig. 11.3 Effects of AMF and compost applied separately or in combination on date palm shoot and root hydrogen peroxide (**a**) and malondialdehyde (**b**) content after 10 and 14 months of cultivation in the absence (0 mM NaCl) and presence (240 mM NaCl) of salinity stress. The bars denoted by a different letter show significant differences as determined by Duncan's test ($P < 0.05$) following the completion of the three-way ANOVA

Table 11.5 Effects of AMF and compost applied separately or in combination on date palm proline and soluble sugar content in shoot and root after 10 and 14 months of cultivation in the absence (0 mM NaCl) and presence (240 mM NaCl) of salinity stress

NaCl level	Treatments	10 months of cultivation				14 months of cultivation			
		Shoot		Root		Shoot		Root	
		Proline (μmol g^{-1} FW)	Soluble sugars (μg mg^{-1} FW)	Proline (μmol g^{-1} FW)	Soluble sugars (μg mg^{-1} FW)	Proline (μmol g^{-1} FW)	Soluble sugars (μg mg^{-1} FW)	Proline (μmol g^{-1} FW)	Soluble sugars (μg mg^{-1} FW)
0 mM	Control	11.61 ± 0.82^{h}	386.31 ± 15.15^{h}	34.53 ± 4.02^{j}	78.85 ± 6.65^{ij}	13.27 ± 1.33^{gh}	383.31 ± 25.01^{h}	37.19 ± 3.45^{j}	75.52 ± 4.29^{j}
	Compost	13.08 ± 0.41^{gh}	443.61 ± 33.47^{g}	51.27 ± 4.43^{fg}	92.73 ± 3.38^{ghi}	14.42 ± 1.23^{efg}	456.53 ± 30.22^{fg}	49.27 ± 1.32^{gh}	88.54 ± 2.76^{hij}
	AMF	16.21 ± 1.95^{de}	555.33 ± 11.87^{c}	53.98 ± 1.70^{f}	138.54 ± 17.82^{cd}	15.54 ± 0.84^{def}	516.61 ± 23.61^{cde}	48.98 ± 1.97^{gh}	140.10 ± 4.18^{cd}
	Compost+AMF	13.53 ± 1.56^{fg}	568.97 ± 1.75^{bc}	44.13 ± 3.15^{i}	127.51 ± 4.61^{de}	12.53 ± 0.47^{gh}	497.97 ± 30.25^{def}	45.80 ± 1.90^{hi}	130.84 ± 9.50^{de}
240 mM	Control	19.13 ± 1.14^{c}	435.89 ± 33.71^{g}	65.19 ± 3.65^{de}	96.75 ± 3.41^{gh}	17.46 ± 0.56^{cd}	467.01 ± 29.79^{efg}	61.85 ± 2.15^{e}	105.38 ± 11.55^{fg}
	Compost	23.35 ± 0.85^{b}	563.30 ± 43.87^{c}	79.25 ± 1.30^{ab}	118.22 ± 1.74^{ef}	22.23 ± 0.95^{b}	536.79 ± 23.62^{cd}	69.25 ± 1.30^{cd}	152.23 ± 8.62^{bc}
	AMF	26.79 ± 1.94^{a}	656.73 ± 14.77^{a}	81.99 ± 1.81^{a}	182.46 ± 10.89^{a}	21.45 ± 1.38^{b}	616.61 ± 23.61^{ab}	70.33 ± 1.46^{c}	190.10 ± 12.15^{a}
	Compost+AMF	23.27 ± 0.88^{b}	649.62 ± 45.15^{a}	75.85 ± 3.09^{b}	165.83 ± 3.42^{b}	21.61 ± 0.89^{b}	858.97 ± 38.80^{a}	64.52 ± 0.79^{e}	188.54 ± 2.76^{a}

The values of each column denoted by a different letter show significant differences as determined by Duncan's test ($P < 0.05$) following the completion of the three-way ANOVA

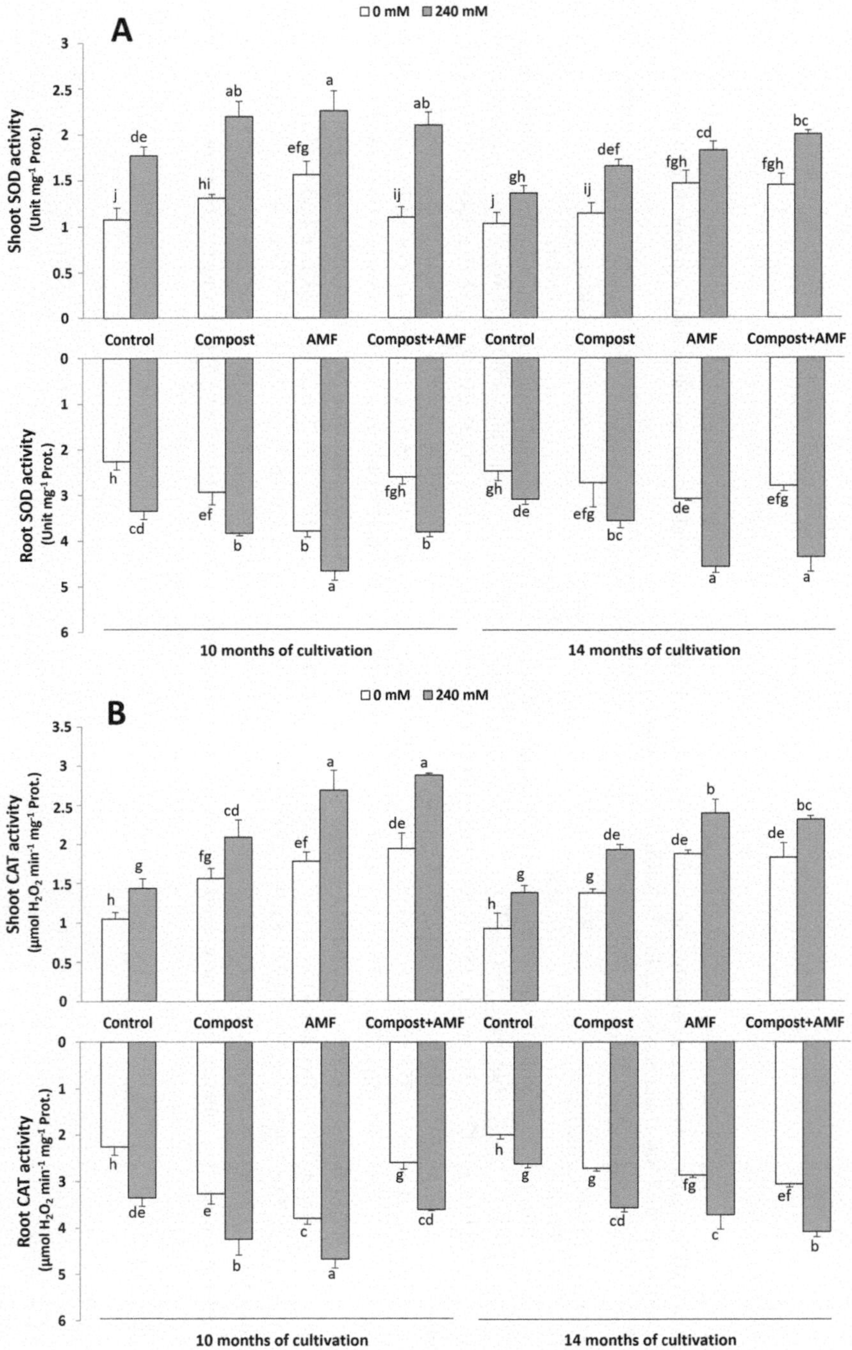

Fig. 11.4 Effects of AMF and compost applied separately or in combination on date palm shoot and root SOD (**a**) and CAT (**b**) antioxidant enzyme activity after 10 and 14 months of cultivation in the absence (0 mM NaCl) and presence (240 mM NaCl) of salinity stress. The bars denoted by a different letter show significant differences as determined by Duncan's test ($P < 0.05$) following the completion of the three-way ANOVA

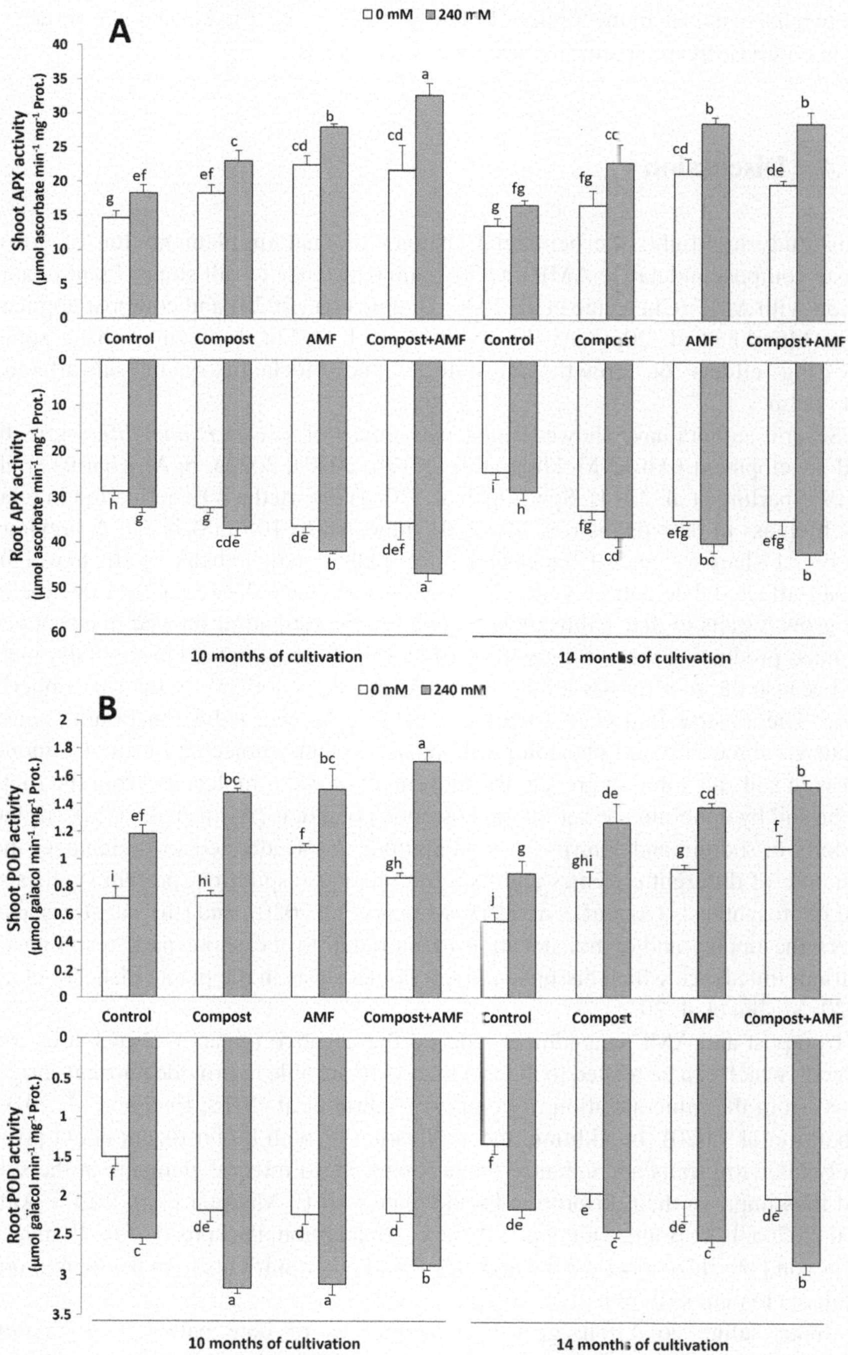

Fig. 11.5 Effects of AMF and compost applied separately or in combination on date palm shoot and root APX (**a**) and POD (**b**) antioxidant enzyme activity after 10 and 14 months of cultivation in the absence (0 mM NaCl) and presence (240 mM NaCl) of salinity stress. The bars denoted by a different letter show significant differences as determined by Duncan's test ($P < 0.05$) following the completion of the three-way ANOVA

the two harvests where the highest decline (25%) was recorded in shoot SOD activity in compost-treated palms between the two harvests.

11.4 Discussion

This study highlights the beneficial changes of dual supplementation of green waste compost and native AMF on date palm resistance to salt stress. Plant inoculation with AMF (Chebaane et al. 2020; Diagne et al. 2020) and compost application (Mbarki et al. 2020; Ait-El-mokhtar et al. 2022a, b) seem to have some boosting effects on growth, physiology, and biochemistry of salt-affected date palm.

Several authors have showed a negative impact of salt on *P. dactylifera* growth and development (Ait-El-Mokhtar et al. 2019b, 2020a, 2022a, b; Al Kharusi et al. 2019; Sperling et al. 2014). Sperling et al. (2014) reported a 40% reduction in total dry biomass in date palms (cv. Medjoul) subjected to 105 mM NaCl. A previous study (Al Kharusi et al. 2019) revealed a clear reduction in both dry and fresh weight of salt-affected date palm (cv. Zabad). The present study showed a clear decline in the growth traits of date palms under stress. Our investigation showed that root dry biomass production was strongly lowered by salt stress compared to shoot dry matter because the root tissues are in contact with excess soil salt in the rhizospheric zone. The adverse impact of excess soil salinity on date palm functioning could occur via three different physiological paths: (i) plants subjected to small osmotic stress in soil salt solution prevent the movement of water molecules from the roots to the soil by upholding lesser internal osmotic potentials (Ayub et al. 2020), (ii) the toxicity of sodium and chlorine ions, ubiquitous in salt-affected soils, damages the structure of different enzymes and therefore disrupt respiration, protein synthesis, and photosynthesis (Ashraf et al. 2018; Munns et al. 2020), and (iii) salt stress also alters the uptake and/or translocation of nutrients to the shoot part, resulting in nutrient imbalance which disrupts nutrient homeostasis in the plant (Elsheery et al. 2020; Mishra et al. 2021).

Compost and AMF amendment reduced the negative impact of NaCl stress on growth, which can be related to the fact that AMF are able to provide nutrients originated from the mineralization of compost (Ashraf et al. 2018; Evelin et al. 2019; Mbarki et al. 2020). In addition, the application of both biofertilizers could positively alter soil traits and enhance the accessibility to mineral elements in the soil and thus improve their absorption by the plant (Ait-El-Mokhtar et al. 2020a; Anli et al. 2020a, b, c; Boutasknit et al. 2020a, b). In addition, the spread of AMF mycelia beyond the rhizospheric zone acquires P and other nutrients to make them more available to plants (Evelin et al. 2019).

Under saline condition, growth performances of date palms were mainly increased with the combined supplementation of AMF and compost, despite the fact that the application of compost and salinity hinders mycorrhizal colonization. Nourinia et al. (2007) revealed that AMF barley root infection was reduced by salinity, likely linked to the toxicity of Na^+ and Cl^- on these fungi. Yarahmadi et al. (2018) revealed the adverse effect of salt on root AMF colonization by inhibiting hyphae development and spore germination or by reducing the number of arbuscles and the spread of AMF colonization. The adverse effects of salt on AMF root colonization may further be linked to the accumulation of H_2O_2 in mycorrhized plants roots, which can cause arbuscules degradation (Hause and Fester 2005). In the same vein, the reduction in AMF root colonization rate in the compost-amended soil has been highlighted by several works (Anli et al. 2020a, c; Boutasknit et al. 2020b, 2021a, b) and may be linked to the increasing delivery of P that may cause a reduction in the percentage of mycorrhizal infection (Ben-Laouane et al. 2021). Furthermore, compost OM includes products of decomposition that could induce AMF inhibition (Gryndler et al. 2008).

The application of salt and the two biofertilizers further affected date palm mineral uptake. In the present investigation, the content of N, P, K^+, and Ca^{2+} and the ratios Ca^{2+}/Na^+ and K^+/Na^+ in *P. dactylifera* were reduced under salt. However, sodium and chlorine contents were increased in date palms exposed to salinity. The imbalance of many nutrients figures among the greatest direct effect of salinity as reported by multiple authors (Ait-El-Mokhtar et al. 2020a; Ben-Laouane et al. 2020). Nutrient hemostasis plays a determining role in several metabolic functions such as photosynthesis (Ashraf et al. 2018). This effect may be due to the synthesis of secondary compounds that boost different plant processes via improving the uptake of essential nutrients (de Oliveira et al. 2019; Ayub et al. 2020). Salt stress disrupts the uptake of micro and macro nutrients since sodium ions present an antagonistic effect on the uptake of these elements (Hashem et al. 2016). Salinity also limits the mobility and transport of Ca^{2+} and K^+ elements to growing plant parts, which affects the integrity of the vegetative and reproductive organs.

Compost and AMF addition has been shown to boost plant metabolic functioning, specifically nutrient absorption, under salinity (Chaichi et al. 2017; Mbarki et al. 2020; Ait-El-Mokhtar et al. 2022a, b). In this study, the increase in mineral elements absorption under salt regime was higher when plants were grown with a dual application of compost and AMF. In a previous study, Chaichi et al. (2017) showed that AMF + compost-treated tomato presents an increase in nutrients acquisition such as K^+, Mg^{2+}, and Ca^{2+}, under saline regime. Furthermore, Sallaku et al. (2019) demonstrated an evident improvement in essential mineral elements uptake in mycorrhized plants, whereas mycorrhizae lowered Na^+ acquisition. In addition, inhibition of sodium ion uptake by the roots of AMF treated plants and/or reduction of sodium ions translocation in composted soil could be another mechanism to promote plant adjustment to salinity (Mbarki et al. 2020; Ait-El-Mokhtar et al. 2022a, b). Compost and AMF application could alleviate the negative impact of toxic ions. Miransari (2010) reported that AMF can store Na^+ and Cl^- in vesicle vacuoles under saline stress. In addition, it has been shown that organic amendments are directly related to Na^+ elimination from the soil, either through its leaching or through its adsorption (Chaganti and Crohn 2015; Drake et al. 2016).

The increased absorption of K^+ and Ca^{2+} is another mechanism that allows the treated plants to show a better tolerance to salinity. K^+ is a key element for plant functioning since it is required for osmotic adjustment, regulates stomatal opening and closing, and performs a pivotal role in protein biosynthesis (Selvakumar et al. 2018). Calcium is considered a second messenger, and under different stresses, its content is increased to initiate signaling pathways. Several authors have highlighted an increase of calcium ion concentrations in AMF treated plants under saline stress (Ait-El-Mokhtar et al. 2019b, 2020a). Furthermore, OM amendments have been shown to increase P and N content of salinity-stressed crops (Mbarki et al. 2020). The large number of saprophyte microorganisms in the applied organic amendment may promote the accessibility to nutrients by plants after OM mineralization (Abud-Archila et al. 2018). The improved resistance of date palms to NaCl stress by enhancing P uptake may be linked to the involvement of P in Na^+ compartmentalization and thus in the preservation of the cell membrane from toxic ions effects (Bothe 2012). This prevents the disruption of metabolism paths by Na^+ ions, thus limiting the adverse impact of salt (Khalid and Teixeira da Silva 2010). Our findings indicate that Ca^{2+}/Na^+ and K^+/Na^+ ratios in mycorrhized date palms grown in amended substrate were higher in the presence of salinity compared to stressed untreated plants. The increased values of these ratios proved to be a significant factor to reduce the salinity effects on the plants due to the antagonism between Ca^{2+} and K^+ ions and Na^+ ion for membrane transporter (Munns et al. 2020). Thereby, greatest ratios preserve photosynthesis components by eliminating Na^+ input and promote cell signaling transduction and hydraulic conductivity (Evelin et al. 2019).

Saline stress damages the photosynthesis components in numerous paths. Certainly, salt-induced osmotic stress hampers g_s, which lowers CO_2 assimilation (Sharma et al. 2020). In addition, salinity negatively alters the functioning of PSII in the photochemical reactions phase (Ait-El-Mokhtar et al. 2020a; Toubali et al. 2020). In this investigation, a decline in photosynthetic pigments content was recorded with a decline in chlorophyll fluorescence in salt environment. This detrimental impact is related to the direct effect of NaCl stress on chloroplast inducing their destruction and disruption in the functioning of many enzymes implicated in the biosynthesis of chlorophyll (Sharma et al. 2020). In addition, the low values of chlorophyll concentrations can be explained by the reduced uptake of nutrients, particularly Mg^{2+}, in saline environment (Ibrahim et al. 2011).

Under salt stress, photosynthetic traits (stomatal conductance, photosynthetic efficiency, and chlorophyll concentrations) were substantially enhanced by AMF + compost application. Improved nitrogen absorption owing to the biofertilizers supplementation may boost photosynthesis as nitrogen figures among the key component of photosynthesis enzymes (Ahanger et al. 2014). Similarly, the release of Mg^{2+} from compost and its possible assimilation by AMF may be considered as a potential mechanism for increased photosynthetic capacity (Sharma et al. 2020). In addition, increased concentrations of proline and glycine betaine in mycorrhized plants could prevent the damage of CO_2 assimilation enzymes including RuBisCO and Rubisco activase and PSII pigment-protein complexes without forgetting their involvement in osmotic adaptation strategies (Marček et al. 2016).

The water status traits were also negatively affected by salinity. In fact, salt-affected palms recorded an evident decline in leaf water potential and relative water content (RWC). The decrease in these parameters could be due in part to the changes induced by NaCl on the electrical potential of the plasma membrane, which disrupts water uptake inducing a water deficit (Munns et al. 2020). Consequently, NaCl prevents water movement from soil into the plant. In contrast, the application of compost and AMF clearly enhanced the above-mentioned traits in salt-stressed palms, since we recorded higher leaf water potential and RWC values. This effect could be related to the key role of AMF in improving water uptake capacity and hydraulic conductivity under salt stress (Evelin et al. 2019). Conversely, compost and AMF may have altered soil structure by inducing the formation of aggregates that improve soil water availability (Anli et al. 2021; Boutasknit et al. 2021a, b).

The increase in H_2O_2 and MDA levels in date palm under 240 mM NaCl was coupled with a decrease in biomass accumulation. Many authors supported the same findings in several plant species such as *Brassica juncea* L. (Hashem et al. 2015), *P. dactylifera* (Ait-El-Mokhtar et al. 2020a; Toubali et al. 2020), and *Medicago sativa* L. (Ben-Laouane et al. 2020). Root tissues showed higher ROS damage than shoot tissues, as roots are in direct contact with high salt concentrations in the rhizospheric soil solution. ROS react with unsaturated membrane lipids, inducing high membrane damage (Evelin et al. 2019). *P. dactylifera* grown in the presence of 240 mM NaCl and treated with AMF + compost recorded lower concentrations of MDA and H_2O_2 compared to untreated stressed date palms, resulting in limited ROS levels and loss of membrane integrity in treated date palm. This result could be related to the improvement of the antioxidant defense system that removes free radicals before reacting with biomolecules and therefore reduces ROS damages (Foyer 2018).

Soluble sugars and proline constitute the most common metabolites involved in the osmotic adjustment under salt stress (Rao and Chaitanya 2016; Fatima and Arora 2023). Our investigation showed increased levels of soluble sugars and proline in palms affected by salinity. These levels were further enhanced in compost and AMF treated date palm seedlings under the same conditions. These changes could promote osmotic adjustment and mitigate free radical effects, which constitute a defensive strategy to improve date palm tolerance to salinity stress (Evelin et al. 2019). In the same vein, Chinsamy et al. (2013) and Talaat and Shawky (2014) corroborate our findings since they reported higher concentrations of soluble sugar and proline in the presence of the two biofertilizers under salt stress.

Antioxidant enzymes enable the elimination of ROS and therefore, the reduction of the damages caused by oxidative stress on many sensitive biomolecules such as DNA, proteins, pigments, and lipids (Evelin et al. 2019). Stimulation of ROS-eliminating enzymes including CAT, POD, SOD, and APX is the most common strategy to scavenge ROS toxicity under stressful conditions (Foyer 2018). As a result, plants tolerance to salinity is closely linked to the boosting of antioxidant enzyme activity and thereby the mitigation of the oxidative damages. Current results demonstrate that the application of salt stress promoted the antioxidant enzyme activities of *P. dactylifera* seedlings. Our results are corroborated with the results of

Elsheery et al. (2020) in *Cucumis sativus*, Ben-Laouane et al. (2020) in *Medicago sativa* L., and Ait-El-Mokhtar et al. (2020a) and Toubali et al. (2020) in *P. dactylifera* L. High antioxidant activity in saline environments helps plants eliminate ROS and maintain their values below harmful levels. SOD is the first defense enzyme and allows the transformation of superoxide radicals into hydrogen peroxide and oxygen. The resulted H_2O_2 is then converted to oxygen and water via APX or CAT (Mittler 2002). The action of POD can also remove H_2O_2 in the apoplast of lignified tissues (Ros-Barceló et al. 2006).

Under saline conditions, applied biofertilizers improved the antioxidant enzymes activity evaluated in date palm seedlings, especially those treated with AMF + compost. The improved antioxidant activities help plants to rapidly eliminate toxic ROS and thus reduce damage to metabolic processes. Our findings showed an enhancement in SOD activity, which was linked to the increase in the activity of H_2O_2-scavenging enzymes, namely CAT, POD, and APX. The same results were reported by Kamanga et al. (2020), Ait-El-Mokhtar et al. (2020a), and Toubali et al. (2020). A previous study indicated that mycorrhized *Lycopersicon esculentum* seedlings grown under salt stress maintained greater SOD, CAT, and POD activities in comparison to uninoculated seedlings affected by salinity (Abouelsaad and Renault 2018). Lucho et al. (2019) also reported an enhancement of antioxidant system defense in wheat colonized by AMF under salinity. In addition, Kamanga et al. (2020) demonstrated that compost application in *Solanum lycopersicum* L. stimulates the activity of the antioxidant system to prevent ROS damages. The induction of antioxidant enzyme activity in stressed AMF + compost-treated plants in comparison to stressed control plants was closely linked to reduced stress marker (H_2O_2 and MDA) levels, and thus limited oxidative damage in treated date palms coupled with an effective ROS scavenging system.

Our findings indicate that the strategies underlying the tolerance of the date palm to soil salt are gradually enhanced over time with the application of biofertilizers and particularly in AMF + compost-treated palms. Between the two harvests, growth traits were significantly improved with nutrient (P, N, and K^+) concentrations increasing and Na^+ and Cl^- contents decline. The substantial improvement in antioxidant enzymes activity in treated palms under salinity after 5 months of salt application possibly run a great decline in H_2O_2 and MDA accumulation. This decline is more remarkable after 9 months of saline stress (14 months of cultivation). The reduced obtained concentrations of the H_2O_2 and MDA may well induce a negative feedback action on antioxidant enzymes activity after 14 months of cultivation through lowering it. The recorded decrease in stress markers levels after 14 months of cultivation may also induce the decline of proline and soluble sugars.

11.5 Future Perspectives

Salinity is a crucial stressor that disrupts date palm performances. Recently, significant advances have been made to unveil the machineries and the role of biostimulants, including AMF, plant growth promoting rhizobacteria (PGPR), and organic amendments, on *P. dactylifera* development and growth and how they help this tree to survive under stressful environment (Meddich et al. 2018, 2020; Ait-El-Mokhtar et al. 2019b, 2020a; Saadaoui et al. 2019; El Kinany et al. 2019; Toubali et al. 2020; Anli et al. 2020a; Chebaane et al. 2020; Mohamed et al. 2020). Under salt stress, date palm performances are boosted by beneficial microorganisms and composts. Indeed, these biostimulants are involved in salinity adjustment mechanisms via inducing stomatal closure, enhancing water use efficiency, osmotic adjustment, reduction of stress markers, and boosting antioxidant defense system (Ait-El-Mokhtar et al. 2019b, 2020a, 2022a, b; Chebaane et al. 2020; Toubali et al. 2020). Although biostimulants have great effect on mitigating salt stress, their signaling pathways can be complex because of the wide diversity in plant physiology traits and stress extent and intensity. Despite recent advances in omics-based research (Jamil et al. 2020; Xu et al. 2021; Zancarini et al. 2021; Doni et al. 2022), these techniques should investigate many gaps especially the mechanisms orchestrating the molecular interactions between biostimulants and date palm at the cellular and transcriptional levels in plants. Therefore, genome-wide studies focusing on the biostimulants' signaling responses in the date palm under challenging environmental conditions are required and will significantly advance our knowledge of the exchanges among the host plant and the biostimulants in the future, enhancing crop production and stress tolerance.

11.6 Conclusion

Salt stress has significantly reduced date palm growth, physiology, and biochemistry characteristics. Plants supplemented with the combination of AMF and compost recorded a great improvement in these traits despite the applied saline stress. The used biofertilizers, applied alone or combined, improved hydric status, photosynthesis attributes, nutrient uptake, osmolytes content, and antioxidant system and reduced stress markers and toxic ions absorption under salt regime. Plants tolerance to salinity varies over time where antioxidant enzyme activity and organic osmolytes were reduced between 10 and 14 months of cultivation as a feedback to the decline in stress markers accumulation, indicating a boosting of plant's tolerance to salinity. Our findings sturdily propose the combined use of compost and AMF to mitigate the negative impact of NaCl stress and to lift crops performances in salt-affected soils in degraded ecosystems such as the Moroccan palm groves.

References

Abouelsaad I, Renault S (2018) Enhanced oxidative stress in the jasmonic acid-deficient tomato mutant def-1 exposed to NaCl stress. J Plant Physiol 226:136–144. https://doi.org/10.1016/j.jplph.2018.04.009

Abud-Archila M, Luján-Hidalgo MC, López-Pérez JM, Ordaz-Rivera J, Ruiz-Valdiviezo VM, Oliva-Llaven MÁ, Gutiérrez-Miceli F (2018) Growth and fruit chemical characteristics of Blackberry (*Rubus Fruticosus*) cultivated with vermicompost, *Glomus Mosseae* and phosphate rock. Compost Sci Util 26:225–231. https://doi.org/10.1080/1065657X.2018.1471367

Aebi H (1984) Catalase *in vitro*. Methods Enzymol 105:121–126. https://doi.org/10.1016/S0076-6879(84)05016-3

Ahanger MA, Hashem A, Abd-Allah EF, Ahmad P (2014) Arbuscular mycorrhiza in crop improvement under environmental stress. In: Ahmad P, Rasool S (eds) Emerging technologies and management of crop stress tolerance. A sustainable approach, vol 2. Elsevier, pp 69–95. https://doi.org/10.1016/B978-0-12-800875-1.00003-X

Ait Rahou Y, Ait-El-Mokhtar M, Anli M, Boutasknit A, Ben-Laouane R, Douira A et al (2020) Use of mycorrhizal fungi and compost for improving the growth and yield of tomato and its resistance to *Verticillium dahliae*. Arch Phytopathol Plant Protect:1–26. https://doi.org/10.1080/03235408.2020.1854938

Ait-El-Mokhtar M, Ben Laouane R, Anli M, Boutasknit A, Fakhech A, Wahbi S et al (2019a) Climate change and its impact on oases ecosystem in Morocco. In: Climate change and its impact on ecosystem services and biodiversity in Arid and semi-Arid zones, pp 217–245. https://doi.org/10.4018/978-1-5225-7387-6

Ait-El-Mokhtar M, Laouane RB, Anli M, Boutasknit A, Wahbi S, Meddich A (2019b) Use of mycorrhizal fungi in improving tolerance of the date palm (*Phoenix dactylifera* L.) seedlings to salt stress. Sci Hortic (Amsterdam) 253:429–438. https://doi.org/10.1016/j.scienta.2019.04.066

Ait-El-Mokhtar M, Baslam M, Ben-Laouane R, Anli M, Boutasknit A, Mitsui T et al (2020a) Alleviation of detrimental effects of salt stress on date palm (*Phoenix dactylifera* L.) by the application of arbuscular mycorrhizal fungi and/or compost. Front Sustain Food Syst 4:131. https://doi.org/10.3389/fsufs.2020.00131

Ait-El-Mokhtar M, Fakhech A, Anli M, Ben-Laouane R, Boutasknit A, Wahbi S et al (2020b) Infectivity of the palm groves arbuscular mycorrhizal fungi under arid and semi-arid climate and its edaphic determinants towards efficient ecological restoration. Rhizosphere 15:100220. https://doi.org/10.1016/j.rhisph.2020.100220

Ait-El-Mokhtar M, Boutasknit A, Ben-Laouane R, Anli M, El Amerany F, Toubali S et al (2021) Vulnerability of oasis agriculture to climate change in Morocco. In: Research anthology on environmental and societal impacts of climate change, vol 4. IGI Global, pp 76–106. https://doi.org/10.4018/978-1-7998-3343-7.ch004

Ait-El-Mokhtar M, Ben-Laouane R, Boutasknit A, Anli M, Fakhech A, Ait-Rahou Y et al (2022a) Evaluation of young date palm tolerance to salinity stress under arbuscular mycorrhizal fungi and compost application. Environ Sci Proc 16(1):15. https://doi.org/10.3390/environsciproc2022016015

Ait-El-mokhtar M, Fakhech A, Ben-Laouane R, Anli M, Boutasknit A, Ait-Rahou Y et al (2022b) Compost as an eco-friendly alternative to mitigate salt-induced effects on growth, nutritional, physiological and biochemical responses of date palm. Int J Recycl Org Waste Agric 11:85–100. https://doi.org/10.30486/ijrowa.2021.1927528.1233

Al Kharusi L, Al Yahyai R, Yaish M (2019) Antioxidant response to salinity in salt-tolerant and salt-susceptible cultivars of date palm. Agriculture 9:8. https://doi.org/10.3390/agriculture9010008

Amako K, Chen GX, Asada K (1994) Separate assays specific for ascorbate peroxidase and guaiacol peroxidase and for the chloroplastic and cytosolic isozymes of ascorbate peroxidase in plants. Plant Cell Physiol 35:497–504. https://doi.org/10.1093/oxfordjournals.pcp.a078621

Anli M, Baslam M, Tahiri AI, Raklami A, Boutasknit A, Symanczik S et al (2020a) Biofertilizers as strategies to improve photosynthetic apparatus, growth, and drought stress tolerance in the date palm. Front Plant Sci 11:516818. https://doi.org/10.3389/fpls.2020.516818

Anli M, Kaoua MEL, Ait-el-Mokhtar M, Boutasknit A, Ben-Laouane R, Toubali S et al (2020b) Seaweed extract application and arbuscular mycorrhizal fungal inoculation: a tool for promoting growth and development of date palm (*Phoenix dactylifera* L.) cv «Boufgous». South African J Bot 132:15–21. https://doi.org/10.1016/j.sajb.2020.04.004

Anli M, Symanczik S, El Abbassi A, Ait-El-Mokhtar M, Boutasknit A, Ben-Laouane R et al (2020c) Use of arbuscular mycorrhizal fungus *Rhizoglomus irregulare* and compost to improve growth and physiological responses of *Phoenix dactylifera* "Boufgouss". Plant Biosyst Int J Deal with all Asp Plant Biol 155:1–14. https://doi.org/10.1080/11263504.2020.1779848

Anli M, Boutasknit A, Ben-Laouane R, Ait-El-Mokhtar M, Ait-Rahou Y, Raho O et al (2021) Evaluating the performance of *Lactuca sativa* under four different organic fertilizers and subsequent impact on the soil health. J Basic Appl Res Int 27:53–66

Anli M, Fakhech A, Boutasknit A, Ait-El-Mokhtar M, Ben-Laoaune R, Ait-Rahou Y et al (2022) Comparing the response of growth and physiologic variables of onion to olive mill wastewater application and arbuscular mycorrhizal fungi inoculation. Gesunde Pflanz 75:655. https://doi.org/10.1007/s10343-022-00731-1

Arnon DI (1949) Copper enzymes in isolated chloroplasts. Polyphenoloxidase in *Beta vulgaris*. Plant Physiol 24:1–15. https://doi.org/10.1104/pp.900074

Ashraf MA, Iqbal M, Rasheed R, Hussain I, Riaz M, Arif MS (2018) Environmental stress and secondary metabolites in plants: an overview. In: Ahmad P, Ahanger MA, Singh VP, Tripathi DK, Alam P, Alyemeni MN (eds) Plant metabolites and regulation under environmental stress. Academic Press, Elsevier, pp 153–167. https://doi.org/10.1016/B978-0-12-812689-9.00008-X

Ayub MA, Ahmad HR, Ali M, Rizwan M, Ali S, Zia ur Rehman M et al (2020) Salinity and its tolerance strategies in plants. In: Tripathi DK, Chauhan DK, Prasad SM, Ramawat N, Singh VP, Sharma S, Dubey NK (eds) Plant life under changing environment. Responses and management. Academic Press, Elsevier, pp 47–76. https://doi.org/10.1016/b978-0-12-818204-8.00003-5

Baker NR (2008) Chlorophyll fluorescence: a probe of photosynthesis in vivo. Annu Rev Plant Biol 59:89–113. https://doi.org/10.1146/annurev.arplant.59.032607.092759

Barrs HD, Weatherly PE (1962) A re-examination of the relative turgidity technique for estimating water deficits in leaves. Aust J Biol Sci 15:413–428

Bates LS, Aldren RP, Teare LD (1973) Rapid determination of free proline for water-stress studies. Plant Soil 39:205–207

Benaffari W, Boutasknit A, Anli M, Ait-El-mokhtar M, Ait-Rahou Y, Ben-Laouane R et al (2022) The native arbuscular mycorrhizal fungi and vermicompost-based organic amendments enhance soil fertility, growth performance, and the drought stress tolerance of Quinoa. Plan Theory 11:1–29. https://doi.org/10.3390/plants11030393

Ben-Laouane R, Baslam M, Ait-El-Mokhtar M, Anli M, Boutasknit A, Ait-Rahou Y et al (2020) Potential of native arbuscular mycorrhizal fungi, Rhizobia, and/or green compost as Alfalfa (*Medicago sativa)* enhancers under salinity. Microorganisms 8:1695. https://doi.org/10.3390/microorganisms8111695

Ben-Laouane R, Ait-El-Mokhtar M, Anli M, Boutasknit A, Ait Rahou Y, Raklami A et al (2021) Green compost combined with mycorrhizae and rhizobia: a strategy for improving Alfalfa growth and yield under field conditions. Gesunde Pflanz 73:193–207. https://doi.org/10.1007/s10343-020-00537-z

Beyer WF, Fridovich I (1987) Assaying for superoxide dismutase activity: some large consequences of minor changes in conditions. Anal Biochem 161:559–566. https://doi.org/10.1016/0003-2697(87)90489-1

Bothe H (2012) Arbuscular mycorrhiza and salt tolerance of plants. Symbiosis 58:7–16. https://doi.org/10.1007/s13199-012-0196-9

Boursier P, Lynch J, Lauchli A, Epstein E (1987) Chloride partitioning in leaves of salt-stressed Sorghum, Maize, Wheat and Barley. Funct Plant Biol 14:463. https://doi.org/10.1071/pp9870463

Boutasknit A, Anli M, Tahiri A, Raklami A, Ait-El-Mokhtar M, Ben-Laouane R et al (2020a) Potential effect of horse manure-green waste and olive pomace-green waste composts on physiology and yield of garlic (*Allium sativum* L.) and soil fertility. Gesunde Pflanz:1–11. https://doi.org/10.1007/s10343-020-00511-9

Boutasknit A, Baslam M, Ait-El-mokhtar M, Anli M, Ben-Laouane R, Douira A et al (2020b) Arbuscular mycorrhizal fungi mediate drought tolerance and recovery in two contrasting carob

(*Ceratonia siliqua* l.) ecotypes by regulating stomatal, water relations, and (in) organic adjustments. Plan Theory 9:80. https://doi.org/10.3390/plants9010080

Boutasknit A, Ait-Rahou Y, Anli M, Ait-El-Mokhtar M, Ben-Laouane R, Meddich A (2021a) Improvement of garlic growth, physiology, biochemical traits, and soil fertility by Rhizophagus irregularis and compost. Gesunde Pflanz 73(2):149–160. https://doi.org/10.1007/s10343-020-00533-3

Boutasknit A, Baslam M, Ait-El-Mokhtar M, Anli M, Ben-Laouane R, Ait-Rahou Y et al (2021b) Assemblage of indigenous arbuscular mycorrhizal fungi and green waste compost enhance drought stress tolerance in carob (*Ceratonia siliqua* L.) trees. Sci Rep 11:1–23. https://doi.org/10.1038/s41598-021-02018-3

Boutasknit A, Anli M, Ben-Laouane R, Fakech A, Ait Rahou Y, Wahbi S et al (2023) Arbuscular mycorrhizal fungi for enhanced production and crop yield under changing climate. In: Ahmed K (ed) Intelligent solutions for optimizing agriculture and tackling climate change: current and future dimensions. IGI Global, pp 115–147. https://doi.org/10.4018/978-1-6684-4649-2.ch005

Chaganti VN, Crohn DM (2015) Evaluating the relative contribution of physiochemical and biological factors in ameliorating a saline–sodic soil amended with composts and biochar and leached with reclaimed water. Geoderma 259:45–55. https://doi.org/10.1016/j.geoderma.2015.05.005

Chaichi MR, Keshavarz-Afshar R, Lu B, Rostamza M (2017) Growth and nutrient uptake of tomato in response to application of saline water, biological fertilizer, and surfactant. J Plant Nutr 40:457–466. https://doi.org/10.1080/01904167.2016.1246567

Chao CT, Krueger RR (2007) The date palm (*Phoenix dactylifera* L.): overview of biology, uses, and cultivation. Hort Science 42:1077–1082

Chebaane A, Symanczik S, Oehl F, Azri R, Gargouri M, Mäder P et al (2020) Arbuscular mycorrhizal fungi associated with *Phoenix dactylifera* L. grown in Tunisian Sahara oases of different salinity levels. Symbiosis 81:173. https://doi.org/10.1007/s13199-020-00692-x

Chinsamy M, Kulkarni MG, Van Staden J (2013) Garden-waste-vermicompost leachate alleviates salinity stress in tomato seedlings by mobilizing salt tolerance mechanisms. Plant Growth Regul 71:41–47. https://doi.org/10.1007/s10725-013-9807-6

de Oliveira VP, Lima MDR, da Silva BRS, Batista BL, da Silva LAK (2019) Brassinosteroids confer tolerance to salt stress in *Eucalyptus urophylla* plants enhancing homeostasis, antioxidant metabolism and leaf anatomy. J Plant Growth Regul 38:557–573. https://doi.org/10.1007/s00344-018-9870-3

Dhindsa RS, Plumbdhindsa P, Thorpe TA (1981) Leaf senescence: correlated with increased levels of membrane-permeability and lipid peroxidation, and decreased levels of superoxide dismutase and catalase. J Exp Bot 32:93–101. https://doi.org/10.1093/jxb/32.1.93

Diagne N, Ndour M, Djighaly PI, Ngom D, Ngom MCN, Ndong G et al (2020) Effect of plant growth promoting rhizobacteria (PGPR) and arbuscular mycorrhizal fungi (AMF) on salt stress tolerance of *Casuarina obesa* (Miq.). Front Sustain Food Syst 4:601004. https://doi.org/10.3389/fsufs.2020.601004

Doni F, Miranti M, Shakirin M, Mohamed Z, Uphoff N (2022) Rhizosphere multi-omics approaches for deciphering the microbial modulation of plants' genetic potentials: what's known and what's next? Rhizosphere 24:100613. https://doi.org/10.1016/j.rhisph.2022.100613

Drake JA, Cavagnaro TR, Cunningham SC, Jackson WR, Patti AF (2016) Does biochar improve establishment of tree seedlings in saline sodic soils? L Degrad Dev 27:52–59. https://doi.org/10.1002/ldr.2374

Dubois M, Gilles KA, Hamilton JK, Rebers PA, Smith F (1956) Colorimetric method for determination of sugars and related substances. Anal Chem 28:350–356. https://doi.org/10.1021/ac60111a017

El Kinany S, Achbani E, Faggroud M, Ouahmane L, El Hilali R, Haggoud A et al (2019) Effect of organic fertilizer and commercial arbuscular mycorrhizal fungi on the growth of micropropagated date palm cv. Feggouss. J Saudi Soc Agric Sci 18:411–417. https://doi.org/10.1016/j.jssas.2018.01.004

Elsheery NI, Helaly MN, Omar SA, John SVS, Zabochnicka-Świątek M, Kalaji HM et al (2020) Physiological and molecular mechanisms of salinity tolerance in grafted cucumber. South African J Bot 130:90–102. https://doi.org/10.1016/j.sajb.2019.12 014

Evelin H, Devi TS, Gupta S, Kapoor R (2019) Mitigation of salinity stress in plants by arbuscular mycorrhizal symbiosis: current understanding and new challenges. Front Plant Sci 10. https://doi.org/10.3389/fpls.2019.00470

Fatima T, Arora NK (2023) Structural, compositional characterization of novel exopolysaccharide from halotolerant strain of Pseudomonas entomophila PE3 and their role in growth promotion of sunflower under saline condition. Env Exp Bot 213:105418

Foyer CH (2018) Reactive oxygen species, oxidative signaling and the regulation of photosynthesis. Environ Exp Bot 154:134–142. https://doi.org/10.1016/j.envexpbot.2018.05.003

Gryndler M, Sudová R, Püschel D, Rydlová J, Janoušková M, Vosátka M (2008) Cultivation of high-biomass crops on coal mine spoil banks: can microbial inoculation compensate for high doses of organic matter? Bioresour Technol 99:6391–6399. https://doi.org/10.1016/j.biortech.2007.11.059

Guangming L, Xuechen Z, Xiuping W, Hongbo S, Jingsong Y, Xiangping W (2017) Soil enzymes as indicators of saline soil fertility under various soil amendments. Agric Ecosyst Environ 237:274–279. https://doi.org/10.1016/j.agee.2017.01.004

Hashem A, Abd Allah EF, Alqarawi AA, Aldubise A, Egamberdieva D (2015) Arbuscular mycorrhizal fungi enhances salinity tolerance of *Panicum turgidum* forssk by altering photosynthetic and antioxidant pathways. J Plant Interact 10:230–242. https://doi.org/10.1080/17429145.2015.1052025

Hashem A, Allah EFA, Alqarawi AA, Shah MA (2016) Induction of osmoregulation and modulation of salt stress in *Acacia gerrardii* benth. by arbuscular mycorrhizal fungi and *Bacillus subtilis* (BERA 71). Biomed Res Int 2016:11

Hause B, Fester T (2005) Molecular and cell biology of arbuscular mycorrhizal symbiosis. Planta 221:184–196. https://doi.org/10.1007/s00425-004-1436-x

Ibrahim AH, Abdel-Fattah GM, Eman FM, Abd El-Aziz MH, Shohr AE (2011) Arbuscular mycorrhizal fungi and spermine alleviate the adverse effects of salinity stress on electrolyte leakage and productivity of wheat plants. Phyton 51(2):261–276

Jamil IN, Remali J, Azizan KA, Nor Muhammad NA, Arita M, Goh HH et al (2020) Systematic multi-omics integration (MOI) approach in plant systems biology. Front Plant Sci 11:944. https://doi.org/10.3389/fpls.2020.00944

Kamanga RM, Echigo K, Yodoya K, Mekawy AMM, Ueda A (2020) Salinity acclimation ameliorates salt stress in tomato (*Solanum lycopersicum* L.) seedlings by triggering a cascade of physiological processes in the leaves. Sci Hortic 270:109434. https://doi.org/10.1016/j.scienta.2020.109434

Khalid KA, Teixeira da Silva JA (2010) Yield, essential oil and pigment content of *Calendula officinalis* L. flower heads cultivated under salt stress conditions. Sci Hortic 126:297–305. https://doi.org/10.1016/j.scienta.2010.07.023

Lahbouki S, Anli M, El Gabardi S, Ait-El-Mokhtar M, Ben-Laouane R, Boutasknit A et al (2021) Evaluation of arbuscular mycorrhizal fungi and vermicompost supplementation on growth, phenolic content and antioxidant activity of prickly pear cactus (*Opuntia ficus-indica*). Plant Biosyst Int J Deal with all Asp Plant Biol 156:1–19. https://doi.org/10.1080/11263504.2021.1947408

Lucho SR, do Amaral MN, Auler PA, Bianchi VJ, Ferrer MÁ, Calderón AA et al (2019) Salt stress-induced changes in in vitro cultured *Stevia rebaudiana* bertoni: effect on metabolite contents, antioxidant capacity and expression of steviol glycosides-related biosynthetic genes. J Plant Growth Regul 38:1341–1353. https://doi.org/10.1007/s00344-019-09937-6

Marček T, Tkalec M, Vidaković-Cifrek Ž, Ježić M, Ćurković-Perica M (2016) Expression of dehydrins, HSP70, Cu/Zn SOD, and RuBisCO in leaves of tobacco (*Nicotiana tabacum* L.) dihaploids under salt stress. In Vitro Cell Dev Biol Plant 52:233–240

Mbarki S, Cerdà A, Zivcak M, Brestic M, Rabhi M, Mezni M et al (2018) Alfalfa crops amended with MSW compost can compensate the effect of salty water irrigation depending on the soil texture. Process Saf Environ Prot 115:8–16. https://doi.org/10.1016/j.psep.2017.09.001

Mbarki S, Skalicky M, Talbi O, Chakraborty A, Hnilicka F, Hejnak V et al (2020) Performance of *Medicago sativa* grown in clay soil favored by compost or farmyard manure to mitigate salt stress. Agronomy 10:1–14. https://doi.org/10.3390/agronomy10010094

McGonigle TP, Miller MH, Evans DG, Fairchild GL, Swan JA (1990) A new method which gives an objective measure of colonization of roots by vesicular-arbuscular mycorrhizal fungi. New Phytol 115:495–501. https://doi.org/10.1111/j.1469-8137.1990.tb00476.x

Meddich A, Hafidi M, Ait-El-Mokhtar M, Boumezzough A (2015a) Characterization of physico-chemical parameters and mycorrhizal potential of salt soils of North-east date palm grove of Marrakesh. J Mater Environ Sci 6(9):2469–2475

Meddich A, Jaiti F, Bourzik W, El A, Hafidi M (2015b) Use of mycorrhizal fungi as a strategy for improving the drought tolerance in date palm (*Phoenix dactylifera*). Sci Hortic 192:468–474. https://doi.org/10.1016/j.scienta.2015.06.024

Meddich A, Elouaqoudi FZ, Khadra A, Bourzik W (2016) Valorization of green and industrial wastes by composting process. Rev des Compos des Mater Av 26:451–469. https://doi.org/10.3166/rcma.26.451-469

Meddich A, El Mokhtar MA, Wahbi S, Boumezzough A (2017) Evaluation of mycorrhizal potential in relation to the physicochemical parameters of the soils of palm groves in Morocco (Marrakech and Tafilalet). Cah Agric 26(4):45012. https://doi.org/10.1051/cagri/2017044

Meddich A, Ait El Mokhtar M, Bourzik W, Mitsui T, Baslam M, Hafidi M (2018) Optimizing growth and tolerance of date palm (*Phoenix dactylifera* L.) to drought, salinity, and vascular *Fusarium*-induced wilt (*Fusarium oxysporum*) by application of arbuscular mycorrhizal fungi (AMF). In: Giri B, Prasad R, Varma A (eds) Root biology. Soil biology, vol 52. Springer, Cham, pp 239–250. https://doi.org/10.1007/978-3-319-75910-4_9

Meddich A, Oufdou K, Boutasknit A, Raklami A, Tahiri A, Ben-Laouane R et al (2019) Use of organic and biological fertilizers as strategies to improve crop biomass, yields and physico-chemical parameters of soil. In: Meena RS (ed) Nutrient dynamics for sustainable crop production. Springer, Singapore, pp 247–288. https://doi.org/10.1007/978-981-13-8660-2_9

Meddich A, Boutasknit A, Anli M, Ait Ahmed M, El Abbassi A, Boutaj H et al (2020) Use of olive mill wastewaters as bio-insecticides for the control of *Potosia opaca* in date palm (*Phoenix dactylifera* L.). In: Basso TP, Basso TO, Basso LC (eds) Biotechnological applications of biomass. IntechOpen. https://doi.org/10.5772/intechopen.93537

Miransari M (2010) Contribution of arbuscular mycorrhizal symbiosis to plant growth under different types of soil stress. Plant Biol 12:563–569. https://doi.org/10.1111/j.1438-8677.2009.00308.x

Mishra P, Mishra J, Arora NK (2021) Plant growth promoting bacteria for combating salinity stress in plants—recent developments and prospects: a review. Microbiol Res 252:126861

Mishra P, Bhattacharya A, Verma P, Bharti C, Arora NK (2022) Plant growth-promoting bacteria as biostimulants of crops in saline agroecosystems. In: Arora NK, Bouizgarne B (eds) Microbial biotechnology for sustainable agriculture. Volume 1. Microorganisms for sustainability, vol 33. Springer, Singapore

Mittler R (2002) Oxidative stress, antioxidants and stress tolerance. Trends Plant Sci 7:405–410. https://doi.org/10.1016/S1360-1385(02)02312-9

Mohamed OZ, Yassine B, Hilali Rania E, El Hassan A, Abdellatif H, Rachid B (2020) Evaluation of compost quality and bioprotection potential against *Fusarium* wilt of date palm. Waste Manag 113:12–19. https://doi.org/10.1016/j.wasman.2020.05.035

Muchate NS, Nikalje GC (2016) Plant salt stress: adaptive responses, tolerance mechanism and bioengineering for salt tolerance. Bot Rev 82:371. https://doi.org/10.1007/s12229-016-9173-y

Munns R, Day DA, Fricke W, Watt M, Arsova B, Barkla BJ et al (2020) Energy costs of salt tolerance in crop plants. New Phytol 225:1072–1090. https://doi.org/10.1111/nph.15864

Nourinia AA, Faghani E, Rejali F, Safarnezhad A, Abbasi MR (2007) Evaluation effects of symbiosis of mycorrhiza on yield components and some physiological parameters of barley genotypes under salinity stress. Asian J Plant Sci 6:1108–1112

Olsen SR, Dean LA (1965) Phosphorus. In: Black CA (ed) Methods of soil analysis Part 2. Wiley

Parida AK, Panda A, Rangani J (2018) Metabolomics-guided elucidation of abiotic stress tolerance mechanisms in plants. In: Ahmad P, Ahanger MA, Singh VP, Tripathi DK, Alam P, Alyemeni MN (eds) Plant metabolites and regulation under environmental stress. Academic Press, Elsevier, pp 89–131. https://doi.org/10.1016/B978-0-12-812689-9.00005-4

Polle A, Otter T, Seifert F (1994) Apoplastic peroxidases and lignification in needles of Norway spruce (*Picea abies* L.). Plant Physiol 106:53–60. https://doi.org/10.1007/BF02234561

Rao DE, Chaitanya KV (2016) Photosynthesis and antioxidative defense mechanisms in deciphering drought stress tolerance of crop plants. Biol Plant 60:201–218. https://doi.org/10.1007/s10535-016-0584-8

Ros-Barceló A, Gómez-Ros LV, Ferrer MA, Hernández JA (2006) The apoplastic antioxidant enzymatic system in the wood-forming tissues of trees. Trees-Struct Funct 20:145–156. https://doi.org/10.1007/s00468-005-0020-8

Saadaoui I, Sedky R, Rasheed R, Bounnit T, Almahmoud A, Elshekh A et al (2019) Assessment of the algae-based biofertilizer influence on date palm (*Phoenix dactylifera* L.) cultivation. J Appl Phycol 31:457–463. https://doi.org/10.1007/s10811-018-1539-6

Sallaku G, Sandén H, Babaj I, Kaciu S, Balliu A, Rewald B (2019) Specific nutrient absorption rates of transplanted cucumber seedlings are highly related to RGR and influenced by grafting method, AMF inoculation and salinity. Sci Hortic 243:177–188. https://doi.org/10.1016/j.scienta.2018.08.027

Scholander PF, Hammel HT, Bradstreet ED, Hemmingsen EA (1965) Sap pressure in vascular plants. Science 148(3668):339–346. https://doi.org/10.1126/science.148.3668.339

Selvakumar G, Shagol CC, Kim K, Han S, Sa T (2018) Spore associated bacteria regulates maize root K^+/Na^+ ion homeostasis to promote salinity tolerance during arbuscular mycorrhizal symbiosis. BMC Plant Biol 18:1–13

Shankar V, Evelin H (2019) Strategies for reclamation of saline soils. In: Giri B, Varma A (eds) Microorganisms in saline environments: strategies and functions. Soil biology, vol 56. Springer, Cham, pp 439–449. https://doi.org/10.1007/978-3-030-18975-4_19

Sharma A, Kumar V, Shahzad B, Ramakrishnan M, Singh Sidhu GP, Bali AS et al (2020) Photosynthetic response of plants under different abiotic stresses: a review. J Plant Growth Regul 39:509–531. https://doi.org/10.1007/s00344-019-10018-x

Smith SE, Read D (2008) The symbionts forming arbuscular mycorrhizas. In: Mycorrhizal symbiosis. Academic Press, Elsevier, pp 13–41. https://doi.org/10.1016/B978-012370526-6.50003-9

Sperling O, Lazarovitch N, Schwartz A, Shapira O (2014) Effects of high salinity irrigation on growth, gas-exchange, and photoprotection in date palms (*Phoenix dactylifera* L., cv. Medjool). Environ Exp Bot 99:100–109. https://doi.org/10.1016/j.envexpbot.2013.10.014

Talaat NB, Shawky BT (2014) Protective effects of arbuscular mycorrhizal fungi on wheat (*Triticum aestivum* L.) plants exposed to salinity. Environ Exp Bot 98:20–31. https://doi.org/10.1016/j.envexpbot.2013.10.005

Toubali S, Tahiri A, Anli M, Symanczik S, Boutasknit A, Ait-El-Mokhtar M et al (2020) Physiological and biochemical behaviors of date palm vitroplants treated with microbial consortia and compost in response to salt stress. Appl Sci 10:8665. https://doi.org/10.3390/app10238665

Toubali S, Ait-El-Mokhtar M, Boutasknit A, Anli M, Ait-Rahou Y, Benaffari W et al (2022) Root reinforcement improved performance, productivity, and grain bioactive quality of field-droughted quinoa (*Chenopodium quinoa*). Front Plant Sci 13:860484. https://doi.org/10.3389/fpls.2022.860484

Trivedi P, Singh K, Pankaj U, Verma SK, Verma RK, Patra DD (2017) Effect of organic amendments and microbial application on sodic soil properties and growth of an aromatic crop. Ecol Eng 102:127–136. https://doi.org/10.1016/j.ecoleng.2017.01.046

Velikova V, Yordanov I, Edreva A (2000) Oxidative stress and some antioxidant systems in acid rain-treated bean plants. Plant Sci 151:59–66. https://doi.org/10.1016/S0168-9452(99)00197-1

Wolf B (1982) Communications in soil science and plant analysis a comprehensive system of leaf analyses and its use for diagnosing crop nutrient status. Commun Soil Sci Plant Anal 13:1035–1059

Xu L, Pierroz G, Wipf HML, Gao C, Taylor JW, Lemaux PG et al (2021) Holo-omics for deciphering plant-microbiome interactions. Microbiome 9:69. https://doi.org/10.1186/s40168-021-01014-z

Yarahmadi MA, Shahsavani S, Akhyani A (2018) Pomegranate growth affected by arbuscular mycorrhizae, phosphorus fertilizer, and irrigation water salinity pomegranate growth affected by arbuscular mycorrhizae. Commun Soil Sci Plant Anal 49:478–488. https://doi.org/10.1080/00103624.2018.1431265

Zancarini A, Westerhuis JA, Smilde AK, Bouwmeester HJ (2021) Integration of omics data to unravel root microbiome recruitment. Curr Opin Biotechnol 70:255–261. https://doi.org/10.1016/j.copbio.2021.06.016

Chapter 12
Suppressive Effect of Different Compost Extracts Against *Fusarium oxysporum* f. sp. *ciceri*: Causal Agent of Fusarium Wilt of Chickpea (*Cicer arietinum*)

Khalid Azim, Yamna Elbakhouch, Ilyass Tabrika, Fouad Elame, Brahim Bouizgarne, Rachid Aboutayeb, and Abdelhi Dihazi

Abstract *Fusarium oxysporum* f. sp. *ciceri* (*Foc*) is the causal agent of fusarium wilt in chickpea (*Cicer arietinum*). One of the main objectives of waste composting and management is to make available products that could contribute to soil health by fertilization and by suppressing plant pathogens. In this work, the suppressive effects of six different compost extracts A, B, C, D, E, and F at four different concentrations (extracts and their dilutions at 25%, 50%, and 75%) used either directly

K. Azim (✉) · F. Elame
Integrated Crop Production Research Unit, Regional Center of Agricultural Research of Agadir, National Institute of Agricultural Research, Rabat, Morocco
e-mail: khalid.azim@inra.ma

Y. Elbakhouch
Integrated Crop Production Research Unit, Regional Center of Agricultural Research of Agadir, National Institute of Agricultural Research, Rabat, Morocco

Laboratory of Agrobiotechnology and Bioengineering, Cadi Ayyad University, Faculty of Sciences and Technology, Marrakech, Morocco

I. Tabrika
Laboratory of Mechanic, Process Energy and Environment, National School of Applied Sciences, Ibn Zohr University, Agadir, Morocco

B. Bouizgarne
Laboratory of Plant Biotechnology, Department of Biology, Faculty of Sciences, Ibn Zohr University (UIZ), Agadir, Morocco

R. Aboutayeb
Environment and Sustainable Management of Natural Resources Research Unit, Regional Center of Agricultural Research of Settat, National Institute of Agricultural Research, Rabat, Morocco

A. Dihazi
Laboratory of Agrobiotechnology and Bioengineering, Cadi Ayyad University, Faculty of Sciences and Technology, Marrakech, Morocco

N. K. Arora, B. Bouizgarne (eds.), *Microbial Biotechnology for Sustainable Agriculture Volume 2*, Microorganisms for Sustainability 34,
https://doi.org/10.1007/978-981-97-2355-3_12

or after sterilization or not against *Foc* growth and sporulation were investigated. The composts were prepared mainly from tomato leaves in mixtures with other components at different ratios (sawdust, poultry manure, sheep manure, and olive pomace). All composts contain 50% tomato leaves and 20% sawdust but differ for other component contents. The results showed that all the compost extracts displayed an inhibitory effect on both *Foc* growth and sporulation by up to 89%. Following compost extracts sterilization, inhibition was reduced to only 53%. The inhibition of *Foc* growth varied according to the type of compost, the composting stage, and the compost extract concentration. The olive pomace free compost type A and type B, containing both sheep and poultry manure, were the most effective. This inhibition was mainly noticed at the second and fourth composting sampling stages corresponding to the 13th and 39th day after mixing the five components for composting. The suppressive effect of the compost type A and type B could be due to their manure content, which could be involved in the development of antagonistic microorganisms against the chickpea fusarium wilt agent.

Keywords Waste management · Sanitation · Compost tea · Fusarium wilt · Inhibition

12.1 Introduction

Legumes, which are the second most important crop after cereals, are a key component of healthy traditional diets worldwide. In Morocco, chickpea (*Cicer arietinum*) is the second major food legume after faba bean. It is mainly grown in rainfed and intermediate favorable regions of the country (Houasli et al. 2020). According to FAOSTAT (2021), the total area under chickpea cultivation in Morocco is estimated to about 63.411 ha with a yield of 1.144 kg.ha^{-1} and a production of about 72.547 T. However, the global yield of legumes is vulnerable and constrained by disease susceptibility (Idrissi et al. 2019). This major constraint in chickpea production worldwide is the vascular wilt disease, caused by *Fusarium oxysporum* f. sp. *ciceri (Foc),* resulting in significant crop loss of up to 100% and reduced quality of the harvested crop (Purohit et al. 2023; Chichaybelu et al. 2021). Chemical fungicides represent the traditional method of disease control, but environmental and health concerns have limited their use (Wu et al. 2023). Alternatively, organic amendments such as compost have been reported to inhibit soil-borne pathogens and improve plant growth (Azim 2019). Compost is a hygienic product rich in humus substances formed due to biochemical conversion by aerobic biodegradation of organic matter (Azim et al. 2017).

Composts have been used to control various plant diseases in greenhouses and field trials. Studies have demonstrated that the application of compost can reduce disease incidence and severity caused by soil-borne pathogens. Among plant diseases controlled by compost, those caused by *F. oxysporum, Laetisaria fuciformis, Microdochium nivale*, *Phytophthora capsici*, *Pythium ultimum*, *Pythium graminicola, Rhizoctonia solani*, *Sclerotinia homoeocarpa Typhula ishikariensis,* and *Verticillium dahlia* are reported (Noble and Coventry 2005; Bellini et al. 2020;

Rasool et al. 2021). It was also suggested that supplementation of beneficial microbiota from compost into conducive soils could render them resistant by increasing suppression against soil-borne pathogens (De Corato 2020). The great antagonistic potential of compost against plant diseases is explained by its chemical composition, the presence of nutrients, its microbial composition and ability to promote microbial activity by microbial introduction and stimulation in the soil (Azim et al. 2022). In addition, the biological and physical properties of compost, such as its ability to enhance soil structure and release organic acids, are also believed to contribute to its suppressive effect (De Corato 2020). The organic matter in compost can provide a source of nutrients for beneficial microorganisms that compete with pathogens for resources (Aguilar-Paredes et al. 2023). Furthermore, compost can stimulate the production of plant defense compounds, which can help to protect against infection (Lutz et al. 2020).

Liquid compost extracts are a potent solution, serving as a valuable resource for organic fertilization as well. It is produced by soaking compost in water and then separating the liquid from the solid material (Fernández-Delgado et al. 2020). Recent studies have shown that compost liquid extract can have a suppressive effect on soil-borne pathogens (Orosz et al. 2021). The extract contains a high concentration of beneficial microorganisms, such as bacteria, fungi, and actinomycetes, which can compete with and inhibit the growth of pathogenic microorganisms in the soil (De Corato 2020). Furthermore, compost liquid extract contains a range of bioactive compounds, such as enzymes, organic acids, and plant growth regulators, which can help to stimulate plant growth and enhance plant defense against pathogens (Liu et al. 2023). When applied to soil, compost liquid extract has been shown to reduce the incidence and severity of diseases caused by soil-borne pathogens, such as *F. oxysporum* and *P. ultimum* (Yatoo et al. 2021). De Corato (2020) in his review has counted several modes of action of compost extract against soil-borne pathogens including: (i) improved microbial activity and release of mineral nutrients during organic matter degradation, (ii) stimulation of competition for nutrients (iii) induction of microbiostasis and hyperparasitism, (iv) release of diffusible antibiotic-like compounds, and (v) induction of systemic disease resistance in host plants that may extend to the aerial parts due to the protective effect provided by the compost amended soil and improvement of the overall condition of the plants. The purpose of this study is to investigate the inhibitory effects of various compost extracts on Fusarium wilt disease of chickpea caused by *F. oxysporum* f. sp. *ciceri*.

12.2 Materials and Methods

12.2.1 F. oxysporum *f. sp.* ciceri*: Causal Agent of Chickpea Wilt*

The pathogen used in this study to assess the suppressive potential of compost extracts was a virulent strain of *F. oxysporum* f. sp. *ciceri* (*Foc*), provided by the laboratory of biotechnology and protection of plant resources of the Faculty of

Sciences Semlalia, Marrakech (Morocco). The multiplication of the fungus was carried out by taking 6 mm plugs from the margin of the colonies from the original culture and inoculating them aseptically in potato dextrose agar (PDA) plates that were subsequently incubated at 28 °C for 8 days.

12.2.2 Compost Preparation

Six water extracts were prepared from six different composts (A, B, C, D, E, and F) in the experimental center of the National Institute of Agricultural Research of Agadir (INRA), Morocco. These composts were obtained mainly from tomato waste mixed with other components. Table 12.1 shows the initial composition of the mixture for each compost. The composting of this horticultural waste was carried out in 220-liter bio-composters (®Biolan 220eco, Finland) during a composting cycle of 40 days.

Composite samples of 200 g for each compost were taken at different stages of composting (1st, 13th, 21st, and 39th days after composting starting up corresponding respectively, to stages 1, 2, 3, and 4). During sampling, the reactors were turned twice manually by emptying the reactors and then mixing the compost and filling them back. Samples were taken to INRA laboratory in order to determine the suppressive effect of the extracts against *Foc*. Table 12.2 represents the different isolates with respect to their macroscopic and microscopic (×40) characterization.

12.2.3 Compost Physicochemical Parameters

Samples of the compost (5 g) were transferred into tubes containing 25 ml of distilled water. pH was determined using 1/5 extract method in distilled water (1:5 w/v). The mixture was then stirred for 1 h at 22 °C ± 3 °C. Then the pH of the suspension was measured using a pH meter. The electrical conductivity (EC) was determined by transferring 5 g of the sample into tubes containing 25 ml of distilled water (1:5; w/v). The mixture was then stirred for 1 h at 22 °C ± 3 °C, then centrifuged for 10 min at 2000 rpm. The EC of the suspension was measured using a conductivity

Table 12.1 Initial composition of different studied composts (w/w)

	Compost					
Composition	A	B	C	D	E	F
Tomato leaves (TL)	50%	50%	50%	50%	50%	50%
Sawdust (SD)	20%	20%	20%	20%	20%	20%
Poultry manure (PM)	20%	10%	10%	–	20%	–
Sheep manure (SM)	10%	20%	10%	–	–	20%
Olive pomace (OP)	–	–	10%	30%	10%	10%

Table 12.2 Macroscopic ([A]×40) and microscopic ([B]×400) characterization of 4 microbial isolates from composts extract

Characteristics	Macroscopic ([A]×40) and microscopic ([B]×400) observation	
Isolate 1 (*Aspergillus* sp. 1): The colonies of this isolate have a blackish color with a white border and a white reverse (A). The conidia are globose, sometimes slightly flattened. The phialides are elongated, presenting several budding sites (B)	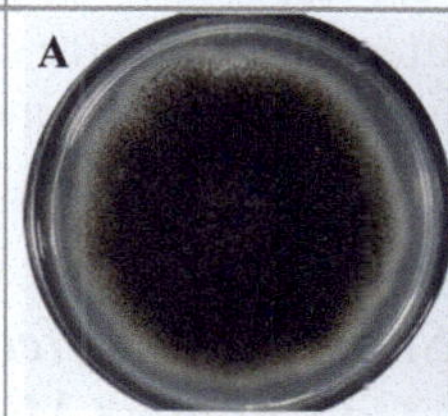	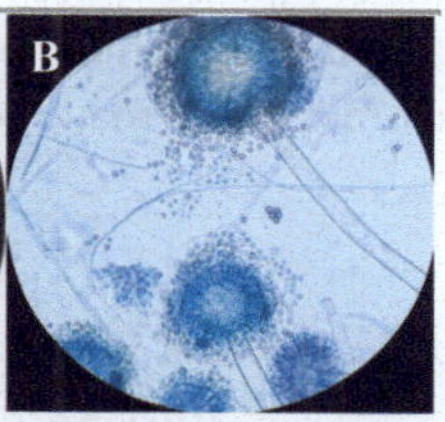
Isolate 2 (*Aspergillus* sp. 2): The colonies of this isolate are powdery green in color and velvety in appearance and colorless on the reverse (A). The conidial heads are blue-green and compact. Short conidiophores with a spatulate-like vesicle on which phialides are densely clustered on its upper half bearing globose to subglobose conidia (B)	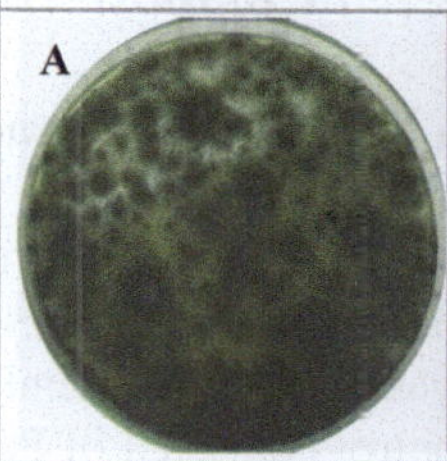	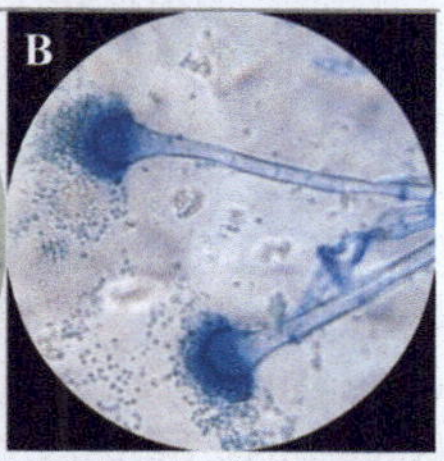
Isolate 3 (*Trichoderma* sp.): The colonies of this isolate are lightly compacted into green clumps. The reverse is colorless (A). The mycelium is composed of septate and branched hyphae. The conidiophores are pyramidal in shape, branched, and bearing conidia (B)	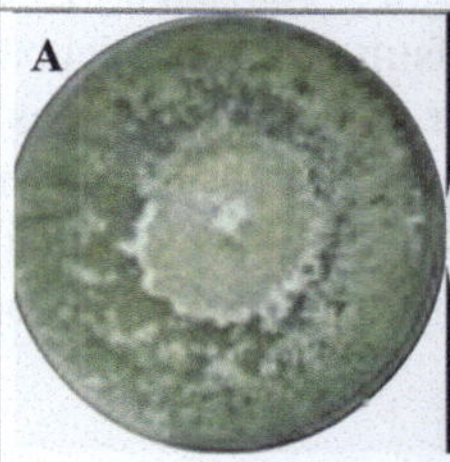	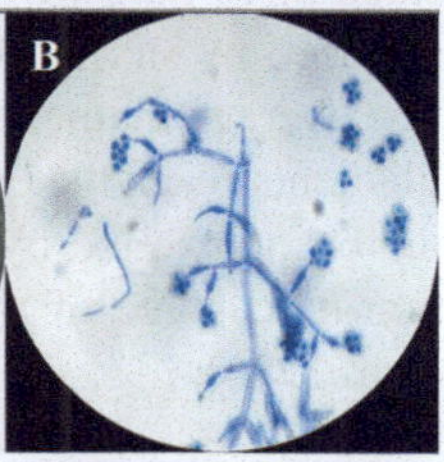
Isolate 4: Unidentified bacterium. The colonies of this bacteria are round, flat with regular edges and small of 0.2 mm (A). The surface is smooth in white color. The bacterium is Gram-negative (B)	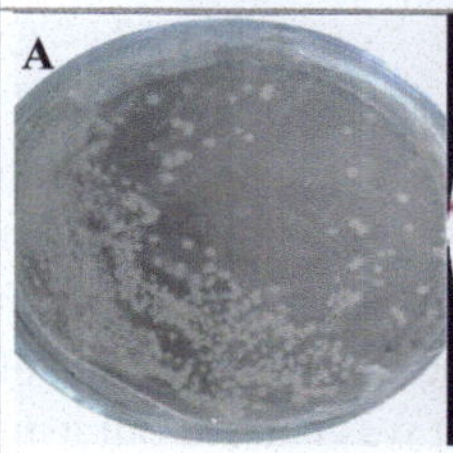	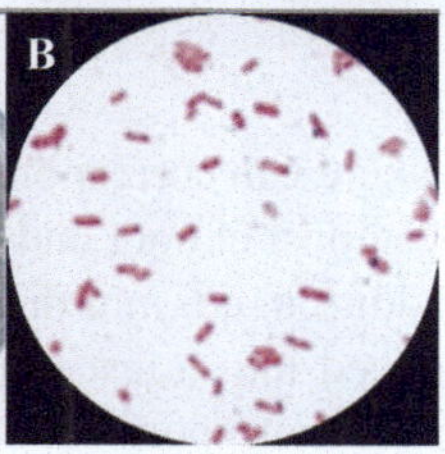

meter. Temperatures were measured daily for the first 2 weeks and thereafter every 3 days using a multiparameter device (Demista instruments, IL, USA) with a 1 m long temperature probe.

12.2.4 Compost Extracts Preparation

The extract was prepared according to the aerobic method developed by Gariglio et al. (2002). It consisted of 100 ml of sterile distilled water added to 50 g of each compost. The mixture was then stirred for 6 h at 25 ° C, subsequently centrifuged at

8000 rpm for 20 min. The supernatant obtained represents the compost extract, a part of it was used directly and the other part was autoclaved for 20 min at 120 °C. Sterilization was performed to determine whether the microorganisms in the compost extract were responsible for the suppressive effect against *Foc*. In addition to the original extract (100%), three dilutions 25%, 50%, and 75% of compost extracts were made for each type of compost.

12.2.5 *In Vitro Suppressive Effect of Compost Extract and Selected Compost Microbial Isolates against* Foc

12.2.5.1 Inhibition by Compost Extracts

The technique used consists of adding 1 ml of the compost extract to PDA medium before pouring it into Petri dishes. After solidification, an agar plug of 6 mm diameter containing *Foc* was placed in the center of each dish. Each treatment was replicated three times. Control consisted of Petri dishes where the compost extract was replaced by sterile distilled water. Incubation was carried out at 28 °C for 8 days. The average diameter of the *Foc* colonies growth was measured when the mycelial filaments reach the periphery of the dish in the control dishes. The evaluation of the inhibition exerted by the compost extract was estimated by calculating the inhibition percentage of mycelial growth according to the following formula (Hmouni et al. 1996):

$$\mathrm{I}(\%) = (1 - \mathrm{Cm} / \mathrm{Co}) \times 100$$

where Cm corresponds to the average diameter of the colonies in the presence of the compost extract and Co represents the average diameter of the control colonies. In addition, a 6 mm diameter mycelial plug was taken from each Petri dish and introduced into a test tube containing 1 ml of sterile distilled water. After mechanical agitation, spore count was carried out using a Malassez cell. The inhibition percentage of *Foc* sporulation was calculated relative to the control.

12.2.6 *Inhibition by Selected Microbial Isolates from Compost (Direct Confrontation Method)*

The antagonistic activity of four isolated microbia was tested in vitro according to the method of Benhamou and Chet (1996), on culture medium, by direct confrontation with the phytopathogenic agent. This technique consists of placing in a Petri dish, containing the PDA medium, two 6 mm diameter agar plugs, one carrying the antagonist agent and the other the pathogen. The two fragments were placed along a diametral axis of 30 mm and equidistant from the center of the box. For the control dishes, a mycelial plug of the pathogen was placed in the center of a dish containing the PDA. Triplicates were carried out per phytopathogenic agent and for the controls. Incubation was carried out at 26 °C for 8 days. The evaluation of the inhibition exerted by the compost extract was estimated by calculating the inhibition percentage of mycelial growth. In addition, after 8 days of incubation, a 6 mm diameter mycelial plug was taken from each Petri dish and introduced into a test tube containing 1 ml of sterile distilled water. After mechanical agitation, spore count was carried out using a Malassez cell. The inhibition percentage of *Foc* sporulation was calculated and compared relative to the control.

12.2.7 Statistical Analysis

The analysis of the variance of all the results was accomplished by the software SPSS version 25. The tests carried out in vitro were conducted by ANOVA 3 (type of compost, stage of compost, and concentration of compost extract). The in vitro antagonist test was analyzed following a completely randomized design by ANOVA 1 (type of antagonist). We conducted a one-way analysis of variance and compared means using Tukey's honest significant difference (HSD) test, applying a significance threshold of 5% ($P < 0.05$) to identify significant difference among treatments.

12.3 Results and Discussions

12.3.1 Compost Physicochemical Parameters

12.3.1.1 Potential of Hydrogen (pH)

Figure 12.1 shows the evolution of the compost's pH during the composting cycle. The pH of the six composts (A, B, C, D, E, and F) increased by more than one unit (from 7.25 to 8.95, 7.04 to 8.79, 7.09 to 8.63, 7.96 to 8.75, 7.06 to 8.93, and 7.09 to 8.85 for the composts A, B, C, D, E, and F respectively) after 39 days of composting. Similar pH variation was described by Rich and Bharti (2015) who reported that a pH range between 6.7 and 9 was favorable for microbial activity during composting. Chen et al. (2015) showed that the optimal pH values for composting are between 5.5 and 8. The rise in pH is explained by an increase in the concentration of potassium (Kalemelawa et al. 2012) and NH_4^+ (Gavilanes-Terán et al. 2016). This increase in pH is likely to prevent the development of certain pathogens, belonging mainly to fungi and confirms the work of Borrero et al. (2009) who explained the suppressive effect of certain composts from agricultural and industrial waste on *Fusarium* wilt of tomato and carnation by increased pH.

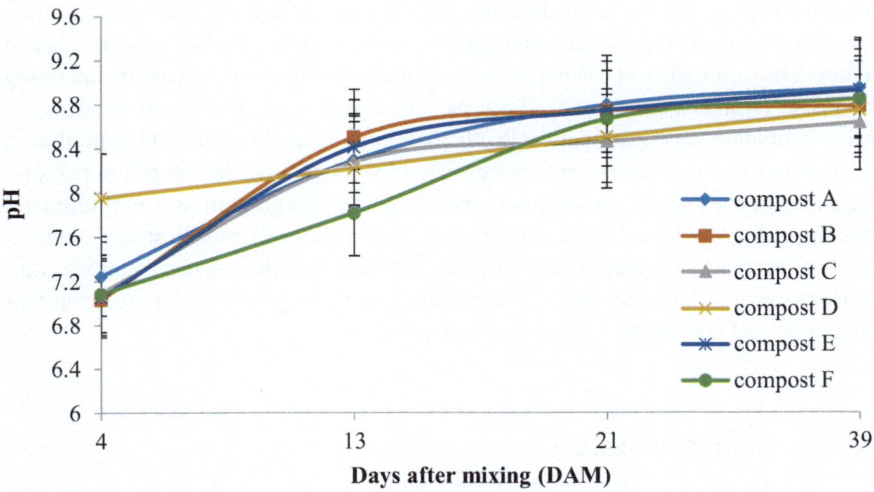

Fig. 12.1 pH evolution during the composting cycle for the six compost treatments (A, B, C, D, E, and F)

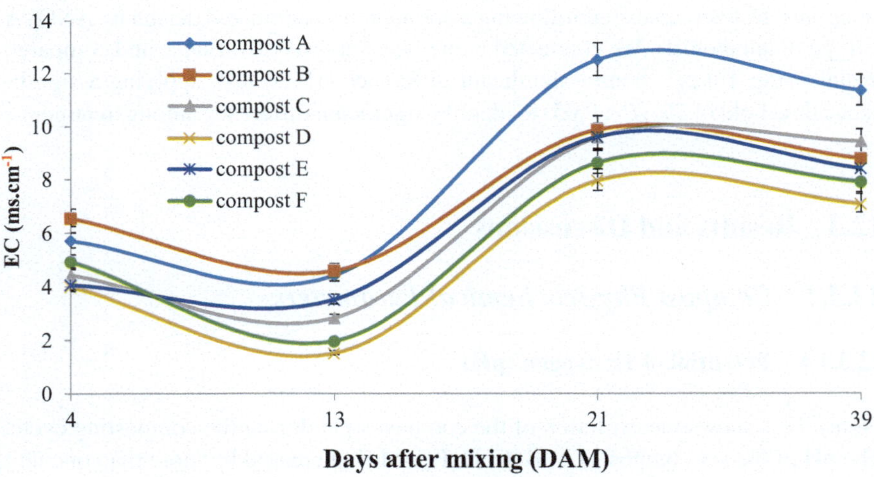

Fig. 12.2 Variation of EC in extracts of the six types of compost (A, B, C, D, E, and F) at different stages of composting (4th, 13th, 21st, and 39th day)

12.3.1.2 Electrical Conductivity (EC)

The results obtained concerning the EC in Fig. 12.2 show a moderate decrease in the first 13 days for all the studied composts. The stage 2 presents the relatively lowest EC values for all the six composts. For example, the EC decreased from 6 to 4.42 ms.cm^{-1} and from 4 to 1.53 ms.cm^{-1} respectively, for composts A and D. Afterwards, the EC increased for all types of compost to reach a maximum at the stage 3 corresponding to the 21st day. For composts A and D, on the 21st day, the EC was 12.5 and 7.98 ms.cm^{-1} respectively. Subsequently, the EC did not show significant variations even if the overall trend was toward a slight decrease. The increase in EC could be linked to the degradation of complex materials into simple compounds with the release of mineral ions such as phosphate, ammonium, or potassium (Azim et al. 2022). In general, high EC values are phytotoxic (Campitelli and Ceppi 2008). In addition, high salinity levels increase susceptibility to phytopathogens and accentuate fungal soil-borne diseases effects in plants as increased EC could have a negative impact on biocontrol agents. For instance, *Trichoderma*, a well-known biocontrol fungal agent was found to be ineffective at high salinity (Bheemaraya et al. 2013). Also, for composts, it was a negative impact of EC on compost effectiveness to combat phytopathogens (Pane et al. 2011; Krause et al. 2001; Otlewska et al. 2020). For instance, it was demonstrated that compost applied at the high rate of 72 t. ha^{-1} failed to suppress root rot of chile peppers caused by *P. capsici*, possibly by increasing the salinity of the soil (Dickerson 1996) The negative impact of the salinity in hampering suppressive effect of compost against phytopathogens was also confirmed by works on *Pythium* and *Phytophthora* diseases (Hoitink and Grebus 1996; Krause et al. 2001). Hoitink and Grebus (1996) reported that highly saline composts do not inhibit diseases by those two pathogens due to stress inflected on plants. This negative effect was observed even when the salinity levels are not toxic to the plants, suggesting that high salinity hampers the suppression by compost (Hoitink 2007).

12.3.1.3 Temperature Evolution

Concerning the temperature variation (Fig. 12.3), the overall trend was the opposite of that recorded for the EC. After 4 days, the temperature rose rapidly in all the reactors until it exceeded 45 °C. This rapid increase in temperature has also been observed

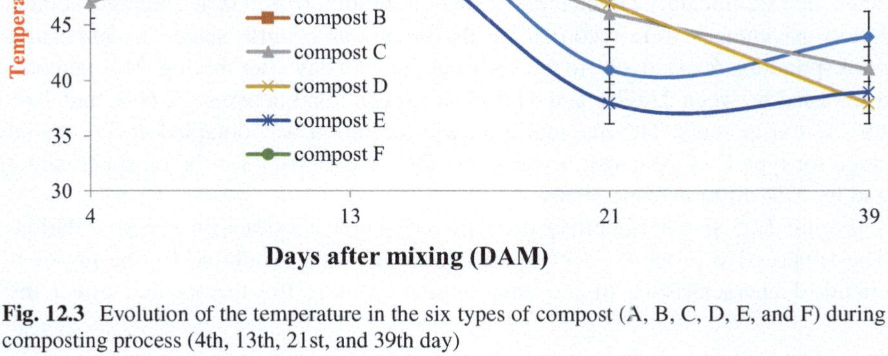

Fig. 12.3 Evolution of the temperature in the six types of compost (A, B, C, D, E, and F) during composting process (4th, 13th, 21st, and 39th day)

during the thermophile phase of composting (El-Mrini et al. 2021), which was maintained during the 39 days of composting. The reactor containing compost B recorded the highest temperature of above 58 °C, while the reactors of composts A, C, D, E, and F recorded temperatures between 50 and 55 °C. The thermophilic phase lasted almost 37 days, which is likely to allow complete sanitization of the compost and maximum reduction of pathogens (Chan et al. 2016). After the 13th day turnaround, a decrease in temperature was observed, due to improved oxygenation and homogenization of the mixture. Reactivation of the composting process with turnings has also been reported in compost from horticultural wastes (Jurado et al. 2014).

12.3.2 *In Vitro Suppressive Effect of Compost Extracts and Selected Microbial Isolates from Compost Against* Foc

12.3.2.1 Inhibition by Compost Extracts

Effect of Sterile Compost Extract on *Foc* Mycelial Growth and Sporulation

In order to understand the role of native compost microorganisms, microorganism-free compost extracts were used. Sterilized compost extracts, were tested in vitro against *Foc*. Results showed that three compost parameters (concentration of the extract, stage of composting, and type of compost) were significantly involved in the inhibition of *Foc* growth and sporulation. Regarding the effect of sterilized compost extract on *Foc* mycelial growth (Fig. 12.4), the treatment with different compost extracts showed significant inhibitory effects on mycelial growth in comparison to the control. The percentages of inhibition recorded ranged from 36% to 50% compared to the controls depending on the type of compost ($p \leq 0.05$). The extracts of composts A and C inhibited *Foc* growth by more than 42%, while those of B, E, F, and D showed an inhibition of *Foc* growth by more than 32%. These inhibition percentages also varied significantly depending on the concentration of the extracts. The original, undiluted extract (100%) and the 75% concentration of the different composts showed the highest inhibition percentages, 51.84% on average for the original extract and 46% on average for the 75% concentration. The composting stage also significantly affected *Foc* growth inhibition ($p \leq 0.05$). The highest inhibition percentages were recorded for the second and fourth stage of composting corresponding, respectively, to the 13th and the 39th day after mixing. This percentage varied between 23.07% and 41.53% for stage 2 and between 22.69% and 40% for the fourth stage. The maximum growth inhibition was obtained at the second stage for type C of compost, it was of 63.46% and 59.61% for the original extract and its 75% dilution respectively.

Figure 12.5 shows the effect of sterilized compost extract on *Foc* sporulation. The suppressive property of compost can be affected or altered by the physicochemical characteristics of the environment. Among the factors that affect the

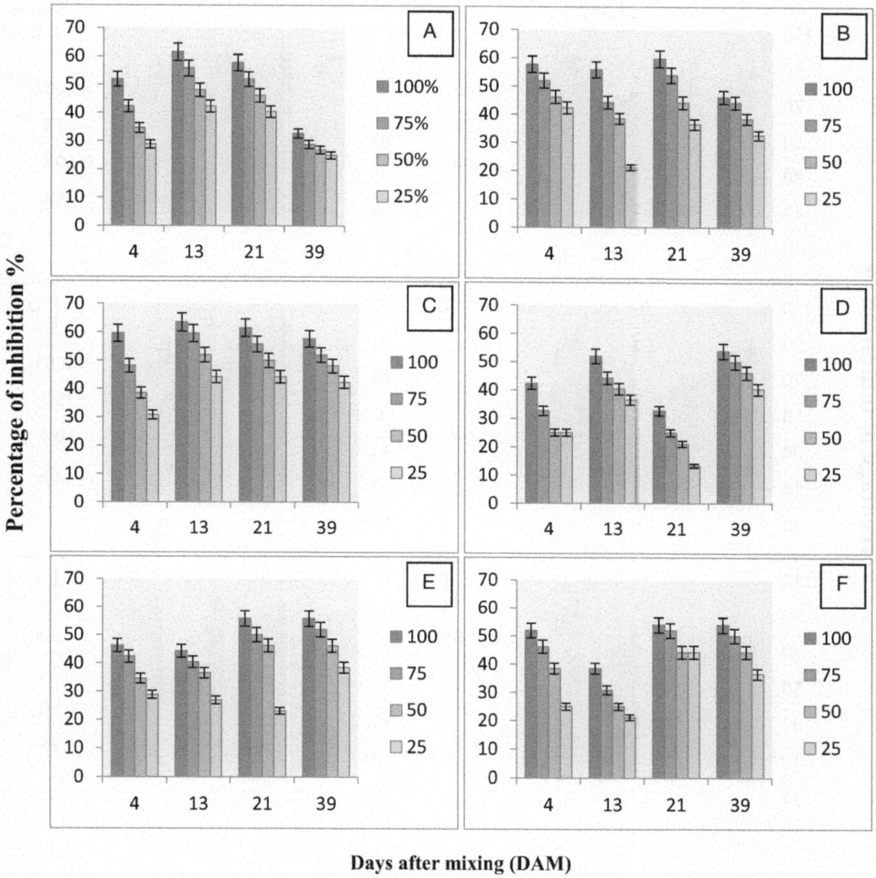

Fig. 12.4 Kinetics of mycelial growth inhibition of *Foc* treated with different concentrations (original non-diluted extract, 75, 50, and 25%) of the five sterilized compost extracts (A, B, C, D, E, and F) at different stages of composting (4th, 13th, 21st, and 39th day)

ability of certain pathogens to grow in compost are pH, temperature, and conductivity (Gurtler et al. 2018). Results showed a significant inhibition of the growth and sporulation of *Foc* by the six different types of compost, especially during stages 2 and 4 of composting. These two stages are characterized by an increase in pH values, which raised from 7.28 to 8.35 at stage 2 and to 8.82 at stage 4. Studies by Borrero et al. (2009) suggested that the increase in pH may explain the suppressive effect of *Fusarium* wilt of tomato by certain composts originating from agricultural and industrial waste. This factor is likely to affect the availability of macronutrients and micronutrients necessary for the growth and sporulation of *F. oxysporum* (Orr and Nelson 2018). Indeed, a depletion in nutrients such as of Fe, Cu, P, Na, Ca, and Zn caused by the high pH of composts and compost–peat mixes was reported (Cotxarrera et al. 2002; Borrero et al. 2004). Regarding the EC, the results obtained are different depending on the stage of composting. An average value of

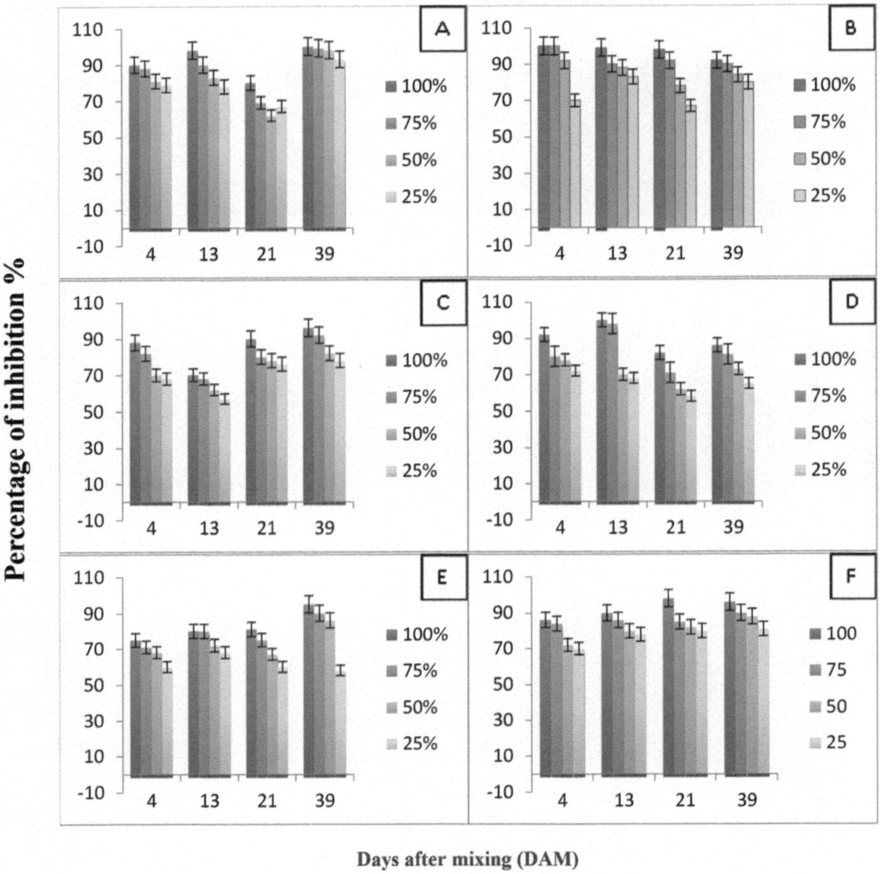

Fig. 12.5 Kinetics of suppression of *Foc* sporulation treated with different concentrations (original non-diluted extract, 75, 50, and 25%) of sterilized compost (A, B, C, D, E, F) extracts at different stages of composting (4th, 13th, 21st, and 25th day)

3.38 $ms.cm^{-1}$ was noted at stage 2 and an average value of 9.15 $ms.cm^{-1}$ at stage 4. These results are very different from stage 1 (CE = 4.97 $ms.cm^{-1}$). Pane et al. (2011) and Otlewska et al. (2020) showed that high EC values can counteract the suppressive effects of compost against pathogens. This may be correlated with the inhibition of the development of *Foc* recorded at stage 2, which showed relatively low EC values but cannot explain the inhibition of *Foc* at stage 4 for which the EC increased sharply. The temperature of the compost also showed significant variations during composting. The maximum temperature values were obtained at stage 2 of composting (56.2 °C as the average value). Soobhany et al. (2017) reported that the increase in temperature during the thermophilic phase is likely to eliminate pathogens and Bernal et al. (2009) recommended temperatures above 55 °C as effective for this purpose. Gao et al. (2018) reported that it is enough for the thermophilic phase to last more than 3 days for the compost to become pathogen-free. All the above mentioned physicochemical factors made it possible to understand *Foc* inhibition generated by the different types of compost but do not explain the causes of distinction of compost C from the others. Compost C differs from all other types by the fact that it is the only one containing all components (tomato leaves, sawdust, poultry manure, sheep manure, and olive pomace) and this distinction can be attributed to the composition of this compost.

Effect of Non-sterile Compost Extract on *Foc* Mycelial Growth and Sporulation

After 8 days of incubation at 28 °C, the results obtained (Fig. 12.6) show that all the non-sterilized extracts of the composts significantly inhibited *Foc* mycelial growth in comparison with the control. The inhibition effects obtained by non-sterilized extracts are higher than those obtained by sterilized extracts with inhibition

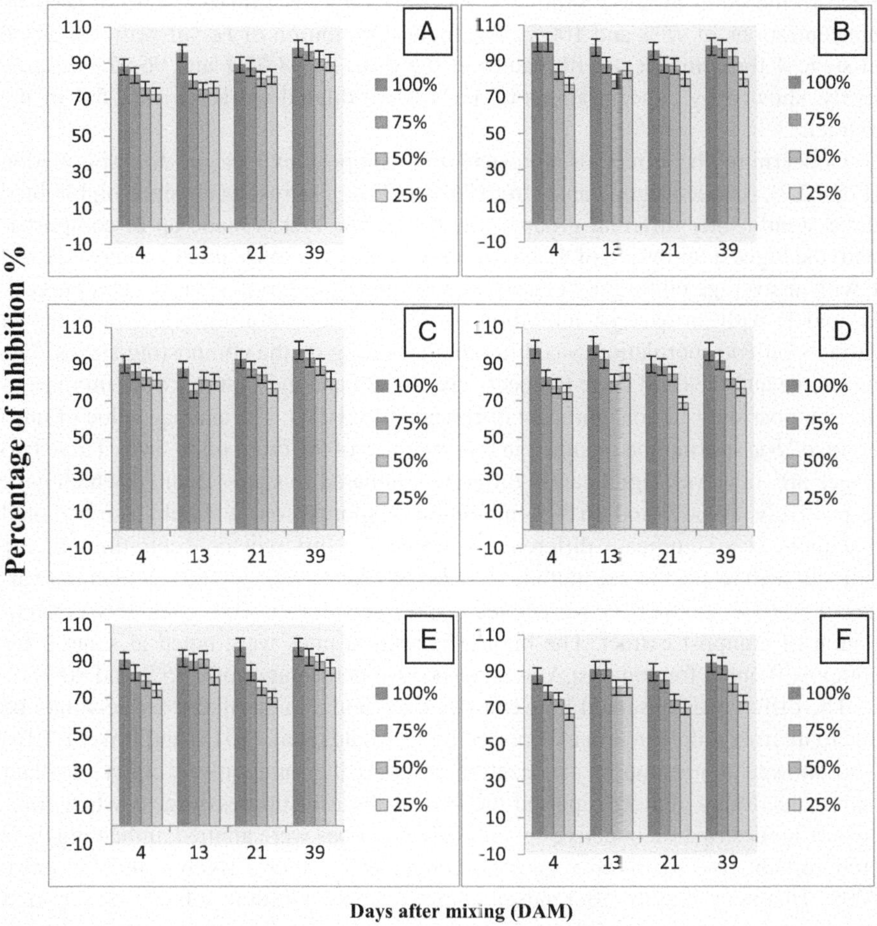

Fig. 12.6 Kinetics of mycelial growth inhibition of *Foc* treated with different concentrations of non-sterile compost extracts (A, B, C, D, E, F) at different stages of composting

percentages fluctuating between 53% and 89%. However, these percentages vary significantly depending on the type of compost, the stage of composting, and the concentration of the extract. Regarding the effect of the type of compost on the inhibition of *Foc* growth, results made it possible to highlight three significantly different groups ($p \leq 0.05$). The first is made up of composts A, C, D, and F with an average inhibition of 85.45%, the second made up of compost B with the maximum inhibition of 89.01%, and the last group made up of compost E with an average inhibition of 82.02%. The statistical analysis showed that there is no significant difference between the first and the third stage, corresponding to 4th and 21st day after mixing ($p = 0.60 > 0.05$); the average value of inhibition of the growth of *Foc* obtained in this case was of the order of 82.67%. However, these stages are significantly different from stages 2 and 4. The highest inhibition percentage was obtained at stage 4 with an average of 89.92%. The effect of different concentrations of compost extracts gave significant differences. The inhibition rate was positively correlated to the concentration with 72% and 97% inhibition respectively, showed by concentrations of 75% and 100%. The highest inhibition of *Foc* growth was noted at stage 4 for compost A with values of the order of 98.03% and 96.07% respectively, showed by concentrations of 100% (non-diluted extracts) and 75% of the extracts.

Concerning the effect of non-sterilized compost extract on *Foc* sporulation (Fig. 12.7), it varied significantly from 75% to 95%. The results obtained highlighted three significantly different groups ($p \leq 0.05$). The first is made up of compost B with the highest inhibition of 87.68, the second group is made up of composts A and F with an average inhibition of 84.68, and the third group is made up of composts C, D, and E with an average inhibition of 77.31. The inhibiting effect of compost extracts on *Foc* sporulation was different depending on the composting stages. The first stage and the third stage (respectively, the 4th and the 21st day after mixing) of the compost showed no significant difference ($p \leq 0.05$). The average value of inhibition of *Foc* sporulation obtained in this case was of the order of 76.58%. These two stages are, however, significantly different compared to stages 2 and 4 which gave respectively, 80.66% and 86.37% inhibition of sporulation. It should also be noted that there is a significant difference between the different concentrations of the extracts with respect to the inhibition of *Foc* sporulation ($p \leq 0.05$). Maximum values of 90.08% and 84.62% were noted respectively, for the 100% and 75% concentration of compost extract. The highest inhibition rates were noted at stage 2 for compost B and 4 for compost A, with respective percentages of 97.5% and 90%.

The differences observed between sterilized and non-sterilized extracts may be due to microbial destruction as reported by El Khaldi et al. (2015) and Raviv (2016) who showed that compost sterilization reverses the antagonistic effect on plant pathogens. Many authors reported the loss of disease suppressiveness when composts after destruction of active antagonistic microbes were applied, indicating their predominant role in disease suppressiveness of composts (Noble and Coventry 2005; Tilston et al. 2005; Berendsen et al. 2012). In addition, it has been reported that compost extracts are rich in *Trichoderma* and *Penicillium* strains (Bellini et al. 2020; Haas et al. 2016). The highest inhibition rate was obtained at stage 2 for

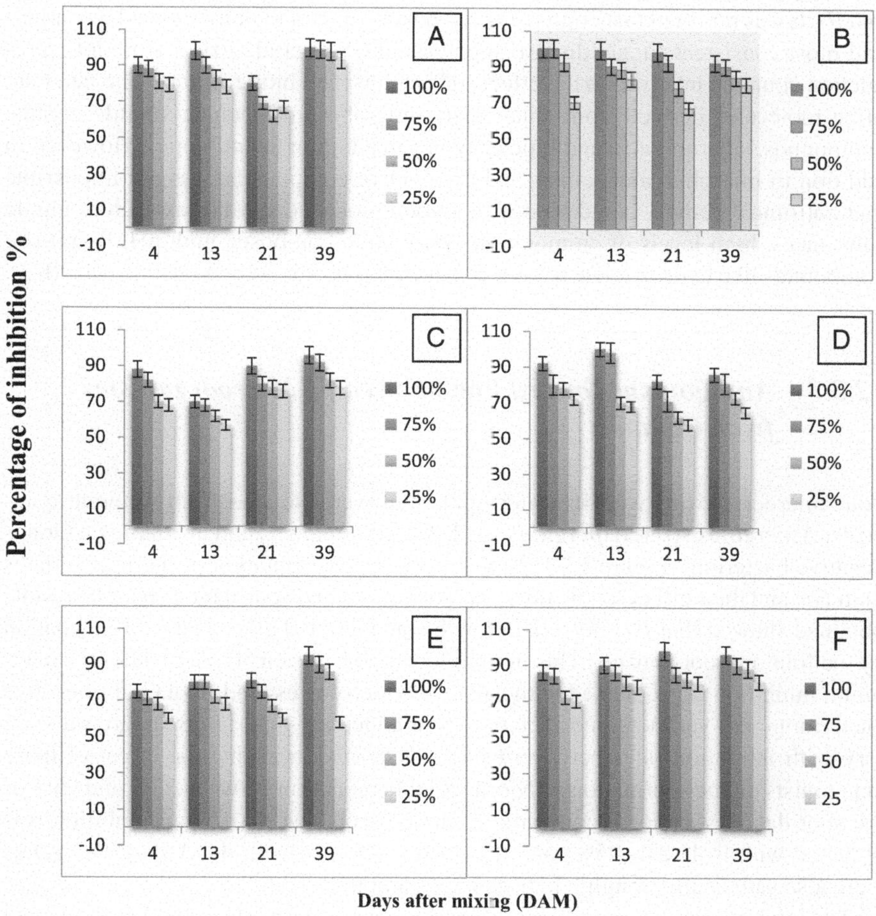

Fig. 12.7 Kinetics of suppression of sporulation of *Foc* treated with different concentrations of extract from non-sterile compost (A, B, C, D, E, F) at different stages of composting

compost B and stage 4 for compost A. These stages are characterized by high temperature values concomitant to replacement of some mesophilic microorganisms by thermophilic ones (Sánchez et al. 2017), which may play an important role in suppressing pathogens through antibiotic production (Azim et al. 2022). Among the thermophilic microorganisms involved in pathogen suppression, actinomycetes, *Pusillimonas*, *Pseudoxanthomonas*, and *Desulfobacter* are frequently reported (Liu et al. 2018). Actinobacteria have been reported to suppress *Pythium* disease (Ebrahimi-Zarandi et al. 2022) and many other plant pathogens (Bouizgarne 2013, 2022). Furthermore, the differences in *Foc* inhibition seen between different composts could be attributed to their composition. A-type and B-type compost contain nutrients that promote growth of microbial antagonists leading to *Foc* suppression. This effect could be related to their high manure content (poultry and sheep manure at a sum of 30%) compared to other composts (0% for compost D and 20% for C, E and F). It was reported that compost derived from animal manure showed the largest and most consistent fungal disease suppression (Pane et al. 2011). Antibiotic production could be involved in the effects of compost and biotic factors. The latter are often molecules derived from antagonistic microbes and include volatile organic compounds, siderophores, and lytic enzymes in addition to antibiotics. However, in addition to microbial antagonists and their active compounds, it is not impossible that antifungal activity could be due to abiotic bioactive substances such as humic substances, high levels of ammonia in non-mature composts, high pH, fungitoxic compounds like hydroxyoleic acids and salts (de Bertoldi 2009; Verrillo et al. 2022).

12.3.3 Antagonism Toward Foc *Showed by Microorganisms from Compost*

Four microorganisms with inhibitory potential were isolated from different composts: *Aspergillus* sp. 1, *Aspergillus* sp. 2, *Trichoderma* sp., and an unknown Gram-negative bacterium. Figure 12.8 shows the direct confrontation of the four isolates with *Foc* and their effects on its mycelial growth (A) and sporulation (B). The results obtained showed that *Foc* mycelial growth was inhibited after 8 days of incubation by the four microorganisms. This inhibition varied between 50% and 98%, the maximum inhibition was obtained with isolate 4 which corresponds to a Gram-negative bacterium—not yet identified. The results obtained by *Aspergillus sp.* are satisfactory with inhibition rates between 62% and 78%. Concerning the effect of these antagonists on the sporulation of *Foc*, a variable reduction between 37% and 95% of the sporulation was noted compared to the control. The maximum inhibition was obtained with isolate 4 (95%) and the results obtained with the *Aspergillus* genus were also satisfactory ranging between 77% and 62%.

The inhibitory effect exerted by these antagonists against *Foc* could be explained by high competition potential resulting from their rapid development by colonizing the shared medium. Theoretically, all fungal agents can use this mode of action to compete with pathogens for niche and nutrients (Ghorbanpour et al. 2018). The ability to have antifungal activity against *Fusarium* sp. has been described as a mode of antagonism of *Trichoderma* sp. and *Penicillium* sp. isolated from compost (Daami-Remadi et al. 2006; Chen et al. 2021). Nigam and Tomar (2022) noted antagonistic potential of *Aspergillus niger* against *F. oxysporum* f sp. *pisi, R. solani,* and *P. ultimum.* The antifungal effect of these antagonists on the pathogen was attributed either to the production of antibiotics (Frisvad et al. 2018) or to the production of lytic enzymes (Aydi Ben Abdallah et al. 2023). The lytic enzymatic activities are diverse; the most described are chitinases, proteases, lipases, cellulases, and β-1,3-glucanases (Ghasemi et al. 2019). These enzymes have the capability to break down the polysaccharides that contribute to the integrity of fungal cell walls (Aydi Ben Abdallah et al. 2023). Another mechanism by which compost could suppress plant pathogen was described as induction of plant resistance (Pharand et al. 2002).

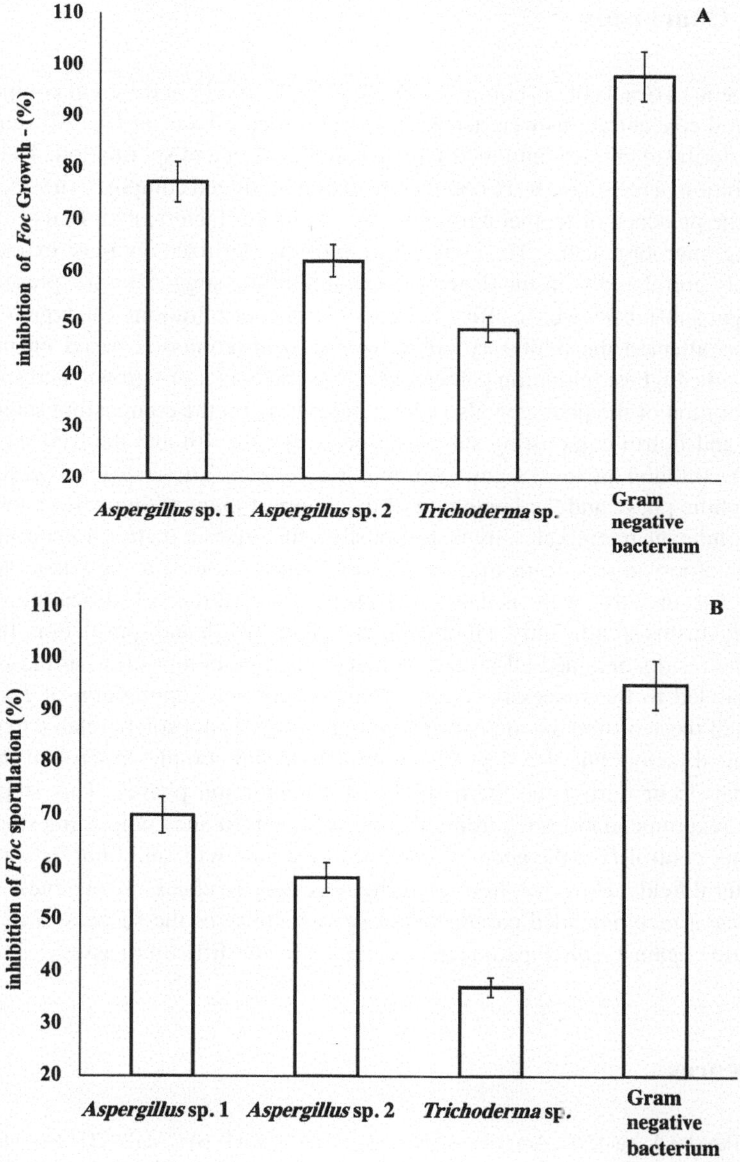

Fig. 12.8 Antagonism effect of the four isolates (*Aspergillus* sp. 1, *Aspergillus* sp. 2, *Trichoderma* sp., and a Gram-negative bacterium) on *Foc* mycelial growth (A) and sporulation (B)

Those mechanisms were shown to be triggered by compost against *F. oxysporum* f. sp. *radicis-lycopersici,* causal agent of tomato crown and root rot (Pharand et al. 2002). Although induced plant resistance to many phytopathogens was reported, to the best of our knowledge, rare are works aiming to apply compost as resistance inducer in order to protect chickpea from *F. oxysporum* f. sp. *ciceri*.

12.4 Conclusion

The aqueous extracts of six composts (A, B, C, D, E, and F) at different composting stages and concentrations were tested in vitro for their effects on *Foc*. All sterilized or non-sterilized extracts inhibited *Foc* mycelial growth and sporulation. The highest inhibition percentages were obtained with non-sterilized compost extracts, likely due to the presence of temperature sensitive biological factors, mainly represented by living microorganisms. The concentration of the sterilized compost extract was positively correlated with the degree of *Foc* inhibition, suggesting the presence of antifungal substances whose effect increased with concentration. The type of compost also affected the inhibitory effect on *Foc*, with composts based on manure showing the highest inhibition percentages. The percentages of growth and sporulation inhibition of the pathogen also varied depending on the composting stage. The second and fourth composting stages (respectively, the 4th and the 21st day after mixing) exhibited the most significant inhibitory effects. These stages represent the thermophilic phase and the beginning of the compost maturation phase, known for their significant microbial activity responsible for organic matter transformation. Some microorganisms from these composts, which showed antagonistic activity against *Foc* in vitro, were isolated and purified for further identification. These microorganisms significantly inhibited in vitro *Foc* growth and sporulation. In summary, the results obtained allowed to conclude that the composition of these composts has led to the successful composting process and suppression of *Foc*. The success of these different composts in eliminating *Foc* is not solely related to physicochemical factors but also depends on numerous and complex microbial interactions that occur during the thermophilic and maturation phases. This study can provide recommendations regarding the type of compost and composting stage that can better control *Foc*, the agent of chickpea Fusarium wilt, and limit its spread in agricultural fields. However, field trials are necessary to check effectiveness of our composts at recommended parameters as predictability of the success of compost application against a phytopathogen is usually a factor difficult to assess.

References

Aguilar-Paredes A, Valdés G, Araneda N, Valdebenito E, Hansen F, Nuti M (2023) Microbial community in the composting process and its positive impact on the soil biota in sustainable agriculture. Agronomy 13:542. https://doi.org/10.3390/agronomy13020542

Aydi Ben Abdallah R, Hassine M, Jabnoun-Khiareddine H, Daami-Remadi M (2023) Exploration of non-phytopathogenic *Aspergillus spp.* isolates recovered from soil and compost as potential source of bioactive metabolites for potato *Fusarium* dry rot control. Braz J Microbiol 54:1–11. https://doi.org/10.1007/s42770-023-00925-3

Azim K (2019) Organic amendments to alleviate plant biotic stress. In: Ansari R, Mahmood I (eds) Plant health under biotic stress. Springer, Singapore. https://doi.org/10.1007/978-981-13-6043-5_8

Azim K, Komenane S, Soudi B (2017) Agro-environmental assessment of composting plants in southwestern of Morocco (Souss-Massa region). Int J Recycl Org Waste Agric 6(2):107–115. https://doi.org/10.1007/s40093-017-0157-7

Azim K, Soudi B, Périssol C, Thami-Alami I, Roussos S (2022) Suppressive effect of root knot nematode *Meloidogyne spp.* during composting of tomato residues. In: Microbial biotechnology for sustainable agriculture Vol 1 (eds. Arora NK and Bouizgarne B), pp 449–469. https://doi.org/10.1007/978-981-16-4843-4_15

Bellini A, Ferrocino I, Cucu MA, Pugliese M, Garibaldi A, Gullino ML (2020) A compost treatment acts as a suppressive agent in *Phytophthora capsici*–*Cucurbita pepo* pathosystem by modifying the rhizosphere microbiota. Front Plant Sci 11:885. https://doi.org/10.3389/fpls.2020.00885

Benhamou N, Chet I (1996) Parasitism of sclerotia of *Sclerotium rolfsii* by *Trichoderma harzianum*: ultrastructural and cytochemical aspects of the interaction. Phytopathology 86:405–416

Berendsen RL, Pieterse CMJ, Bakker PAHM (2012) The rhizosphere microbiome and plant health. Trends Plant Sci 17:478–486. https://doi.org/10.1016/j.tplants.2012.04.001

Bernal MP, Alburquerque JA, Moral R (2009) Composting of animal manures and chemical criteria for compost maturity assessment. A review. Bioresour Technol 100:5444–5453. https://doi.org/10.1016/j.biortech.2008.11.027

Bheemaraya PMB, Ramesh YST, Amaresh YS, Naik MK (2013) Salinity stress tolerance in native *Trichoderma* isolates. Environ Ecol 31:727–729

Borrero C, Trillas MI, Ordovás J, Tello JC, Avilés M (2004) Predictive factors for the suppression of *Fusarium* wilt of tomato in plant growth media. Phytopathology 94:1094–1101. https://doi.org/10.1094/PHYTO.2004.94.10.1094

Borrero C, Trillas I, Avilés M (2009) Carnation *Fusarium* wilt suppression in four composts. Eur J Plant Pathol 123:425–433. https://doi.org/10.1007/s10658-008-9380-4

Bouizgarne B (2013) Bacteria for plant growth promotion and disease management. In: Maheshwari D (ed) Bacteria in agrobiology: disease management. Springer, Berlin/Heidelberg, pp 15–47. https://doi.org/10.1007/978-3-642-33639-3_2

Bouizgarne B (2022) Phosphate-solubilizing actinomycetes as biofertilizers and biopesticides: bioformulations for sustainable agriculture. In: Arora NK, Bouizgarne B (eds) Microbial biotechnology for sustainable agriculture, vol 1. Springer, Singapore. https://doi.org/10.1007/978-981-16-4843-4_13

Campitelli P, Ceppi S (2008) Chemical, physical and biological compost and vermicompost characterization: a chemometric study. Chemometr Intellig Labo Sys 90:64–71. https://doi.org/10.1016/j.chemolab.2007.08.001

Chan MT, Selvam A, Wong JW (2016) Reducing nitrogen loss and salinity during 'struvite' food waste composting by zeolite amendment. Bioresour Technol 200:838–844. https://doi.org/10.1016/j.biortech.2015.10.093

Chen R, Wang Y, Wang W, Wei S, Jing Z, Lin X (2015) N_2O emissions and nitrogen transformation during Windrow composting of dairy manure. J Environ Manag 160:121–127. https://doi.org/10.1016/j.jenvman.2015.06.021

Chen J, Zhou L, Din IU, Arafat Y, Li Q, Wang J, Wu T, Wu L, Wu H, Qin X, Pokhrel GR, Lin S, Lin W (2021) Antagonistic activity of *Trichoderma* spp. against *Fusarium oxysporum* in rhizosphere of radix pseudostellariae triggers the expression of host defense genes and improves its growth under long-term monoculture system. Front Microbiol 12. https://doi.org/10.3389/fmicb.2021.579920

Chichaybelu M, Girma N, Fikre A, Gemechu B, Mekuriaw T, Geleta T, Chiche W, Rubyogo JC, Akpo E, Ojiewo CO (2021) Enhancing chickpea production and productivity through stakeholders' innovation platform approach in Ethiopia. In: Akpo E, Ojiewo CO, Kapran I, Omoigui LO, Diama A, Varshney RK (eds) Enhancing smallholder farmers' access to seed of improved legume varieties through multi-stakeholder platforms. Springer, Singapore, pp 97–111. https://doi.org/10.1007/978-981-15-8014-7_7

Cotxarrera L, Trillas-Gay M, Steinberg C, Alabouvette C (2002) Use of sewage sludge compost and *Trichoderma asperellum* isolates to suppress *Fusarium* wilt of tomato. Soil Biol Biochem 34:467–476

Daami-Remadi M, Hibar K, Jabnoun-Khiareddine H, Ayed F, El Mahjoub M (2006) Effect of two *Trichoderma* species on severity of potato tuber dry rot caused by *Tunisian fusarium* complex. Int J Agric Res 1:432–441. https://doi.org/10.3923/ijar.2006.432.441

de Bertoldi M (2009) Production and utilization of suppressive compost: environmental, food and health benefits. In: Insam H, Franke-Whittle I, Goberna M (eds) Microbes at work. Springer, Berlin, pp 153–170. https://doi.org/10.1007/978-3-642-04043-6_8

De Corato U (2020) Disease-suppressive compost enhances natural soil suppressiveness against soil-borne plant pathogens: a critical review. Rhizosphere 13:100192. https://doi.org/10.1016/j.rhisph.2020.100192

Dickerson GW (1996) Compost dressing helps Chile peppers. Bio Cycle 37:80–87

Ebrahimi-Zarandi M, Saberi Riseh R, Tarkka MT (2022) Actinobacteria as effective biocontrol agents against plant pathogens, an overview on their role in eliciting plant defense. Microorganisms 10:1739. https://doi.org/10.3390/microorganisms10091739

El Khaldi R, Daami-Remadi M, Cherif M (2015) Biological control of stem canker and black scurf on potato by date palm compost and its associated fungi. J Phytopathol 164:40–51. https://doi.org/10.1111/jph.12423

El-Mrini S, Aboutayeb R, Azim K, Zouhri A (2021) Co-composting process assessment of three-phase olive mill pomace and Turkey manure in Morocco. J Southwest Jiaotong Univ 56:764–778. https://doi.org/10.35741/issn.0258-2724.56.6.68

FAOSTAT (2021) Food and Agriculture Organization of the United Nations (FAO), Rome. Available https://www.fao.org/faostat/fr/#data/QCL. Accessed 14 May 2023

Fernández-Delgado M, Amo-Mateos E d, Lucas S, García-Cubero MT, Coca M (2020) Recovery of organic carbon from municipal mixed waste compost to produce fertilizers. J Clean Prod 265:121805. https://doi.org/10.1016/j.jclepro.2020.121805

Frisvad JC, Møller LL, Larsen TO, Kumar R, Arnau J (2018) Safety of the fungal workhorses of industrial biotechnology: update on the mycotoxin and secondary metabolite potential of *Aspergillus niger*, *Aspergillus oryzae*, and *Trichoderma Reesei*. Appl Microbiol Biotechnol 102:9481–9515. https://doi.org/10.1007/s00253-018-9354-1

Gao Y, Yu H, Liu P, Ma C, Li Q, Jiang W (2018) Ending composting during the thermophilic phase improves cultivation substrate properties and increasing winter cucumber yield. Waste Manag 79:260–272. https://doi.org/10.1016/j.wasman.2018.07.048

Gariglio NF, Buyatti MA, Pilatti RA, Russia DE, Acosta MR (2002) Use of a germination bioassay to test compost maturity of willow (*Salix* sp.) sawdust. N Z J Crop Hortic 30:135–139. https://doi.org/10.1080/01140671.2002.9514208

Gavilanes-Terán I, Jara-Samaniego J, Idrovo-Novillo J, Bustamante MA, Moral R, Paredes C (2016) Windrow composting as horticultural waste management strategy—a case study in Ecuador. Waste Manag 48:127–134. https://doi.org/10.1016/j.wasman.2015.11.026

Ghasemi S, Safaie N, Shahbazi S, Shams-Bakhsh M, Askari H (2019) Enhancement of lytic enzymes activity and antagonistic traits of *Trichoderma harzianum* using γ-radiation induced mutation. J Agric Sci Technol 21:1035–1048

Ghorbanpour M, Omidvari M, Abbaszadeh-Dahaji P, Omidvar R, Kariman K (2018) Mechanisms underlying the protective effects of beneficial fungi against plant diseases. Biol Control 117:147–157. https://doi.org/10.1016/j.biocontrol.2017.11.006

Gurtler JB, Doyle MP, Erickson MC, Jiang X, Millner P, Sharma M (2018) Composting to inactivate foodborne pathogens for crop soil application: a review. J Food Prot 81:1821–1837. https://doi.org/10.4315/0362-028X.JFP-18-217

Haas D, Lesch S, Buzina W, Galler H, Gutschi AM, Habib J, Pfeifer B, Luxner J, Reinthaler FF (2016) Culturable fungi in potting soils and compost. Medic Mycol 54:825–834. https://doi.org/10.1093/mmy/myw047

Hmouni A, Hajlaoui MR, Mlaiki A (1996) Résistance de *Botrytis cinerea* aux benzimidazoles et aux dicarboximides dans les cultures abritées de tomate en Tunisie. EPPO Bull 26:697–705. https://doi.org/10.1111/j.1365-2338.1996.tb01513.x

Hoitink HAJ (2007) Compost use for disease suppression. Ohio State University, Wooster

Hoitink H, Stone A, Grebus M (1996) Suppression of plant disease by composts. In: De Bertoldi M (ed) The science of composting. Chapman & Hall, London, pp 373–381

Houasli C, Idrissi O, Nsarellah N (2020) Chickpea genetic improvement in Morocco: state of the art, progress and prospects. Moroccan J Agric Sci 1(1)

Idrissi O, Sahri A, Houasli C, Nsarellah N (2019) Breeding progress, adaptation, and stability for grain yield in Moroccan lentil improved varieties. Crop Sci 59:925–936. https://doi.org/10.2135/cropsci2018.07.0431

Jurado MM, Suárez-Estrella F, Vargas-García MC, López MJ, López-González JA, Moreno J (2014) Evolution of enzymatic activities and carbon fractions throughout composting of plant waste. J Environ Manag 133:355–364. https://doi.org/10.1016/j.jenvman.2013.12.020

Kalemelawa F, Nishihara E, Endo T, Ahmad Z, Yeasmin R, Tenywa MM, Yamamoto S (2012) An evaluation of aerobic and anaerobic composting of banana peels treated with different inoculums for soil nutrient replenishment. Bioresour Technol 126:375–382. https://doi.org/10.1016/j.biortech.2012.04.030

Krause MS, Hoitink HAJ, Han DY (2001) Spectrum and mechanisms of plant disease control with composts. In: Stofella PJ, Kahn BA (eds) Compost utilization in horticultural cropping systems. Lewis Publishers, Boca Raton, pp 263–274

Liu L, Wang S, Guo X, Zhao T, Zhang B (2018) Succession and diversity of microorganisms and their association with physicochemical properties during green waste thermophilic composting. Waste Manag 73:101–112. https://doi.org/10.1016/j.wasman.2017.12.026

Liu Z, de Souza TSP, Holland B, Dunshea F, Barrow C, Suleria HAR (2023) Valorization of food waste to produce value-added products based on its bioactive compounds. Processes 11:840. https://doi.org/10.3390/pr11030840

Lutz S, Thuerig B, Oberhaensli T, Mayerhofer J, Fuchs JG, Widmer F, Freimoser FM, Ahrens CH (2020) Harnessing the microbiomes of suppressive composts for plant protection: from metagenomes to beneficial microorganisms and reliable diagnostics. Front Microbiol 11:1810. https://doi.org/10.3389/fmicb.2020.01810

Nigam R, Tomar J (2022) Bioefficacy of *Aspergillus niger* isolates against soil borne pathogens of field pea. Ann Agric Res 43:195–198. https://epubs.icar.org.in/index.php/AAR/article/view/127777

Noble R, Coventry E (2005) Suppression of soil-borne plant diseases with composts: a review. Biocontrol Sci Tech 15:3–20. https://doi.org/10.1080/09583150400015904

Orosz V, Tomócsik A, Demeter I, Aranyos TJ, Makádi M (2021) Control of plant pathogen *Fusarium* spp. with compost, compost tea application—a review. COLUMELLA–J Agric Environ Sci 8:55–71. https://doi.org/10.18380/SZIE.COLUM.2021.8.2.55

Orr R, Nelson PN (2018) Impacts of soil abiotic attributes on fusarium wilt, focusing on bananas. Appl Soil Ecol 132:20–33. https://doi.org/10.1016/j.apsoil.2018.06.019

Otlewska A, Migliore M, Dybka-Stępień K, Manfredini A, Struszczyk-Świta K, Napoli R, Białkowska A, Canfora L, Pinzari F (2020) When salt meddles between plant, soil, and microorganisms. Front Plant Sci 11:553087. https://doi.org/10.3389/fpls.2020.553087

Pane C, Spaccini R, Piccolo A, Scala F, Bonanomi G (2011) Compost amendments enhance peat suppressiveness to *Pythium Ultimum*, *Rhizoctonia Solani* and *Sclerotinia minor*. Biol Control 56:115–124. https://doi.org/10.1016/j.biocontrol.2010.10.002

Pharand B, Carisse O, Benhamou N (2002) Cytological aspects of compost-mediated induced resistance against fusarium crown and root rot in tomato. Phytopathology 92:424–438. https://doi.org/10.1094/PHYTO.2002.92.4.424

Purohit A, Ghosh S, Chaudhuri RK, Chakraborti D (2023) Biological control of Fusarium wilt in legumes. In: Plant stress mitigators. Elsevier, pp 435–454. https://doi.org/10.1016/b978-0-323-89871-3.00019-7

Rasool M, Akhter A, Soja G, Haider MS (2021) Role of biochar, compost and plant growth promoting rhizobacteria in the management of Tomato early blight disease. Sci Rep 11(1):6092. https://doi.org/10.1038/s41598-021-85633-4

Raviv M (2016) Compost as a tool to suppress plant diseases: established and putative mechanisms. Acta Hortic 1146:11–24. https://doi.org/10.17660/ActaHortic.2016.1146.2

Rich N, Bharti A (2015) Assessment of different types of in-vessel composters and its effect on stabilization of MSW compost. Int Res J Eng Technol 2:37–42. https://www.irjet.net/archives/V2/i3/Irjet-v2i308.pdf

Sánchez ÓJ, Ospina DA, Montoya S (2017) Compost supplementation with nutrients and microorganisms in composting process. Waste Manag 69:136–153. https://doi.org/10.1016/j.wasman.2017.08.012

Soobhany N, Mohee R, Garg VK (2017) Inactivation of bacterial pathogenic load in compost against vermicompost of organic solid waste aiming to achieve sanitation goals: a review. Waste Manag 64:51–62. https://doi.org/10.1016/j.wasman.2017.03.003

Tilston EL, Pitt D, Fuller MP, Groenhof AC (2005) Compost increases yield and decreases take-all severity in winter wheat. Field Crop Res 94:176–188. https://doi.org/10.1016/j.fcr.2005.01.003

Verrillo M, Salzano M, Savy D, Di Meo V, Valentini M, Cozzolino V, Piccolo A (2022) Antibacterial and antioxidant properties of humic substances from composted agricultural biomasses. Chem Biol Technol Agric 9:28. https://doi.org/10.1186/s40538-022-00291-6

Wu P-H, Chang H-X, Shen Y-M (2023) Effects of synthetic and environmentally friendly fungicides on powdery mildew management and the phyllosphere microbiome of cucumber. PLoS One 18:e0282809. https://doi.org/10.1371/journal.pone.0282809

Yatoo AM, Niamat AM, Baba ZA, Birjees H (2021) Sustainable management of diseases and pests in crops by vermicompost and vermicompost tea, a review. Agron Sustain Dev 41(1). https://doi.org/10.1007/s13593-020-00657-w

Chapter 13
Nutrient and Pest-Managing Microbial Inoculants for Sustainable Production of Rice

Upendra Kumar, Megha Kaviraj, Snehasini Rout, Archana Singha, and Rajeev Padbhushan

Abstract Rice production is mostly dependent upon chemical fertilizers and pesticides across the globe; however, extensive use of these in rice crop has shown adverse effects on environment and human health. Besides, continuous application of these chemicals in rice cultivation has also deteriorated the soil physical conditions and quality. Achieving higher sustainable yield through microbial technology could be one of the better options under rice cultivation system. Rice-specific biofertilizers and biopesticides are one of the possible substitutes of dangerous chemical products as these are host-specific, eco-friendly, and cost-effective. These are actually organic formulations containing living microorganisms like bacteria, fungi, and algae that are capable of exhibiting various plant growth promoting (PGP) traits. Numerous biofertilizers such as nitrogen (N) fixing microorganisms (*Azolla*, *Cyanobacteria*, *Azotobacter*, *Azospirillum*, etc.), phosphate, potassium, zinc solubilizing microorganisms (*Pseudomonas*, *Bacillus*, *Aspergillus*, *Arbuscular mycorrhiza*, etc.), and sulfur oxidizing microorganisms are being used in rice cultivation. Heterotrophic N-fixing microbes fix nitrogen from 15.6 to 22 kg N ha^{-1} $year^{-1}$ in rice field. Moreover, cyanobacteria and *Azolla* could supplement the N need of rice crop by replacing 30–50% of the requisite chemical N, whereas phosphate and potassium solubilizing bacteria may reduce chemical P and K fertilizer applications by 15–30% without any significant reduction in rice yield. Besides biofertilizers, microbial pesticides (bacteria, fungi, and viruses) could perform as antagonists and entomopathogens against various rice pests. About 23 bacterial antagonists from the genera *Bacillus*, *Pseudomonas*, *Serratia*, and *Erwinia* have been found, which prevent rice diseases like sheath blight, stem and sheath rot, brown spot, and blast. Similarly, the most popular entomopathogens used mainly to manage rice insects

U. Kumar (✉) · M. Kaviraj · S. Rout · A. Singha
ICAR-National Rice Research Institute, Cuttack, Odisha, India

R. Padbhushan
Department of Soil Science and Agricultural Chemistry, Bihar Agricultural University, Sabour, Bihar, India

N. K. Arora, B. Bouizgarne (eds.), *Microbial Biotechnology for Sustainable Agriculture Volume 2*, Microorganisms for Sustainability 34,
https://doi.org/10.1007/978-981-97-2355-3_13

(leaf folder, brown leaf hopper, yellow stem borer, etc.) are *Metarhizium anisopliae, Beauveria bassiana,* and *Bacillus thuringiensis*. Overall, researchers are interested in PGP microbes as biofertilizers and biocontrol bioinoculants and therefore, various such microbes are now commercialized for several crops including rice. This chapter describes the biofertilizers and microbial pesticide formulations exclusively used in rice to enhance its production and productivity by exhibiting their desirable mechanisms and PGP functions.

Keywords Rice · Biofertilizers · Biocontrol · Microbial formulations

13.1 Introduction

Rice (*Oryza sativa* L.) is one of the major cereal crops consumed by majority of human population (Kumar et al. 2020). Asia has 60% of the world population that produces and consumes 90% of rice in the world. In terms of rice production, India has ranked second after China, whereas Chhattisgarh is the largest rice producer state followed by Andhra Pradesh and West Bengal in India. At present the fertilizer consumption has increased four-fold than that has been applied during 1990s without any significant yield difference (Biswas and Sharma 2008). Use of chemical fertilizers of nitrogen (N), phosphorous (P), and potassium (K) over a period of time has shown harmful effect as they deteriorate the physico-chemical and biological properties of the soil, as well as cause ground water pollution by eutrophication of water bodies and are also harmful to human health (Bhardwaj et al. 2014; Dębska et al. 2016; Naher et al. 2016; Sao and Samuel 2018).

By 2025, heavy dose of multi-nutrients as well as pesticides would be required to produce 40% more rice to feed the growing population (Nayak et al. 2019). However, injudicious application of agrochemicals has resulted in soil health deterioration, alteration in quality of water, and decline in population of useful microorganisms and insects in the soil. In turn, this has resulted in decline of microbial diversity along with suppression of ecosystem function and vulnerability of crops towards attack of several diseases (Kumar et al. 2023a, b). Moreover, the imbalance that is created in the overall microbial structure causes irreversible damage to the agricultural system (Vance 2001; Jha and Mishra 2005; Kumar et al. 2017, 2018a, b, 2019a, 2020; Arora 2023). Hence, the reduced dependency on chemical fertilizers is important to accomplish the goal of agriculture towards sustainable production system so that environmental health may not be compromised with production. In order to maintain the higher grain yield of rice, microbial resources have to be given more attention. It is also evident that any regular deceleration in the yield could be mitigated by incorporating soil microbes in the agriculture production.

In recent years, biofertilizers (microbial inoculants) are being used, which are eco-friendly, cost-effective, and a substitute to chemical fertilizers (Rao 1982; Rajput et al. 2020). Biofertilizers are organic formulations containing living microorganisms like bacteria, fungi, and algae that are capable of exhibiting various plant growth promoting (PGP) activities (Govindasamy et al. 2009) (Fig. 13.1). In addition, they keep the soil fertile for long period of time (Myo et al. 2016; Yasin et al. 2012; Aggani 2013; Mohammadi and Sohrabi 2012). There are wide range of biofertilizers that are used in the cultivation of rice crop, like formulations containing N-fixing microorganisms (*Azolla, Cyanobacteria, Azotobacter, Azospirillum,* etc.), P, K, zinc (Zn) solubilizing microorganisms, and sulfur (S) oxidizing microorganisms (Mishra et al. 2023). The interaction of microbes with plant, particularly under rhizosphere and endosphere can be valuable, unbiased, adaptable, or not injurious for plant development. These microbes are helpful in plant growth and termed as plant growth-promoting microbes (PGPM) (Kumar and Nayak 2019; Kumar et al. 2018c, 2019a, b; Kumar 2017).

Among all the nutrients, NPK are essential for plant growth as well as microbial survival. These nutrients perform several physiological roles, which accumulate and release energy during cellular metabolism. N is the major essential nutrient for plant survival and it is also responsible for causing significant changes in plant growth and development (Kumar et al. 2019b; Sharma et al. 2019). Another essential element is P, which is deficit in most of the natural soils. If soils are inoculated with the phosphate solubilizing microorganisms (PSB), then plants may not face the problem of P deficiency. PSB have been successfully used for P nutrition, improving growth and yield of crops (Pattnaik et al. 2019). K is another essential nutrient involved in growth and metabolism of plant. It is also called as quality nutrient as it helps in maintaining shape, size, colour, taste, shelf life, fibre quality, and other factors of plant.

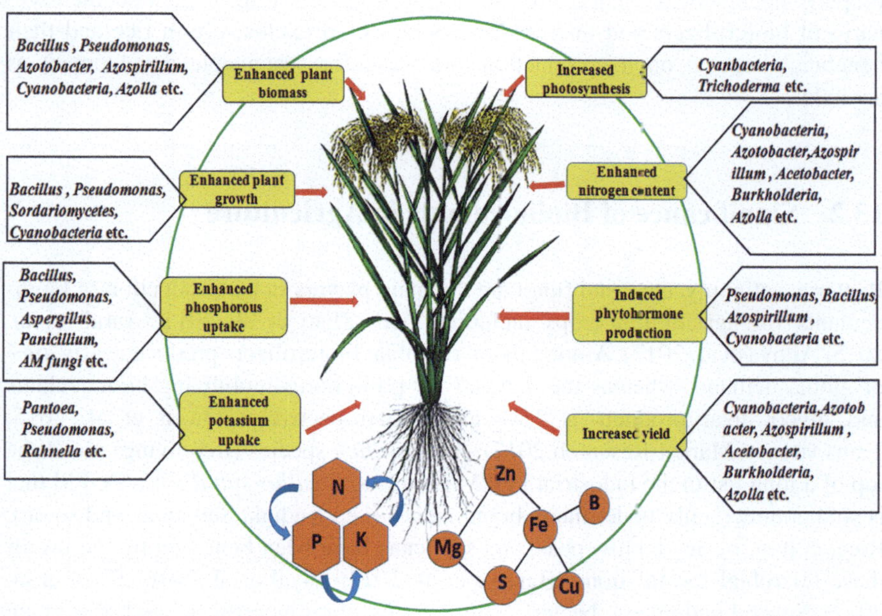

Fig. 13.1 Group of agriculturally important microbial bioinoculants and their role in plant growth-promotion in rice

Microbial pesticides may significantly control various rice pests besides minimizing the environmental pollution arising due to application of chemical pesticides (Kirubakaran et al. 2014; Kumar et al. 2018d; Dangar et al. 2019). Microbial pesticides such as biopesticides and biocontrol agents (BCAs) have been recognized to be highly valued species and eco-friendly in nature that has led to their adaptation as pest management strategies around the globe (Kumar et al. 2013). The extent of market of microbial pesticides is about 92% of the total biopesticide market, and there is adequate opportunity to use their potential in agriculture, though there are several problems associated with handling. More than 452 BCAs have been used for production of more than 2000 commercial products at global level (Dangar et al. 2019). BCAs include microorganisms (149), natural products (89), macro-organisms (140), and semi-chemicals (74). India is one of the leading producer and consumer of BCAs. Nearly 16 commercial preparations have been made with *Bacillus thuringiensis* (Bt), among fungal formulations, about 38 made up of *Trichoderma*, *Metarhizium,* and *Beauveria* and in case of virus-based formulations about 45 made with *Helicoverpa* and *Spodoptera* (Dangar et al. 2019). These agents have potential to substitute approximately 20% of the total chemical pesticides (Kumar et al. 2018d). For supplying BCAs across the different regions of the country, 128 units including 80 private companies are playing an important role (Dangar et al. 2019). Besides, ICAR institutes (8), State Agricultural Universities (10), Central Integrated Pest Management (30) centres, and parasitoid-producing laboratories (4) are also providing BCAs (Wahab 2003, 2004, 2009).

Microbes used to manage nutrient problems and pest management in rice are not properly documented. Hence, this chapter is an attempt to provide a detailed scenario of biofertilizers and microbial pesticides used exclusively in rice and their beneficial impacts on its production and yield for sustainable development as a whole.

13.2 Significance of Bioinoculants in Agriculture

Bacteria, actinomycetes, and fungi are the main players in the production of bioinoculants for agricultural crops including rice (Zhao et al. 2011; Aarab et al. 2015; Adhya et al. 2017). Among them, rhizobial biofertilizers produce nearly 79% of global demand, whereas the demand for phosphorus-mobilizing bioinoculants and mycorrhizal products is 15% and 7% respectively (Owen et al. 2015; Transparency Market Research 2017). *Azospirillum* species (free living) are at the top of a long list in the industrial PGPM products. *Bacillus subtilis* is a PGPM that is sold commercially under many brand names (i.e., Kodiak, Serenade, and so on). Rice, cotton, beans, lentils, peas, and soybeans have been benefited by the use of these microbial formulations (Banayo et al. 2012; Suyal et al. 2016; Khan et al. 2017). Several endophytic bacteria, fungi, arbuscular mycorrhizae, and blue green algae are also known to show significant effects on plant growth.

13.3 Biofertilizer Application in Rice

13.3.1 N-Fixing Microorganisms

The process of biological N-fixation (BNF) plays a significant role in replacing the commercially available chemical N fertilizer and thus, reducing the environmental problems to some extent (Sao and Samuel 2018). *Azolla*, *Cyanobacteria*, *Azospirillum*, and *Azotobacter* are some of the major N-fixing bioinoculants of rice (Tables 13.1 and 13.2).

13.3.1.1 *Azolla*

Azolla is an aquatic floating fern and a known biofertilizer for rice crop, which is symbiotically associated with N-fixing cyanobionts. Taxonomically, seven distinct species of *Azolla*, namely *A. filiculoides*, *A. caroliniana*, *A. mexicana*, *A. microphylla*, *A. rubra*, *A. pinnata*, and *A. nilotica* are reported so far and these are categorized into two sections, namely Euazolla and Rhizosperma on the basis of reproductive structures. N-fixing potentiality of *Azolla* was standardized and found that the fresh weight of *Azolla* augmented growth by 2–6 folds in a week with N-fixation of 75 mg g^{-1} dry weight day^{-1}. The total N content was 3–5% on dry weight basis and 0.2–0.3% based on fresh weight. Most importantly, within a year, *Azolla* biomass multiplied very fast and produced around 347 t fresh weight ha^{-1}, which contains around 868 kg N (Kumar and Nayak 2019).

Table 13.1 Estimated amounts of fixed N_2 by different N_2-fixing systems in rice

N_2-fixing systems	Estimated amounts of fixed N_2	Increases in rice grain yield		References
		Amount	(%)	
Azolla-anabaena symbiosis	48.2 kg ha^{-1}	1.5 t ha^{-1}	50	Mian (2002)
Cyanobacteria	24.2 kg ha^{-1}	1.4 t ha^{-1}	29	Hashem (2001)
Azotobacter sp.	11–20 kg ha^{-1}	0.4–0.9 t ha^{-1}	7–20	Yanni and El-Fattah (1999); Kumar et al. (2018c)
Azospirillum lipoferum	58.9% Ndfa	6.7 g $plant^{-1}$	81	Mirza et al. (2000)
Herbaspirillum spp.	38.1–58.2% Ndfa*	3.7–7.5 g $plant^{-1}$	45–90	Mirza et al. (2000)
Burkholderia vietnamiensis	Data not available	0.5–0.8 t ha^{-1}	13–22	Tran Van et al. (2000)
Rhizobium leguminosarum	23–31 mg pot^{-1}	0.6–7.9 g pot^{-1}	2–22	Biswas et al. (2000)

**Ndfa* nitrogen derived from atmosphere

Table 13.2 Microbial bioinoculants used in rice for nutrient management

Inoculant	Rice cultivar	Effect	Yield/biomass/germination	References
Azolla	PSBRc 54	Increase in tiller number up to 64.5%, reduced urea up to 41.1%	Grain yield (20–41%) more than control	Macale et al. (2004)
Azolla filiculoides	–	High quantities of nitrogen accumulation (2.58%) in the leaf	Increase in plant biomass	Bocchi and Malgioglio (2010)
Anabaena variabilis, *Nostoc muscorum*, *Aulosira fertilissima*, and *Tolypothrix tenuis*	–	Improved N uptake	3.8% increase in the yield	Bhooshan et al. (2018)
Nostoc BI42, *Anabaena* BI46, and *Calothrix* BI 22	El Paso 144 and INIA Tacuar	Increased N efficiency (5–12%)	Grain yield (6401 ± 381 kg ha^{-1})	Irisarri et al. (2007)
Azospirillum and *Trichoderma*	BU Dhan 1	Increased plant height, tiller number per hill, panicle length and chlorophyll content in leaf tissue	Grain yield (29.84 g $plant^{-1}$), nitrogen efficiency: *Azospirillum* + *Trichoderma* (41%) and *Trichoderma* (40%)	Khan (2018)
Azospirillum brasilense TPS05	ADT 36	Increased root number, root length, numbers of leaf, leaf length, shoot length	(59.5 g $plant^{-1}$) grain weight	Kanimozhi and Panneerselvam (2015)
Azospirillum lipoferum and *A. brasilense*	–	Increased grain yield	Improved growth and better Grain yield (kg ha^{-1})	Sahoo et al. (2014)
A. brasilense	CO43, ASD16, ASD18, AS89044, ADT36, ADT37, and IR64	Increased shoot and root length	Increased seed germination CO43 (93.3%), ASD16 (78.3%), ASD18 (88.5%), AS89044 (90.5%), ADT36 (76.5%), ADT37 (76.3%), and IR64 (72%)	Kannan and Ponmurugan (2010)
Azospirillum and BGA	ADT 37, IET 1444, CO 47, white ponni, and china ponni	Supplement for N, P, K in soil	Highest yield (2088 kg $acre^{-1}$ in ADT 37)	Subashini et al. (2007)

Burkholderia and *Acinetobacter*	Zn responsive (NDR 359), Zn nonresponsive (PD 16)	Increased dry matter yield pot^{-1} (12.9%), productive tillers plant^{-1} (15.1%), number of panicles plant^{-1} (13.3%), number of grains panicle^{-1} (12.8%), grain yield (17.0%), and straw yield (12.4%)	52.5% of Zn uptake	Vaid et al. (2014)
Acinetobacter sp. (TM56) and *Serratia* sp. (TM9)	MR219	Increased plant height, higher chlorophyll content, greater root length and volume	22–49%	Othman et al. (2018)
Staphylococcus epidermidis, Erwinia tasmaniensis, Pseudomonas aeruginosa, Bacillus subtilis	–	Improved growth of both shoot and root length and also increased biomass as compared to the control	Significantly higher germination index was observed with the application of PSB	Duarah et al. (2011)
Herbaspirillum strains (H18, ZA15), *Burkholderia vietnamiensis* strain (AR114)	IR42	Enhanced N-use efficiency of the ^{15}N-labelled fertilizer	Increased rice grain yield from 18 to 47%	Estrada et al. (2012)
Pseudomonas sp. strain PAC, *Serratia* sp. strain CMR165, and *A. brasilense* strain FT326	INTA Puitá CL variety	Increased growth of both shoot and root length	Concentration of soluble P was significantly higher in shoots of inoculated plants	Nico et al. (2012)
Burkholderia thailandensis, Sphingomonas pituitosa, and *Burkholderia seminalis*	MR 219	Improved root volume and seedling dry weight of the inoculated plants	–	Panhwar et al. (2014)
Rahnella aquatilis, Enterobacter sp., *Pseudomonas fluorescens,* and *Pseudomonas putida*	Shiroodi, Tarom, and Tarom Hashemi	Increased total number of stems hill^{-1}, number of panicles hill^{-1}, and plant height	Increased grain yield by 8.50–26.9% compared to the control	Bakhshandeh et al. (2015)
Pantoea ananatis (KM977993), Rahnella aquatilis (KM977991), and Enterobacter sp. *(KM977992)*	Tarom Hashemi	Increased plant height, stem diameter, root length, leaf area, and biomass dry weight	–	Bakhshandeh et al. (2017)

(continued)

Table 13.2 (continued)

Inoculant	Rice cultivar	Effect	Yield/biomass/germination	References
Pseudomonas aeruginosa, Ralstonia pickettii, Burkholderia cepacia, and Klebsiella pneumoniae	Rice (Jabalpur, Madhya Pradesh, India)	Aminocycloproparane-1-carboxylic acid Utilization, exopolysaccharide production, phosphate and potassium solubilisation	Zinc solubilisation (62.48 mg l^{-1}) in broth assay	Mishra et al. (2017)
Acinetobacter sp.	Rice (Cuddalore district, Tamil Nadu, India)	Increased plant height, no. of grains per plants, depletion of Zn	–	Gandhi and Muralidharan (2016)
Trichoderma harzianum strain S2	Sahbhagi Dhan	Increased plant height, tiller number	4.91 t ha^{-1}	Zaidi et al. (2018)
T. harzianum strain S2	Swarna-Sub1	Increased plant height, tiller number	4.81 t ha^{-1}	Zaidi et al. (2018)
Arbuscular mycorrhizae (*Glomus intraradices*), *Rhizobium*	N-L-19, FARO 44	Increased grain weight, higher panicle	6090 t ha^{-1} and 100% higher than control	Oladele and Awodun (2014)
ZOB-1 (*Anabaena variabilis*, *Chlorella vulgaris*, and *Azotobacter* sp.) and ZOB-2 (*Nostoc calcicola*, *Chlorella vulgaris*, and *Azotobacter* sp.)	–	Higher growth rate and photosynthetic activity	90% seed germination	Zayadan et al. (2014)

Cyanobacteria (*Aulosira fertilissima*, *Anabaena sphaerica*, *Nostoc hatei*, *Cylindrospermum majus*, and *Westiellopsis prolifica*), *Azospirillum*	–	Increased in fertilizer use efficiency (5–10%)	Increased rice grain yield up to 35%	Jha et al. (2013)
Bacillus megaterium, *Bacillus licheniformis*	KDML 105	Increased the shoot wet-dry weight, number of grains per panicle	–	Sapsirisopa et al. (2009)
Trichoderma viride, *B. megaterium*	MR220	Increased root length, seedlings height	20% more than control	Al-Taweila et al. (2010)
A. brasilense	Sakha 101	Increased the productivity of straw, grain plant^{-1} and total weight of 1000 grains	–	Shaban et al. (2009)

Incorporating fresh *A. pinnata* (20 t ha^{-1}) and its compost as powdered form (25 kg ha^{-1}) is known to improve soil and plants P content and number of tillers, but no significant effect was found in the contents of organic carbon, soil N, plant N content, and yield of rice crop (Setiawati et al. 2018). Rice crop inoculated with *Azolla* without inorganic N fertilizer but with moderate N fertilizer considerably minimized methane (CH_4) emissions by 12.3% and 25.3% over the conventional rice cropping system with general recommended N fertilizer dose (Xu et al. 2015). Dual cropping of *Azolla* has a major impact on soil redox potential and dissolved oxygen, which are most important factors for CH_4 emission. The potentiality of *Azolla*-cyanobionts as biofertilizers (to supplement as source of N) is mentioned in Table 13.1, while various studies related to *Azolla* are reflected in Table 13.2.

13.3.1.2 *Cyanobacteria*

Cyanobacteria (blue green algae, BGA) are the oldest known N-fixing biofertilizers for rice crop (Sao and Samuel 2018). The genera, namely *Anabaena, Nostoc, Aulosira, Cylindrospermum, Gloeotrichia,* and *Aphanothece* are available as major BGA for rice crop. Most of the BGA are prevailing in moist soil and *Aphanothece* is found in all the water regimes, whereas *Aulosira* is favoured to grow in fresh clean standing water. BGA could be inoculated with crop at the quantity of 50–100 kg fresh weight or 6–10 kg dry weight ha^{-1} after 7 days of planting (Selvaraj and Balasubramanian 2020). Application of combined strains of several BGA was more superior to the use of single strain inoculum. Fresh BGA inoculum grew better than dry BGA and local native strain achieved better than the exotic strains. BGA grows very well with addition of superphosphate at the rate of 20–40 kg P_2O_5 ha^{-1} and carbofuran at 75–90 g ha^{-1}. Several field studies reported that application of BGA results in yield at par with application of 20–25 kg N ha^{-1} as chemical fertilizers and increased yield by 15–60%. Evaluation of BGA biomass in planted paddy fields showed that N content varied from 4 to 32 kg, 4 to 28 t fresh wt ha^{-1}, and 99 to 822 kg dry wt ha^{-1}. The acetylene reduction assay (ARA) estimates and range varied from 189 to 259 nm of ethylene mg^{-1} chl h^{-1} for *Aphanothece* sp. and 249 to 357 nm ethylene mg^{-1} chl h^{-1} for *Aulosira* sp. Studies reported that the maximum yield was obtained in *Aulosira* sp. among different BGA tested.

Response to algalization is superior mainly in dry season and for long duration rice varieties. The combined application of BGA and chemical fertilizers up to 30–40 kg N ha^{-1} is recommended because algal growth decreased at high dosages of these chemical fertilizers. *Azolla* and BGA significantly increased rice yield and have a great potential as biofertilizer for rice. On an average, BGA could supply 20–25 kg N $ha^{-1}$$season^{-1}$, whereas *Azolla* can supply 20–40 kg N ha^{-1} in 20–75 days. Jochum et al. (2018) revealed that N-fixing cyanobacteria (*Nostoc muscorum* UTEX 2209S, *Anabaena* sp. UTEX 2576) and a polyculture of *Scenedesmus dimorphus* (UTEX 1237) and *Chlorella vulgaris* (UTEX 2714) enhanced the rice plant heights and growth, which are similar to or better than the chemical urea treatment (Table 13.2).

13.3.1.3 *Azospirillum*, *Azotobacter*, and Other Heterotrophic N-Fixing Microbes

Besides *Azolla* and *Cyanobacteria*, *Azospirillum* and *Azotobacter* are frequently used N-fixing bioinoculants for rice crop. *Azospirillum* increases rice yield by up to 30% as compared to the chemical fertilizers because it exhibits a wide range of PGP activities (fixing atmospheric N and by producing growth promoting substances) that also impact the root growth, leaf length and breadth, inhibit pathogenic fungi and other bacteria in the rhizospheric region (Khan 2018; Kecskés et al. 2015). *Azospirillum* along with *Trichoderma* enhanced the plant performances in rice variety BU Dhan 1 and showed better compared to 25% less inorganic N with *Trichoderma* (Khan 2018). Highest level of N-fixation is done by *A. brasilense* TPS05, which fixes N at a rate of 11.2 mg N g^{-1} and inoculation of *A. brasilense* TPS05 on paddy ADT 36 in pot culture produced the maximum number of tillers, roots, leaves, leaf length and breadth, shoot and root length and grain yield (Kanimozhi and Panneerselvam 2015). Rhizospheric *Azospirillum* inoculation improved the rice yield by 4% and native endophytic *Azospirillum* strain *AzA* increased yield by 6% in winter and 10% in summer season over the untreated plots in field experiment (de Salamone et al. 2010).

Recently, *Azotobacter* is focussed as N-fixing rhizospheric bioinoculant for rice crop (Chen et al. 2018; Sahoo et al. 2014; Yadav et al. 2014) and its endophytic behaviour is also established with rice as it forms a biofilm within the root (Banik et al. 2016; Kumar et al. 2019b). The formed biofilm may improve the N-fixation because the biofilm declines the oxygen tension from nitrogenase (oxygen sensitive enzyme) in the system and traps the nutrients (Wang et al. 2017; Vyas 2018; Rathnathilaka et al. 2023). Nevertheless, better performance of indigenous rice root endophytic *Azotobacter* strain suggests that they should be chosen compared to the rhizobacteria, for inoculating and developing a successful inoculant. *Azotobacter* sp. had differential effects on development of rice. It assists overall growth and physiological aspects of the rice crop (Banik et al. 2019; Wei et al. 2018). Field evaluation of *A. chroococcum* Avi2 (endophytic nitrogen fixer) and *A. vinelandii* SRIAz3 (free living nitrogen fixer) was performed in different rice cultivars, which indicated that application of these with 75% N + full dose of P and K recorded yield at par with recommended dose of fertilizers and thus, it is revealed that the application of the liquid formulations of Avi2 and SRIAz3 could save approximately 12–25% of N without compromising the rice yield (Kumar et al. 2018c; Banik et al. 2019).

Xanthomonas sp., *Acetobacter* sp., *Alcaligenes* sp., *Klebsiella* sp., *Enterobacter* sp., *Pseudomonas* sp., *Burkholderia* sp., *Herbaspirillum* sp., *Paenibacillus* sp., *Microbacterium* sp. are also reported as N-fixing bioinoculants for rice crop (Arshad and Frankenberger 1998; Leveau and Lindow 2005) (Table 13.2). Besides, some siderophore producing N-fixing bioinoculants interfere with the detrimental microbial communities of rice, as siderophores can chelate iron, zinc, nickel, and copper (co-factor of the enzymes), thereby cutting their availability to other organisms and suppressing plant–pathogen interactions (Dimkpa et al. 2008; Sun et al. 2010; Wang et al. 2014). Altogether, 47% of rice ecologies harboured microbes showed the wider extent siderophore production (Islam et al. 2009). Similarly, in Korean rice cultivars diverse kind of PGP activities of endophytes has been found (Ji et al. 2014), which may enhance rice production in natural condition.

13.3.2 Phosphate Solubilizing Microorganisms

Phosphate solubilizers are a group of microbes that include bacterial (*Bacillus*, *Pseudomonas*, and *Rhizobium*), fungal (*Penicillium* and *Aspergillus*), actinomycetes genera, and arbuscular mycorrhizal fungi, having capability of converting insoluble phosphate to soluble forms by producing various metabolites like organic acids, siderophores, and various antibiotics so that the plants can uptake it or is protected from pathogens (Kalayu 2019; Panhwar et al. 2012; Rodríguez and Fraga 1999). Some of the prominent phosphate solubilizing microbes for rice are mentioned in Table 13.2.

13.3.2.1 Arbuscular Mycorrhizal Fungi

Arbuscular mycorrhiza (AM), a common symbiotic association of plants and fungi (Chen et al. 2018), have a wide range of PGP activities including conversion of inorganic orthophosphate to soluble forms so that plant can uptake it, suppress disease caused by the pathogens, form water-stable aggregates in soil through which soil health is maintained (Bernaola et al. 2018; Panneerselvam et al. 2018a, 2019a), tolerate drought as well as heavy metals (Bholay et al. 2018). Narwal et al. (2018) have shown to have highest response towards mycorrhizal colonization (between 26.60% and 55.40%) in ten upland rice varieties in the fields of IARI, New Delhi, India. Rajeshkannan et al. (2009) conducted a pot experiment by using mycorrhizal and rice seedling where they observed the root colonization varied between 17.4% and 37.2% over the time period of 7–70 days.

AM fungal relationship was investigated in 72 rice cultivars, including two low P tolerant controls, Kasalath and Dular, which are grown in P deficient soil (6.0–8.0 ppm). The AM fungal root colonization ranged from 20 to 90%, whereas it was 80–90% in Kasalath and Dular cultivars. The dominating unique kind of vesicle-forming AM colonization in these two types was not detected in numerous low P tolerant cultivars. This observation clearly demonstrates that some AM fungal taxa may select a specific rice genotype. It is generally known that Kasalath and Dular have the protein kinase gene; *pstol* for phosphorus-deficiency tolerance, hence, these cultivars with a unique type of AM root colonization in P deficient soil require additional research (Panneerselvam et al. 2018a).

13.3.2.2 Phosphate Solubilizing Bacteria

The efficacy of phosphate-solubilizing bacteria (PSB) in rice was determined by various researchers in past (Panhwar et al. 2014). During incubation PSB isolates could solubilize P by up to 43.7% after 72 h of incubation. Estrada et al. (2012) isolated 49 PSB isolates, among all, three strains *Burkholderia thailandensis*, *Sphingomonas pituitosa,* and *Burkholderia seminalis* were found to be effective in pot grown rice under two sources of P, i.e., tricalcium phosphate and simple superphosphate (SSP). Application of two PSB, namely *Burkholderia vietnamiensis* strain (AR114) and *Herbaspirillum* strains (H18, ZA15) improved rice grain yield by 33–47% with TCP and 18–44% with TSS, respectively (Table 13.2).

13.3.3 Potassium Solubilizing Microbes

Due to more use of N and P fertilizers by farmers, K level in soil is declining (Sindhu et al. 2016) and as a result potassium solubilizing bacteria (KSB) are getting more importance in modern agriculture (Ahmad et al. 2016). KSB include diverse microbes like *Pseudomonas*, *Burkholderia*, *Acidithiobacillus ferrooxidans*, *Bacillus mucilaginosus*, *Bacillus edaphicus*, and *Bacillus circulans* (Sindhu et al. 2016), which are known to release K from minerals by various mechanisms including productions of different types of organic acids (citric, oxalic, and tartaric acids) and also by producing capsular polysaccharides (Shanware et al. 2014). Khanghahi et al. (2017) isolated KSB in northern Iran from 40 different paddy cultivated fields of Mazandaran province and screened for their K-solubilizing capability on K specific Aleksandrow media. They selected three most promising potassium solubilizing isolates identified as *Pantoea agglomerans*, *Rahnella aquatilis,* and *Pseudomonas orientalis*. The pot experiment revealed that inoculums of all three potential KSB improved the amount of rice grain yield, K uptake, IAA production, and tolerance to different environment stresses and obtained higher value when KSB inoculums integrated with inorganic K at half dosage of recommendation.

13.3.4 Zinc Solubilizing Microbes for Rice

Due to excessive use of chemical fertilizers, Zn deficiency occurs in soil due to conversion of insoluble forms like zinc hydroxide [$Zn(OH)_2$] at high soil pH, zinc carbonate ($ZnCO_3$) in calcium rich alkali soils, zinc phosphate [$Zn_3(PO_4)_2$] in alkali soils (Mishra et al. 2017). But this can be overcome by the application of Zn solubilizing microbes (ZSM) such as *Thiobacillus thiooxidans*, *Thiobacillus ferrooxidans*, *Acinetobacter*, *Bacillus*, *Gluconacetobacter,* and *Pseudomonas* (Gandhi and Muralidharan 2016), because they are able to solubilize numerous ores such as zinc oxide (ZnO), $ZnCO_3$, $Zn_3(PO_4)_2$. Altogether 234 bacterial isolates were obtained from the roots of basmati-385 and basmati super rice varieties grown in clay loam and saline soil from Punjab, India. Among them 27 isolates were Zn solubilizers from Zn containing ores (Zn_3 $(PO_4)_2$, $ZnCO_3$ and ZnO). The strain SH-10 showed highest Zn solubilization zone (24 mm) from $Zn_3(PO_4)_2$ and grain yield was increased by 22–49% and strain SH-17 showed largest Zn solubilization zone (14–15 mm) from ZnO and $ZnCO_3$ and grain yield was increased by 18–47%. The strains SH-10 and SH-17 on identification were found to be *Bacillus* sp. and *B. cereus*, respectively. The bacterial strains improved Zn translocation (Shakeel et al. 2015). Under xenobiotic conditions, effect of ZSB such as *Acinetobacter* sp. (TM56) and *Serratia* sp. (TM9), showed promising colonization in rice (Othman et al. 2018). Gandhi and Muralidharan (2016) showed that, among 143 ZNB, the promising Zn solubilizing bacteria was *Acinetobacter* sp. AGM3 (Table 13.2).

13.3.5 Silicon Solubilizing Bacteria

Silicon (Si) is one of the important metalloids present as silicate minerals in the soil. It is a beneficial element for the plant (Cooke and Leishman 2011). The role of Si is to expand the photoassimilation of carbon and improvement of the deposition of carbon in the panicle of rice crop (Lin et al. 2002). Si has an inter-relation with P and helps in the sustenance of P. Together with above facts, Si has the growth promoting effects when applied to several crops (Sheng and Huang 2002; Gong et al. 2005; Liang et al. 2005; Hamayun et al. 2010; Kim et al. 2011). Si when applied in the dissolved silicate form in low P containing soil resulted in release of P from the adsorbed sites and increased the P content in the soil (Rusel 1973). Si present in soil solution plays an important role by providing strength to the cell wall by absorbing silicic acid with the help of plant roots and further move to the active growing plant and then reach cell wall. At the cell wall, this Si is combined with organic compounds and makes the plants stronger and healthier. Insoluble Si added with *Burkholderia eburnean* improves the shoot length, root length, shoot weight, root weight, and chlorophyll content and causes significant improvement in rice growth. Si-solubilizing bacteria facilitate Si availability to the plants, thus improving several plant growth parameters (Avakyan et al. 1984; Malinovskaya et al. 1990; Vandevivere et al. 1994; Liu et al. 2006). Si not only promotes crop growth but also protects the rice plant from diseases (Hamayun et al. 2010; Liu et al. 2006).

13.4 Microbial Inoculants Against Rice Plant Disease

13.4.1 Biocontrol Agents for Rice Pathogens

The research on biocontrol of pathogens started to take shape worldwide in the 1980s. The studies on the biocontrol agents are still focused on the identification, evaluation, and formulation of these agents (Panneerselvam et al. 2019b). We already know that diseases in rice plants are caused by fungi, bacteria, virus, nematode, and mycoplasma like organisms. Some of the major known rice diseases and their causal agents are blast (*Pyricularia grisea*), brown spot (*Bipolaris oryzae*), sheath blight (*Rhizoctonia solani*), stem rot (*Sclerotium oryzae*), sheath rot (*Sarocladium oryzae*), bacterial blight (*Xanthomonas oryzae pv. oryzae*), and rice tungro (rice tungro virus). Due to these diseases the production of rice at optimum level is a serious constraint. The biocontrol agent of diverse groups exists in nature. Among these agents, bacterial antagonists are measured as ultimate candidates due to their rapid growth, easy handling, and aggressive colonizing character (Panneerselvam et al. 2019b). *Pseudomonas* and *Bacillus* are the major bacterial antagonists and key players for biocontrol. *Bacillus* are Gram-positive endospore-producing bacteria that are heat tolerant and their desiccation is suitable for field use. The pseudomonads, Gram-negative rods are excellent colonizers and require simple and minimum nutrients. They are mostly dominant in the rice rhizosphere. Several antagonists' bacteria (fluorescent and non-fluorescent strains) are present in association with rice rhizospheric soils (lowland and upland) and observed effective against *R. solani* (sheath blight). Antagonistic bacteria belonging to the genera *Bacillus, Pseudomonas, Serratia,* and *Erwinia* inhibit mycelial growth of *R. solani*, and some of them also inhibit growth of other fungal pathogens like *S. oryzae* (stem and sheath rot), *B. oryzae* (brown spot), *P. grisea* (blast), and *Fusarium fujikuroi* (bakanae). Different researchers documented bacterial strains with disease controlling activities in rice. Till date 40 bacterial isolates antagonistic to the rice sheath blight pathogen have been reported. The isolates belonging to fungal antagonists reported were *Trichoderma* spp., *Penicillium, Myrothecium verrucaria, Chaetomium globosum,* and *Laetisaria arvalis*. The list of important biocontrol agents of major rice diseases is shown in Fig. 13.2 and Table 13.3.

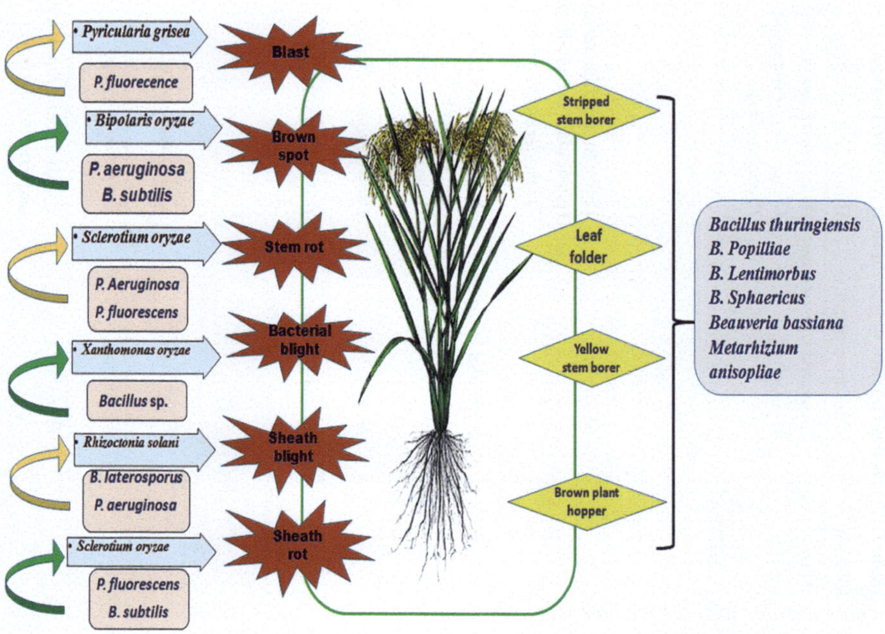

Fig. 13.2 Group of biocontrol agents used to control rice pests

Table 13.3 Disease control in rice through microbial inoculants

Disease	Causative agent	Biocontrol microbes	Mode of action	Inhibition	References
Blast	*Magnaporthe oryzae* B. Coch.	*Acidovorax, Sphingomonas, Chryseobacterium,* and *Pseudomonas*	*Acidovorax* and *Sphingomonas* sp. produced calcium-dependent metalloproteinases, *Chryseobacterium* sp. produced calcium/zinc-dependent metalloproteinases. Decreased the adhesion of germling of *M. oryzae* to rice leaves	–	Shimoi et al. (2010)
	M. oryzae	Different rhizobacteria (particulars not mentioned)	Produces peroxidase, b-1,3-glucanase, and chitinase	Reduces the disease by 90–95%	Sperandio et al. (2017)
		Trichoderma harzianum		Reduces 10–25% disease intensity	Singh et al. (2015)
	Pyricularia oryzae	*Bacillus firmus* E65, *Serratia marcescens* E31, *Pseudomonas aeruginosa* C32b, *Bacillus cereus* II.14	Reduces the mycelial growth by producing chitinase and glucanase	Inhibits fungal mycelia by 73–85%	Suryadi et al. (2013)
	M. oryzae	*Pseudomonas, Bacillus*	Produces HCN	Inhibits fungal growth by 76%	Spence et al. (2014)
	M. oryzae	*Bacillus subtilis*	Produces antibiotics (surfactin, iturin, and fengycin), toxins, biosurfactants, and extracellular cell wall degrading enzymes such as chitinase and β-1,3-glucanase	73.5–83.5%	Sha et al. (2016)
	P. oryzae Cav.	Endophytic bacteria	Causes mycelia malformation	Inhibits fungal mycelia by 50%	Widiantini et al. (2017)
	M. oryzae	*Streptomyces* spp.	Possesses antagonistic activities against *M. oryzae*	Reduces the disease up to 88.3%	Law et al. (2017)
	P. oryzae	*Pseudomonas fluorescens* and *Trichoderma* sp.	Produces protease and chitinase	–	Subhalakshmi and Devi (2017)

	P. oryzae	*Penicillium* sp. GM15, *Fusarium* sp. LM1, *Curvularia* sp. GM11, *Penicillium* sp. GT1, GM20, GM7, GT19, and GT27	Produces endoglucanase	–	Paraiso and Capacao (2019)
	M. oryzae	*B. subtilis* 5, *B. cereus* 3S5, and *P. fluorescens* 10S2	Produces peroxidases	Controls disease up to 83.9%	Chen et al. (2018)
Sheath blight	*Rhizoctonia solani*	*Trichoderma viride*	Produces alpha-glucosidase which inactivates RS toxin	–	Shanmugam et al. (2001)
	R. solani Kuhn	*B. subtilis* AUBS1	Produces two β-1-3-glucanases and chitinase	–	Jayaraj et al. (2004)
	R. solani	*P. fluorescens*	Produces chitinase, beta-1,3-glucanase, siderophores, salicylic acid (SA), and hydrogen cyanide (HCN) and inhibits the mycelial growth of *R. solani*	–	Saikia et al. (2006)
	Pyricularia grisea and *R. solani*	*B. subtilis* NSRS 89–24	Produces glucanase enzyme that inhibits the growth of hyphal growth of both fungi	–	Leelasuphakul et al. (2006)
	R. solani AG1-IA	*Burkholderia* sp.	*RecA* gene supresses the fungal growth	–	Cuong et al. (2011)
	R. solani Kuhn	*B. subtilis* var. *amyloliquefaciens* (FZB24)	Produces peroxidase, polyphenol oxidase, and phenylalanine ammonia-lyase	–	Nagendran et al. (2014)
	R. solani	*Streptomyces aurantiogriseus* VSMGT1014	Produces antibiotics, lytic enzymes, volatile compounds, and siderophore that inhibits the mycelial growth	–	Harikrishnan et al. (2014)
	R. solani	*Trichoderma asperellum* T12			Chen et al. (2015)

(continued)

Table 13.3 (continued)

Disease	Causative agent	Biocontrol microbes	Mode of action	Inhibition	References
	R. solani	*P. fluorescens* PF-08 and *T. harzianum* UBSTH-501	Causes significant morphological and physiological changes in Rhizoctonia hyphae, including swelling, knotting, crumpling, flattening, shrivelling, bursting, and cytoplasm leaking	–	Singh et al. (2016)
	R. solani	*P. fluorescens* 4aYN11		–	Yua et al. (2017)
Bacterial blight	*Xanthomonas oryzae*	*P. aeruginosa*	Produces siderophores (1-hydroxy-phenazine, pyocyanin, and pyochelin), rhamnolipids, and a series of 4-hydroxy-2-alkylquinolines (HAQs) as well as novel 2,3,4-trihydroxy-2-alkylquinolines and 1,2,3,4-tetrahydroxy-2-alkylquinolines and even induces defence related enzymes in the rice plant	Suppressed diseased leaf area 3.7%	Yasmin et al. (2017)
	X. oryzae	*Pseudomonas* sp.E227, E233, Rh323, *Serratia* sp. Rh269, and *Bacillus* sp. Rh219	Produced siderophores and defence related enzymes including phenylalanine ammonia-lyase and polyphenol oxidase	Reduced diseased leaf area by 80%	Yasmin et al. (2016)
	X. oryzae	*Paenibacillus polymyxa* Sx3	–	Inhibited 25–81%	Abdallah et al. (2019)

	X. oryzae	*Streptomyces toxytricini*	Inhibits by producing bioactive compounds	Reduced 30% of infected crops	Duong et al. (2014)
Brown sheath rot	*Sarocladium oryzae*	*Pseudomonas fuscovaginae*	–	–	Kakar et al. (2014)
Brown spot	*Bipolaris oryzae*	*Trichoderma virens*, *Trichoderma atroviride*, and *T. harzianum*	Produces volatile and nonvolatile metabolites having inhibitory effect against the pathogen	Inhibited 100%	Khalili et al. (2012)
	B. oryzae	*T. viride* and *T. harzianum*	–	Inhibited 60%	Harish et al. (2008)
	H. oryzae	*Bacillus* sp. (Gram positive)	Produces antimicrobial compounds, such as antibiotics (e.g., bacilysin, iturin, mycosubtilin) and siderophores	–	Jatoi et al. (2016)
					Venkateswarlu et al. (2015)

The potentiality of biocontrol agents in disease suppression rests on the suitable form for application on the crop (Kaari et al. 2023). It can be added either by pouring seeds inside the culture medium, aerial spray or dispersing the seeds in furrows by drip system (direct inoculation) or using different solid-phase inoculants. Vasudevan et al. (2002) explained the effect of the biocontrol agents on the pathogens and plants as mentioned below:

1. Bacterium *Pseudomonas fluorescens* application reduces a variety of diseases by 20–42% and grain yield increased by 3–160%.
2. Bakanae disease (*F. fujikuroi*) is reduced by 72–96% by using antagonistic bacteria for seed treatment.
3. Bacterial blight pathogen is suppressed by up to 59% and enhanced yield by twice on seed treatment with different species of *Bacillus*.
4. *P. fluorescens* strains of Pf 7-14 and PpV14i can cause about 90% mortality of rice tungro virus disease vector *Nephotettix virescens* treated on rice leaves for 7 days.

Al-Taweila et al. (2010) studied the effect of microbial inoculants against sheath spot disease caused by *Rhizoctonia oryzae*. They isolated *Trichoderma* sp. and *Bacillus megaterium* from soil and inoculated separately into soil as suspension. They observed that the linear growth of *Trichoderma* sp. was more rapid than of *R. oryzae* and was able to inhibit 59.5% after 8 days of inoculation. Treatments with both microorganisms (*Trichoderma* sp. and *B. megaterium*) also significantly affect positive growth parameters like grain and stem dry matter, panicles, and harvest index comparing to control plants. Yua et al. (2017) conducted an experiment for improving a stable biocontrol agent against sheath blight disease caused by *R. solani* Kuhn. They isolated 14 different strains and tested their biocontrol efficacy against *R. solani*, which differed from 36.23% to 88.24%, but, among all, *P. fluorescens* 4aYN11 showed a comparatively stable control efficacy of 32.26–78.79% and growth promotion of 18.43%. Harikrishnan et al. (2014) isolated 132 actinomycetes from diverse rice rhizospheric soil ecosystem of Tamil Nadu, India. They found that among all isolates, 57 showed antagonism towards *R. solani* and other fungal pathogens such as *Fusarium oxysporum*, *Fusarium udum, Macrophomina phaseolina*, and *Alternaria alternata* with a detectable zone of inhibition. Based on the morphological, physiological, biochemical characteristics and 16S rRNA sequence analysis, the most potential bacterial isolate was identified as *Streptomyces aurantiogriseus* VSMGT1014. It was able to produce lytic enzymes, siderophore, secondary metabolites, indole acetic acid, and some volatile substances; the crude metabolites of VSMGT1014 showed better antagonistic activity against *R. solani*.

Gupta et al. (2018) used dual culture techniques to test *T. viride*, *T. harzianum*, *P. fluorescens,* and *B. subtilis* against *Bipolaris oryzae* (causes brown spot of rice) by taking three rice varieties: basmati 370, Jaya, and PC-19. They observed maximum inhibition of mycelial growth of *B. oryzae* by *T. harzianum* (65.33%), followed by *T. viride* (57.76%), and *P. fluorescens* (32.68%). They also observed the foliar application of *T. harzianum* exhibited maximum disease reduction (64.00%)

and also resulted in increased yield in PC-19 (8%) as compared to their respective control.

13.4.2 Microbial Inoculants Against Entomopathogens of Rice

Predation, parasitism, pathogenesis are different types of biological control methods, which are practised against the insect pests (Fig. 13.2) (Poinar and Thomas 1984; Joung et al. 2000). Microorganisms, such as viruses, bacteria, fungi, and protozoa are the pathogens or antagonists of all types of living organisms. The diseases of various pests have been used as biocontrol agents. This group of pathogens can control various pests, viz. vectors of human, animal and plant diseases, and insect pests of agricultural crops (Elad et al. 1980; Whipps 2001; Compant et al. 2005; Das and Dangar 2008; Raddadi et al. 2009; Dangar et al. 2019; Sahoo et al. 2014; Ghosh et al. 2017). Biological control of insects is a complex phenomenon and is dependent on the virulence of the pathogen, resistance of the host, and effects of the environment. Virulence of the pathogens is assessed by LD_{50} or LC_{50} and LT_{50} values (Poinar and Thomas 1984; Lacey 1997).

Rice crop is attacked by numerous insect pests. Among them, rice leaf folder (LF) is caused by *Cnaphalocrocis medinalis.* LF is successfully reared in the laboratory with *Drosophila* and fruit fly diets, which could be exploited for mass assay of entomopathogens including strains of *Skermanella* and *Serretia* (Panneerselvam et al. 2018b). Six entomopathogens, of which four bacterial, namely *B. thuringiensis* (Bt) strain TB 160, 161, 261, and 263 and two fungal strains, namely *Beauveria bassiana* strain TF6 and *Metarhizium anisopliae* strain TF19 are identified to manage the LF at NRRI, Cuttack and formulations of these potential strains are also filed for Indian patents with following numbers 264/KOL/2015, 263/KOL/2015, 261/KOL/2015, 262/KOL/2015, 260/KOL/2015, 265/KOL/2015, respectively. Out of three native *Bt* isolates, one is more effective than the commercial formulations Bactospeine, Bioasp, Biolep, and one equally effective with Bactospeine. These indigenous isolates could replace the commercial formulations. Fourteen *Bt* isolates with Cyt protein toxic to dipteran insects, i.e., mosquitoes and 24 organisms with Cry5 protein toxic to lepidopteran (butterflies and moths) and coleopteran (beetles) insects are identified. *P. aeruginosa* and *P. fluorescens* caused epizootic of LF in the field. The epizootic causing two *Pseudomonas* sp. had the LC_{50} $2.3–2.5 \times 10^3$ bacteria ml^{-1}, which showed that they would cause natural epizootic of the pest.

Bt strain was 100% virulent to third instar stripped borer (*Chilo suppressalis*) larvae in the laboratory and more effective than Biolep. *B. bassiana, N. rileyi,* and *M. anisopliae* survived 3–16 months more at −20 °C but all of them died within 8 months at 27 °C. Lepidoptera specific *cry* toxin gene amplification of 40 rhizoplane and rhizosphere Bt detected *cry1, 2, 3, 4,* and 11 genes from 10, 10, 6, 4, and 1 Bt, respectively. Presence of more than one *cry* gene, viz. *cry1, 2* in TB 113, 114, 118, and 155, *cry1, 3* in TB 116, 118, and 121, and *cry 2, 4* in TB 117 explained that the concerned Bt might be virulent against wide hosts. Out of 381 indigenous Bt isolates, the 27 kDa Cyt protein toxic to dipteran insects, i.e., mosquitoes and the 81 kDa Cry 5 protein toxic to lepidopteran (butterflies and moths) and coleopteran (beetles) insects were identified from 14 and 24 organisms, respectively. The *cry1, cry10, cry3,* and *cry1Ac* genes were identified from 31, 4, 16, and 15 Bt, respectively, which were effective against the lepidopteran, dipteran, and coleopteran insects. The δ-endotoxins, Cry2 (69.1 kDa) toxic to lepidopteran/dipteran insects, Cry4C toxic to dipteran insects, Cry5 (78.8 kDa) toxic against lepidopteran/coleopteran insects, Cry3 (74.2 kDa) toxic to coleopteran insects, and Cry4C (78.8 kDa) toxic to dipteran insects were identified from 67 Bt isolates.

Reports indicated that a bioformulation combining *P. fluorescens* along with *B. bassiana* showed 36.05% of *C. medinalis* larval mortality (Karthiba et al. 2010) and similarly, Kirubakaran et al. (2014) made an oil-formulation with *B. bassiana* and *M. anisopliae* showing 50% lethal percentage at 9.09×10^4 conidia ml^{-1} and 6.08×10^5 conidia ml^{-1}, respectively (Table 13.4). Study also revealed that *Skermanella* sp. and *Serratia* sp. affect both pink stem borer (*Sesamia inferens*) and leaf folder (*C. medinalis*) (Panneerselvam et al. 2018b). Unlike *Skermanella* sp. and *Serratia* sp. some other microbes are effective on more than one causative insect. *B. bassiana* is effective against leaf folder, rice water weevil, and brown plant hopper, whereas Bt is effective on causative insects like yellow stem borer and leaf folder. Most of the microbes attack these causative insects in their larval stage but two of the major exception are rice bug (*Leptocorisa oratorius* and *Leptocorisa acuta)* and brown plant hopper (*N. lugens)* where microbial biocontrol inoculants attack in their nymph stage (Table 13.4).

13.5 Global Biofertilizers and Biopesticides Market

13.5.1 Global Market of Biofertilizers

Soils with low organic matter content as well as N status have increased the demand of biofertilizers use in the agriculture (FAO 2005). Application of biofertilizers can help in upbringing sustainable agriculture production system via soil health improvement as well as producing quality products (Tawate et al. 2018). The increase in demand brought the opportunity to increase its production and marketability (Röös et al. 2018). Global market of biofertilizer has reached up to USD 2.0 billion in the year 2019 and further projected to achieve the value of USD 3.8 billion by 2025 (Biofertilizer market 2019). The market of biofertilizers is determined based on increased organic farmland and the awareness among the farmers for the use of biofertilizers in their farmland. Carrier or liquid-based preparations of biofertilizers support plant growth-promotion and nutrition (Motsara et al. 1995) (Table 13.5). Among the both forms of biofertilizers, the market of liquid

Table 13.4 Microbial intervention for pest management in rice

	Causative pest (scientific name)	Mortality/lethality	References
B. thuringiensis	Yellow stem borer (*Scirpophaga incertulas*)	6.9% mortality after 21 days of second application B.t.	Chakraborty (2011) Singh et al. (2015)
Skermanella sp. and *Serratia* sp.	Pink stem borer (*S. inferens*)	50% larval mortality of *S. inferens* at 3.80×10^4 CFU ml^{-1} *Skermanella* sp. and 2.29×10^5 CFU ml^{-1} *Serratia* sp.	Panneerselvam et al. (2018)
Skermanella sp. and *Serratia* sp.	Leaf folder (*C. medinalis*)	50% larval mortality at 2.95×10^3 and 5.88×10^3 CFU ml^{-1} for *Skermanella* sp. and *Serratia* sp.	Panneerselvam et al. (2018)
Bioformulations combining *P. fluorescens* Migula strains Pf1 and AH1 and *B. bassiana* (Balsamo) Vuill. isolate B2	Leaf folder (*C. medinalis*)	Larval mortality of 36.05% in Pf1 + AH1 + B2 treatment	Karthiba et al. (2010)
B. thuringiensis strain TS 110	Leaf folder (*C. medinalis*)	The field assay of the bacterial pathogen Bt TS110 against third, fourth, and fifth instar larvae of *C. medinalis* showed LC50 (×10 cfu ml^{-1}) values of 3.77, 5.29, 4.83 and 4.93, 4.42, 4.72 in dry and wet season, respectively	Ghosh et al. (2017)
B. bassiana and *M. anisopliae*	Leaf folder (*C. medinalis*)	The mortality of the larvae was dose-dependent and the oil-formulated *B. bassiana* MTCC7690 conidia was highly virulent isolate with lower LC50 of 9.09×10^4 conidia ml^{-1} followed by oil-formulated *M. anisopliae* MTCC4101 conidia with LC50 of 6.08×10^5 conidia ml^{-1}	Kirubakaran et al. (2014)
B. bassiana strain YS03	Rice water weevil (*Lissorhoptrus oryzophilus*)	The control effect was about 40–60%, the yield loss rate of return was 18.75%, by using YS03 strain	Xu et al. (2015)
Lecanicillium saksenae	Rice bug (*L. oratorius* and *L. acuta*)	LT50 of *L. saksenae* was 18.58 h and 17.58 h for adults and nymphs rice bug	Hari Sankar (2017)

(continued)

Table 13.4 (continued)

	Causative pest (scientific name)	Mortality/lethality	References
M. anisopliae	Brown plant hopper (*Nilaparvata lugens*)	BPH-derived *M. anisopliae* isolates from the Philippines (ARSEF456) and Indonesia (ARSEF576) killed >50% of the nymphs. Their LC50 values were estimated as 731 and 1124 conidia mm^{-2} on day 7 and fell to 284 and 306 conidia mm^{-2}, respectively, on day 10	Jin et al. (2008)
B. bassiana	Brown plant hopper (*Nilaparvata lugens*)	Mortality percentage of brown planthoppers was 85.0–83.3%	Lee et al. (2015)

Table 13.5 The global key market players of biofertilizers for agricultural crops

Company name	Country of origin	Products
Novozymes	Denmark	Bio AG
IFFCO	India	ZSB, KMB, *Azotobacter*, Liquid consortia NPK, *Azospirillum*, PSB, *Acetobacter,* Sagarika
T Stanes & Company Ltd	India	SYMBION-N, SYMBION-N, SYMBION-N, SYMBION-P, SYMBION-K, SYMBION-S, CONSORT NPK, SYMBION-VAM PLUS and GROW CARE
SOM Phytopharma Ltd	India	Agri Life Nitrofix AC Bio Fertilizer, Agri Life Nitrofix AV Bio Fertilizer, Agri Life Nitrofix AS Bio Fertilizer, Agri Life Nitrofix RZ Bio Fertilizer, Agri Life P Sol B Bio Fertilizer, Agri Life K Sol B Bio Fertilizer, Agri Life Fe Sol B Bio Fertilizer, Agri Life S-Sol B Bio Fertilizer, Agri Life BioHume SL Fertilizer, Agri Life Biohume Granule, Agri Life Aminocid SL Fertilizer, Agri Life Aminocid GR Bio Fertilizers, Agri Life Oxyrich Fertilizer, Agri Life SomZyme SL Fertilizer, Agri Life Stimula Bio Fertilizer, Agri Life Rootam Bio Fertilizer, Agri Life Stoma EW Bio Fertilizer, Agri Life Nitrofix AC Bio Fertilizer, Agri Life Agri VAM Fertilizer, Agri Life Aminocid Granules
International Panaacea Limited	India	BioNPK (Gtranular & ltr.), Cane Master, Potas activa (Granular & ltr.), Premium Azospi, Premium PHOSPHOFIX, Premium Phoster, Premium Rhizo, Premium VAM Shakti, Premium Zinc Activator, Premium Zinko, soil gold ltr, Soil Shakti & Zinc extra
Kan Biosys	India	Taba R, Vitormone R, TabaDrip R, Bioplin R, B-seepel-RP R, Phosfert R, Taba G R, TeamBio-3 (Fertidose), TeamBio-2 (Fertidose), Potak R, B:seepel (PSB), BioNit G R, TeamBio-3 (Granular), TeamBio-2 (Granular), Elicitomone&Mycozoots
Madras Fertilizers Limited	India	VIJAY Biofertilizers
Gujarat State Fertilizers and Chemicals Ltd	India	Liquid Biofertilizers and VAM–Mycorrhiza
National Fertilizers Limited	India	*Rhizobium* (symbiotic), *Azotobacter* (non-symbiotic), phosphate-solubilizing bacteria, zinc solubilizing bacteria
Rashtriya Chemicals and Fertilizers Ltd	India	Biola
Varsha Bioscience and Technology India Pvt. Limited	India	Nitromax, Phospomax, Potaz, Varsha VAM, Rhizodyne, Nanozinc
UPL	India	Opteine, Pilatus, Copio, Macarena, neo root

(continued)

Table 13.5 (continued)

Company name	Country of origin	Products
Kiwa Biotech	China	Biological organic fertilizer
China Bio-Fertilizer AG	China	–
Symborg	Spain	VITASOIL, MBB, QLIMAX
Ficosterra	Spain	Prebiotic—soluble fertilizer, marine amino acids, liquid biostimulants Probiotic—ficosagro, ficosturf
Rizobacter Argentina S.A	Argentina	Biological fertilizers for phosphorus
Mapleton Agribiotech	Australia	TwinN, NitroGuard, NitroGuard Defender, pastureN, Catapult Superfine
ASB Greenworld,	Australia	–
Australian Bio Fert Pvt. Ltd	Australia	–
Kula Bio	USA	Nitrogen biofertilizers
Earnest Earth	USA	Vermicompost biofertilizers
Fertoz	Canada	Phosphorus biofertilizers
Lallemand Inc.	Canada	Biostimulants
Groundwork Bio Ag	Israel	Rootella (fungi biofertilizers)
Vegalab SA	Switzerland	Charged Bioboost, MycoBioBoost

biofertilizer is edged over the carrier-based biofertilizers globally. The reason for the more acceptances of liquid biofertilizers is shelf life, which is more than the carrier-based biofertilizer (Sahoo et al. 2014). The liquid biofertilizers are more convenient in use and affordable in carrying for the farmers of developing countries with a life expectancy of 2–3 years. Thus, in terms of value, the liquid biofertilizers are going to catch the global market in near future. N-based biofertilizers are being used in leguminous as well as non-leguminous crops like rice and sugarcane. Hence, the N-fixing biofertilizers are estimated to grow in the global market (Biofertilizer market 2019) (Table 13.5).

Global biofertilizer market is influenced by several factors. One of the foremost constraints is low capacity of R&D and testing laboratories, which has hindered the growth of the global biofertilizer market. It is also important to have a quality management system which provides comparatively good microbial products to the farmers. The management of quality product is critical and hence must be handled properly. Another factor that affects the growth of biofertilizer market is lack of skilled labourers. Thus, if these factors are over-ruled, then the production of quality biofertilizer will not have any constraints at global level. Hence, a proper module is required to design and implement quality control systems for efficient use of biofertilizers.

13.5.2 Global Market of Biopesticides

Global biopesticides market has reached at USD 3.06 billion in the year 2018 and it is estimated to reach USD 6.42 billion in 2023 to USD 13.9 billion by 2028 (Biopesticide market 2019). The biopesticide market is mainly determined through the growth in the production of different crops like paddy, wheat, corn, and soybean. The development of the biopesticides is dependent on the different compounds produced from microbes that are being developed to survive under specific conditions, in that way creating a niche in the market is quite demanding. Bayer, Syngenta, and BASF are the important microbial pesticides providers at global market and the foliar biopesticides are the dominated sector. The foliar biopesticides are used conveniently in all crops and hence preferred among farmers

Biopesticides improve the overall quality of the applied crops due to better pest management mode and thereby provide higher price of their produce to the farmers who are using them (Takore 2006). Biopesticides also offer social benefits for managing pests that cannot be controlled with the use of insecticides (Popp et al. 2013). Farmers from the developing countries are economically weaker can easily afford these due to lower cost over the chemical pesticides. The biopesticides are prepared using the compounds/microbes obtained from the natural sources and hence minimize the chance of strict regulatory control. The raw materials required for biopesticides formulation are also easily available at very low cost or can be developed onsite. Furthermore, getting regulatory approval for the biopesticides product is convenient and rapid due to their sustainable environmental benefits. Different market players/organizations are involved to regulate the global market of biopesticides actively. This helps to support new organizations/players to enter this market and be part of this production system. The health hazards and pollution potentials of conventional agrochemicals have changed the thoughts towards natural biopesticides for user and policy makers as well as experts globally (Biopesticide market 2019).

13.6 Conclusion and Future Prospects

Biofertilizers and biopesticides application in rice crop have emerged as potential eco-friendly agricultural inputs and are gaining priority over chemicals, which are known to have many negative effects upon ecosystem and human health. Biofertilizers make the nutrients available through fixation, solubilization, and mobilization to the rice crops and they can be part of integrated nutrient management system in the fields. The sustainable production for rice can be achieved by the use of these eco-friendly, non-harmful, and omnipotent microbes. Moreover, these are economically viable for weaker sections of farming community who are not in a position to afford costly chemicals from market. Though several biofertilizers and biopesticides are commercially available in market but these are not specific to rice crop, hence, there is need to develop rice-specific bioinoculants and entomopathogens. Besides N and P as nutrients, rice crop also requires K, S and Zn for its growth and development. But microbial-based formulations of these nutrients for rice crop are still lacking. Similarly, entomopathogens to control crop pests are commercially available but these are restricted to only few major rice pests. Therefore, finding microbial agents and their suitable formulation against other rice diseases (Bakanae, false smut, leaf streak, tungro, etc.) and insect pests (rice swarming caterpillar, rice caseworm, rice gall midge, common cutworm, rice skipper, etc.) would be important and will help in taking a step towards the evergreen revolution in rice production in a sustainable manner.

Acknowledgement The authors are grateful to the Director, ICAR-National Rice Research Institute (NRRI), Cuttack, Odisha, India, for encouraging and giving fruitful suggestions for writing this book chapter. The first author is also thankful to research associates working under different external aided projects such as EAP200, EAP 275, and EAP 350.

References

Aarab S, Ollero FJ, Megías M, Laglaoui A, Bakkali M, Arakrak A (2015) Isolation and screening of bacteria from rhizospheric soils of rice fields in Northwestern Morocco for different plant growth promotion (PGP) activities: an in vitro study. Int J Curr Microbiol App Sci 4(1):260–269

Abdallah Y, Yang M, Zhang M, Masum MM, Ogunyemi SO, Hossain A, Li B (2019) Plant growth promotion and suppression of bacterial leaf blight in rice by *Paenibacillus polymyxa* Sx3. Lett Appl Microbiol 68(5):423–429

Adhya TK, Mishra BB, Annapurna K, Verma DK, Kumar U (2017) Advances in soil microbiology: recent trends and future prospects. In: Arora NK (ed) Microorganisms for Sustainability, vol 4. Springer, Singapore

Aggani SL (2013) Development of bio-fertilizers and its future perspective. Sch Acad J Pharm 2(4):327–332

Ahmad M, Nadeem SM, Naveed M, Zahir ZA (2016) Potassium-solubilizing bacteria and their application in agriculture. Springer, India

Al-Taweila HI, Osmana MB, Abdulhamida A, Mohammadb NH, Yussof WMW (2010) Microbial inoculants for enhancing rice growth and sheath spots disease suppression. Arch Agron Soil Sci 56(6):623–632

Arora P (2023) Life below land: the need for a new sustainable development goal. Front Environ Sci 11:1215282. https://doi.org/10.3389/fenvs.2023.1215282

Arshad M, Frankenberger WT (1998) Plant growth-regulating substances in the rhizosphere: microbial production and functions. Adv Agronom 62:46–152

Avakyan ZA, Belkanova NP, Karavaiko GT, Piskunov VP (1984) Silicon compounds in solution during bacterial quartz degradation. Mikrobiologiya 54:301–309

Bakhshandeh E, Rahimian H, Pirdashti H, Nematzadeh GA (2015) Evaluation of phosphate-solubilizing bacteria on the growth and grain yield of rice (Oryza sativa L.) cropped in northern Iran. J Appl Microbiol 119(5):1371–1382

Bakhshandeh E, Pirdashti H, Lendeh KS (2017) Phosphate and potassium-solubilizing bacteria effect on the growth of rice. Ecol Eng 103:164–169

Banayo NPM, Cruz PCS, Aguilar EA, Badayos RB, Haefele SM (2012) Evaluation of biofertilizers in irrigated rice: effects on grain yield at different fertilizer rates. Agriculture 2:73–86

Banik A, Mukhopadhyay SK, Sahana A, Das D, Dangar TK (2016) Fluorescence resonance energy transfer (FRET) based technique for tracking of endophytic bacteria in rice roots. Biol Fert Soils 52:277–282

Banik A, Dash GK, Swain P, Kumar U, Mukhopadhyay SK, Dangar TK (2019) Application of rice (*Oryza sativa* L.) root endophytic diazotrophic *Azotobacter* sp. strain Avi2 (MCC 3432) can increase rice yield under green house and field condition. Microbiol Res 219:56–65

Bernaola L, Cosme M, Schneider RW, Stout M (2018) Belowground inoculation with arbuscular mycorrhizal fungi increases local and systemic susceptibility of rice plants to different pest organisms. Front Plant Sci 9:747

Bhardwaj D, Ansari MW, Sahoo RK, Tuteja N (2014) Biofertilizers function as key player in sustainable agriculture by improving soil fertility, plant tolerance and crop productivity. Microb Cell Factories 13:66

Bholay AD, Barbosa S, Jadhav K (2018) In vitro cultivation of vesicular-arbuscular mycorrhizal fungi and its biological efficacy. Int J Curr Microbiol Appl Sci 7(3):931–941

Bhooshan N, Pabbi S, Singh A, Sharma A, Chetan, Jaiswal A, Kumar A (2018) Impact of blue green algae (BGA) technology: an empirical evidence from northwestern Indo-Gangetic Plains. 3 Biotech 8:1–8

Biofertilizer market (2019) Biofertilizer market report code: AGI 2569. https://www.marketsandmarkets.com/Market-Reports/compound-biofertilizers-customized-fertilizers-market-856.html

Biopesticides market (2019) Biopesticides market growth, trends, and forecast (2020-2025) https://www.mordorintelligence.com/industry-reports/global-biopesticides-market-industry

Biswas PP, Sharma PD (2008) A new approach for estimating fertilizer response ratio-the Indian scenario. Indian J Fert 4(7):59

Biswas JC, Ladha JK, Dazzo FB (2000) Rhizobia inoculation improves nutrient uptake and growth of lowland rice. Soil Sci Soc Am J 64(5):1644–1650

Bocchi S, Malgioglio A (2010) Azolla-Anabaena as a biofertilizer for rice paddy fields in the Po Valley, a temperate rice area in Northern Italy. Int J Agron. https://doi.org/10.1155/2010/152158

Chakraborty K (2011) Assessment of the efficacy of some bio-rational pesticide formulations for the management of yellow stem borer, Scirpophaga incertulas Wlk. in paddy field. J Biopest 4(1):75–80

Chen LH, Zhang J, Shao XH, Wang SS, Miao QS, Mao XY, Zhai YM, She DL (2015) Development and evaluation of Trichoderma asperellum preparation for control of sheath blight of rice (Oryza sativa L.). Biocontrol Sci Tech 25(3):316–328

Chen SL, Tsai MK, Huang YM, Huang CH (2018) Diversity and characterization of Azotobacter isolates obtained from rice rhizosphere soils in Taiwan. Ann Microbiol 68:17–26

Compant S, Reiter B, Sessitsch A, Nowak J, Clément C, Barka EA (2005) Endophytic colonization of *Vitis vinifera* L. by plant growth-promoting bacterium *Burkholderia* sp. strain PsJN. Appl Environ Microbiol 71(4):1685–1693

Cooke J, Leishman MR (2011) Is plant ecology more siliceous than we realize. Trends Plant Sci 16:61–68

Cuong ND, Nicolaisen MH, Sørensen J, Olsson S (2011) Hyphae-colonizing Burkholderia sp.—a new source of biological control agents against sheath blight disease (Rhizoctonia solani AG1-IA) in rice. Microb Ecol 62:425–434

Dangar TK, Bag MK, Kumar U, Prakash A (2019) Microbial pesticides. In: Biopesticides in Indian agriculture, vol 4. AZRA Publications, pp 163–168

Das J, Dangar TK (2008) Microbial population dynamics, especially stress tolerant *Bacillus thuringiensis*, in partially anaerobic rice field soils during post-harvest period of the Himalayan, Island, brackish water and coastal habitats of India. World J Microb Biot 24(8):1403–1410

de Macale MA, Vlek PL (2004) The role of Azolla cover in improving the nitrogen use efficiency of lowland rice. Plant Soil 263(1):311–321

de Salamone IEG, Di Salvo LP, Ortega JSE, Sorte PMB, Urquiaga S, Teixeira KR (2010) Field response of rice paddy crop to *Azospirillum* inoculation: physiology of rhizosphere bacterial communities and the genetic diversity of endophytic bacteria in different parts of the plants. Plant Soil 336:351–362

Dębska A, Bieniek J, Fabrykowski E, Molendowski F, Romański L (2016) Exploitation assessment of New Holland CX780 combine harvester. Agricul Eng 20:43

Dimkpa C, Svatos A, Merten D, Buchel G, Kothe E (2008) Hydroxamate siderophores produced by *Streptomyces acidiscabies* E13 bind nickel and promote growth in cowpea (*Vigna unguiculata* L.) under nickel stress. Can J Microbiol 5:163–172

Duarah I, Deka M, Saikia N, Deka Boruah HP (2011) Phosphate solubilizers enhance NPK fertilizer use efficiency in rice and legume cultivation. Biotech. https://doi.org/10.1007/s13205-011-0028-2

Duong VH, Hoa PP, Nguyen DQ, Phan HT, Trinh HH, Nguyen VH, Vu TT (2014) Biological control of *Xanthomonas oryzae* pv. oryzae causing rice bacterial blight disease by *Streptomyces toxytricini* VN08-A-12, isolated from soil and leaf-litter samples in Vietnam. Biocontrol Sci 19(3):103–111

Elad Y, Chet I, Katan J (1980) *Trichoderma harzianum*: a biocontrol agent effective against *Sclerotium rolfsii* and *Rhizoctonia solani*. Phytopathology 70(2):119–121

Estrada GA, Baldani VLD, DMde O, Urquiaga S, Baldani JI (2012) Selection of phosphate-solubilizing diazotrophic *Herbaspirillum* and *Burkholderia* strains and their effect on rice crop yield and nutrient uptake. Plant Soil 369:115–129

Food and Agriculture Organization of United Nations, Fertilizer use by crops in India (2005). Available ftp://ftp.fao.org/agl/agll/docs/fertuseindia.pdf

Gandhi A, Muralidharan G (2016) Assessment of zinc solubilizing potentiality of *Acinetobacter* sp. isolated from rice rhizosphere. Eur J Soil Biol 76:1–8

Ghosh TS, Chatterjee S, Azmi SA, Mazumdar A, Dangar TK (2017) Virulence assay and role of *Bacillus thuringiensis* TS110 as biocontrol agent against the larval stages of rice leaffolder *Cnaphalocrocis medinalis*. J Parasit Dis 41(2):491–495

Gong H, Zhu X, Chen K, Wang S, Zhang C (2005) Silicon alleviates oxidative damage of wheat plants in pots under drought. Plant Sci 169(2):313–321. https://doi.org/10.1016/j.plantsci.2005.02.023

Govindasamy V, Senthilkumar M, Kumar U, Annapurna K (2009) PGPR-biotechnology for management of abiotic and biotic stresses in crop plants. In: Potential microorganisms of sustainable agriculture. IK International Publishing, pp 27–48

Gupta V, Shamas N, Razdan VK, Mahajan S, Fatima K, Sharma S, Rai PK (2018) Management of Brown Spot of Rice (*Oryza sativa* L.) caused by *Bipolaris oryzae* by bio-control agents. Int J Curr Microbiol Appl Sci 7(4):3472–3477

Hamayun M, Sohn EY, Khan SA, Shinwari ZK, Khan AL, Lee IJ (2010) Silicon alleviates the adverse effects of salinity and drought stress on growth and endogenous plant growth hormones of soybean (*Glycine max* L.). Pak J Bot 42(3):1713–1722

Hari Sankar SS (2017) Efficacy of chitin enriched formulations of *Lecanicillium* spp against sucking pests of rice *Oryza sativa* L. Doctoral dissertation,. Department of Agricultural Entomology, College of Agriculture, Vellayani

Harikrishnan H, Shanmugaiah V, Balasubramanian N, Sharma MP, Kotchoni SO (2014) Antagonistic potential of native strain *Streptomyces aurantiogriseus* VSMGT1014 against sheath blight of rice disease. World J Microbiol Biotechnol 30:3149–3161

Harish S, Duraiswamy S, Ramalingam R, Ebenezar EG, Seetharaman K (2008) Use of plant extracts and biocontrol agents for the management of brown spot disease in rice. BioControl 53:555–567

Hashem MA (2001) Problems and prospects of cyanobacterial biofertilizer for rice cultivation. Funct Plant Biol 28(9):881–888

Irisarri P, Gonnet S, Deambrosi E, Monza J (2007) Cyanobacterial inoculation and nitrogen fertilization in rice. World J Microbiol Biotechnol 23:237–242

Islam MR, Madhaiyan M, Deka BH, Yim W, Saravanan VS, Fu Q, Hu H, Sa T (2009) Characterization of plant growth-promoting traits of free-living diazotrophic bacteria and their inoculation effects on growth and nitrogen uptake of crop plants. J Microbiol Biotechnol 19:1213–1222

Jatoi GH, Bhatti AG, Hidayatullah SM, Bhatti MU, Abro MA, Memon S (2016) Rhizobacteria associated from rhizospheric soil and there antifungal activity against Brown spot fungus (*Helminthosporium Oryzae*). Sci Int-Lahore 28(4):4119–4122

Jayaraj J, Yi H, Liang GH, Muthukrishnan S, Velazhahan R (2004) Foliar application of Bacillus subtilis AUBS1 reduces sheath blight and triggers defense mechanisms in rice/ Blattapplikation von Bacillus subtilis AUBS1 reduziert die Blattscheidendürre und induziert Abwehrmechanismen in Reis. J Plant Dis Protect 1:115–125

Jha MN, Mishra SK (2005) Biological responses of cyanobacteria to insecticides and their insecticide degrading potential. Bull Environ Contam Toxicol 75(2):374–381

Jha MN, Chaurasia SK, Bharti RC (2013) Effect of integrated nutrient management on rice yield, soil nutrient profile, and cyanobacterial nitrogenase activity under rice–wheat cropping system. Commun in Soil Sci Plant Anal 44(13):1961–1975

Ji SH, Gururani MA, Chun SC (2014) Isolation and characterization of plant growth promoting endophytic diazotrophic bacteria from Korean rice cultivars. Microbiol Res 169:83–98

Jin SF, Feng MG, Chen JQ (2008) Selection of global Metarhizium isolates for the control of the rice pest *Nilaparvata lugens* (Homoptera: Delphacidae). Pest Manag Sci 64(10):1008–1014

Jochum M, Moncayo LP, Jo YK (2018) Microalgal cultivation for biofertilization in rice plants using a vertical semi-closed airlift photobioreactor. PLoS One 13(9):e0203456

Joung JK, Ramm EI, Pabo CO (2000) A bacterial two-hybrid selection system for studying protein–DNA and protein–protein interactions. Proc Natl Acad Sci 97(13):7382–7387

Kaari M, Manikkam R, Annamalai KK, Joseph J (2023) Actinobacteria as a source of biofertilizer/ biocontrol agents for bio-organic agriculture. J Appl Microbiol 134(2):lxac047

Kakar KU, Duan YP, Nawaz Z, Sun G, Almoneafy AA, Hassan MA, Xie GL (2014) A novel rhizobacterium Bk7 for biological control of brown sheath rot of rice caused by *Pseudomonas fuscovaginae* and its mode of action. Eur J Plant Pathol 138(4):819–834

Kalayu G (2019) Phosphate solubilizing microorganisms: promising approach as biofertilizers. Int J Agrondoi 2019:1. https://doi.org/10.1155/2019/4917256

Kanimozhi K, Panneerselvam A (2015) Isolation and characterization of *Azospirillum* sp. from paddy field soil, Thanjavur District, Tamil Nadu. Int J Sci Res 6(5)

Kannan T, Ponmurugan P (2010) Response of paddy (Oryza sativa L.) varieties to *Azospirillum brasilense* inoculation. J Phytology 2(6)

Karthiba L, Saveetha K, Suresh S, Raguchander T, Saravanakumar D, Samiyappan R (2010) PGPR and entomopathogenic fungus bioformulation for the synchronous management of leaffolder pest and sheath blight disease of rice. Pest Manag Sci 66(5):555–564

Kecskés ML, Choudhury ATMA, Casteriano AV, Deakera R, Roughleya RJ, Lewinb L, Ford R, Kennedy IR (2015) Effects of bacterial inoculant biofertilizers on growth, yield and nutrition of rice in Australia, J Plant Nutr. http://www.tandfonline.com/loi/lpla20

Khalili E, Sadravi M, Naeimi S, Khosravi V (2012) Biological control of rice brown spot with native isolates of three *Trichoderma* species. Braz J Microbiol 43(1):297–305

Khan HI (2018) Appraisal of biofertilizers in rice: to supplement inorganic chemical fertilizer. Rice Sci 25(6):357–362

Khan MMA, Haque E, Paul NC, Khaleque MA, Al-Garni SM, Rahman M, Islam MT (2017) Enhancement of growth and grain yield of rice in nutrient deficient soils by rice probiotic bacteria. Rice Sci 24(5):264–273

Khanghahi MY, Pirdashti H, Rahimian H, Nematzadeh G, Sepanlou MG (2017) Potassium solubilising bacteria (KSB) isolated from rice paddy soil: from isolation, identification to K use efficiency. Symbiosis 76:13–23

Kim YH, Khan AL, Hamayun M, Kang SM, Beom YJ, Lee IJ (2011) Influence of short-term silicon application on endogenous physiohormonal level of *Oryza sativa* L. under wounding stress. Biol Trace Elem Res 144:1175–1185. https://doi.org/10.1007/s12011-011-9047-4

Kirubakaran SA, Sathish-Narayanan S, Revathi K, Chandrasekaran R, Senthil-Nathan S (2014) Effect of oil-formulated *Metarhizium anisopliae* and *Beauveria bassiana* against the rice leaf-folder *Cnaphalocrocis medinalis* Guenée (Lepidoptera: Pyralidae). Arch Phytopathol Plant Prot 47(8):977–992

Kumar U (2017) Diazotrophic microbes in rice: a boon to save nitrogen fertilizers. EC Microbiol 6(1):1–3

Kumar U, Nayak AK (2019) *Azolla* germplasms at NRRI: conservation, characterization and utilization. NRRI Research Bull 19:68

Kumar U, Agrawal C, Paul S, Annapurna K (2013) Endophytes as biocontrol agents of plant pathogens and insects. Kavaka 41:92–95

Kumar U, Shahid M, Tripathi R, Mohanty S, Kumar A, Bhattacharyya P, Lal B, Gautam P, Raja R, Panda BB, Jambhulakar NN, Shukla AK, Nayak AK (2017) Variation of functional diversity of soil microbial community in sub-humid tropical rice-rice cropping system under long-term organic and inorganic fertilization. Ecol Indic 73:536–543

Kumar U, Nayak AK, Shahid M, Gupta VVSR, Panneerselvam P, Mohanty S, Kaviraj M, Kumar A, Chatterjee D, Lal B, Gautam P, Tripathi R, Panda BB (2018a) Continuous application of inorganic and organic fertilizers over 47 years in paddy soil alters the bacterial community structure and its influence on rice production. Agric Ecosyst Environ 262:65–75

Kumar U, Banik A, Panneerselvam P, Annapurna K (2018b) Lower frequency and diversity of antibiotic-producing fluorescent pseudomonads in rhizosphere of Indian rapeseed-mustard (*Brassica juncea* L. Czern.). Proc Natl Acad Sci India: Biol Sci 88:579–586

Kumar U, Pannereselvam P, Nayak AK (2018c) Endophytic and rhizospheric nitrogen-fixing bio-inoculants for rice crop. NRRI News Lett 39(1):15

Kumar U, Panneerselvam P, Kaviraj M, Raut S, Narayan H (2018d) Can biopesticides alone capable to replace chemical pesticides in agriculture? In: Rajput AS, Singla A, Kumar U (eds) Compendium on thirty days certificate course. Regional Centre of Organic Farming Bhubaneswar, Odisha, p 107

Kumar U, Nayak AK, Panneerselvam P, Kumar A, Mohanty S, Shahid M, Sahoo A, Kaviraj M, Priya H, Jambhulkar NN, Dash PK (2019a) Cyanobiont diversity in six *Azolla* spp. and relation to *Azolla*-nutrient profiling. Planta 25:1–3

Kumar U, Kaviraj M, Panneerselvam P, Priya H, Chakraborty K, Swain P, Chatterjee SN, Sharma SG, Nayak PK, Nayak AK (2019b) Ascorbic acid formulation for survivability and diazotrophic efficacy of *Azotobacter chroococcum* Avi2 (MCC 3432) under hydrogen peroxide stress and its role in plant-growth promotion in rice (*Oryza sativa* L.). Plant Physiol Biochem 139:419–427

Kumar U, Behera S, Saha S, Das D, Guru PK, Kaviraj M, Munda S, Adak T, Nayak AK (2020) Non-target effect of bispyribac sodium on soil microbial community in paddy soil. Ecotoxicol Environ Saf 189:110019

Kumar U, Shelake RM, Singh R (eds) (2023a) Soil-plant-microbe interactions: an innovative approach towards improving soil health and plant growth. Frontiers Media SA, Lausanne. https://doi.org/10.3389/978-2-83251-919-6

Kumar U, Kaviraj M, Kundu S, Rout S, Priya H, Nayak AK (2023b) Microbial alleviation of abiotic and biotic stresses in rice. In: Sustainable agriculture reviews 60: microbial processes in agriculture. Springer Nature, Cham, pp 243–268

Lacey LA (1997) Manual of techniques in insect pathology. Academic Press

Law JW, Ser HL, Khan TM, Chuah LH, Pusparajah P, Chan KG, Goh BH, Lee LH (2017) The potential of Streptomyces as biocontrol agents against the rice blast fungus, Magnaporthe oryzae (Pyricularia oryzae). Front Microbiol 8:215552

Lee SJ, Yu JS, Nai YS, Parker BL, Skinner M, Kim JS (2015) *Beauveria bassiana* sensu lato granules for management of brown planthopper, *Nilaparvata lugens* in rice. Bio Control 60(2):263–270

Leelasuphakul W, Sivanunsakul P, Phongpaichit S (2006) Purification, characterization and synergistic activity of β-1, 3-glucanase and antibiotic extract from an antagonistic Bacillus subtilis NSRS 89-24 against rice blast and sheath blight. Enzym Microb Technol 38(7):990–997

Leveau JH, Lindow SE (2005) Utilization of the plant hormone indole-3-acetic acid for growth by *Pseudomonas putida* strain 1290. Appl Environ Microbiol 71:2365–2371

Liang Y, Wong JW, Wei L (2005) Silicon-mediated enhancement of cadmium tolerance in maize (*Zea mays* L.) grown in cadmium contaminated soil. Chemosphere 58(4):475–483. https://doi.org/10.1016/j.chemosphere.2004.09.034

Lin QM, Rao ZH, Sun YX, Yao J, Xing LJ (2002) Identification and practical application of silicate dissolving bacteria. Agri Sci China 1:81–85

Liu W, Xu X, Wu S, Yang Q, Luo Y, Christie P (2006) Decomposition of silicate minerals by *Bacillus mucilaginosus* in liquid culture. Env Geochem Health 28:133–140. https://doi.org/10.1007/s10653-005-9022-0

Malinovskaya IM, Kosenko LV, Votselko SK, Podgorskii VS (1990) Role of *Bacillus* mucilaginosus polysaccharide in degradation of silicate minerals. Microbiology 59:49

Mian MH (2002) Azobiofer: a technology of production and use of Azolla as biofertiliser for irrigated rice and fish cultivation. In: Biofertilisers in action. Rural Industries Research and Development Corporation, Canberra, pp 45–54

Mirza MS, Rasul G, Mehnaz S, Ladha JK, So RB, Ali S, Malik KA (2000) Beneficial effects of inoculated nitrogenfixing bacteria on rice. The Quest for Nitrogen Fixation in Rice, pp 191–204

Mishra IG, Sapre S, Tiwari S (2017) Zinc solubilizing bacteria from the rhizosphere of rice as prospective modulator of zinc biofortification in rice. Rhizosphere 3:185 https://doi.org/10.1016/j.rhisph.2017.04.013

Mishra P, Mishra J, Arora NK (2023) Biofortification revisited: addressing the role of beneficial soil microbes for enhancing trace elements concentration in staple crops. Microbiol Res 275:127442

Mohammadi K, Sohrabi Y (2012) Bacterial biofertilizers for sustainable crop production: a review. ARPN J Agric Biol Sci 7(5)

Motsara MR, Bhattacharyya P, Srivastava B (1995) Biofertilizer technology, marketing and usage—a source book-cum-glossary. Fertilizer Development and Consultation Organisation, New Delhi

Myo EM, Myint M, Oo ZK (2016) Use of the effective microorganisms as biofertilizer for rice growth and production. Inter J Tech Res Appl 4(4):94–99

Nagendran K, Karthikeyan G, Mohammed Faisal P, Kalaiselvi P, Raveendran M, Prabakar K, Raguchander T (2014) Exploiting endophytic bacteria for the management of sheath blight disease in rice. Biol Agric Hortic 30(1):8–23

Naher UA, Panhwar QA, Othman R, Ismail MR, Berahim Z (2016) Biofertilizer as a supplement of chemical fertilizer for yield maximization of rice. J Agri Food Dev 2:16–22

Narwal E, Annapurna K, Choudhary J, Sangwan S (2018) Effect of arbuscular mycorrhizal fungal colonization on nutrient uptake in rice aerobic conditions. Inter J Curr Microbiol App Sci 7(4):1072–1093

Nayak AK, Shahid M, SAHA S, Tripathi R, Mohanty S, Kumar U, Bhattacharyya P, Munda S, Guru PK, Debnath M, Chatterjee D, Priyadarshini S (2019) Production technologies of rice: activities, achievements and aspirations. ICAR-National Rice Research Institute, Cuttack 72–120

Nico M, Ribaudo CM, Gori JI, Cantore ML, Curá JA (2012) Uptake of phosphate and promotion of vegetative growth in glucose-exuding rice plants (Oryza sativa) inoculated with plant growth-promoting bacteria. Appl Soil Ecol 61:190–195

Oladele S, Awodun M (2014) Response of lowland rice to biofertilizesr inoculation and their effects on growth and yield in Southwestern Nigeria. J Agric Environ Sci 3(2):371–390

Othman NMI, Othman R, Saud HM, Wahab PEM (2018) Effects of root colonization by zinc-solubilising bacteria on rice plant (*Oryza sativa* MR219) growth. Agric Nat Resour 51:532–537

Owen D, Williams AP, Griffith GW, Withers PJ (2015) Use of commercial bio-inoculants to increase agricultural production through improved phosphorus acquisition. Appl Soil Ecol 86:41–54

Panhwar QA, Othman R, Rahman ZA, Meon S, Ismail MR (2012) Isolation and characterization of phosphate-solubilizing bacteria from aerobic rice. Afr J Biotechnol 11(11):2711–2719

Panhwar QA, Naher UA, Jusop S, Othman R, Latif MA, Ismail MR (2014) Biochemical and molecular characterization of potential phosphate-solubilizing bacteria in acid sulfate soils and their beneficial effects on rice growth. PLoS One 9(10):e97241. https://doi.org/10.1371/journal.pone.0097241

Panneerselvam P, Kumar U, Sugitha TC, Parameswaran C, Sahoo S, Binodh AK, Jahan A, Anandan A (2018a) Arbuscular mycorrhizal fungi (AMF) for sustainable rice production. In: Advances in soil microbiology: recent trends and future prospects. Springer, Singapore, pp 99–126

Panneerselvam P, Kumar U, Sahu S, Mohapatra SD, Dangar TK, Parameswaran C, Govindharaj GPP (2018b) Larvicidal potential of *Skermanella* sp. against rice leaf folder (*Cnaphalocrocis medinalis* Guenee) and pink stem borer (*Sesamia inferens* Walker). J Invertebr Pathol 157:74–79

Panneerselvam P, Kumar U, Senapati A, Parameswaran C, Anandan A, Kumar A, Jahan A, Padhy SR, Nayak AK (2019a) Influence of elevated CO_2 on arbuscular mycorrhizal fungal community elucidated using Illumina MiSeq platform in sub-humid tropical paddy soil. App Soil Ecol 145:103344

Panneerselvam P, Ansuman, Kumar U, Lepcha LSP, Prabhukarthikeyan SR, Jahan A, Parameshwaran, Pandi GP, Lenka S, Nayak PK, Mitra D, Sagarika, Sugitha T, Sivakumar (2019b) Antagonistic and plant-growth promoting novel *Bacillus* species from long-term organic farming soils from Sikkim, India. 3 Biotech, 9, 416

Paraiso L, Capacao R (2019) Fungal epiphytes for biological control of rice blast fungus (Pyricularia oryzae). Plant Pathol Quarantine 1:64–71

Pattnaik S, Mohapatra B, Kumar U, Pattnaik M, Samantaray D (2019) Microbe-mediated plant growth promotion: a mechanistic overview on cultivable plant growth-promoting members. In: Biofertilizers for sustainable agriculture and environment. Springer, Cham, pp 435–463

Poinar GO, Thomas GM (1984) A fossil entomogenous fungus from Dominican amber. Experientia 40(6):578–579

Popp J, Peto K, Nagy J (2013) Pesticide productivity and food security. A review. Agron Sustain Dev 33:243–255

Raddadi N, Belaouis A, Tamagnini I, Hansen BM, Hendriksen NB, Boudabous A, Daffonchio D (2009) Characterization of polyvalent and safe *Bacillus thuringiensis* strains with potential use for biocontrol. J Basic Microbiol 49(3):293–303

Rajeshkannan V, Sumathi CS, Manian S (2009) Arbuscular mycorrhizal fungi colonization in upland rice as influenced by agrochemical application. Rice Sci 16(4):307–313

Rajput AS, Singla A, Kumar U (2020) Organic package and practices for rice and vegetables of Odisha. Regional Centre of Organic Farming (RCOF), Ghatikia, Bhubaneswar

Rao NSS (1982) Biofertilizers. Interdisciplin. Sci Rev 7(3):220–229

Rathnathilaka T, Premarathna M, Madawala S, Pathirana A, Karunaratne K, Seneviratne G (2023) Biofilm biofertilizer application rapidly increases soil quality and grain yield in large scale conventional rice cultivation: a case study. J Plant Nutri 46:1220–1230

Rodríguez H, Fraga R (1999) Phosphate solubilizing bacteria and their role in plant growth promotion. Biotechnol Advan 17:319–339

Röös E, Mie A, Wivstad M et al (2018) Risks and opportunities of increasing yields in organic farming. A review. Agron Sustain Dev 38:14

Rusel EW (1973) Soil condition and plant growth, 10th edn. Longman, London, p 756

Sahoo RK, Ansari MW, Pradhan M, Dangar TK, Mohanty S, Tuteja N (2014) Phenotypic and molecular characterization of efficient native *Azospirillum* strains from rice fields for crop improvement. Protoplasma 251:943–953

Saikia R, Kumar R, Arora DK, Gogoi DK, Azad P (2006) Pseudomonas aeruginosa inducing rice resistance against Rhizoctonia solani: production of salicylic acid and peroxidases. Folia Microbiol 51:375–380

Sao S, Samuel K (2018) Study of *Cyanobacteria* as biofertilizer from the rice field. World J Pharm Res 4(3):1696–1706

Sapsirisopa S, Chookietwattana K, Maneewan K, Khaengkhan P (2009) Effect of salt-tolerant Bacillus inoculum on rice KDML 105 cultivated in saline soil. Asian J Food Ag-Ind 2:S69–S74

Selvaraj B, Balasubramanian S (2020) Formulations of BGA for paddy crop. In: Agroecosystems–very complex environmental systems, IntechOpen, p 137

Setiawati MR, Damayani M, Herdiyantoro D, Suryatmana P, Anggraini D, Khumairah FH (2018) The application dosage of *Azolla pinnata* in fresh and powder form as organic fertilizer on soil chemical properties, growth and yield of rice plant. AIP Conf Proc:1927. https://doi.org/10.1063/1.5021210

Sha Y, Wang Q, Li Y (2016) Suppression of Magnaporthe oryzae and interaction between Bacillus subtilis and rice plants in the control of rice blast. Springerplus 5:1–3

Shaban KA, Mahmoud AA, Mansour A, Abd El-Kader MG (2009) Biofertilizer and organic manure affects rice productivity in newly reclaimed saline soil. Dyn Soil Dyn Plant 3(1):55–60

Shakeel M, Afroz Rais A, Hassan MA, Hafeez FY (2015) Root associated *Bacillus* sp. improves growth, yield and zinc translocation for basmati rice (*Oryza sativa*) varieties. Front Microbiol 6:1286. https://doi.org/10.3389/fmicb.2015.01286

Shanmugam V, Sriram S, Babu S, Nandakumar R, Raguchander T, Balasubramanian P, Samiyappan R (2001) Purification and characterization of an extracellular α-glucosidase protein from Trichoderma viride which degrades a phytotoxin associated with sheath blight disease in rice. J Appl Microbiol 90(3):320–329

Shanware AS, Kalkar SA, Trivedi MM (2014) Potassium solublisers: occurrence, mechanism and their role as competent biofertilizers. Int J Curr Microbiol App Sci 3(9):622–629

Sharma S, Padbhushan R, Kumar U (2019) Integrated nutrient management in rice–wheat cropping system: an evidence on sustainability in Indian subcontinent through meta-analysis. Agron 9(2):71

Sheng XF, Huang WY (2002) Study on the conditions of potassium release by strain NBT of silicate bacteria scientia. Agri Sinica 35(6):673–677

Shimoi S, Inoue K, Kitagawa H, Yamasaki M, Tsushima S, Park P, Ikeda K (2010) Biological control for rice blast disease by employing detachment action with gelatinolytic bacteria. Biol Control 55(2):85–91

Sindhu SS, Parmar P, Phour M, Sehrawat A (2016) Potassium-solubilizing microorganisms (KSMs) and its effect on plant growth improvement. Springer, India

Singh P, Singh R, Dhaka SS, Kumar D, Kumar H, Kumari N (2015) Bioefficacy of insecticides and bio-pesticides against yellow stem borer, *Scirpophaga incertulas* (walk.) and their effect on spiders in rice crop. South Asian J Food Technol Environ 1:179–183

Singh UB, Malviya D, Singh S, Pradhan JK, Singh BP, Roy M, Imram M, Pathak N, Baisyal BM, Rai JP, Sarma BK (2016) Bio-protective microbial agents from rhizosphere eco-systems trigger

plant defense responses provide protection against sheath blight disease in rice (*Oryza sativa* L.). Microbiol Res 192:300–312
Spence C, Alff E, Johnson C, Ramos C, Donofrio N, Sundaresan V, Bais H (2014) Natural rice rhizospheric microbes suppress rice blast infections. BMC Plant Biol 14:1–7
Sperandio EM, do Vale HM, de Souza Reis M, de Carvalho Barros Cortes MV, Lanna AC, de Filippi MC (2017) Evaluation of rhizobacteria in upland rice in Brazil: growth promotion and interaction of induced defense responses against leaf blast (*Magnaporthe oryzae*). Acta Physiol Plant 39:1
Subashini HD, Malarvannan S, Kumaran P (2007) Effect of biofertilizers (N-fixers) on the yield of rice varieties at Puducherry, India, pp 146–150
Subhalakshmi T, Devi IS (2017) Blast of rice in Manipur and its biocontrol by *Pseudomonas fluorescens* and *Trichoderma* sp. Int J Curr Microbiol App Sci 6(6):1619–1634
Sun LN, Zhang YF, He LY, Chen ZJ, Wang QY, Qian M, Sheng XF (2010) Genetic diversity and characterization of heavy metal-resistant-endophytic bacteria from two copper-tolerant plant species on copper mine wasteland. Bioresour Technol 101:501–509
Suryadi Y, Susilowati DN, Priyatno TP, Samudra IM, Kadir TS, Mubarik NR (2013) Prospect of using bacterial bioformulation to suppress bacterial leaf blight of rice: a case study in Cianjur, West Java. J Penelit Pertanian Tanaman Pangan 32(2):123937
Suyal DC, Soni R, Sai S, Goel R (2016) Microbial inoculants as biofertilizer. In: Microbial inoculants in sustainable agricultural productivity. Springer, New Delhi, pp 311–318
Takore Y (2006) Industrial biotechnology, vol 2. Gen Publishing Inc., A Mary Ann Liebert Inc. Company, pp 194–208
Tawate S, Gupta K, Jain K (2018) Technology commercialization in bio-fertilizer firm: an Indian case. Int J Glob Bus Competitiveness 13:65–74
Trân Van V, Berge O, Ngô Kê S, Balandreau J, Heulin T (2000) Repeated beneficial effects of rice inoculation with a strain of *Burkholderia vietnamiensis* on early and late yield components in low fertility sulphate acid soils of Vietnam. Plant Soil 218:273–284
Transparency Market Research (2017) Biofertilizers market (product-nitrogen fixing, phosphate mobilizing, potassium mobilizing; application-cereals & grains, fruits & vegetables, and oil seeds & pulses)—global industry analysis, size, share, growth, trends, and forecast 2017–2025. https://www.transparencymarketresearch.com/biofertilizers-market.html
Vaid SK, Kumar B, Sharma A, Shukla AK, Srivastava PC (2014) Effect of Zn solubilizing bacteria on growth promotion and Zn nutrition of rice. J Soil Sci Plant Nutr 14(4):889–910
Vance CP (2001) Symbiotic nitrogen fixation and phosphorus acquisition. Plant nutrition in a world of declining renewable resources. Plant Physiol 127(2):390–397
Vandevivere P, Welch SA, Ullman WJ, Kirchman DJ (1994) Enhanced dissolution of silicate minerals by bacteria at near neutral pH. Microbiol Ecol 27:241–251. https://doi.org/10.1007/BF00182408
Vasudevan P, Kavitha S, Priyadarisini VB, Babujee L, Gnanamanickam SS (2002) Biological control of rice diseases. Biological control of crop diseases, vol 468. Marcel Dekker Inc., New York, pp 11–32
Venkateswarlu N, Sireesha O, Aishwayra S, Vijaya T, Sriramulu A (2015) Isolation, screening of rhizosphere fungi antagonistic to rice stem rot disease pathogen *Sclerotium oryzae* catt. Asian J Pharm Clin Res 8(5):54–57
Vyas P (2018) Endophytic microorganisms as bio-inoculants for sustainable agriculture. In: Singh J, Sharma D, Kumar G, Sharma N (eds) Microbial bioprospecting for sustainable development. Springer, Singapore. https://doi.org/10.1007/978-981-13-0053-0_3
Wahab S (2003) Biotechnological approaches in plant protection. In: Koul O, Dhaliwal GS, Marwaha SS, Arora JK (eds) Biopesticides and pest management, vol 1. Campus Books International, New Delhi, pp 113–127
Wahab S (2004) Biotechnological approaches in the management of plant pests, diseases and weeds for sustainable agriculture. In: Deep roots, open skies: new biology in India. Narosa Publishing House, New Delhi, pp 113–129

Wahab S (2009) Biotechnological approaches in the management of plant pests, diseases and weeds for sustainable agriculture. J Biopest 2(2):115–134

Wang W, Qiu Z, Tan H, Cao L (2014) Siderophore production by actinobacteria. Biol Metals 27:623–631

Wang D, Xu A, Elmerich C, Ma LZ (2017) Biofilm formation enables free-living nitrogen-fixing rhizobacteria to fix nitrogen under aerobic conditions Inter Soc Microb Ecol J 11:1602–1613

Wei H, Meng T, Li X, Dai Q, Zhang H, Yin X (2018) Sink-source relationship during rice grain filling is associated with grain nitrogen concentration. Field Crop Res 215:23–38

Whipps JM (2001) Microbial interactions and biocontrol in the rhizosphere. J Exp Bot 52:487–511

Widiantini F, Herdiansyah A, Yulia E (2017) Biocontrol potential of endophytic bacteria isolated from healthy rice plant against rice blast disease (Pyricularia oryzae Cav.). KnE Life Sci 26:287–295

Xu J, Yang M, Shi P, Zou X (2015) Field control effects of *Beauveria bassiana* strain YS03 on rice water weevil. Southwest China J Agri Sci 28(4):1630–1633

Yadav J, Verma JP, Jaiswal DK, Kumar A (2014) Evaluation of PGPR and different concentration of phosphorus level on plant growth, yield and nutrient content of rice (*Oryza sativa*). Ecol Eng 62:123–128

Yanni YG, ABD EL-Fttah FK (1999) Towards integrated biofertilization management with free living and associative dinitrogen fixers for enhancing rice performance in the Nile delta. Symbiosis 27(3):319–331

Yasin M, Ahmad K, Mussarat W, Tanveer A (2012) Bio-fertilizers, substitution of synthetic fertilizers in cereals for leveraging agriculture. Crop Environ 3(1–2):62–66

Yasmin S, Zaka A, Imran A, Zahid MA, Yousaf S, Rasul G, Mirza MS (2016) Plant growth promotion and suppression of bacterial leaf blight in rice by inoculated bacteria. PLoS One 11(8):e0160688

Yasmin S, Hafeez FY, Mirza MS, Rasul M, Arshad HM, Zubair M, Iqbal M (2017) Biocontrol of bacterial leaf blight of rice and profiling of secondary metabolites produced by rhizospheric *Pseudomonas aeruginosa* BRp3. Front Microbiol 8:1895

Yua YY, Jianga CH, Wanga C, Chen LJ, Li HY, Xu Q, Guoa JH (2017) An improved strategy for stable biocontrol agents selecting to control rice sheath blight caused by *Rhizoctonia solani*. Microbiol Res 203:1–9

Zaidi NW, Singh M, Kumar S, Sangle UR, Singh R, Prasad R, Singh SS, Singh S, Yadav AK, Singh A, Waza SA (2018) Trichoderma harzianum improves the performance of stress-tolerant rice varieties in rainfed ecologies of Bihar, India. Field Crop Res 220:97–104

Zayadan BK, Matorin DN, Baimakhanova GB, Bolathan K, Oraz GD, Sadanov AK (2014) Promising microbial consortia for producing biofertilizers for rice fields. Microbiology 83:391–397

Zhao J, Shan T, Mou Y, Zhou L (2011) Plant-derived bioactive compounds produced by endophytic fungi. Mini Rev Med Chem 11(2):159–168

Chapter 14
Augmenting of Bioremediation Through Quorum Sensing and Biofilms

Nileshwariba Jadeja, Ketaki Ramani, Disha Changela, Kruti Dangar, and Mihir Raval

Abstract The amplified boost in the rate of pollution, potentially due to anthropogenic activities, has compelled us to search for proficient strategies that can be executed for environmental restoration. The key dilemma with traditional processes used for environmental cleaning of pollutants is incompetence and soaring economic costs. Bioremediation is a budding technology with profound potential for cleaning of pollutants and has turned out to be a ray of hope for environmental safety and its long-lasting survival. Quorum sensing involves cell–cell communication through signalling molecules and is based on population density. Quorum sensing regulates the formation of biofilms which has been implicated in competent bioremediation procedures. The important feature of biofilm is its stress resistance nature, which can be utilized for competent metabolism of toxic pollutants. This chapter throws light on the association of quorum sensing and formation of biofilms in relation to bioremediation of heavy metals, organic pollutants, and plastics.

Keywords Quorum sensing · Biofilms · Bioremediation · Heavy metals · Organic pollutants · Plastics

---

N. Jadeja · D. Changela · K. Dangar (✉)
Department of Biochemistry, Saurashtra University, Rajkot, Gujarat, India

K. Ramani
Department of Microbiology, School of Science, RK University, Rajkot, Gujarat, India

M. Raval
Department of Pharmaceutical Sciences, Sardar Patel University, Vallabh Vidhyanagar, Gujarat, India

N. K. Arora, B. Bouizgarne (eds.), *Microbial Biotechnology for Sustainable Agriculture Volume 2*, Microorganisms for Sustainability 34,
https://doi.org/10.1007/978-981-97-2355-3_14

14.1 Introduction

Bacteria communicate with one another to swap/exchange information using chemical signal molecules called autoinducers, which help in synchronizing or co-ordinating the activities of large bacterial groups of cells (Waters and Bassler 2005; Hmelo 2017). Bacteria make use of these information exchange systems with help of autoinducers to perform various processes, ranging from metabolic and developmental mechanisms to virulence, which is estimated to have about 5–25% of the genes encoded in a bacterial genome. This information exchange process in bacteria is known as quorum sensing (QS) (Helman and Chernin 2015). The function of signalling molecules is to regulate gene expression in same bacterium or surrounding bacterial cells of the similar or different bacterial communities, which control various physiological processes, including virulence, motility, luminescence, biofilm formation, sporulation, fluorescence and production of toxins, proteolytic enzymes, hemolysins, and siderophores (Moghaddam et al. 2014; Tiwary et al. 2017; Asfour 2018).

Autoinducers are mainly of two types, i.e. type-1 or AI-1 and type-2 or AI-2, and are known to be as small diffusible signalling molecules or peptides, which provide responses after achieving threshold value concentration of itself, outside the bacterial cell, to effectively work as quorum sensing system (QSS). Autoinducers type-1 or AI-1 are species-specific and work for intraspecies communication. Autoinducers type-2 participate in interspecies communication, as well as enable bacteria to communicate with microorganisms such as fungi and protozoa to harmonize their own behaviour in a population of microorganisms. AI-1 is formulated as acyl-homoserine lactone ring (AHL), whereas AI-2 is a furanosyl borate diester (Waters and Bassler 2005; Mangwani et al. 2012; Moghaddam et al. 2014; Grandclément et al. 2016).

In Gram-negative bacteria, quorum-dependent mechanisms are regulated by the luxI/luxR gene family. LuxI signals for AHL formation, and then AHL binds LuxR, which is a transcriptional activator of lux operon. S-adenosyl-methionine (SAM) is the precursor molecule for synthesizing AHLs by LuxI synthases, which in association with the LuxR family of transcriptional regulators trigger gene expression in response to autoinducers. In Gram-positive bacteria, QS is usually peptide-based in which post-transcriptionally processed small peptides serve as signalling molecules. These autoinducing peptides (AIPs) are synthesized on ribosomes and express a range of linear and cyclic structures. Autoinducing peptides (AIPs) are ribosomally synthesized and exhibit a variety of linear and cyclic structures (Mangwani et al. 2012; Koh et al. 2013; Sibanda et al. 2018; Grandclément et al. 2016).

The exponential rise in toxic waste effluents and recalcitrant compounds due to industrialization over the past few decades has resulted in the contamination of the environment (Soares et al. 2008; Rizzo et al. 2019), especially that of water and soil (Ghosh et al., 2019). A wide variety of persistent toxic organic pollutants, heavy metals, and plastics, and their immense repositioning in the ecosystem, are critical matters of the hour (Birch et al. 2001; Oyelola and Babatunde 2008). These pollutants can harm humans as well as animals by disrupting the cell membrane, denaturing the DNA, altering various enzymatic activities, which may cause liver and kidney failure, skin lesions, and mental and physical retardation as well (Edwards and Kjellerup 2013; Shukla et al. 2014; Kumar et al. 2019). Mounting understanding of the harmful effects of environmental pollution has led to an upgrading of the strategies that can be employed for environmental clean-up and restoration.

The biological remediation processes are more preferable than physico-chemical processes, as these are environmentally safe, techno-economically feasible, and generate less or null toxic byproducts (Molnár et al. 2007; Shukla and Subba Rao 2017). Bioremediation, classified under biological remediation process, is one of the most emerging/uprising environment-friendly and economically favourable biotechnological innovation to remove persistent organic pollutants (Bala et al. 2022). Bioremediation process is generally classified into two different types depending upon which place the pollutant is being treated. This process can be in situ (at the site of contamination) or ex situ (at a specially prepared place) (Vogt and Richnow 2013; Sharma 2021). Microbial bioremediation is the application of microbes to clean up a contaminated site resulting in degradation of pollutants into less harmful products that are environmentally safe, with economic feasibility, and maintain sustainability (Molnár et al. 2007; Shukla and Subba Rao 2017; Sharma 2021; Mishra et al. 2022). Use of microorganisms can be an ideal choice for remediation due to its adaptability and being ecofriendly (Bala et al. 2022).

14.2 Quorum Sensing in Bioremediation

Researchers have proven and are still investigating the contribution of QS systems with a wide range of applications by using various biotechnological protocols, such as bioremediation of oil and heavy metal contaminated soils, prevention of biofouling and biocorrosion (Liao and Li 2013). The few fabrications with a quorum sensing system, which stimulates the production of engineered biofilms with enhanced degradation kinetics, can be one of the effective ways to reduce pollutants in the natural environment (Mangwani et al. 2016; Bala et al. 2022).

Varieties of environments have been investigated for bacteria with QS mechanisms. The community structure and function necessary for bioremediation have been regulated by QS (Feng et al. 2013; Mangwani et al. 2016). The molecules that assist in QS are autoinducers and hence in the environmental and bioremediation applications, QS mediated via AIs plays a crucial role. From the environmental and bioremediation perspective, the processes regulated by QS are synthesis of biofilms, biosurfactants, and exopolysaccharides. In addition to these, catabolic gene expression, motility, and horizontal gene transfer also occur (López et al. 2010; Huang et al. 2013).

14.3 Biofilm Formation and Its Role in Bioremediation

Biofilm is an outcome of the QSS in bacteria that modulates factors such as antimicrobial resistance, virulence factors, toxins production, and many more (Chan et al. 2015; Bahari et al. 2017). Living tissues, natural aquatic systems, indwelling medical equipment, and industrial machine devices are a broad range of areas where biofilms can be formed. The most ideal environment meant for the attachment as well as growth is between a surface and an aqueous medium, i.e. solid–liquid interface (Donlan 2002). Bacteria attain an irreversible attachment, which can shear forces and retain a persistent grip with the surface. With time, after the maturation of biofilm, there is active exchange of nutrients and products between the microorganisms, which play a crucial role in overall preserving the biofilm structure as well as providing favourable environment to the microorganisms (Kostakioti et al. 2013). Biofilm formation, especially mixed biofilm (a variety of microorganisms involved in forming biofilm) (Galie et al. 2018), is a long-lasting survival strategy in the surrounding environment produced by the bacteria, which results in spreading infections such as dental plaque, respiratory tract infection in cystic fibrosis, osteomyelitis, etc. (Coenye et al. 2020), and is mostly due to the higher resistance in antibiotic treatment (van Tilburg et al. 2015).

Biofilm formation, as mentioned in Fig. 14.1a, b, is a coalition of microbes where they produce a matrix sheath, mushroom like structure (Singh et al. 2020), of extracellular polymeric substance on the surface by sticking to each other, which becomes accessible within biofilm structure to transport the nutrients (Muhsin et al. 2015). The initial step involves the bacterial attachment to either biotic or abiotic surfaces, where they mature and get enclosed in a self-produced sheath of extracellular polymeric substance (EPS). EPS shapes and gives three-dimensional architectural form and stabilization to biofilms produced by a combination of proteins, polysaccharides (homopolysaccharides or heteropolysaccharides), in some cases lipids as well as nucleic acids also participate (De Carvalho 2018) along with other biopolymers. The EPS sheath helps in several ways, including protection from biotic and abiotic stresses (Fatima and Arora 2023). The sheath helps in binding with water molecules and supports during drought and temperature variations. Conversely, cells are also protected from toxins, antimicrobials, pollutants, etc. Biofilm formation also helps in exchange of nutrients and genetic material between the population. From time to time, some cells detach from their planktonic stage, become free, and begin the formation of biofilm at other sites or locations (Annous et al. 2009; Lambert et al. 2014; Liaqat et al. 2019; Singh et al. 2020).

QS plays an important role in controlling monospecies as well as multispecies biofilm formation and a few other physiological behaviours including virulence, symbiosis, toxin production, genetic competence, programmed cell death (Parsek and Greenberg 2005). Biofilm formation can be initiated by a single bacterial species, referred to as monofilms or monoculture, and if multiple species of bacteria are involved, they are called multifilms or multiculture (Maksimova 2014). Monofilms are crucially involved in controlling the processes necessary enough for increasing the yield and quality of a desired product, whereas multispecies are helpful for

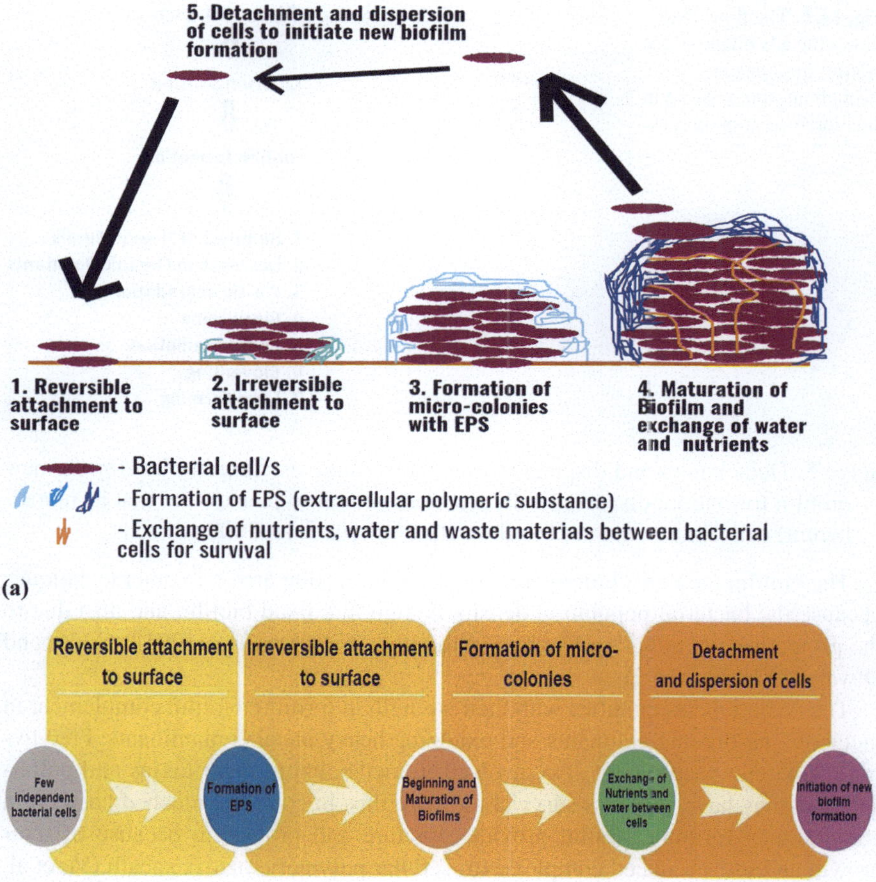

(a)

(b)

Fig. 14.1 (**a**) and (**b**) depict steps involved in the biofilm formation

multistage processes like as in biogas production and pharmaceutical formulations (Maksimova 2014; Liaqat et al. 2019). Multispecies biofilms are bacteria forming biofilms with other microorganisms, which are responsible for macrofouling (De Carvalho 2018).

Figure 14.1 shows the following steps in brief regarding biofilm formation mechanism:

Stage 1: Independent bacteria involves in reversible attachment to a surface,

Stage 2: Irreversible attachment mediated by the formation of exopolymeric material,

Stage 3: Formation of microcolonies and the beginning of biofilm maturation,

Stage 4: A 3-dimensional structure of a mature biofilm having cells packed in clusters with channels in between them, allowing the transport of water and nutrients and waste removal, and finally,

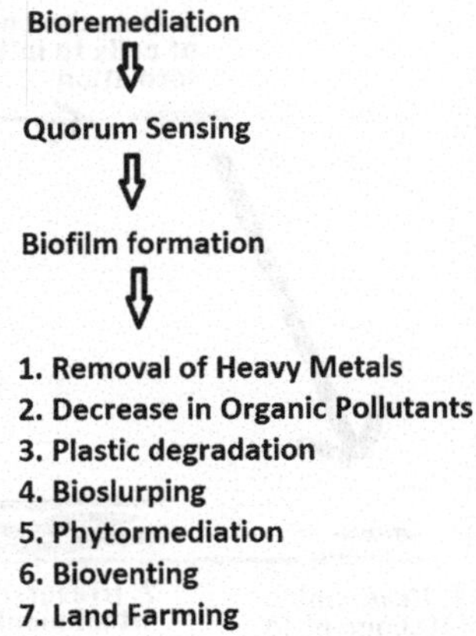

Fig. 14.2 The flow chart shows the advantage of the biofilm mechanism through quorum sensing in the bioremediation process

Stage 5: Detachment and dispersion of cells from the existing biofilm for the new biofilm formation; dispersed cells are more similar to planktonic (that is, nonadherent) cells than to mature biofilm cells (Annous et al. 2009).

The biofilm mediated bioremediation is better in comparison to other techniques because the bacterial population density is high in a fixed biofilm and also due to the protective and adhesive features of the film (Nilanjana et al. 2012; Gupta and Diwan 2017) (Fig. 14.2).

The bacterial communities with their strength in forming biofilm complement in metabolizing organic pollutants and oxidizing heavy metal contaminants. Free living planktonic bacteria are also involved in metabolizing these toxins and pollutants, but they have very less survival rates. To this, in contrast, attached bacteria in communities forming biofilm provide structure and protection because of their growth and self-produced complex extracellular polymeric matrix sheath (Vu et al. 2009; Santos et al. 2018).

Microbial biofilms, in particular those developing in harsh and polluted environments can be a very useful source of resilient microbes. Such microbes and the biofilms formed by them can be utilized for bioremediation and biotransformation of even recalcitrant pollutants (Mohapatra et al. 2020). Enhanced bioremediation of pollutants by the biofilm is attributed to the evolution of microbes within the film due to the efficient gene transfer and also because of efficient chemotaxis involving QS (Singh et al. 2006).

14.4 Quorum Sensing and Biofilms in Bioremediation of Heavy Metals

Heavy metal contamination has risen extremely high in the soil, sediment, and aquatic environments due to industries and unplanned waste management (Hossain and Rao 2014). Numerous techniques are reported for removal of metal ions but currently bioremediation using bacterial biofilm has drawn major attention (Shukla et al. 2017). Biofilm formation by QS has been visualized as an innovative methodology to entrap microbes possessing the ability to catabolise pollutants and thus has been exploited for bioremediation (Arora 2018; Mohapatra et al. 2020). Usually, there is a wide area all over the world having heavy metal pollution, which is generated through various sources such as agricultural, domestic, industrial, etc. The use of microbial surfactants can be a magnificent choice for metal contaminated sites for bioremediation. For heavy metal remediation, biofilm forming capability has been considered as promising application (Singh et al. 2006; Mani 2020). The importance of biofilms in removal of heavy metals from wastewater was highlighted in a study by Azizi et al. (2016). In this study, a modified packed bed biofilm reactor (PBBR) was capable of removing the industrial contamination of heavy metals like copper, cadmium, zinc, and nickel from wastewater. Shukla and Subba Rao (2017) signified the prospective use of biofilm of recombinant *Deinococcus radiodurans* in the remediation of heavy metals cobalt and nickel. The study showed that in the presence of 25 mM Ca^{2+}, the *D. radiodurans* (Kan^r) biofilm showed 35% and 25% removal of Co^{2+} and Ni^{2+}, respectively. Grujić et al. (2017) have presented the comparison of results for heavy metals (cadmium, zinc, and nickel) tolerance and removal efficiency of the *Rhodotorula mucilaginosa* planktonic cells and biofilm. It was found that the tolerance of biofilms was higher than the planktonic cells. Abinaya and Gautam (2017) reported studies on the biofilm produced by *Pseudomonas aeruginosa* grown in different metal fatty acid salt (MFAS) media. *P. aeruginosa* could form biofilms in the presence of different metal fatty acid salts. Thus, the study underlined a proof-of-concept of using *P. aeruginosa* in the biodegradation of MFAS. Meliani and Bensoltane (2016) presented a promising application of biofilms of *Pseudomonas* in heavy metal uptake. The study investigated if the isolates *P. aeruginosa* (P8) and *Pseudomonas fluorescens* (P4, P9, P10) could withstand metal toxicity and simultaneously assessed the interaction between heavy metals and biofilm formation.

14.5 Quorum Sensing and Biofilms in Bioremediation of Organic Pollutants

The exponential rise in industries has led to a profound increase in organic pollutants. These are widespread pollutants produced in elevated quantities due to anthropogenic activities. Because of their hydrophobicity and high molecular weight, their degradation is highly compromised. The bioremediation of a variety of organic compounds can be mediated by biofilms. Organisms like *Pseudomonas*, *Rhodococcus*, *Alcaligenes*, *Sphingomonas*, and *Methylosinus* can produce biofilms and can degrade diverse organic pollutants (Singh et al. 2006). Mangwani et al. (2015) isolated *P. aeruginosa* N6P6 from seawater, which was responsible for forming biofilms and acyl-homoserine lactone (AHL) synthase. In the study, it was found that *P. aeruginosa* N6P6 could degrade phenanthrene and pyrene but the rate of degradation was affected by biofilm growth and lasI expression. In the presence of QS inhibitors, degradation of PAHs and biofilm formation were reduced, which established the fundamental role of QS in biodegradation of PAHs. Recently, in a study by Chen and Huang (2020) it was indicated that in *Novosphingobium pentaromativorans* US6-1, QS negatively affects phenanthrene removal efficiency by bringing alteration in cell surface properties and thus concluding QS as a target for controlling bioremediation. Yu et al. (2020) used *Croceicoccus naphthovorans* PQ-2 and elucidated that the consequence of inactivation of the QS system was a noteworthy drop off in PAHs degradation. The cell surface hydrophobicity and cell morphology of bacteria were changed in the QS-deficient mutant as the QS system positively organized the expression of three PAH-degrading genes (*ahdA1e*, *xylE*, and *xylG*) and a regulatory gene *ardR*. Thus the study threw light on how the QS system swayed the degradation of PAHs and hence assisted the advances of new insights for the bioremediation of PAH. Chicca et al. (2020) isolated *Pseudomonas putida* strain (AQ8) from the soil of an oil refinery and characterized its ability to degrade BTEX. The strain was found to possess genes involved in BTEX degradation and QS. When the QS signal molecule, acyl-homoserine lactone was cleaved with lactonase, which was expressed from a plasmid, the strain reduced only 10% of BTEX concentration.

14.6 Quorum Sensing and Biofilms in Bioremediation of Plastics

Non-biodegradable plastic-based resources are extensively employed for numerous purposes of domestic and industries. Plastics or synthetic polymers such as polyethylene used in food packaging, plastic bags, milk bottles; polystyrene used in disposable cups, packing material, laboratory wares; polypropylene used in making bottle caps, drinking straws, disposable syringe, etc. are an environmentally hazardous threat as they are difficult to degrade, leading towards the accumulation of increased solid waste in the natural environment (Shah et al. 2008; Morohoshi et al. 2018). Marine pollution, generally due to microplastics, having a size of 5 mm, radiates diverse chemicals and toxins attached to their surfaces, which can be harmful to marine life. Overall, such a state of affairs results in land as well as marine pollution. Such plastic polymer squander wastes accumulate in the environment posing a global constant increasing ecological threat to terrestrial and marine wildlife (Pathak 2017; Morohoshi et al. 2018). Plastic biodegradation is a method to transform physical properties with the help of microbes, such as change in tensile strength, colour, chemical structure, shape, and the molecular weight (Shah et al. 2008; Haben 2019). Some of the species of bacterial genera like *Streptococcus*, *Staphylococcus*, *Micrococcus*, *Bacillus*, *Azotobacter*, *Moraxella*, and *Pseudomonas,* and species of fungi like *Aspergillus glaucus* and *Aspergillus niger* are involved in biodegradation mechanisms (Haben 2019). The investigation by researchers focuses on microbial transformation on the organic contaminants occurrence, as these contaminants can provide carbon as well as energy to microorganisms; carbon being basic building blocks of new cells and electrons providing energy to them (Devi et al. 2016). In a latest study by Miao et al. (2020), it was found that biofilms growing on plastics demonstrated elevated biomasses as compared to natural substrates. Thus they concluded that plastic wreckage is a new microbial niche which could shape carbon cycling in aquatic ecosystems. In an another recent study by Faheem et al. (2020), the impact of periphytic biofilms (epiphyton and epixylon) to bioremediate polyethylene (PE) microplastic, lead (Pb^{2+}), and cadmium (Cd^{2+}) was studied. The findings concluded that epiphyton had more potential than epixylon in lead and cadmium removal and PE biodegradation. Delacuvellerie et al. (2019) illustrated the capability of *Alcanivorax borkumensis* to form thick biofilms particularly on low density polyethylene (LDPE) and its potential to degrade it. They illustrated that plastisphere hosted microbial communities were entirely plastic nature dependent. The findings pointed out that *Alcanivorax*, *Marinobacter*, and *Arenibacter* genera were the potential players in plastic degradation. Table 14.1 shows the bioremediation of heavy metals, organic pollutants, and plastics mediated by biofilms.

Table 14.1 Bioremediation process of heavy metals, organic pollutants, and plastics mediated by respective microbial biofilms

Microorganisms	Bioremediation of material	References
Pseudomonas putida	Polyethylene terephthalate	Vague et al. (2019)
Bacillus cereus	Polyethylene terephthalate	Vague et al. (2019)
Proteobacteria	Microplastics	Urbanek et al. (2018)
Clostridium botulinum	Polycaprolactone	Dobretsov et al. (2013)
Streptomyces	Polyvinyl chloride, polyhydroxybutyrate	Dang et al. (2018)
Deinococcus radiodurans R1	Cobalt, nickel	Shukla and Subba Rao (2017)
Rhodotorula mucilaginosa	Mercury, copper, lead	Grujić et al. (2017)
Staphylococcus aureus	Uranium	Shukla et al. (2020)
Cyanobacteria, Gammaproteobacteria, Alphaproteobacteria	Cadmium, lead, polyethylene (PE) microplastic	Faheem et al. (2020)
Rhodotorula taiwanensis MD1149	Mercury, chromium	Tkavc et al. (2018)
Deinococcus radiodurans	Uranium	Manobala et al. (2019)
Serratia marcescens S14	Cadmium	Xu et al. (2020)
Methylobacter, Methylosinus, Methylocystis	Cadmium, lead, chromium	Faheem et al. (2020)
Bacillus sp.	Lead	Pepi et al. (2016)
Nostoc muscorum	Cadmium	Raghavan et al. (2020)

(continued)

Table 14.1 (continued)

Microorganisms	Bioremediation of material	References
Acinetobacter calcoaceticus strain, GSN3	Phenol	Irankhah et al. (2019)
Stenotrophomonas acidaminiphila NCW702, Alcaligenes faecalis NCW402, Pseudomonas mendocina NR802, Pseudomonas aeruginosa N6P6, Pseudomonas pseudoalcaligenes NP103	Polycyclic aromatic hydrocarbons (PAHs)	Mangwani et al. (2017)
Hydrocarbon contaminated soil and water microbes	PAHs	Demeter et al. (2017)
Halobacterium salinarum	Phenanthrene	Di Meglio et al. (2020)
Pseudomonas migulae AN-1	Aniline	Zhao et al. (2016)

14.7 Future Perspectives Regarding Bioremediation

There is an alarming disturbance in the ecological balance and also threat to human health because of increase in concentration of persistent organic pollutants, heavy metals, pesticides, etc. due to their toxic effect (Mishra et al. 2022). One of the most useful techniques used to remove waste from contaminated sites and areas is considered to be bioremediation. Microbial remediation is the most prominent area gained by omics in a variety of industries for maintaining and controlling environmental pollution (Bala et al. 2022). There are various approaches to achieve bioremediation such as designing synthetic microbial communities, nanotechnological methods by using effective microbes, genetic and metabolic engineered nanoparticles, etc.

A wide range of various bioinformatics databases, such as databases of chemical structure and composition, catalytic enzymes, microbial degradation pathways, RNA/protein expression, etc. of a particular organism can be used to alter genes to achieve potential results for bioremediation. Use of various tools based on omics like genomics, transcriptomics, metabolomics, and proteomics can be beneficial in bioremediation studies. Scarcity of data regarding certain areas like factors affecting growth control and metabolism of microbes results in limited bioremediation application (Bala et al. 2022).

Using nanotechnological methods can be useful in removing many toxic substances. Using nanotechnological methods can be useful in removing many toxic substances. Wastewater can be treated using Effective Microbes (EM) technology, which can make water effective enough to be used for irrigation. Genetic as well as metabolic engineering targets the specific site of gene editing by knock-in and knock-out effects using CRISPRs, TALENs, and ZFNs methods. These can be examined for usage in bioremediation analysis by targeting particular DNA sequence, specific DNA-binding element of host DNA, etc. There has been successful achievement of bioremediation of hexachlorocyclohexane and methyl parathion by genetically engineered microorganism *Sphingobium japonicum* and *Pseudomonas* sp. WBC-3 (Dangi et al. 2019; Bala et al. 2022). Also, through the development of an artificial microbiome with the model studying for particular functions and structural characteristics of specific species, it can be possible to create a synthetic microbial community. This approach can be a bit difficult and time-consuming to identify the species that participate in the bioremediation process in a natural bacterial community. Organic pollutants such as petroleum hydrocarbons can reduce solubility as well as mobility of soil, when absorbed by soil. Such can be treated by engineered polymeric nanoparticles, e.g., chains of poly-(ethyl) glycol-modified urethane acrylate polymeric nano-network particles are used to improve and maintain the properties of contaminated soil areas in the presence of diverse range of bacterial communities (Kumar 2021; Sharma and Shukla 2022; Bala et al. 2022). Figure 14.3 depicts the various future aspects of QS.

One of the most innovative approaches to confront the unsolicited biofilms, EPS matrix can be targeted through application of enzymes or by activating natural dispersion mechanisms to break off the biofilm (Coenye et al. 2020). Certain specific molecules, as well as extracted or released from microorganisms, can be used to lower or inhibit certain function/s (such as swarming motility or swimming motility) of other microorganisms such as biofilm formation, without affecting the growth rate. One such molecule, lactone, extracted from *Delisea pulchra* interfered with AHL-mediated system as well as swarming motility in *Serratia liquefaciens*, bioluminescence in *Vibrio harveyi* and *Vibrio fischeri* and also caused abnormal swarming motility by *Proteus mirabilis* without affecting the growth rate. There has been structural similarity between family of furanones and core structure of AHLs that permits furanones to competitively bind to proteins, which are involved in inhibiting the complete activation of AHL mediated processes (Armstrong et al. 2000). Molina et al. (2003) used wild type soil bacterium *Bacillus* sp. strain A24 having AHL-degrading capability in plant pathogens such as *Erwinia carotovora* and *Agrobacterium tumefaciens* resulting in biocontrol of plant disease in potato and tomato. Hence, such efforts can be made as there are a broad range of microorganisms available in nature.

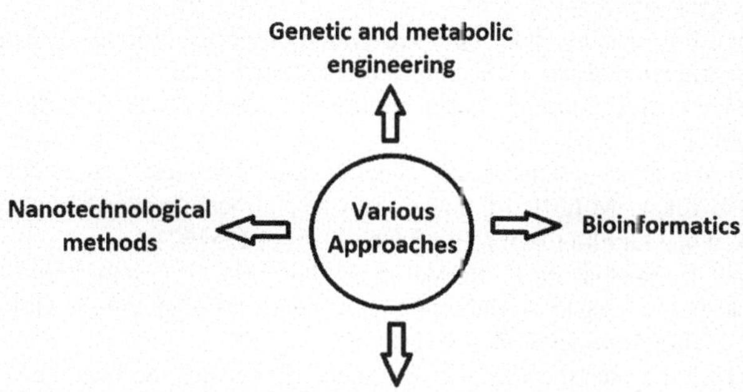

Fig. 14.3 Glimpse of various future aspects through quorum sensing for several bioremediation targets

14.8 Conclusion

Pollutants exert harm on humans as well as other living organisms, so environmental pollution is a serious and severe issue that needs to be addressed. Both, chemical and physical methods for contaminants removal can be costly as well as may result in increased site disruption and pollution. Hence remediation techniques using chemical and physical methods can give results but are not eco-sustainable. Contrary to these techniques, bioremediation is a recommendable solution to eradicate or lessen various persistent contaminants mainly relying on biological processes like microorganisms. Currently, the investigation on QS with perspective on the environment is getting a tremendous boost. Biofilms formed through QS cell–cell communication process have gained attention for the bioremediation of impending pollutants. The studies implied that biofilms have the potential to be engaged in the bioremediation process, as they guard microbes from stringent environmental conditions. Biofilms are capable of immobilizing and solubilizing numerous pollutants and provide a most favourable niche for the growth of diverse microbes. Thus their stress resistance and metabolic dynamicity made them the most suitable candidates for bioremediation of environmental pollutants. Although the mechanism of QS and biofilms has been studied at length, yet their successful application in bioremediation is quite limited. Thus more comprehension on QS molecules and biofilm formation will assist in consortium development with improved bioremediation potential.

References

Abinaya SP, Gautam P (2017) Studies on the biofilm produced by *Pseudomonas aeruginosa* grown in different metal fatty acid salt media and its application in biodegradation of fatty acids and bioremediation of heavy metal ions. Can J Microbiol 63(1):61–73

Annous BA, Fratamico PM, Smith JL (2009) Scientific status summary: quorum sensing in biofilms: why bacteria behave the way they do. J Food Sci 74(1):R24–R37

Armstrong E, Boyd KG, Burgess JG (2000) Prevention of marine biofouling using natural compounds from marine organisms. Biotechnol Annu Rev 6:221–241

Arora NK (2018) Bioremediation: a green approach for restoration of polluted ecosystems. Environ Sustain 1(4):305–307

Asfour HZ (2018) Anti-quorum sensing natural compounds. J Microsc Ultrastruct 6(1):1

Azizi S, Kamika I, Tekere M (2016) Evaluation of heavy metal removal from wastewater in a modified packed bed biofilm reactor. PLoS One 11(5):e0155462

Bahari S, Zeighami H, Mirshahabi H, Roudashti S, Haghi F (2017) Inhibition of *Pseudomonas aeruginosa* quorum sensing by subinhibitory concentrations of curcumin with gentamicin and azithromycin. J Glob Antimicrob Resist 10:21–28

Bala S, Garg D, Thirumalesh BV, Sharma M, Sridhar K, Inbaraj BS, Tripathi M (2022) Recent strategies for bioremediation of emerging pollutants: a review for a green and sustainable environment. Toxics 10(8):484

Birch GF, Taylor SE, Matthai C (2001) Small-scale spatial and temporal variance in the concentration of heavy metals in aquatic sediments: a review and some new concepts. Environ Pollut 113(3):357–372

Chan KG, Liu YC, Chang CY (2015) Inhibiting N-acyl-homoserine lactone synthesis and quenching *Pseudomonas quinolone* quorum sensing to attenuate virulence. Front Microbiol 6:1173
Chen A, Huang Y (2020) Acyl homoserine lactone based quorum sensing affects phenanthrene removal by *Novosphingobium pentaromativorans* US6-1 through altering cell surface properties. Int Biodeterior Biodegradation 147:104841
Chicca I, Becarelli S, Dartiahl C, La China S, De Kievit T, Petroni G, Levin DB (2020) Degradation of BTEX mixture by a new *Pseudomonas putida* strain: role of the quorum sensing in the modulation of the upper BTEX oxidative pathway. Environ Sci Pollut Res 27:36203–36214
Coenye T, Kjellerup B, Stoodley P, Bjarnsholt T (2020) The future of biofilm research–report on the '2019 biofilm bash'. Biofilms 2:100012
Dang TCH, Nguyen DT, Thai H, Nguyen TC, Hien Tran TT, Le VH, Nguyen QT (2018) Plastic degradation by thermophilic *Bacillus sp.* BCBT21 isolated from composting agricultural residual in Vietnam. Adv Nat Sci Nanosci Nanotechnol 9(1):015014
Dangi AK, Sharma B, Hill RT, Shukla P (2019) Bioremediation through microbes: systems biology and metabolic engineering approach. Crit Rev Biotechnol 39(1):79–98
De Carvalho CC (2018) Marine biofilms: a successful microbial strategy with economic implications. Front Mar Sci 5:126
Delacuvellerie A, Cyriaque V, Gobert S, Benali S, Wattiez R (2019) The plastisphere in marine ecosystem hosts potential specific microbial degraders including *Alcanivorax borkumensis* as a key player for the low-density polyethylene degradation. J Hazard Mater 380:120899
Demeter MA, Lemire JA, Mercer SM, Turner RJ (2017) Screening selectively harnessed environmental microbial communities for biodegradation of polycyclic aromatic hydrocarbons in moving bed biofilm reactors. Bioresour Technol 228:116–124
Devi RS, Kannan VR, Natarajan K, Nivas D, Kannan K, Chandru S, Antony AR (2016) The role of microbes in plastic degradation. Environ Waste Manage 341:341–370
Di Meglio LG, Busalmen JP, Pegoraro CN, Nercessian D (2020) Biofilms of *Halobacterium salinarum* as a tool for phenanthrene bioremediation. Biofouling 36(5):564–575
Dobretsov S, Abed RM, Teplitski M (2013) Mini-review: inhibition of biofouling by marine microorganisms. Biofouling 29(4):423–441
Donlan RM (2002) Biofilms: microbial life on surfaces. Emerg Infect Dis 8(9):881
Edwards SJ, Kjellerup BV (2013) Applications of biofilms in bioremediation and biotransformation of persistent organic pollutants, pharmaceuticals/personal care products, and heavy metals. Appl Microbiol Biotechnol 97:9909–9921
Faheem M, Shabbir S, Zhao J, Kerr GP, Ali S, Sultana N, Jia Z (2020) Multifunctional periphytic biofilms: polyethylene degradation and $Cd2^+$ and $Pb2^+$ bioremediation under high methane scenario. Int J Mol Sci 21(15):5331
Fatima T, Arora NK (2023) Structural, compositional characterization of novel exopolysaccharide from halotolerant strain of *Pseudomonas entomophila* PE3 and their role in growth promotion of sunflower under saline condition. Environ Exp Bot 213:105418
Feng L, Wu Z, Yu X (2013) Quorum sensing in water and wastewater treatment biofilms. J Environ Biol 34(2 suppl):437
Galie S, García Gutiérrez C, Miguélez EM, Villar CJ, Lombó F (2018) Biofilms in the food industry: health aspects and control methods. Front Microbiol 9:9–898
Ghosh S, Qureshi A, Purohit HJ (2019) Microbial degradation of plastics: Biofilms and degradation pathways. *In* Contaminants in agriculture and environment: health risks and remediation 25(1):184–99. https://doi.org/10.26832/AESA-2019-CAE-0153-014
Grandclément C, Tannières M, Moréra S, Dessaux Y, Faure D (2016) Quorum quenching: role in nature and applied developments. FEMS Microbiol Rev 40(1):86–116
Grujić S, Vasić S, Radojević I, Čomić L, Ostojić A (2017) Comparison of the *Rhodotorula mucilaginosa* biofilm and planktonic culture on heavy metal susceptibility and removal potential. Water Air Soil Pollut 228:1–8
Gupta P, Diwan B (2017) Bacterial exopolysaccharide mediated heavy metal removal: a review on biosynthesis, mechanism and remediation strategies. Biotechnol Rep 13:58–71

Haben FM (2019) Degradation of plastic materials using microorganisms: a review. Public Health Open J 4:57–63
Helman Y, Chernin L (2015) Silencing the mob: disrupting quorum sensing as a means to fight plant disease. Mol Plant Pathol 16(3):316–329
Hmelo LR (2017) Quorum sensing in marine microbial environments. Annu Rev Mar Sci 9:257–281
Hossain K, Rao AR (2014) Environmental change and its affect. Eur J Sustain Dev 3(2):89–89
Huang Y, Zeng Y, Yu Z, Zhang J, Feng H, Lin X (2013) In silico and experimental methods revealed highly diverse bacteria with quorum sensing and aromatics biodegradation systems–a potential broad application on bioremediation. Bioresour Technol 148:311–316
Irankhah S, Abdi Ali A, Mallavarapu M, Soudi MR, Subashchandrabose S, Gharavi S, Ayati B (2019) Ecological role of *Acinetobacter calcoaceticus* GSN3 in natural biofilm formation and its advantages in bioremediation. Biofouling 35(4):377–391
Koh CL, Sam CK, Yin WF, Tan LY, Krishnan T, Chong YM, Chan KG (2013) Plant-derived natural products as sources of anti-quorum sensing compounds. Sensors 13(5):6217–6228
Kostakioti M, Hadjifrangiskou M, Hultgren SJ (2013) Bacterial biofilms: development, dispersal, and therapeutic strategies in the dawn of the postantibiotic era. Cold Spring Harbo Perspect Med 3(4):a010306
Kumar V (ed) (2021) Rhizomicrobiome dynamics in bioremediation. CRC Press, Taylor Francis Group, Boca Raton
Kumar V, Kumar R, Singh J, Kumar P (2019) Contaminants in agriculture and environment: health risks and remediation. Agro Environ Media, Haridwar
Lambert G, Bergman A, Zhang Q, Bortz D, Austin R (2014) Physics of biofilms: the initial stages of biofilm formation and dynamics. New J Phys 16(4):045005
Liao NQ, Li HM (2013) Conceivable bioremediation techniques based on quorum sensing. In: Applied mechanics and materials. Trans Tech Publications Ltd 295:39–44
Liaqat I, Liaqat M, Ali S, Ali NM, Haneef U, Mirza SA (2019) Biofilm formation, maturation and prevention: a review. J Bacteriol Mycol 6(1):1092
López D, Vlamakis H, Kolter R (2010) Biofilms. Cold Spring Harbor Perspect Biol 2(7):a000398
Maksimova YG (2014) Microbial biofilms in biotechnological processes. Appl Biochem Microbiol 50:750–760
Mangwani N, Dash HR, Chauhan A, Das S (2012) Bacterial quorum sensing: functional features and potential applications in biotechnology. Microb Physiol 22(4):215–227
Mangwani N, Kumari S, Das S (2015) Involvement of quorum sensing genes in biofilm development and degradation of polycyclic aromatic hydrocarbons by a marine bacterium *Pseudomonas aeruginosa* N6P6. Appl Microbiol Biotechnol 99:10283–10297
Mangwani N, Kumari S, Das S (2016) Bacterial biofilms and quorum sensing: fidelity in bioremediation technology. Biotechnol Genet Eng Rev 32(1–2):43–73
Mangwani N, Kumari S, Surajit DAS (2017) Marine bacterial biofilms in bioremediation of polycyclic aromatic hydrocarbons (PAHs) under terrestrial condition in a soil microcosm. Pedosphere 27(3):548–558
Mani I (2020) Biofilm in bioremediation. In: Pandey VC, Singh V (eds) Bioremediation of pollutants. Elsevier, pp 375–385
Manobala T, Shukla SK, Subba Rao T, Kumar DM (2019) A new uranium bioremediation approach using radio-tolerant *Deinococcus radiodurans* biofilm. J Biosci 44:1–9
Meliani A, Bensoltane A (2016) Biofilm-mediated heavy metals bioremediation in PGPR *Pseudomonas*. J Bioremed Biodegr 7(5):2
Miao L, Wang C, Adyel TM, Wu J, Liu Z, You G, Hou J (2020) Microbial carbon metabolic functions of biofilms on plastic debris influenced by the substrate types and environmental factors. Environ Int 143:106007
Mishra S, Huang Y, Li J, Wu X, Zhou Z, Lei Q, Chen S (2022) Biofilm-mediated bioremediation is a powerful tool for the removal of environmental pollutants. Chemosphere 294:133609
Moghaddam MM, Khodi S, Mirhosseini A (2014) Quorum sensing in bacteria and a glance on *Pseudomonas aeruginosa*. Clin Microbial 3(4):156

Mohapatra RK, Behera SS, Patra JK, Thatoi H, Parhi PK (2020) Potential application of bacterial biofilm for bioremediation of toxic heavy metals and dye-contaminated environments. In: Yadav MK, Singh BP (eds) New and future developments in microbial biotechnology and bioengineering: microbial biofilms. Elsevier, pp 267–281

Molina L, Constantinescu F, Michel L, Reimmann C, Duffy B, Défago G (2003) Degradation of pathogen quorum-sensing molecules by soil bacteria: a preventive and curative biological control mechanism. FEMS Microbiol Ecol 45(1):71–81

Molnár M, Gruiz K, Halász M (2007) Integrated methodology to evaluate bioremediation potential of creosote-contaminated soils. Period Polytech Chem Eng 51(1):23–32

Morohoshi T, Oi T, Aiso H, Suzuki T, Okura T, Sato S (2018) Biofilm formation and degradation of commercially available biodegradable plastic films by bacterial consortiums in freshwater environments. Microbe Environ 33(3):332–335

Muhsin J, Ufaq T, Tahir H, Saadia A (2015) Bacterial biofilm: its composition, formation and role in human infections. J Microbiol Biotechnol 4:1–14

Nilanjana D, Basak LVG, Salam JA, Abigail MEA (2012) Application of biofilms on remediation of pollutants-an overview. J Microbiol Biotechnol Res 2(5):783–790

Oyelola OT, Babatunde AI (2008) Effect of municipal solid waste on the levels of heavy metals in Olusosun dumpsite soil, Lagos State, Nigeria. Int J Pure Appl Sci 2(1):17–21

Parsek MR, Greenberg EP (2005) Sociomicrobiology: the connections between quorum sensing and biofilms. Trends Microbiol 13(1):27–33

Pathak VM (2017) Review on the current status of polymer degradation: a microbial approach. Bioresour Bioprocess 4:1–31

Pepi M, Borra M, Tamburrino S, Saggiomo M, Viola A, Biffali E, Casotti R (2016) A *Bacillus sp.* isolated from sediments of the Sarno River mouth, Gulf of Naples (Italy) produces a biofilm biosorbing Pb (II). Sci Total Environ 562:588–595

Raghavan PS, Potnis AA, Bhattacharyya K, Salaskar DA, Rajaram H (2020) Axenic cyanobacterial (Nostoc muscorum) biofilm as a platform for Cd (II) sequestration from aqueous solutions. Algal Res 46:101778

Rizzo L, Malato S, Antakyali D, Beretsou VG, Đolić MB, Gernjak W, Fatta-Kassinos D (2019) Consolidated vs new advanced treatment methods for the removal of contaminants of emerging concern from urban wastewater. Sci Total Environ 655:986–1008

Santos ALSD, Galdino ACM, Mello TPD, Ramos LDS, Branquinha MH, Bolognese AM, Roudbary M (2018) What are the advantages of living in a community? A microbial biofilm perspective! Mem Inst Oswaldo Cruz 113:e180212

Shah AA, Hasan F, Hameed A, Ahmed S (2008) Biological degradation of plastics: a comprehensive review. Biotechnol Adv 26(3):246–265

Sharma S (2021) Biofilm: used as a brand new technology in bioremediation. Vidya Vol. 16(2), 2021, pp. 99–116

Sharma B, Shukla P (2022) Futuristic avenues of metabolic engineering techniques in bioremediation. Biotechnol Appl Biochem 69(1):51–60

Shukla SK, Subba Rao T (2017) The first recorded incidence of *Deinococcus radiodurans* R1 biofilm formation and its implications in heavy metals bioremediation. Bio Rxiv 234781

Shukla SK, Mangwani N, Rao TS, Das S (2014) Biofilm-mediated bioremediation of polycyclic aromatic hydrocarbons. In: Das S (ed) Microbial biodegradation and bioremediation. Elsevier, pp 203–232

Shukla SK, Mangwani N, Karley D, Rao TS (2017) Bacterial biofilms and genetic regulation for metal detoxification. In: Das S, Das HR (eds) Handbook of metal-microbe interactions and bioremediation. CRC Press, pp 317–332

Shukla SK, Hariharan S, Rao TS (2020) Uranium bioremediation by acid phosphatase activity of *Staphylococcus aureus* biofilms: can a foe turn a friend? J Hazard Mater 384:121316

Sibanda S, Moleleki LN, Shyntum DY, Coutinho TA (2018) Quorum sensing in gram-negative plant pathogenic bacteria. Adv Plant Pathol 10

Singh R, Paul D, Jain RK (2006) Biofilms: implications in bioremediation. Trends Microbiol 14(9):389–397

Singh MP, Singh P, Li HB, Song QQ, Singh RK (2020) Microbial biofilms: development, structure, and their social assemblage for beneficial applications. In: Yadav MK, Singh BP (eds) New and future developments in microbial biotechnology and bioengineering: microbial biofilms. Elsevier, pp 125–138

Soares A, Guieysse B, Jefferson B, Cartmell E, Lester JN (2008) Nonylphenol in the environment: a critical review on occurrence, fate, toxicity and treatment in wastewaters. Environ Int 34(7):1033–1049

Tiwary BK, Ghosh R, Moktan S, Ranjan VK, Dey P, Choudhury D, Chakraborty R (2017) Prospective bacterial quorum sensing inhibitors from Indian medicinal plant extracts. Lett Appl Microbiol 65(1):2–10

Tkavc R, Matrosova VY, Grichenko OE, Gostinčar C, Volpe RP, Klimenkova P, Daly MJ (2018) Prospects for fungal bioremediation of acidic radioactive waste sites: characterization and genome sequence of *Rhodotorula taiwanensis* MD1149. Front Microbiol 8:8–2528

Urbanek AK, Rymowicz W, Mirończuk AM (2018) Degradation of plastics and plastic-degrading bacteria in cold marine habitats. Appl Microbiol Biotechnol 102:7669–7678

Vague M, Chan G, Roberts C, Swartz NA, Mellies JL (2019) *Pseudomonas* isolates degrade and form biofilms on polyethylene terephthalate (PET) plastic. BioRxiv 647321

van Tilburg BE, Lewenza S, Reckseidler-Zenteno S (2015) Current research approaches to target biofilm infections. Postdoc J 3(6):36

Vogt C, Richnow HH (2013) Bioremediation via in situ microbial degradation of organic pollutants. In: Schippers A, Glombitza F, Sand W (eds) Geobiotechnology II. Advances in biochemical engineering/biotechnology, vol 142. Springer, Berlin/Heidelberg, pp 123–146

Vu B, Chen M, Crawford RJ, Ivanova EP (2009) Bacterial extracellular polysaccharides involved in biofilm formation. Molecules 14(7):2535–2554

Waters CM, Bassler BL (2005) Quorum sensing: cell-to-cell communication in bacteria. Annu Rev Cell Dev Biol 21:319–346

Xu S, Xing Y, Liu S, Luo X, Chen W, Huang Q (2020) Co-effect of minerals and Cd (II) promoted the formation of bacterial biofilm and consequently enhanced the sorption of Cd (II). Environ Pollut 258:113774

Yu Z, Hu Z, Xu Q, Zhang M, Yuan N, Liu J, Yin J (2020) The LuxI/LuxR-type quorum sensing system regulates degradation of polycyclic aromatic hydrocarbons via two mechanisms. Int J Mol Sci 21(15):5548

Zhao Y, Qu D, Zhou R, Yang S, Ren H (2016) Efficacy of forming biofilms by Pseudomonas migulae AN-1 toward in situ bioremediation of aniline-contaminated aquifer by groundwater circulation wells. Environ Sci Pollut Res 23:11568–11573

GPSR Compliance

The European Union's (EU) General Product Safety Regulation (GPSR) is a set of rules that requires consumer products to be safe and our obligations to ensure this.

If you have any concerns about our products, you can contact us on ProductSafety@springernature.com

In case Publisher is established outside the EU, the EU authorized representative is:

Springer Nature Customer Service Center GmbH
Europaplatz 3
69115 Heidelberg, Germany

The manufacturer's authorised representative in the EU is Springer Nature Customer Service Centre GmbH, Europaplatz 3, 69115 Heidelberg, Germany. If you have any concerns regarding our products, please contact ProductSafety@springernature.com

Printed and bound by CPI Group (UK) Ltd, Croydon, CR0 4YY
01/10/2026
02225857-0012